Polymers
Properties and Applications

4

A. Hebeish · J. T. Guthrie

The Chemistry and Technology of Cellulosic Copolymers

With 91 Figures

Springer-Verlag
Berlin Heidelberg New York 1981

Professor Dr. Ali Hebeish
National Research Center
Dokki, Cairo, Egypt

Dr. James Thomas Guthrie
University of Leeds
Leeds, U.K.

This volume continues the series *Chemie, Physik und Technologie der Kunststoffe in Einzeldarstellungen*, which is now entitled *Polymers/Properties and Applications.*

ISBN 3-540-10164-0 Springer-Verlag Berlin · Heidelberg · New York
ISBN 0-387-10164-0 Springer-Verlag New York · Heidelberg · Berlin

Library of Congress Cataloging in Publication Data:
Hebeish, Ali, 1936 — The chemistry and technology of cellulosic copolymers. (Polymers; v. 4). Includes bibliographical references and index. I. Guthrie, James Thomas, 1943 — joint author. II. Title.
QD321.H42 547.7′82 80-19442

Printed in GDR

Bookbinding: Konrad Triltsch, Graphischer Betrieb, Würzburg.
2152/3020-543210

To our families

A. H./J. T. G.

Preface

Many excellent volumes have been written on the chemistry of cellulose and its derivatives. Judging by the number of conferences which have been assembled to deal with the topic, cellulose and its derivatives continue to arouse great scientific interest. Matching this interest has been the development in copolymer science and technology. In both instances the driving force has been the search for products having useful, new or interesting properties. It appeared inevitable that these two concepts would be brought together at some time in the research and development of cellulosic copolymers. That time has arrived.

In assembling this text our aim was to present an informative account of the chemistry and technology of cellulosic copolymers. As such, we intended that the contents be of interest to all those concerned with the production and use of cellulosic products whether in academic or industrial circles. Sections of the text should be of value in undergraduate and post-graduate teaching, provided the student is given guidance in following the text. The volume is divided into eight chapters, each dealing with factors which are relevant to an understanding of cellulosic copolymers. Each chapter carries its own bibliography and is reasonably self-contained.

In assembling the text we have drawn on the experience of many individuals and organizations whose interest we acknowledge. We are very grateful to various publishers and authors for granting permission to use their diagrams in this text. We extend our thanks to Mrs. Leisel Carter and her colleagues for their efficient typing, and we are indebted to Mr. A. Kazlauciunas for his interest and enthusiasm in reproducing the figures contained in the text. We acknowledge the help given by the publishers in preparing the manuscript in its final form as well as in the removal of ambiguities or errors.

However, there may be some errors which have been too well hidden for our detection. We would be indebted if these were brought to our attention by readers.

Cairo and Leeds, October 1980 A. Hebeish, J. T. Guthrie

Table of Contents

1. The Homopolymeric Species

1.1. Introduction

A consideration of the chemical structure of natural high polymers such as cotton, wool, silk, starch, gums, and rubber reveals that in all these substances, a very large number of identical or similar atoms or groups of atoms are linked by primary valency forces to form macromolecules. Thus, each of the high polymers is composed of certain fundamental repeat units in the molecular structure whereby it acquires some useful properties. Such structural features common to all high polymers impart similar physical properties to these substances, even if they are chemically different. The fibrous structure of asbestos and cellulose, the elastic properties of poly(phosphonitrilic chloride) and rubber, the optical properties of inorganic and organic glasses, and the optical anisotropy of cellulose and nylon are examples.

Some of the high polymers described above are fiber-forming while the others are not. One difference between these two classes of high polymers is that fiber forming polymers, though they are of the same size as ordinary molecules in two directions (breadth and depth), are hundreds or even thousands of times as long in the third dimension (length), i.e., they are similar in shape to the fibers they form (very long when compared with their diameter), whereas in the non-fiber-forming high polymers, the molecules are roughly of the same dimensions in each direction. This may be explained by giving an analogy. If a sugar molecule is enlarged to the size of the head of an ordinary match stick, a cellulose molecule will, on the same scale, correspond to a match stick 6 meters long.

All natural fiber forming substances consist of macromolecules built up from a large number of fragments of smaller molecules of non-fiber forming substances.

Though textile fibers share the property of being fibrous, they differ in many other important properties. The differences lie largely in physical properties, such as moisture uptake, tensile strength, elasticity, and luster. Differences in chemical properties include their dye-ability and their resistance to various chemicals commonly employed in bleaching and other processes. Thus, cotton, wool, acetate rayon, and terylene have to be dyed by different classes of dyes having entirely different chemical properties. Wool is quite resistant to fairly high concentrations of acids but is easily damaged by boiling with even dilute caustic liquors, while the reverse is the case with cotton. In fact, cotton acquires useful properties, such as increased luster and tensile strength, on treatment with caustic soda (mercerization).

Examination of the properties of natural high polymers useful as textile fibers has yielded some important observations, which have led to a greater appreciation of the prime requirements of fiber-forming polymers. A high polymer should have at least some of the following characteristics if it is to serve as a fibrous material:

(1) The polymer must have a comparatively high molecular mass. The average length of its molecular chain should be on the order of 1×10^{-7} m or more. This requirement may be relaxed depending on the chemical nature of the polymer and on the arrangement of the polymer molecules.

(2) The macromolecules of the polymers must be linear in structure, and possess a high degree of symmetry so as to allow for good orientation of the molecules and the build-up of an ordered structure. The molecules should not contain bulky side chains.

(3) A high degree of orientation of the molecules in the polymer is a prerequisite for producing good tensile strength.

(4) The chemical nature of the materials should be such that the polymer has a high melting point and the ability to maintain orientation. To ensure these requisites, the molecular chains should have a sufficient number of sites of attraction (certain chemical groups such as hydroxyl groups), having pronounced secondary valency forces which allow the material to acquire a high degree of intermolecular cohesion. The arrangement of such macromolecules in the fiber produces ordered regions and amorphous regions, often better considered as the inaccessible and accessible regions respectively.

Synthetic fibers have come to occupy a very important position in the textile field. Some synthetic polymers have better properties than the natural fibers in several respects. The main advantages of synthetic fibers, as compared to cellulosic fibers, may be summarized as follows: (i) high resistance to chemicals, (ii) high resistance to micro-organisms and moulds, (iii) low inflammability, (iv) high elasticity and, hence, higher resistance to distortion and creasing, and (v) high resistance to abrasion. Nowadays, there is no doubt that with the introduction of various synthetic fibers, which can be made to suit many end uses, all natural fibers are facing stiff competition. Statistics from the 1960's and 1970's show this to be so, though recent figures show the natural fibers to be more than holding their ground. Carothers has shown that synthetic polymers can be classified according to the chemistry of their preparation. He identified two broad categories: namely, polymers made via (a) a step-growth reaction and (b) an addition reaction. Our interest is mainly in addition polymerization methods.

1.2. Cellulose

The term cellulose in the strict scientific sense applies only to the plant cell materials consisting of macromolecules of at least several hundred to several thousand anhydroglucose units. It is the carbohydrate part of the cell wall of plants, formed out of only glucose molecules condensed and linked together linearly by means of 1,4-β-glucosidic bonds.

1.2.1. Structural Features of Cellulose

The nature of the building units and their linking, toether with the average molecular length and its range, are of primary importance in the establishment of the structure of macromolecules. These aspects have been thoroughly studied for cellulose.

On hydrolosis with inorganic acids, cellulose gives almost a quantitative yield of glucosé (90.7%). Completely acetylated cotton on methanolysis gives an equilibrium mixture of methyl α- and β-glucosides, which accounts for 98.1% of cellulose. As the products of reaction are only glucosides, and as the reaction mixture fails to give the furfural test for pentosans, this is taken as good evidence to show that cellulose is built up only of glucose. The lower-than-theoretical yields are attributed to reversion and/or decomposition of glucose by strong acids. Purified cotton and cellulose acetate are hydrolised with 40% HCl to glucose, which is estimated as CO_2 by catalytic oxidation with $FeCl_3$. Results show 99.1% of glucose in the original cellulose in both cases. Sugar identification by chromatography shows that very pure forms of cellulose, such as ramie, yield on hydrolysis only glucose. All the above tests show that the basic unit of cellulose is glucose.

The reaction of one α- or β-glucose form with an alcoholic group of another glucose molecule gives α- or β-bonded dimers, trimers or higher polymers bonded through the acetal group. Maltose is the α(1–4)-linked dimer of glucose. This is the repeating unit found in starch dextrins and amylose. Likewise, cellulose is the β(1–4)-linked dimer of glucose and the polymeric chain built up of cellobiose residues in cellulose.

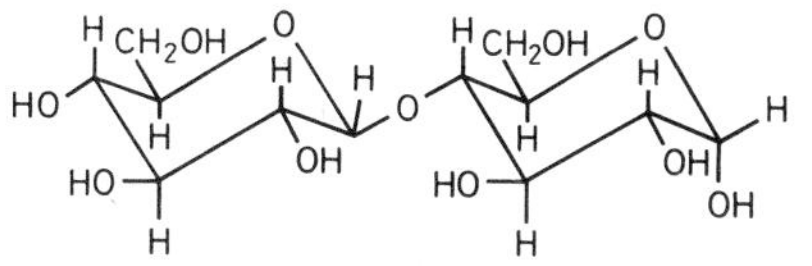

Maltose (α-anomer)

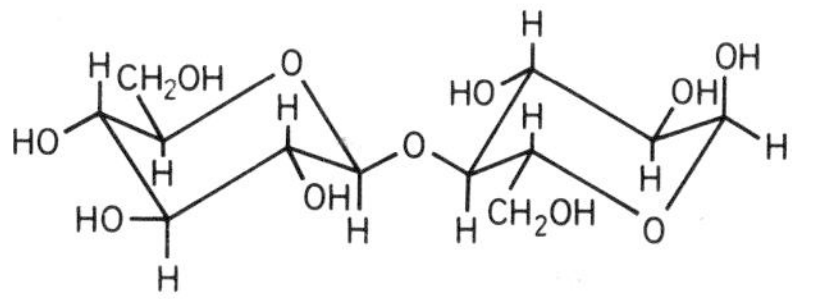

Cellobiose (α-form)

The crystalline and molecular structure of cellobiose, one of the major products of the hydrolysis of cellulose under acid conditions, has been reviewed by Wunderlich et al.[1)]. Cellulose itself has, therefore, been shown to be poly-(1–4)-β-D-glucopyranose.

Carefully isolated cellulose shows very little reducing power, but it develops this property on hydrolysis. This fact, as well as the production of a nearly theoretical yield of cellobiose, indicates that the bond is glucosidic in nature. The bond involves the potential aldehyde group of one glucose molecule and a hydroxyl group of another.

Cellobiose octaacetate, obtained by hydrolysis of fully acetylated cellulose, has been shown by Skraup and Konig[2)] to resist the yeast ferment maltase, which readily hydrolyses the α-glucosidic bonds in the starch degradation product, maltose. It is interesting to note that cellobiose may be hydrolysed to glucose by emulsin, which establishes the disaccharide linkage in the beta configuration. Work on the infrared absorption of cellulose and starch compared with known oligomers has reconfirmed the beta configuration as the only interglucosidic valence bond in cellulose[3].

Cellotriose, cellotetraose, cellopentaose, cellohexaose, and celloheptaose have been identified as products of hydrolysis of cellulose. From a comparison of the physical data obtained from compounds of increasing anhydroglucose (glucopyranose) content, and from the fact that they showed no chemical differences, it became evident that their formulae should extrapolate to cellulose when the number of anhydroglucose units was assumed to be very large.

Methylated cellulose, obtained by treating sodium hydroxide-soaked cotton with dimethyl sulphate, yields upon hydrolysis only the 2,3,6 trimethyl ether of glucose. It was therefore established that the three free hydroxyl groups in the cellulose occupy the 2, 3 and 6 positions. The groups have decreasing acidic properties in the order 2, 3 and 6. The primary hydroxyl group at position 6 is sterically the most unhindered.

Methylation of cellobiose gave the crystalline heptamethyl derivative, which upon acid hydrolysis produced 2,3,4,6-tetra-0-methyl-β-D-glucopyranose in equimolecular amounts. Cellobiose could therefore only be 0-β-D-glucopyranosyl-(1–4) or (1–5)-β-D-glucopyranose. The fact that no 5-methyl derivative was found in either of the two fractions supported the formula already established theoretically, that the 5 position in the molecule was inaccessible for chemical reactions. This was finally proved when the D-glucose-2,3,5,6-tetramethyl ether was obtained with 2,3,4,6-tetramethyl ether from an octamethyl-cellobionic acid. This acid was prepared by eliminating the cyclic structure in the reducing half of the cellobiose by oxidation with bromine water. Cellobiose was thus established to be 0-β-D-gluco-pyranosyl-(1–4)-β-D-glucopyranose.

Four types of evidence for the uniformity of the linkage in cellulose have been seen. This is accurate up to an extent of about 99%. Chemical evidence is derived from methylation studies of glucose, other oligomers, and cellulose. Polarimetric evidence is based on optical rotation of methyl cellulose in suitable solvents. The values show very good agreement with the theoretical value based on considering the chain to be built up of only β-glucopyranose units. Studies on cellopentaose, cellohexaose, and celloheptaose further confirm the results. Kinetic evidence is obtained from the change in rate, optical rotation, and reducing power during acid hydrolysis. These can be quantitatively accounted for by assuming that all the hydrolysable links in a uniform chain are equal and equivalent except for the reducing terminal units which can be neglected when the chains are infinitely long: otherwise, a correction can be made for the faster rate of the cleavage of the bonds adjacent to the reducing end of the chain molecules. Static evidence consisted in the assessment of the actual quantities of oligosaccharides formed during hydrolysis. By assuming all bonds to be equivalent, Kuhn[4] explained, by a mathematical treatment, the low yields of intermediate compounds. The higher yield of cellobiose was attributed to its ready crystallisation, which prevented further breakdown.

As cellulose chain molecules are very long, it is very difficult to detect a small number of bonds other than the β-glucosidic, if present. Even the best evidence applies only to 99% of the bonds. Further work incorporating advanced techniques and mathematics in the study has reduced the extent of error in the proof to about 0.1%. The uncertainty about the nature of this 0.1% of bonds in cellulose itself has for some time resulted in a heated controversy over the existence of weak bonds in cellulose. Certain studies indicate the absence of such bonds in native cellulose, but it is possible that there can be a few β-glucosidic bonds which are sensitized to certain reactions, such as acid hydrolysis by induction effects or physical strains in the molecules.

The chain molecules in natural cellulose are not of the same length. The number of glucose units in different chains varies. This is revealed by different samples of cellulose of no detectable chemical difference giving different alkali solubilities and viscosities. A given sample represents a molecular homologous series in which there is no molecular heterogeneity. We have to deal with averages, such as average molecular mass and average chain length. The degree of polymerization (D.P.) of unopened cotton has been reported at 15,300; this value decreases rapidly to 8,100 on exposure to the atmosphere. Bast fibers have an average D.P. of 9,900, while wood species vary between 7,500 and 10,500.

The long cellulose molecule effectively camouflages the presence of the two end groups in chemical analysis. Upon methylation the non-reducing end group should give a D-glucose-2,3,4,6-tetramethyl ether; this has been obtained under very careful reaction conditions.

The reducing end, on the other hand, could never be isolated, presumably because of the fast demethylation of the C_1 methyl group during acid hydrolysis.

Six-membered rings can assume either a boat (a) or chair (b) conformation according to their energy level. The pyranose ring assumes a chair form in

a b

preference to the boat form because of internal strain in the latter. In cellulose, the chair form will fit the periodicity of 1.03 nm. Two possible chair conformations will exist in the case of the pyran ring; these are chain (c) and (d) for glucose.

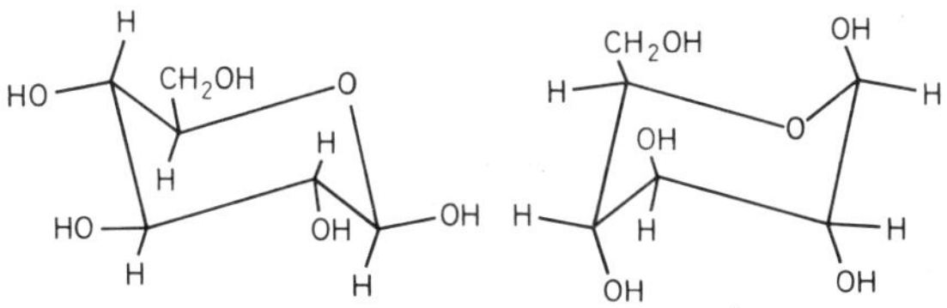

c d

In (c) the oxygen-bearing substituents lie mainly in the same plane as the pyranose ring (equatorial), and the hydrogen atoms stand away from the ring ('axial'); the situation is reversed in (d). Reeves[5] showed that D-glucopyranosides exist and react in the chair conformation. The transition into a boat form, under certain strains or activating-energy influences, might be the cause of differences in physical structure and chemical reactivity of cellulose under these conditions. Similar conclusions have been obtained from *x*-ray diffraction patterns and from infrared spectroscopy.

Whenever the distance between the various oxygen and hydrogen atoms in the cellulose molecule reaches 0.3 nm or less, they interact with each other to form intramolecular and intermolecular hydrogen bonds. Infrared spectroscopy has verified the existence of these hydrogen bonds. The intramolecular hydrogen bridges anchor the anhydroglucose units to a very limited region of free play around the acetal linkage. Thus, they impart a certain stiffness to the cellulose molecule:

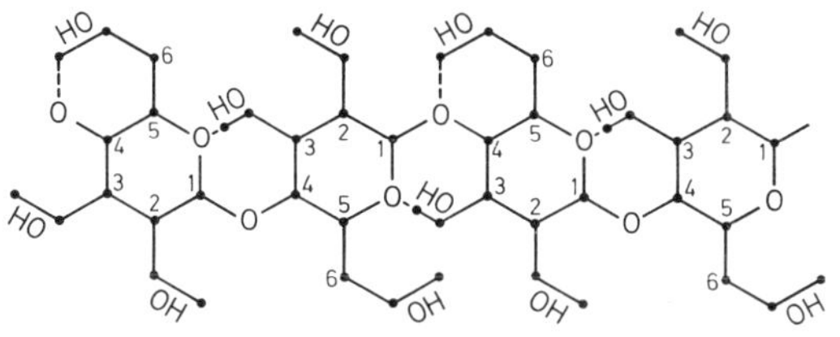

Intramolecular hydrogen bonds in the cellulose molecule

This, and the fact that the (1–4)-β-bond demands a rotation of 180° of each subsequent glucose unit to fit the β-configuration of the connecting hemiacetal linkage, gives the cellulose molecule a rod-like chain structure:

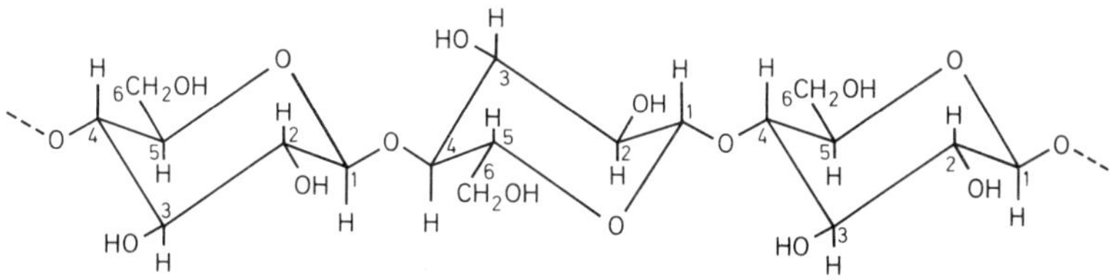

The β-glucosidic linkage in cellulose and the resulting intramolecular hydrogen bonds render the cellulose molecule straight and stiff.

On the other hand, in starch the glucose units can be arranged in a helix-like chain molecule:

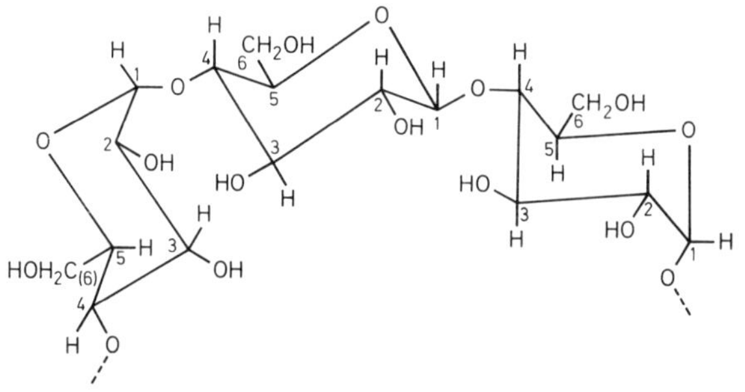

The involvement of the hydroxyl groups in hydrogen bonding, as well as general dispersion forces, determined by the proximity of neighbouring atoms, impart a different reactivity to the three hydroxyl groups available for chemical reactions. Esterification and etherification studies have shown that the C-6 group is esterified ten times faster than the other groups, whereas on etherification the C-2 group is etherified twice as fast as the C-3. The primary alcoholic group at C-6 is distinguished from the other two secondary alcoholic groups in that it has an axis of free rotation around the C-5 to C-6 bond[6], which is, however, somewhat restricted by the hydrogen bonds. It has been observed by infrared spectroscopy[7] that rotational isomers must exist especially in the alkali-swollen cellulose. The reactivity of the primary alcoholic groups seems to be related to this isomerisation. Structural differences between cotton, wood cellulose and mercerised cellulose appear to be due to this rotational isomerisation. Equally relevant are aspects of cellulose reactivity.

1.2.2. Reactivity of Cellulose

Cellulose reacts as a trihydric alcohol with one primary and two secondary hydroxyl groups per glucose unit. The relative reactivity of the hydroxyl groups of both low molecular mass carbohydrates and cellulose has been studied. In the former, the 2- and 6-hydroxyl groups are usually the most reactive. With cellulose, certain data indicate the preferential reactivity of the 2-hydroxyl and others of the 6-hydroxyl group.

The manifold reactions of cellulose may be conveniently divided into two main kinds: those involving the hydroxyl groups and those comprising a degradation of the chain molecules. The former includes the following reactions: (1) Esterification: nitration, acetylation and xanthation. (2) Etherification: alkylation and benzylation. (3) Replacement of —OH by $—NH_2$ and halogen. (4) Replacement of —H in —OH by Na. (5) Oxidation of $—CH_2OH$ to —COOH. (6) Oxidation of secondary —OH groups to aldehyde and carboxyl. (7) Formation of addition compounds with acids, bases, and salts.

These reactions taking place without breakdown of the chain may have only a local effect, e.g., causing change in the terminal groups or in individual members of the chain, or they may affect all, or the majority of, the members of the chain simultaneously. In the former case it is exceedingly difficult to detect the changes analytically in high molecular products, for which reagents of the utmost sensitivity are required. Sometimes, however, these reactions are manifested indirectly. Changes in the cellulose molecule resulting from oxidation in an acid medium affect only a few members of the chain and are scarely to be detected by direct means; yet, later on they become clearly noticeable in that the chain splits up at the affected parts upon subsequent contact with alkaline liquids. There are many chemical reactions — the esterification and etherification of the hyroxyl groups in particular — which are liable to take place over the entire chain more or less uniformly, with often little difference in reactivity of the —OH groups in positions 2, 3 and 6, though occasionally probable distinctions have been made.

Of importance among the degradative reactions are those brought about by hydrolysis of the glucosidic bonds and by oxidation. Hydrolytic breakdown takes place in the presence of acids, while oxidation may occur in an alkaline, acid or neutral medium.

As hydrolytic degradation involves the scission of the cellulose acetal bonds (i.e., the β-glucosidic bonds) by acids, the resulting increase in reducing power, the decrease in D.P. of cellulose, and the extent of oligosaccharide formation provide methods to study the kinetics of this reaction. Hydrolysis of cellulose can be either homogeneous or heterogeneous, i.e., intracrystalline or intercrystalline, being dependent upon the swelling capacity of the acid used. In the case of homogeneous hydrolysis, the end products are the monomer (glucose) and some reversion products of the gentiobiose 1–6 linked type.

With heterogeneous hydrolysis, the rate of reaction decreases as the reaction progresses to regions of high lateral order. There is an initial rapid rate which gradually decreases to a leveling-off D.P. stage. The first effect of heterogeneous acid hydrolysis is the formation of hydrocelluloses.

When the cellulose is oxidized, the chain usually breaks down, probably as the result of opening and cleavage of the monomeric rings. Side by side with this, other reactions not interfering with chain length may occur, such as oxidation of the primary hydroxyl groups in the C-6 position to aldehyde or carboxyl groups, oxidation of the secondary hydroxyl groups at C-2 and C-3 to ketone groups, oxidative opening of rings to form two aldehyde or carboxyl groups, etc. According to the conditions of oxidation, the resulting products have either a marked acidic or equally pronounced reducing character. While the former exhibit intensified sorption towards basic dyes, the latter show an increased copper number. Mixed types also occur.

The various possible types of oxidized groups formed in the cellulose molecule are shown below: —

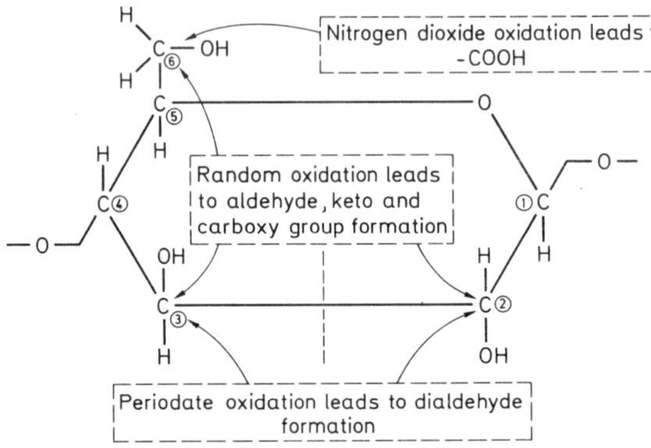

The formation of reducing groups is determined by conventional copper number and by several other methods. Carboxyl groups are determined by their ion exchange properties or by infrared analysis.

Heuser[8] has classified the main reactions of cellulose into four 'possibilities' which are derived from the two-phase crystalline-amorphous structural concept, and the physical and chemical reactivities of the hydroxyl groups. These possibilities are:

(i) Reaction takes place exclusively with one of the two types of hydroxyl groups (either primary or secondary), and is topochemical (micellar/heterogeneous). Duality of the submicroscopic structure determines the rate, e.g., periodic acid oxidation.

(ii) Reaction takes place preferentially with one of the two types of hydroxyl groups and is permutoid (quasi-homogeneous). The difference in reactivity of the hydroxyl types determines the rate, e.g., tritylation.

(iii) Reaction takes place equally with hydroxyls of the two types, and the reaction is topochemical. Duality of the submicroscopic structure determines the rate of reaction, e.g., water sorption.

(iv) Reaction takes place equally with the two types of hydroxyl group, and the reaction is permutoid. The rate of reaction is uniform throughout, e.g., nitration.

One might be tempted to regard cellulose as a trihydric alcohol similar to sugars in the type of its reactions. This, however, has been definately shown not to be the case when reactions with cellulose take place, with all the properties of cellulose as a fiber-forming, high polymeric substance then coming into play. The following division accounts for most cellulose reactions, although it naturally represents a considerable simplification of the extremely complex state of the matter: (A) Heterogeneous reactions [which include surface reactions, macroheterogeneous reactions, micellarheterogeneous reactions and permutoid (or quasi-homogeneous) reactions] and (B) Homogeneous reactions.

Surface reactions are limited to the fiber surface, with the capillary system of pores and cracks remaining closed throughout the reaction due to the absence of any form of swelling, e.g., treatment of unswollen fibres with diazomethane or thallous ethylate and alkylation of trisodium cellulose. In other cases the molar volume of the reagent may be too great to allow the penetration of the latter into the pores and interstices, for example, in the case of isoalkyl halides[9].

Macroheterogeneous reactions are characterized by the fact that the conversion starts on the substrate surface and then proceeds through the fiber from layer to layer. Two different processes may occur: (a) If the secondary wall is penetrable but the primary wall is only partially swelled, then the reaction will start at the points of the fiber surface at which the cuticle has been damaged. A typical topochemical type of reaction is thus obtained. The capillaries, which may regarded as submicroscopical reaction rooms, serve as supplying canals for the reagent. If they contain a liquid, immiscible with the reagents, the penetration of the latter will naturally be opposed, and a topochemical macroheterogeneous reaction will be obtained. The acetylation of fibrous cellulose with acetic anhydride in benzene furnishes a typical example. Another is the denitration of nitrocellulose in alcohol. (b) If the primary wall is swelled but the secondary wall is only slightly penetrable, the reaction will begin simultaneously all over the fiber surface and then proceed from layer to layer in the secondary wall. An example of this type is the denitration of nitrocellulose in aqueous media.

If both the primary and the secondary wall are swelled in a cellulose reaction, the reaction is of the microheterogeneous type. Microheterogeneous reactions in turn consist of two very different types:
(i) micellar heterogeneous reactions, i.e., they involve a rapid and complete conversion of the micellar surface, after which the micelle itself undergoes a reaction layer after layer. An inner core of unchanged cellulose remains until the very end; (ii) permutoid reactions, which refer to the ability of the permutites to exchange quantitatively and rapidly ions or molecules in the lattice with other ions or molecules, the crystal structure still being retained.

The general concept of inter- and intramicellar types of reaction was first presented in 1925. Intercrystalline or intermicellar reactions take place in the amorphous regions or on the surface of crystallites, intracrystalline or intramicellar reactions within the crystallites. The former involve no changes in the X-ray diagram of the sample in question, whereas the latter lead to the formation of a new pattern, the two types being distinguished in this way.

Reactions of cellulose in dispersion conditions produce a very uniform distribution of the entering groups along the chain molecules, as indicated by the good solubility properties of the product obtained. Several cases are known in which cellulose is converted quite homogeneously. An example is the acetylation of cellulose dispersed in phosphoric acid. The reaction of cellulose derivatives, dissolved in pyridine, with tosyl chloride is a homogeneous esterification reaction. All kinds of cellulose ethers can also be prepared homogeneously with the help of quaternary ammonium bases.

Swelling

A solid is said to swell when it takes up a liquid while at the same time (a) it does not lose its apparent homogeneity, (b) its dimensions are enlarged, and (c) its cohesion is diminished. This swelling is not to be confused with capillary imbibition, which is due to the fine structure of cellulose with its capillaries and interstices. As the dimensions of the latter are partially submicroscopic, there is a continuous transition from swelling to capillary imbibition.

The swelling phenomena of cellulose are suitably subdivided through intermicellar or intramicellar swelling, each with either limited or unlimited swelling.

Intermicellar swelling is restricted to the amorphous domains of the fiber, and in X-ray diagrams the diffuse halo of the swelling agent is therefore superimposed upon the original cellulose pattern. If only a limited amount of reagent is absorbed, the swelling will be limited; otherwise it would be unlimited. In the case of intramicellar swelling, the reagents penetrate the crystallites. If the swelling is limited, formation of an addition (swelling) compound takes place between the cellulose and the swelling agent. This compound has a distinct X-ray pattern. In the case of unlimited swelling, the uptake of the reagent continues until the crystallites are completely dispersed.

Cellulose undergoes swelling in solutions of acids, bases, and salts as well as in some organic solvents. Swelling generally involves breaking of intermolecular bonding of cellulose and, in many cases, formation of new bonds with the swelling agents to give swelling compounds. As will be seen later, such swelling is an important feature of cellulose modification through graft polymerization whether to cellulose or one of its derivatives.

Various reasonably recent texts dealing with aspects of cellulose accessibility and reactivity are available[10-13]. These are significant since they provide information relevant to more efficient grafting reactions and to novel cellulosic products.

1.3. Synthetic Polymers

1.3.1. Radical Addition Polymerization

Addition (chain) polymerization consists simply of the union of several molecules with a redistribution of the valency bonds, but with no splitting off of simple by-products as in step-growth polymerization. Addition polymers can be formed from unsaturated compounds or through ring opening procedures. Many unsaturated compounds, particularly ethylene and its derivatives, will polymerize under suitable conditions.

The simplest case, in theory, is the polymerization of ethylene to give polyethylene, which may be illustrated as follows:

$H_2H{:}CH_2 + H_2C{:}CH_2 + H_2C{:}CH_2 \ldots + H_2C{:}CH_2 \rightarrow$
ethylene
$—CH_2—CH_2—CH_2—CH_2—CH_2—CH_2— .$
polyethylene

This reaction may be brought about by applying high pressure and high temperature, usually in the presence of a catalyst.

Polymerization of vinyl compounds of the general formula $H_2C{=}CHX$ involves three stages: initiation, propagation (growth), and termination.

1.3.1.1. Free Radical Initiation

A free radical is a molecular fragment carrying an unpaired electron. It is usually unstable and ready to combine to a more complete state. Free radicals can be produced by the homolytic cleavage of chemical bonds. Such homolytic cleavage may be assisted using ultraviolet light or high energy irradiation, catalysts of the peroxide type, the persulfate type, redox systems, or thermally unstable aliphatic azo compounds. When the free radical comes in contact with a vinyl compound, it disturbs the electronic structure of the carbon atoms at either end of the double bond, breaking the π bond. This is represented by saying that the double bond opens. If $R^{\cdot}$ is the free radical and $CH_2{=}CHX$ is a vinyl monomer:

$$R^{\cdot} + CH_2 = CHX \rightarrow RCH_2\dot{C}HX .$$

The two have combined together to form another free radical, which is itself capable of further growth; the reaction has been initiated.

1.3.1.2. Propagation

In a step-growth polymerization the small polymers that are formed early in the reaction, such as the dimers and trimers, react together. As a result, none of the original monomer is left after a short time, even though the average degree of polymerization may still be very low. This is not the case with addition polymerization, in which a partially polymerized mixture will consist of high molecular mass polymer and of unchanged monomer with virtually no constituents at intermediate stages of growth; some individual polymer molecules grow to maturity, while most of the monomer molecules remain unchanged. A given molecule of very high D.P. (degree of polymerisation) is formed by consecutive steps of a single chain process initiated by a free radical. The propagation step can be illustrated, using a free radical $R^{\cdot}$ as the initiator, as follows:

$$
\begin{aligned}
&RCH_2{-}\dot{C}HX^{\cdot} + CH_2 = CHX \rightarrow RCH_2{-}CHX{-}CH_2{-}\dot{C}HX^{\cdot} \\
&RCH_2{-}CHX{-}CH_2{-}\dot{C}HX^{\cdot} + CH_2 = CHX \rightarrow \\
&RCH_2{-}CHX{-}CH_2{-}CHX{-}CH_2{-}\dot{C}HX^{\cdot}
\end{aligned}
$$

and so on. Ultimately, the chain may contain several thousand monomer units.

1.3.1.3. Termination

When the free radical at the end of one chain meets that at the end of another chain, the two combine. Through such combination termination the long molecule is completed. It cannot grow any more unless reinitiated through chain transfer. Combination termination may be represented as

$$
\begin{aligned}
&R{-}(CH_2{-}CHX)_y{-}CH_2\dot{C}HX^{\cdot} + R^1{-}(CH_2{-}CHX)_z{-} \\
&{-}CH_2{-}\dot{C}HX^{\cdot} \rightarrow R{-}(CH_2{-}CHX)_{y+1}{-}(CHX{-}CH_2)_{z+1}{-}R^1
\end{aligned}
$$

Alternatively, instead of joining together, the two active chain ends may disproportionate, one giving up a β hydrogen atom to the other, so that both molecules are deactivated and can grow no more, unless reinitiated by chain transfer. Disproportionation termination, which is much less common than combination termination, may be shown as

$$
\begin{aligned}
&R{-}(CH_2{-}CHX)_y{-}CH_2{-}\dot{C}HX^{\cdot} + R^1{-}(CH_2{-}CHX)_z{-} \\
&{-}CH_2CHX^{\cdot} \rightarrow R{-}(CH_2{-}CHX)_y{-}CH = CHX + R^1{-} \\
&{-}(CH_2{-}CHX)_z{-}CH_2{-}CH_2X\,.
\end{aligned}
$$

Termination of the propagating chain in radical polymerizations may also occur by chain transfer. Chain transfer reactions provide many subtle means of termination which may be unpredictable. In cellulose modification through grafting, chain transfer can sometimes be used to advantage. Chain transfer reactions are usually hydrogen abstraction reactions which may involve transfer of the radical activity to the solvent, monomer, dead homopolymer, or the substrate. The transfer of radical activity to cellulosic substrates through hydrogen abstraction reactions is one important feature of grafting to cellulosics. Most of the vinyl and acrylic graft polymerizations onto cellulose, which have provoked interest either in research or industry, involve a free radical mechanism, thus, the provision of detailed information on various aspects of radical polymerization is called for. For the more basic aspects of radical and ionic addition polymerization, the reader is referred to one or more of the excellent relevant texts on this subject[14-19].

1.3.2. Kinetics of Polymerization

The kinetics of vinyl polymerization of the various monomers vary considerably with the nature of the monomers and the conditions of polymerization. Discussion of the kinetics will be confined here to homogeneous systems, since these systems form a basis which can be extended to heterogeneous conditions.

1.3.2.1. Initiation

As previously pointed out, the first step in initiation is the decomposition of the initiator, I, into two radicals, $R^{\cdot}$,

$$I \rightarrow 2\,R^{\cdot}\,. \tag{1}$$

If k_i is the reaction constant for this reaction, the rate of production of radicals may be expressed as

$$\frac{d[R^{\cdot}]}{dt} = 2k_i[I]\,. \tag{2}$$

The factor 2 is introduced to take care of the fact that the radicals must be produced in pairs. The radical $R^{\cdot}$ is capable of initiating the reaction chain,

$$R^{\cdot} + M \rightarrow M_1^{\cdot}\,; \qquad M_1^{\cdot} + M_1 \rightarrow M_{11}^{\cdot} \tag{3}$$

in homopolymerization, where M represents the monomer and $M_1^{\cdot}$ is the first member of the propagation chain. The rate of initiation, R_i, of chain radicals can be written in the form

$$R_i = \frac{d[M^{\cdot}]}{dt} = 2fk_i[I] \tag{4}$$

where f is a factor which represents the fraction of primary radicals $R^{\cdot}$, which initiate chains. This is tantamount to saying that reaction (3) is fast, for, if it is not, the factor f will be dependent on the monomer concentration. However, a large fraction of the primary radicals may react with the other molecules in the system to yield an inactive product. Such a reaction will make f

less than unity. Another possibility is that the radicals which are formed in pairs can recombine before they escape from one another. This "cage" effect is a first order combination and will again make f less than unity and a function of monomer concentration.

It is possible, in a case such as the polymerization of acrylonitrile with redox catalysts, to show that the radical concentration is maintained constant when the rate-determining step is as shown by reaction (3), giving the velocity of initiation as

$$R_i = \frac{d[M^{\cdot}]}{dt} = k_i[R^{\cdot}]\,[M]\,. \tag{5}$$

This state of affairs corresponds to the factor f being linearly dependent on the monomer concentration.

The initiator efficiency, f, may be determined by a number of methods:

(a) by comparison of the amount of initiator decomposed with the number of initiator fragments which start the polymer chain;

(b) by determining the number average molecular mass of the polymer and hence calculating the number of molecules formed. If the reaction involves no transfer and termination is by coupling of radicals, the number of combined primary radicals is twice the number of molecules;

(c) by using inhibitors which can react stoichiometrically with the chain radicals giving products of very short chain lengths, one molecule of inhibitor being used per chain initiated. In the inhibition period, the number of chains started is assumed to be equivalent to the number of inhibitor molecules used and, hence, from the knowledge of the rate of decomposition, the efficiency may be calculated.

Polymerization may also be initiated by means of high energy irradiation, a complex process with many side reactions. It has been widely studied in reference to grafting upon cellulose, as will be discussed in Chapter 3.

When polymerization is initiated by irradiation of the monomer solution with light, absorption of light by a monomer molecule will give rise to a radical of some kind and may be formally expressed as

$$M + h\nu \rightarrow M^{\cdot}\,.$$

The rate of production of the radicals will depend on the amount of light energy absorbed. The quantum yield now corresponds to the efficiency, f. If irradiation of intensity I_0 passes through the solution of thickness x, with concentration of monomer [M], the energy absorbed is given according to the Beer/Lambert law by $I_0(I - e^{-\varepsilon[M]x})$ where ε is the extinction coefficient. The rate of radical production will be

$$\frac{d[M^{\cdot}]}{dt} = 2fI_0(I - e^{-\varepsilon[M]x})\ 2fI_0\varepsilon x[M]\,. \tag{6}$$

Initiation in such a system is thus dependent on the monomer concentration.

Polymerization by photochemical initiation is another method worth mentioning. In this method a photo-chemical initiator, which releases free radicals at a wavelength where the monomer is unaffected, is put into the system. Such a substance may be benzoin or azobisisobutyronitrile, which is decomposed to free radicals on irradiation with light from a mercury arc in the region of 360 nm. In this case, the velocity of initiation will depend upon the intensity, the absorption of light by initiator, and the concentration of the initiator. The reaction may be represented as:

$$\mathrm{I} + h\nu \rightarrow 2\,\mathrm{R}^{\cdot}\,,$$
$$\mathrm{R}^{\cdot} + \mathrm{M} \rightarrow \mathrm{M}^{\cdot}\,.$$

As previously indicated, photochemical techniques may be performed at low temperatures when thermal polymerization is absent. Further, photochemical initiation is useful in fundamental studies because the production of free radicals may be instantaneously controlled by variation of the light intensity. Photochemically induced grafting is considered later.

1.3.2.2. Propagation

The reaction of the monomer molecules with the reactive centers at the end of the chains may be written as:

$$\mathrm{M}_1^{\cdot} + \mathrm{M} \rightarrow \mathrm{M}_2^{\cdot}\,,$$
$$\mathrm{M}_2^{\cdot} + \mathrm{M} \rightarrow \mathrm{M}_3^{\cdot}\,,$$
$$\mathrm{M}_3^{\cdot} + \mathrm{M} \rightarrow \mathrm{M}_4^{\cdot}\,,$$
$$\mathrm{M}_n^{\cdot} + \mathrm{M} \rightarrow \mathrm{M}_{n+1}^{\cdot}\,,$$

where $\mathrm{M}_n^{\cdot}$ refers to a chain radical containing n units.

Each of these steps could be described by means of the rate of loss of monomer.

$$\left(\frac{\mathrm{d}[\mathrm{M}]}{\mathrm{d}t}\right)_1 = -k_1[\mathrm{M}_1^{\cdot}][\mathrm{M}]\,,$$

$$\left(\frac{\mathrm{d}[\mathrm{M}]}{\mathrm{d}t}\right)_2 = -k_2[\mathrm{M}_2^{\cdot}][\mathrm{M}]\,, \text{ etc.}$$

where $((\mathrm{d}[\mathrm{M}])/\mathrm{d}t)_n$ refers to the change monomer concentration due to reaction of $\mathrm{M}_n^{\cdot}$ with monomer. The overall rate of change of monomer concentration can be measured, and these equations can be summed to give the velocity of propagation, R_p

$$\mathrm{R}_p = \frac{\mathrm{d}[\mathrm{M}]}{\mathrm{d}t} = -[\mathrm{M}]\{k_1[\mathrm{M}_1^{\cdot}] + k_2[\mathrm{M}_2^{\cdot}]\ldots\}\,. \tag{7}$$

The loss of monomer is a measure of the rate of polymerization or the velocity of polymerization.

To render the kinetic treatment possible, it is assumed that the reaction constants are identical, i.e., $k_1 = k_2 \dots = k_n = k_p$, the propagation constant. Equation 7 then becomes:

$$\frac{d[M]}{dt} = -k_p[M]\{[M_1^{\cdot}] + [M_2^{\cdot}] + \dots\}. \tag{8}$$

The quantity in the braces is the total concentration of the radicals taking part in the reaction and may be written simply as $[M^{\cdot}]$. Equation 8 then becomes:

$$\frac{d[M]}{dt} = -k_p[M][M^{\cdot}]. \tag{9}$$

1.3.2.3. Termination

On the assumption that the termination of the reaction is the union of two radical ends to annihilate both, the equation can be written:

$$M_n^{\cdot} + M_n^{\cdot} \rightarrow M_{n+m}$$

and the loss in radicals is given by

$$\frac{d[M^{\cdot}]}{dt} \rightarrow -2k_{mn}[M_n^{\cdot}][M_m^{\cdot}] \tag{10}$$

where k_{mn} is the termination constant.

Again, a kinetic scheme in which each coupling reaction has a different constant would be incapable of solution. The reaction between two radicals is fast; the rate will be determined by the number of times per second that the reactive heads are in the correct position. It is difficult to calculate the rate at which small radicals collide in a liquid, and any calculation with a polymer radical will be further complicated by the fact that each segment of the polymer chain is capable of free rotation around the single bonds comprising the backbone of the polymer. This means that the mobility of the radical head will be much greater than for a rigid molecule, and the mobility may not be much less than that of a small radical. Thus, the rate of termination may be taken to be independent of molecular mass, and equation 10 may be written in terms of a single termination constant, k_t;

$$\frac{d[M^{\cdot}]}{dt} = -2k_t[M^{\cdot}]^2. \tag{11}$$

This step would seem justified on the basis of the fit of the overall kinetics to the data.

However, the rate at which the growing chain ends collide need not, under all conditions, be independent of molecular mass. Since the diffusion of large radicals may be slow, it is to be expected that a point may be reached when the termination step is diffusion-controlled. A reduction in the rate of the termination step would give more time for the radicals to grow and hence give a product of higher molecular mass. Since the diffusion coefficient is inversely proportional to the viscosity, the effect is to be expected in bulk polymerization when a considerable quantity of monomer is converted and the reaction mass becomes very viscous. This is referred to as the gel effect. The reaction is also diffusion-controlled when the mobility of the active head is reduced by causing the growing molecule to coil itself up. This shields the radical end and may be carried out by conducting the reaction in the presence of a precipitant such as methanol. A reduction of the rate of termination would be accompanied by an increase in the overall rate of polymerization. It must be pointed out that the termination step differs from the propagation step by the fact that in the termination step both the reacting species are large molecules, i.e., growing polymer chains. With the propagation reaction, the supply of monomer units is governed by the diffusion of small molecules even if the movement of the large ones is restricted.

1.3.2.4. Steady State

In irradiation polymerization, for example, the formation of radicals occurs at a constant rate R_i. When irradiation is stopped, the radicals formed will combine with the monomer. As a result, the growing polymer chains increase and so will their rate of destruction until a stage is reached at which the rate of production of radicals is approximately equal to their rate of destruction. Under such conditions, the concentration of radicals and the rate of reaction vary little with time except for effects brought about by changes in the concentration of reactants. The rate of change of radical concentration is given by the difference between the rate of formation and the rate of destruction of the radicals, i.e.,

$$\frac{d[M^\cdot]}{dt} = R_i - 2k_t[M^\cdot]^2 \,. \tag{12}$$

Since R_i is constant, Eq. 12 may be integrated to give

$$[M^\cdot] = \left(\frac{R_i}{2k_t}\right)^{1/2} \mathrm{Tanh}\,(2k_tR_i)^{1/2}\, t \tag{13}$$

provided that $[M^\cdot] = 0$ when $t = 0$.

Now Tanh x is a function which increases rapidly from zero and approaches unity asymptotically. When t approaches infinity, $[M^\cdot]$ approaches a maximum value, $[M^\cdot]_s$. $[M^\cdot]$ approaches within 1% of this value when $(2k_t\, R_i)^{1/2}\, t$ is $\geqq 3$. When this situation is reached, the concentration of $[M^\cdot]$ does not substantially change and the system may be described as being in the "steady state".

The steady state principle or hypothesis states that a reaction involving a transient species is characterised by having rates of formation and disappearance of this species, which are equal. The previous discussion would indicate that this is an approximation, but one which has been amply justified by the consistency of the results obtained. The condition for this stationary state is $d[M^{\cdot}]/dt = 0$ and, hence, from Eq. (12),

$$[M^{\cdot}]_s = \left(\frac{R_i}{2k_t}\right)^{1/2}. \qquad (14)$$

In accepting the stationary state principle, one must realize that the rate of destruction must be fast.

The rate of radical decay can be determined by stopping the initiation. R_i now equals zero, and Eq. (12) becomes:

$$\frac{d[M^{\cdot}]}{dt} = -2k_t[M^{\cdot}]^2. \qquad (11)$$

This equation integrates to

$$2k_t t = \frac{1}{[M^{\cdot}]} - \frac{1}{[M^{\cdot}]_0} \qquad (15)$$

where $[M^{\cdot}]_0$ is the radical concentration at the time when R_i becomes zero and equals $[M^{\cdot}]_s$ if the stationary concentration had been attained prior to switching off the light. A knowledge of k_t again shows that the change of $[M^{\cdot}]$ is rapid. Thus, in this system the radical concentration is rapidly built up and will decay rapidly when the source of initiating radicals is removed.

In addition, it can be shown that the steady state concentrations of the individual radicals will be constant. The rate of change of the concentration of the radical $M_n^{\cdot}$ is given by:

$$\frac{d[M_n^{\cdot}]}{dt} = k_p[M_{n-1}^{\cdot}][M] - k_p[M_n^{\cdot}][M] - k_t[M_n^{\cdot}]\sum[M_m^{\cdot}]. \qquad (16)$$

Equation 16 may be equated to 0 for all values of n.

When this steady state is reached, the concentration of radicals in the system is given by Eq. (14). Substituting this into Eq. (9) and using the value of R_i in Eq. (4) for the radical-initiated polymerization, the rate of polymerization becomes:

$$\frac{d[M]}{dt} = \frac{f^{1/2}k_i^{1/2}}{k_t^{1/2}} k_p[I]^{1/2}[M]. \qquad (17)$$

This equation reveals that the rate of build-up of the polymer (or loss of monomer) is first order with respect to the monomer concentration but only of half order with respect to the concentration of initiator.

1.3.2.5. Degree of Polymerization

In order to known how many units of monomer are used up (on an average) for each radical which starts a chain, the ratio of the rate at which the monomer is consumed to the rate at which the chains are started must be determined. This quantity is called the kinetic chain length, usually written as v, i.e.,

$$v = R_p/R_i \,.$$

At the steady state, the velocity of initiation is identical to that of termination, so that

$$v = R_p/R_t \,.$$

Substituting the value from Eqs. (9) and (11) gives:

$$v = \frac{k_p[M][M^{\cdot}]}{2k_t[M^{\cdot}]^2} = \frac{k_p[M]}{2k_t[M^{\cdot}]} \,. \tag{18}$$

Since it is the steady state which is considered, the radical concentration may be eliminated using Eqs. (4) and (14) to give:

$$v = \frac{k_p}{2(fk_ik_t)^{1/2}} \frac{[M]}{[I]^{1/2}} \tag{19}$$

or in terms of the rate of propagation (Eq. 17) to give

$$v = \frac{k_p^2}{2k_t} \frac{[M]^2}{R_p} \,. \tag{20}$$

It is therefore possible to calculate the kinetic chain length, provided that the constants as well as the concentrations of both monomer and initiator are known. Furthermore, the number average degree of polymerization, P_n, is equal to the rate of polymerization/rate of termination of radicals. Hence, the kinetic chain length is related to P_n by $P_n = 2v$. According to Eq. (20), a straight line should result if the degree of polymerization, D.P., were plotted against the polymerization rate, R_p, for a constant monomer concentration, on the assumption that termination occurs by coupling.

Results for methyl methacrylate were found to be satisfactorily in agreement with this theory for some initiators and not for others. Generally, P_n was reported to be less than $2v$. A lower D.P. than that calculated may arise if termination occurs by disproportionation. When all radicals are terminated by this method, two molecules will be produced for every reaction so that the D.P. will be equal to the kinetic chain length, i.e., $P_n = v$. Thus, depending on the proportion of mutual termination to disproportionation, P_n lies between $2v$ and v. Termination of a growing polymer chain by disproportionation does not alter the kinetics since the reaction may be written as:

$$M_n^{\cdot} + M_m^{\cdot} \xrightarrow{k_d} M_n + M_m \,. \tag{21}$$

This process is essentially a bimolecular one, i.e.,

$$-\frac{d[M^\cdot]}{dt} = 2k_d[M^\cdot]^2 \tag{22}$$

assuming k_d to be independent of the size of the molecule.

1.3.2.6. Transfer Reactions

In a polymerization system, propagation occurs when a growing radical collides with a monomer. Alternatively, a hydrogen atom may be transfered to a monomer molecule ending one chain and producing a radical which will initiate another; thus

$$\text{—}CH_2\text{—}\dot{C}HX + CH_2{=}CHX \rightarrow \text{—}CH{=}CHX + CH_3\dot{C}HX\,.$$

In the presence of a solvent, transfer of an atom from the solvent to the growing radical may take place. Also, a radical may react with the initiator, thereby ending one chain and starting another. These reactions may be represented as follows:

$$M_n^\cdot + M \xrightarrow{k_{trM}} M_n + M^\cdot\,, \qquad \text{transfer to monomer, M,}$$

$$M_n^\cdot + S \xrightarrow{k_{trS}} M_n + S^\cdot\,, \qquad \text{transfer to solvent, S,}$$

$$M_n^\cdot + I \xrightarrow{k_{trI}} M_n + I^\cdot\,, \qquad \text{transfer to initiator, I.}$$

The effect of any of these reactions is to increase the number of polymer molecules, thereby decreasing the D.P. However, the number of radicals in the system remains unaltered provided the radicals formed in these reactions set new chains going. Consequently, the rate of polymerization is unchanged.

Termination reactions of the type

$$S^\cdot + M_n^\cdot \rightarrow M_nS\,,$$

$$S^\cdot + S^\cdot \rightarrow S_1\,,$$

are usually considered to play little part and may be neglected. However, each transfer reaction will certainly terminate a chain so that, for the purpose of calculating the rate of chain termination, each is making its contribution. The average degree of polymerization, $\bar{P}_n$, is given by the ratio of the rate of propagation to the rate of termination by all processes thus:

$$\bar{P}_n = \frac{k_p[M][M^\cdot]}{k_t[M^\cdot]^2 + k_{trM}[M][M^\cdot] + k_{trS}[M^\cdot][S] + k_{trI}[M^\cdot][I]}\,, \tag{23}$$

where $[M^\cdot]$ is the concentration of all the radicals. If all transfer reactions are ignored except that of solvent, Eq. 23 may be written as:

$$\frac{I}{\bar{P}_n} = \frac{k_t[M^\cdot]}{k_p[M]} + \frac{k_{trS}}{k_p}\,\frac{[S]}{[M]}\,. \tag{24}$$

For purposes of simplification, the first term on the right-hand side may be maintained constant provided $[I]^{1/2}$ [M] is kept constant. Under these conditions, $I/\bar{P}_n$, may be plotted against [S]/[M] to give a straight line of slope k_{tr_S}/k_p. The literature contains examples in which the symbols C_M, C_S and C_I are used to denote the ratio of the rate constant for transfer (to monomer, solvent, and initiator respectively) and the constant for propagation.

1.3.2.7. Effect of Temperature

The dependence of the D.P. on the reaction constants and transfer reactions has already been dicussed. The D.P. might also be expected to depend on the temperature, since the reaction constants are temperature-dependent. As will be seen below, the effect of increasing the temperature is to reduce the length of the chain.

The changes in rate constants with temperature are given by the Arrhenius equation

$$k = A\,e^{-E/RT}\,, \tag{25}$$

where

A = the collision frequency factor
E = the energy of activation
T = the absolute temperature
R = the gas constant.

The equation for the D.P. is:

$$\bar{P}_n = \left[\frac{k_p^2}{k_t}\frac{[M]^2}{R_p}\right] = \left[\frac{k_p}{k_t^{1/2}k_i^{1/2}f^{1/2}}\frac{[M]}{[I]^{1/2}}\right] \tag{26}$$

the value of $k_p/k_t^{1/2}$ may be obtained from eqs. (19) and (26). Both these constants obey Eq. (25) so that, in terms of the activation energies for the propagation and termination reactions (E_p and E_t), the relationship of this quantity with temperature may be written as

$$k_p/k_t^{1/2} = A_p/A_t^{1/2}\exp\left[-(E_p - E_t/2)/RT\right]\,. \tag{27}$$

Thus, a plot of $\log(k_p/k_t^{1/2})$ against I/T gives the values of $(E_p - E_t/2)$ and these turn out to be 21–25 kJ/mol, which corresponds to an increase in $k_p/k_t^{1/2}$ of around 30–35% for a temperature rise of 10 °C. Photochemical polymerizations, having initiation reactions which are independent of temperature, have a temperature dependence which is determined by the changes in k_p and k_t.

In radical-initiated reactions, on the other hand, the reaction velocity is determined by $(k_i/k_t)^{1/2}\,k_p$, and hence, the overall activation energy is given by $(E_i/2 + E_p - E_t/2)$ where E_i is the activation energy for the decomposition of the initiator.

The D.P. given in Eq. 26 may be written in terms of activation energies as:

$$\bar{P}_n = \frac{A_p}{(A_i A_t f)^{1/2}} \frac{[\mathrm{M}]}{[\mathrm{I}]^{1/2}} \exp - \frac{\left[E_p - \frac{(E_i + E_t)}{2}\right]}{RT} \tag{28}$$

so that the changes in D.P. are dependent on the value of $E_p - [(E_i + E_t)/2]$. Since this quantity is negative, increases in temperature are accompanied by a decrease in D.P.

With some polymers, depolymerization occurs through a stepwise loss of monomer units if the temperature is high enough. At room temperature, this process is normally so slow that it can be ignored, but, since the activation energy is high, the rate of depolymerization becomes faster, with rising temperature, than that of polymerization. A temperature (the ceiling temperature) may be reached at which the two rates are equal. Above this temperature, the growing polymer chains depolymerize instead of increasing in size.

1.3.2.8. Rate Constants

For a polymerization system with initiation, propagation, and mutual termination, Eqs. (19) and (26) may be used to obtain knowledge of the initiation rate constants k_i, but the propagation rate constant, k_p, and the termination rate constant, k_t, cannot be obtained separately since both equations are dependent on $k_p/k_t^{1/2}$. Although numerical values of this ratio may be assigned for any reaction, a further measurement is necessary to obtain the reaction constants themselves. One way to meet this need is to measure the average life time, τ, of the active center, i.e., the period over which the radical exists from initiation to termination. The value of τ will be equal to the concentration of radicals divided by the rate of their destruction, i.e.,

$$\tau = \frac{[\mathrm{M}^\cdot]}{2k_t[\mathrm{M}^\cdot]^2} = \frac{\mathrm{I}}{2k_t[\mathrm{M}^\cdot]} = \frac{k_p}{2k_t} \frac{[\mathrm{M}]}{\mathrm{R}_p} \quad \text{at steady rate.} \tag{29}$$

τ is the time during which the radical concentration $[\mathrm{M}^\cdot]$ disappears, so that $R_i\tau = [\mathrm{M}^\cdot]$. Since τ is the average lifetime of the kinetic chain, during time τ each radical consumes v monomer molecules. Hence v is related to τ,

$$v = \frac{\tau}{[\mathrm{M}^\cdot]} \frac{d[\mathrm{M}]}{dt}. \tag{30}$$

τ itself is a quantity which must be deduced from a knowledge of $[\mathrm{M}^\cdot]$ and is not directly measurable. Direct measurements of the concentration of radicals by techniques such as paramagnetic resonance are unfortunately not practical as yet and, since the concentrations involved are extremely small, $[\mathrm{M}^\cdot]$ must be deduced from sector disc techniques.

Thus, the three measurements give k_i, $k_p/k_t^{1/2}$ and k_p/k_t; hence, the individual rate constants may be calculated. The method most amenable to the determination of τ is photochemical. Irradiation of monomer systems has the advantage that, during a period of illumination, radicals are being created and their creation can be controlled at will.

1.3.2.9. Initiation Rates

It is possible to determine k_i, provided that the rate of polymerization and the D.P. under steady state conditions are known. Independent values for k_i may also be found from observing the kinetic chain length using the equation $v = R_p/R_i$. The measurements of v could be made by determining the number of initiator fragments in the polymer. One of the simplest experimental methods is to label the catalyst with a radioactive atom and to measure the degree of radioactivity in the polymer.

Moreover, determination of k_i may be accomplished by means of inhibitors such as p-benzoquinone and di-phenylpicrylhydrazyl.

These substances, when added to a monomer, suppress the formation of polymer until they are used up, when the rate of propagation becomes normal. It is assumed that inhibitors react with initial radicals to give products incapable of initiating polymerization. The rate of initiation is given by the inhibitor consumption, provided it is known how many radicals react with each inhibitor molecule. The weakness of this method often lies in lack of knowledge of this number and indeed of the actual reaction of the radical with the inhibitor.

1.4. Vinyl Copolymerization

A very wide range of uses of individual vinyl polymers is attained because polymer properties are dependent on the functional groups contained in the chain backbone. However, there are some applications which demand a combination of properties not available in any single homopolymer. It might seem possible to develop a material with a combination of properties by combining different homopolymers. Unfortunately, the mixing together of two dissimilar homopolymers usually does not result in a single compatible material, but rather it gives a biphasic mixture. The properties of such a mixture are frequently unattractive because of defects arising from weakness, opacity and so on.

It is possible to circumvent this problem by polymerizing a mixture of the two monomers in order to obtain a mixture of the functional groups along the polymer chain. This process, called copolymerization, results in a copolymer with properties which are dependent on its composition and are often intermediate between those of the individual homopolymers.

1.4.1. Copolymer Formation

When two monomers A and B are polymerized together, the following types of copolymers are obtained, depending on the technique used and on the reactivities of the monomers:

Random copolymer

... AAABBBBBAABBBBAAAABBBBBAAABBAA ...

Alternating Copolymer

... ABABABABABABABABABABABABABABABAB ...

Block Copolymer

... AAABBB ... BBBAAA ... or ... AAAAABBBBB ...

Graft Copolymer

```
... AAAAAAAAAAAAAAAAAAAAAAAAAAAAAA ...
          |     |              |
... BBBBBBBB   BBBB ...       BBBBB ...
```

Crosslinked Structure

```
        AAAAA          AAAA ...
            |           |
            ABBBB ... BBBBA
            |           |
    ... AAAAA          AAAA ...
```

Although commercial free radical copolymerization leads to random copolymers, it is also possible to use radical reactions to prepare copolymers consisting of long sequences of the individual monomers, i.e., block and graft copolymers. The sequence of 'A' units in the graft copolymer shown is referred to as the backbone, and the sequence of branches of 'B' units is termed the grafted polymer.

In general, the formation of graft copolymers can be brought about by either (i) initiation of chain growth of monomer B on a prepolymer molecule formed from monomer A, or (ii) termination of chain growth of polymer B by a prepolymer molecule formed from monomer A. Block copolymers can be produced either by linking reactive end groups of two different polymers or by initiating the polymerization of another molecule on the reactive end groups of a polymer chain. Thus, initiating the polymerization of, for example, monomer B with polymeric radicals derived from polymer A, where the free radicals are situated at the ends of the pre-polymer chain or at random positions along the prepolymer backbone, would give rise to block or graft copolymer respectively. The exact structure, or distribution of structures, found in the resultant copolymer depends on two main factors:

(1) the mechanism by which the polymeric radicals are formed, and

(2) the termination reaction in the polymerization of B.

When the production of the initiating radicals for the second stage, i.e., polymerization of monomer B, is by simple fission of a weak link at the end of, or on the side chain of, the prepolymer, two initiating radicals may be formed, a polymeric radical and a small radical derived from the end group or side chains. The polymeric radicals induce polymerization and yield the desired block or graft copolymer, while the small radicals lead to formation of homopolymer of the second monomer unattached to a prepolymer chain. However, homopolymer formation of the second monomer may be avoided if the initiation mechanism can be modified to prevent the formation of small radicals. A mixture of block and graft copolymers may be formed if the initiating radicals are generated simultaneously at the ends of prepolymer chains as well as along the prepolymer backbone. Initiating radicals can also be produced by random main-chain scission of the pre-polymer, leading exclusively to polymeric radicals with the radical in the terminal positions. The formation of homopolymer of the second monomer is avoided and only block copolymers are produced.

Polymerization of the second monomer can be initiated by a conventional initiator in the presence of prepolymer, depending on transfer to either reactive terminal groups or side chains in the prepolymer, to produce polymeric radicals. Incorporation of a divinyl monomer or diene in the formation of the prepolymer leads to vinyl groups pendant to the main prepolymer chain. Formation of nonrandom copolymers then depends on the incorporation of these vinyl groups during the polymerization of the second monomer, effectively giving rise to two branches from the same point on the pre-polymer chain. Alternatively, the unsaturated links may be in the backbone of the prepolymer, as in cis-1,4-polyisoprene, when radicals used to initiate the second stage of the polymerization may add to the double to produce a polymeric radical resulting in a branched structure.

Incorporation of weak linkages in the terminal positions of prepolymer molecules may be achieved by using dihydroperoxides and other multifunctional initiators, or peroxy radicals, in the formation of pre-polymer. For instance, peroxy linkages have been introduced into the main chain of the prepolymer by copolymerization of monomer A with oxygen and into the side groups of the prepolymer by a variety of techniques. It is advantageous to use hydroperoxides at the terminal positions or on the side chains of the prepolymer. In this way, homopolymer formation in the second stage can be avoided when redox initiation is employed.

Polymerization of monomer A in the presence of trichlorobromomethane[20)] or certain sulphur compounds[21)] gives rise to a prepolymer with photosensitive terminal groups, while formation of the prepolymer in the presence of triethylamine introduces terminal amino groups. These terminal groups act as strong transfer agents during the polymerization of the second monomer[22)].

Side chain radicals of the prepolymer can be introduced without simultaneous formation of small radicals in certain cases. These include the abstraction of chlorine atoms from suitable chlorine containing polymers, e.g., poly(vinyltrichloroacetate), by the use of a metal in its zero oxidation state[23)], as in metal carbonyls, hexaphenylisocyanides, and similar derivatives. Cerium^{IV} ions may also be used to produce such polymeric radicals from polymers containing oxidizable side groups[24)]. An advantage of employing these techniques is that they can be used to study termination processes in free radical polymerizations.

The γ-Irradiation of prepolymers usually leads to the random formation of free radicals. Even in the absence of a second monomer, this method can be used directly to produce branched and crosslinked structures. Trapped radicals obtained by irradiation can be used to initiate polymerizations by introducing the monomer. Surface grafting may be achieved by use of a second monomer which does not swell the prepolymer. Radiation grafting may be carried out by irradiation of a prepolymer dissolved in or swollen by the second monomer. Other techniques include the use of spark discharges, ultrasonics, mastication, vapor phase swelling, and bond scission arising from repeated freezing and thawing.

1.4.2. Kinetics of Copolymerization

Copolymerization in its simplest form consists of four propagation reactions where chains ending in units of monomer 1, ($M_1^{\cdot}$) or monomer 2, ($M_2^{\cdot}$) add M_1 or M_2:

$$M_1^{\cdot} + M_1 \xrightarrow{k_{11}} M_1M_1^{\cdot}\,, \tag{3}$$

$$M_1^{\cdot} + M_2 \xrightarrow{k_{12}} M_1M_2^{\cdot}\,, \tag{31}$$

$$M_2^{\cdot} + M_1 \xrightarrow{k_{21}} M_2M_1^{\cdot}\,, \tag{32}$$

$$M_2^{\cdot} + M_2 \xrightarrow{k_{22}} M_2M_2^{\cdot}\,. \tag{33}$$

Each of these reactions will have its own reaction rate constant. These have been written with two suffixes, the former of which refers to the kind of reacting radical, the latter to the kind of monomer being added to the chain.

Reactions 3 and 33 are analogous to homopolymerization, while the "crossover" reactions 31 and 32 represent the tendency for alteration in the chain structure. Statistical reasoning demands that the rates of reactions 31 and 32 be the same:

$$k_{12}[M_1^{\cdot}]\,[M_2] = k_{21}[M_2^{\cdot}]\,[M_1]\,. \tag{34}$$

This leads to the composition equation

$$F_1 = (r_1f_2^2 + f_1f_2)/(r_1f_1^2 + 2f_1f_2 + r_2f_2^2) = 1 - F_2\,. \tag{35}$$

F_1 and f_i are the mol fractions of monomer 1 in the polymer and in the feed respectively. F_2 and f_2 are the values of mol fractions of monomer 2. The reactivity ratios, r_1 and r_2, may be defined as $r_1 = k_{11}/k_{12}$ and $r_2 = k_{22}/k_{21}$. These reactivity ratios represent the relative tendency of a given chain end to add its own type of monomer rather than to add a different monomer. A reactivity ratio less than unity denotes a tendency to alteration; one greater than unity denotes a tendency to formation of blocks of similar monomer units along the chain. The implications pertaining to physical-organic chemistry of particular r values for given monomer pairs have often been discussed.

The reactivity ratio product

$$\pi = r_1r_2 \tag{36}$$

is used to characterize alteration. If $\pi \ll 1$, there is a tendency toward alteration. If $\pi = 1$, the copolymerization is said to be "ideal", since Eq. 35 becomes simply:

$$F_1/F_2 = r_1(f_1/f_2) = (1/r_2)\,(f_1/f_2)\,. \tag{37}$$

In free radical copolymerization, π is frequently very much less than unity and only rarely greater than unity. In ionic copolymerization, π is often, but not always, found close to unity.

Values of reactivity ratios have been tabulated for many hundreds of monomer pairs[25].

The copolymerization process is, in its mathematical essence, a formal analogue of an coin-tossing game where the previous events do not influence the succeeding one. Probability considerations are therefore very useful in describing copolymerization progress. Let us define the conditional probability, P_{12}, that a chain ending in monomer 1 will add monomer 2:

$$P_{12} = K_{12}[M_1^{\cdot}]\,[M_2]/(K_{12}[M_1^{\cdot}]\,[M_2] + K_{11}[M_1^{\cdot}]\,[M_1])\,, \tag{38}$$

and

$$P_{11} = 1 - P_{12}\,. \tag{39}$$

Given the symmetry of Eqs. (3) and (31) to (35) with respect to interchange of the subcripts in Eqs. (3) and (31), we can write immediately

$$P_{21} = 1/(1 + r_2 f_2/f_1)\,, \tag{40}$$

$$P_{22} = 1 - P_{21}\,. \tag{41}$$

It can be shown from Eqs. (37), (38) and (40) that

$$F_1/F_2 = P_{12}/P_{21}\,. \tag{42}$$

Furthermore, the sequence of monomer units along the chain is well described by these probabilities. Actually, there the distribution of sequences will be such that we must describe them by average values. For example, there are three types of pairs of diads: —M_1M_1—, —M_1M_2—, and —M_2M_2—. (It should be noted that formation of diad —$M_1M_2^{\cdot}$— at the chain end can be distinguished from —$M_2M_1^{\cdot}$, but not its presence in the completed chain). The mol fraction of diads, then, is given as follows[26]:

$$\begin{aligned} f_{11} &= P_{21}P_{11}/(\text{normalizing factor})\,, \\ f_{12} = f_{21} &= P_{21}P_{12}/(\text{normalizing factor})\,, \\ f_{22} &= P_{12}P_{22}/(\text{normalizing factor})\,. \end{aligned} \tag{43}$$

The normalizing factor is the sum of the numerators $P_{21}P_{11} + 2P_{21}P_{12} + P_{12}P_{22}$. Using Eqs. (39) and (41), the normalizing factor becomes $(P_{12} + P_{21})$. If one wishes to consider sequences of three units, i.e., triads, they can be described by equations such as

$$f_{111} = f_{11}P_{11}\,; \qquad f_{121} = f_{12}P_{21}\,, \qquad \text{etc.} \tag{44}$$

Diad, triad and higher sequences are measurable[27)] for some copolymers by a variety of wet analytical and instrumental techniques, the most common of which is nuclear magnetic resonance (N.M.R.). If adequate information on intra- and interchain potentials is available, then sequence distribution can, in principle, be used to describe copolymer chain physics and, therefore, properties.

It is important to note that the kinetic model in Eqs. (3) and (31)–(35) is sometimes violated by real systems for reasons other than experimental problems associated with carrying out a copolymerization, such as poor mixing or poor heat transfer. Great care must be taken to recognize those copolymerizations which do not conform to the simple models. Examples of causes of deviations include:

a) Penultimate unit effects;
b) Non-steady-state behavior;
c) Complex formation;
d) Reversibility of propagation.

Penultimate unit effects occur when the rate constant for propagation is a function not only of the monomer and the chain end unit, but also of units preceding the chain end.

$$\sim M_1^{\cdot} + M_2 \xrightarrow{K_{12}} \sim M_1M_2^{\cdot} \tag{31}$$

$$\sim M_1M_1^{\cdot} + M_2 \xrightarrow{K_{112}} \sim M_1M_1M_2^{\cdot} \tag{45}$$

$$\sim M_2M_1^{\cdot} + M_2 \xrightarrow{K_{212}} \sim M_2M_1M_2^{\cdot} . \tag{46}$$

$$K_{112} \text{ is not equal to } K_{212} . \tag{47}$$

Kinetic equations have been developed for these situations to describe copolymer composition as a function of monomer feed. It is important to point out that sequence distribution is more sensitive to penultimate unit effects than is composition[28)].

Non-steady-state behavior occurs when equation (34) is not satisfied. This is especially important in the case of low molecular mass copolymers, where the influence of the initiation reaction may affect the short chain composition. It is sometimes important in the case of ionic copolymerization; a good example is the anionic copolymerization of styrene and methyl methacrylate where a methyl methacrylate chain end cannot add a styrene monomer.

Complex formation between chain ends and monomer is quite commonly used as an explanation for some anomalous ionic copolymerization results[29)]. Charge transfer complex formation between monomers has been the subject of work by Gaylord and Takahashi[30)], which has revolutionized some of our thinking on the general topic of copolymerization.

The reversibility of copolymerization would also be expected to be quite important when considering copolymerization at or near the ceiling temperature of one or both of the comonomers being polymerized.

1.5. Molecular Mass Distribution

During the course of polymerization, a radical may undergo a number of reactions: propagation, transfer and termination. The sequence of reactions of individual radicals will differ. As a result, the polymer obtained will consist of molecules of various molecular masses. The actual distribution is determined by the reaction kinetics prevailing during the polymerization reaction. A change in the reaction conditions definitely influences the molecular mass distribution of the polymer which, in turn, alters the properties of the polymer.

1.6. Polymer Structure

The polymer structure depends mainly on the mode of addition in the propagation reaction, for which there are four possible reactions:

$$\text{—CH}_2\text{—}\dot{\text{C}}\text{XY} + \text{CH}_2 = \text{CXY} \rightarrow \text{—CH}_2\text{—CXY—CH}_2\text{—}\dot{\text{C}}\text{XY}\,, \quad (48)$$

$$\text{—CH}_2\text{—CXY}^{\cdot} + \text{CXY} = \text{CH}_2 \rightarrow \text{—CH}_2\text{—CXY—CXY—CH}_2^{\cdot}\,, \quad (49)$$

$$\text{—CXY—CH}_2^{\cdot} + \text{CXY} = \text{CH}_2 \rightarrow \text{—CXY—CH}_2\text{—CXY—CH}_2^{\cdot}\,, \quad (50)$$

$$\text{—CXY—CH}_2^{\cdot} + \text{CH}_2 = \text{CXY} \rightarrow \text{—CXY—CH}_2\text{—CH}_2\text{—CXY}^{\cdot}\,. \quad (51)$$

Reactions (48) and (50) give rise to a regular structure, termed the head-to-tail structure, while reactions (49) and (51) lead to the incorporation of head-to-head and tail-to-tail linkages. All processes functioning simultaneously would bring about a random structure in the polymer.

A number of reports show that free radical polymerization leads to predominantly head-to-tail addition. Destructive distillation of polystyrene yielded a series of products with phenyl groups on alternate carbon atoms; no products with phenyl groups on successive carbon atoms could be detected. Oxidation of poly(vinyl alcohol), derived from poly(vinyl-acetate), with periodic acid was used to detect the glycol structure. No reduction in the periodic acid concentration could be detected, thereby demonstrating that the polymer contains a predominantly 1,3 structure. Futhermore, the viscosity of an aqueous solution of poly(vinyl alcohol) decreases on reaction with periodic adic, or periodate ions. These cause chain scission of 1,2-glycol units and a reduction in the molecular mass of the polymer corresponding to 1% of the total number of 1,2 units in the polymer.

Evidence for head-to-head structure in free radical polymerized poly(vinyl fluoride) and poly(vinylidene fluoride) and their copolymers has been obtained from nuclear magnetic resonance (NMR) studies. In poly(vinylidene fluoride) it was estimated that at least 5% of the monomers are reversed.

In head-to-tail addition reactions, reaction (48) is more likely than (50) due to higher stability of the radical produced by the former reaction. This arises from the fact that the radical will be stabilized by the substituent groups on the carbon atom associated with the unpaired electron. Direct evidence that propagation occurs by reaction (48) has been obtained by e.s.r. techniques.

In head-to-tail addition polymers, every alternate carbon atom in the backbone is asymmetric. In free radical polymerization, these asymmetric centers will be controlled by steric and electronic interactions of the substituent groups on the monomer. According to the definition of Natta et al., an isotactic chain is one in which all the asymmetric carbon atoms have the same configuration (all d- or l-); in a syndiotatic chain the configurations are alternately d- and l-; and in an atactic chain the arrangement is random. On the basis of these definitions, polymers prepared by a free radical mechanism are atactic.

To obtain a detailed knowledge of the polymer structure, it is necessary to define tacticity in terms of local structure. Natta described tacticity in terms of the bonding between adjacent units in the chain and considered two units to be isotactically bonded if their asymmetric carbon atoms were in the same configuration, and syndiotactically bonded if in the opposite configuration. Tacticity has been defined in terms of triads of monomer units. An isotactic sequence is defined as ddd or lll, when the central unit is an isotactic unit. Syndiotactic sequences are defined as ldl and dld, and heterotactic sequences as ddl, dll, lld, or ldd. This definition is the most convenient for NMR studies of the substituents on the α-carbon atom, and the relative proportions of these sequences can be determined experimentally for certain polymers. For studies on the methylene resonance it is more convenient to consider the chain structure in terms of diads.

On the basis of polymers, for which a detailed structural analysis has been achieved, it may be concluded that, in free radical vinyl polymerization and in the absence of overiding electronic interactions, there is some preference for syndiotactic addition, particularly at low temperatures. It has been pointed out that the relative proportions of iso- and syndiotactic linkages in polymers prepared by free radical polymerization and their temperature dependence can be predicted from a consideration of preferred conformation of the propagating radical, which results from hindered rotation about the terminal carbon-carbon bond.

References

1. Jackson, E., Wunderlich, J. A., Lipscomb, W. N.: Acta Cryst. *14*, 598 (1961)
2. Skraup, L. H., Konig, J.: Monatsh. *22*, 1011 (1901)
3. Barker, S. A., Bourne, E. J., Stacey, M., Whiffen, D.: J. Chem. Soc. *171* (1954)
4. Kuhn, W.: Ber. *63*, 1503 (1930)
5. Reeves, R. E.: Advan. Carbohydrate Chem. *6*, 107 (1951)
6. Skrigan, A. I., Shishko, M. A., Zhbanov, R. G.: Vesti Akad, Nauk Belarusk, SSR, Ser. Fiz Tekhn. Nauk *61*, 4 (1962): Through Abstr. Bull. Inst. Paper Chem. *33*, 8492 (1963)
7. Stepanov, B. I., Zhbanov, R. G., Marupov, R.: Vysokomol Soedin *3*, 1633 (1961)
8. Heuser, E.: Textile Res. J. *20*, 828 (1950)
9. Timell: Studies on Cellulose Reactions, Handlingar N.R. 205 (1950) P. 105
10. Turbak, A. F., ed.: Cellulose Technology Research, A.C.S. Symposium Series, No. 10., Amer. Chem. Soc., Washington: D.C. 1975
11. Goldstein, I. S., ed.: Wood Technology; Chemical Aspects, A.C.S. Symposium Series. No. 43, Amer. Chem. Soc., Washington: D.C. 1977
12. Arthur, J. C. Jr. ed.: Cellulose Chemistry and Technology, A.C.S. Symposium Series No. 48, Amer. Chem. Soc., Washington: D.C. 1977

13. Arthur, J. C. Jr., ed.: Textile and Paper Chemistry and Technology A.C.S. Symposium Series, No. 49, Amer. Chem. Soc., Washington: D.C. 1977
14. Allen, P. E. M., Patrick, C. R.: Kinetics and Mechanisms of Polymerization Reactions, Ellis Herwood 1974
15. Jenkins, A. D., Ledwith, A. eds.: Reactivity, Mechanism and Structure in Polymer Chemistry, New York: Wiley 1974
16. Odian, G.: Principles of Polymerization, McGraw-Hill 1970
17. Blackley, D. C.: Emulsion Polymerization, Applied Science 1975
18. Schildknecht, C. E., Skeist, I., eds.: Polymerization Processes, Wiley — Interscience 1977
19. Ceresa, R. J., ed.: Block and Graft Copolymerization, Vols 1 & 11, New York: Wiley 1977
20. Dunn, A. S., Stead, B. D., Melville, H. W.: Trans, Faraday Soc. *50*, 279 (1954)
21. Otsu, T.: J. Polymer Soc. *26*, 236 (1957)
22. Bamford, C. H., White, E. F. T.: Trans. Faraday Soc. *52*, 716 (1956)
23. Bamford, C. H., Eastmond, G. C., Robinson, V. J.: Trans. Faraday Soc. *60*, 751 (1964)
24. Hebeish, A., Kantouch, A., El-Rafie, M. H.: J. Appl. Polymer Sci. *15*, 1921 (1971)
25. Young, L. J. in: Polymer Handbook, Brandruß, J., Immergut, E. H. (eds.): Ind. Edition, 1974, P. II–105
26. Fischer, T., Kinsinger, J. B., Wilron, C. W. III: J. Polymer Sci. *B 4*, 379 (1966)
27. Harwood, H. J.: Angew. Chem. (Intl. Ed.) *4* 394 and 1051 (1965)
28. Berger, M., Kuntz I.: J. Polymer Sci. *A 2*, 1687 (1964)
29. O'Driscoll, K., Patsiga, R.: J. Polymer Sci. *A 3*, 1037 (1965)
30. Gaylord, N. G., Takahashi, A.: J. Polymer Sci. *B 6*, 743 and 749 (1968)

2. Vinyl Graft Copolymerization onto Cellulose

2.1. Introduction

Cellulose, in particular cotton, is the most widely used textile fiber. This preeminence is due to a happy combination of properties, such as abundance, low cost, fine cross-section, high strengh and durability, high thermal stability, good mechanical properties, ability to absorb moisture, easy dyeability, wearing comfort, etc. However, cellulosic fibers have some inherent drawbacks which can be briefly enumerated as follows: (a) poor solubility in common solvents, which makes improvements in fibers and yarns through spinning processes almost impossible; (b) poor crease resistance, which makes garments made from cellulosic fibres crumple easily during wear; (c) lack of thermoplasticity, which is a requirement for heat setting and shaping of garments; and (d) poor dimensional stability, resulting in distortion of the garments during laundering and ironing. These drawbacks, and the fact that cellulose has encountered stiff competition from synthetic fibers, have directed attention toward improving the properties of cellulose. To achieve such improvements, the physical and/or chemical structure of cellulose must be changed.

The physical structure of cellulose can be changed either by swelling or by regeneration. Cellulose can be swollen in a suitable swelling agent and then partially deswollen by removal of the swelling agent. There is practically no change in the chemical structure of the cellulose, whether fiber, crumb, or film, but there are considerable changes in the physical fine structure, resulting in an enhancement of strength, luster, and reactivity. In regeneration, the native cellulose is treated to render it soluble and then is regenerated from solution. This process, which is the basis of rayon and cellophane manufacture, gives rise to a transformation of the crystal structure from a Cellulose 1 lattice to a Cellulose 11 lattice and results in a corresponding modification of the physical structure.

The chemical structure of cellulose can be changed in several ways:

a) By substitution of the cellulose hydroxyls whereby the cellulose molecules are altered through introducing side groups, usually by an etherification or by an esterification reaction. The effect of this is to eliminate or at least to reduce some of the outstanding drawbacks of the cellulose, such as flammability, susceptibility to rot and mildew, swellability, etc., and is dependent on the type and extent of the reaction as well as on the nature of the substituent groups (see Chapter 6).
b) By reacting cellulose with bi- or polyfunctional compounds, which results in the production of cross-links or resinification products in the cellulose, thereby stabilizing its structure. Such treatments impart crease resistance, shrink resistance, and a number of other properties to the cellulose[1].

c) By preparing a branched cellulose or what is called a graft copolymer of cellulose, a process in which attempts have been made to combine synthetic polymers with cellulose to produce material with the best properties of both. This process is known as grafting, usually done by modifying the cellulose molecules through creation of branches of synthetic polymers that confer certain desirable properties on the cellulose without destroying its intrinsic properties.

The amount of grafted monomers is usually taken as the increase in weight after solvent-extraction of homopolymer and drying. The percent grafting is defined as the ratio of the increase in weight to the initial cellulose weight as follows:

$$\%\,\text{Grafting} = \left(\frac{\text{Final cellulose weight} - \text{Initial cellulose weight}}{\text{Initial cellulose weight}}\right) \times 100. \tag{52}$$

The efficiency of grafting is defined as the ratio of the amount of grafted monomer to that of polymerized monomer.
Efficiency of Grafting =

$$\left(\frac{\text{Final cellulose weight} - \text{Initial cellulose weight}}{\text{Total polymer weight after reaction} - \text{initial cellulose weight}}\right) \times 100. \tag{53}$$

During synthesis of cellulosic graft copolymers, block copolymers may form if the initiating centers are formed at the end of cellulose molecules or at broken chain ends of the cellulose. The block and graft copolymers to be considered here are defined by the schematic structure shown in Figure 1. The block copolymer is composed of a cellulose chain to which synthetic polymer segments are chemically linked to form a linear molecule. It is believed that though block copolymers of cellulose would usually be expected to resemble the schematic structure A, schematic structure B cannot be ruled out. The graft copolymer is comprised of a linear backbone polymer, i.e., cellulose, containing branches or "grafts" of the synthetic polymer, C. These branches may vary considerably in length depending on particular requirements and copolymerization conditions.

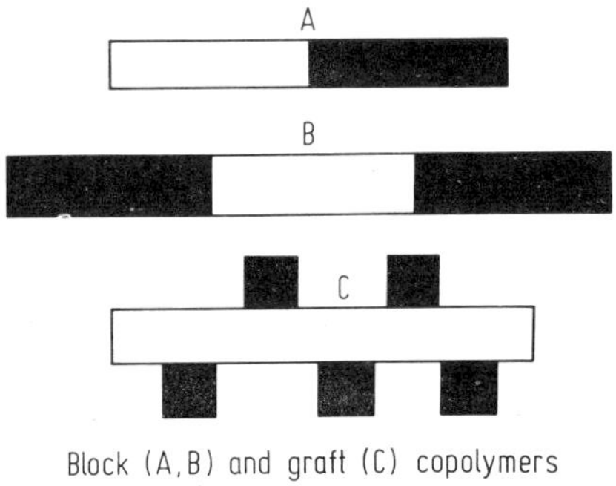

Fig. 1. Schematic representation of block and graft copolymer (white and black segments represent the cellulose backbone and the synthetic polymer, respectively).

2.2. Conditions Affecting Grafting

2.2.1. Chemical Conditions

The preparation of cellulose graft copolymers differs from the preparation of fully synthetic graft copolymers because cellulose is insoluble in all common organic solvents. Grafting onto cellulose is therefore a heterogeneous reaction in which the physical structure and the state of aggregation of the cellulose play a significant role. There are, however, two approaches to homogeneous grafting onto cellulose. One involves the use of soluble cellulose derivatives[2]; the other, the preparation of "inclusion celluloses"[3]. In the latter, the monomer to be grafted is introduced into the cellulose lattice by a series of consecutive solvent exchanges.

Regardless of the type of reaction involved in grafting, the synthetic polymer is generally formed in the presence of cellulose, usually with the intention of linking the synthetic polymer chains to the cellulose by primary valence bonds. Whether or not chemical bonding is actually achieved, the synthetic polymer is usually so intimately associated with the cellulose that the components cannot be separated by any of the normal fractionation techniques. Thus, in the conventional method of grafting, cellulose, initiator, monomer, and solvent are all present. In a complex system like this, several reactions take place simultaneously and the ultimate graft formation depends on the influence of the following factors: (a) solution of the monomer in the solvent and its diffusion from the solution phase to the cellulose, (b) adsorption of the monomer on the cellulose (c) initiation of the active sites on the cellulose backbone, (d) formation and propagation of graft on the cellulose, (e) termination of the active sites on the cellulose backbone, and (f) homopolymerization in solution.

The first four factors cause increased graft formation; the last two, a decrease. In general, these factors are favored by higher concentrations, longer times, and greater temperatures, and there are usually optimum conditions for maximal grafting. The reason for this is that as the grafting reaction proceeds, the monomer and initiator concentrations decrease, as does the number of available grafting sites[4].

To achieve increased amounts of graft formation, various approaches may be undertaken. These approaches are based on the fact that two competitive reactions occur simultaneously during the grafting process, i.e., grafting and homopolymerization. Depressing the latter would be expected to promote the graft formation. Among these approaches are (i) carrying out the initiation (reaction of cellulose with the initiator to produce activated cellulose) and grafting (reaction of the activated cellulose with the monomer) separately, (ii) allowing the monomer to be absorbed on the substrate first (to increase availability of the monomer for grafting), (iii) portion-wise addition of the monomer and initiator during the grafting operation, and (iv) inclusion of the monomer in the cellulose lattice by a series of solvent exchanges prior to activation for grafting.

It is important to note, however, that modification of the technique of grafting as given above depends on the reactivity of monomer and type of initiator, for not all monomers respond to all types of initiators under all conditions. The nature of the activated center depends also on the initiator. Hence, the choice of a modification of the grafting technique requires an understanding of the nature of monomer and initiator.

It is also of importance to point out that whether grafting takes place near the site of production of primary radicals or through some chain-transfer process after the growth of the macroradical, will depend on the catalyst system concerned, the nature of the cellulose material, and the nature of the growing macroradical. The various macroradicals are known to have different activities with regard to abstracting hydrogen atoms from carbohydrates or other materials.

Differences in stability of macroradicals initiated on cellulose and starch are to be expected because of significant differences in their physical structure. However, one cannot be certain that creation of such radicals will lead to initiation of polymerization, since these radicals may be destroyed by other processes. Such processes might involve coupling to form crosslinks in cellulose, abstraction of hydrogen from water, or reaction with impurities that may be present in the grafting medium. Thus, the chemical composition of the starting material has to be well defined; that is to say, the presence of traces of inorganic impurities, the presence of transition metal cations, or the presence of organic contaminants with reducing properties, must be avoided if the kinetics and mechanism of the grafting system are to be studied.

Any foreign matter might act as either a promotor or an inhibitor for the polymerization system, affecting the rate and extent of grafting considerably.

2.2.2. Physical Conditions

Although favorable chemical conditions and suitable techniques are necessary for promoting grafting, they are not the only factors of importance. Most of the earlier attempts to achieve grafting onto cellulose gave poor results. The reasons for the greater success of the more recent work are related to modifications to physical rather than chemical conditions. Most of the work has been done with fibers and films, and it is now realized that swelling is an important prerequisite to making the cellulosic material accessible to the monomer, thus enhancing diffusion processes. With the relatively higher concentration of cellulose in swollen fibers or films compared to that of monomer, the cellulose-monomer reaction, i.e., grafting, will be preferred over a monomer-monomer reaction, i.e., homopolymerization. Indeed, studies of the effect of the cellulose fine structure on the grafting reaction show that the graft yield is a function of the cellulose accessibility[5-7]. Nevertheless, the possible role of swelling and diffusion will often render quantitative interpretation of grafting studies a very difficult task. Even if none of the reactions is diffusion-controlled, the composition of the fiber of film changes gradually as grafting proceeds, and the solubility of the compounds may also change correspondingly. This has been noted in the grafting of cellulose and modified celluloses with vinyl and acrylic monomers[8].

Another point which may be made is that the physical characteristics of grafting onto different cellulosic substrates are not necessarily comparable. For example, grafting onto hydrocellulose, where most of the amorphous portion of the cellulose is removed, is more of a surface phenomenon, whereas grafting onto swollen cottons and viscose takes place inside the fiber as well at the surface. The cross-sectional area of native cotton, cotton treated with sodium hydroxide, and viscose was found to be enlarged and round after grafting when compared with ungrafted samples. With hydrocellulose, on the other hand, it was difficult to distinguish between the cross-sections of untreated and grafted hydrocellulose[9]. In their work on graft polymerization of acrylonitrile to cellophane film using the photo-initiation technique, Geacintov et al.[10] reported that grafting is predominantly a surface reaction and that the amount of grafting is relatively independent of the thickness of the film. Their results indicate that the amount of grafting varies directly with the surface area of the film, the amount of graft per unit area remaining constant. In another communication, the same outhors[11] report that this is only approximately true and that the experimental evidence obtained so far is not sufficient to conclude that grafting is exclusively a surface reaction. The rate of the grafting reaction was also found to vary with the physical state of the cellulose substrate. As can be seen from Figure 2, there is a considerable difference in the rates of grafting obtained with fibers, yarn, and fabric. This difference is unequivocally due to the physical restriction on swelling introduced when the fibers are twisted into yarns and woven into fabrics[12].

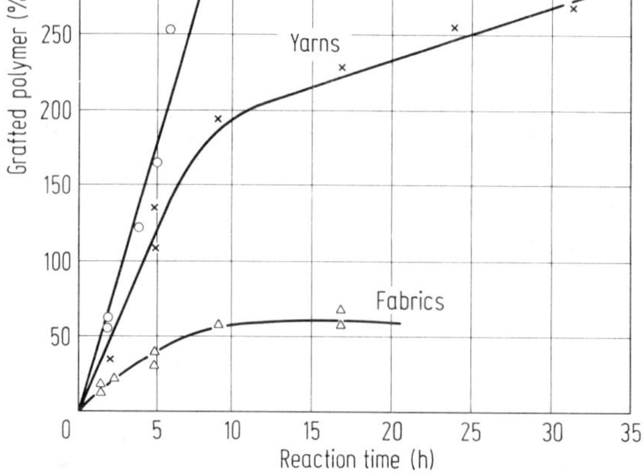

Fig. 2. Post-irradiation grafting of acrylonitrile in 75% aqueous $ZnCl_2$ to cotton fibers, yarns, and fabrics at about 23 °C. Irradiation conditions: dry and in nitrogen, 10^6 R. Reprinted from Ref. [12], p. 312, by courtesy of Textile Research Institute.

2.2.3. Nature of Substrate

Cellulose is composed of anhydroglucose units joined through a hemiacetal linkage into long, linear, chain-like molecules held together laterally by hydrogen bonds. X-ray analysis has shown that cellulose fibers contain both amorphous (disordered) and crystalline (ordered) regions. For cellulose to give an X-ray diffraction pattern, the anhydroglucose units must be arranged in a regular and repeated manner. Hence, the fibers are thought to consist of alternating regions of amorphous and crystalline structure, but with certain molecular chains running through both regions to give an irregular flexible network. There is no sharp boundary between the crystalline and disordered areas. Owing to this heterogeneous nature, graft copolymerization, particularly when performed in an aqueous medium, would be expected to be affected by the pore size of cellulose, the crystalline-amorphous ratio, the degree of orientation, the size of crystallites and, in the case of chemically modified celluloses, the nature, amount and distribution of the substituent groups.

During a copolymerization reaction, grafting starts on the amorphous parts of the fiber. Localization of the grafted polymer in those areas causes the more regular areas to become more porous; the grafting reaction then proceeds in these regions.

Since cellulose fibers possess a high structural regularity, the cellulose acts as a matrix under whose influence a directed graft polymerization of several synthetic polymers can occur. This process is not confined to grafting in the gas phase where there is no disorientation because of the solvent effect. It has been shown[13)] by X-ray analysis and the determination of dichroism in the infrared region, that grafting acrylonitrile and vinyl chloride onto hydrated cellulose fibers produces an orientation of the grafted chain. Such orientation takes place if the polymer formed does not dissolve in its own monomer. Conversely, the presence of monomer always brings about disorientation of the grafted chain.

The degree of orientation of a grafted crystallizable polymer is always greater than that of a non-crystallizable polymer. A number of polymers, for instance, poly(methyl methacrylate), do not generally crystallize on grafting. Obviously, the degree of orientation of the grafted chain is greater when the degree or regularity (orientation) of the structural elements of the cellulose fiber is greater. This effect is, therefore, more clearly shown in the last stages of the grafting process when grafting occurs on the oriented structural elements of the fiber[14)].

The intercrystalline swelling of cellulose by water occurs predominantly in the accessible, less ordered regions of cellulose. Diffusion, and, therefore, the grafting of a monomer, is probably restricted to these regions. Sakurada et al.[15)] have shown that no change in the X-ray diffraction pattern was observed in poly(vinyl alcohol) grafted with large amounts of methyl methacrylate and reported that grafting probably occurred only in the disordered regions. That grafting occurs exclusively in the accessible regions of cellulose has recently been supported by conductivity data and dielectric constant measurements of irradiated celluloses[16)]. Figure 3 shows the graft yields obtained with native cotton, viscose, ramie, hydrocellulose, and cottons treated with sodium hydroxide (24%, w/w), ethylenediamine, EDA (78%, w/w), or zinc chloride (70%, w/w)

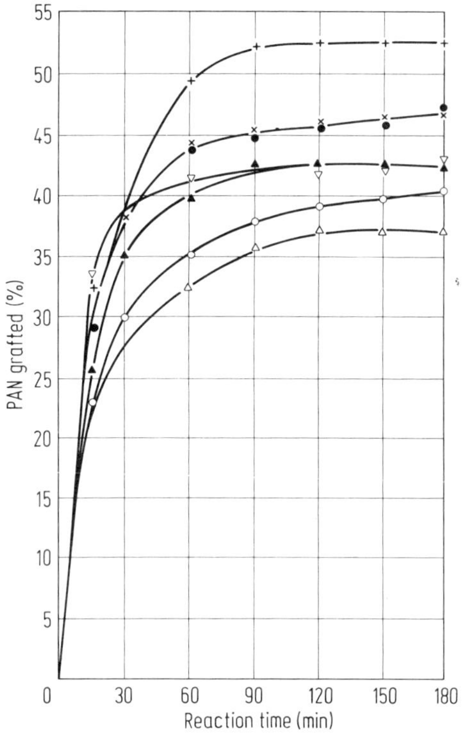

Fig. 3. Grafting of polyacrylonitrile on different cellulosic materials with CAN at 40 °C: (○) native cotton; (+) NaOH-cotton; (▲) EDA-cotton; (×) $ZnCl_2$-cotton; (△) hydrocellulose; (▽) viscose; (●) ramie; $[Ce^{IV}]$ = 0.006 M in 1% HNO_3; cellulose, 1 g; acrylonitrile, 1 cm^3; liquor ratio 30. Reprinted from Ref. [6]; p. 1629, by courtesy of John Wiley and Sons, Inc.

using Ce^{IV} ammonium nitrate (CAN) and acrylonitrile (AN) as initiator and monomer respectively. It can be seen that less grafting was obtained with native cotton, ramie, and hydrocellulose than with the three swollen cottons. This was expected, since swollen cottons have higher accessibility and, therefore, greater reactivity than native cotton, hydrocellulose, and ramie[6].

With viscose, the grafting curve (Fig. 3) shows a higher initial rate than that of many other substrates. However, this levels off to give a maximum graft yield which is less than those obtained with swollen cottons. With viscose, grafting appears to occur at the point of maximum swelling in the Ce^{IV}-acrylonitrile-water system. Beyond this, the polyacrylonitrile probably forms a diffusion barrier and inhibits the diffusion of acrylonitrile monomer. It is well known that the average pore size in cotton is considerably larger than in viscose. Huang and Rapson[17] have reported that radiation grafting of cotton yarn in styrene solution containing methanol, ethanol, and acetic acid was found to proceed without the presence of water, whereas with viscose no grafting took place under these conditions. Thus, smaller pore sizes and the rapid slowing down of diffusion of initiator (Ce^{IV} ions) and monomer, i.e., acrylonitrile as grafting proceeds favors the formation of homopolymer in solution rather than grafting. This may explain the low value of maximum graft yield obtained with viscose.

It is interesting to point out, in this connection, that when treatments of celluloses with CAN, namely, initiation followed by exposure to monomer to

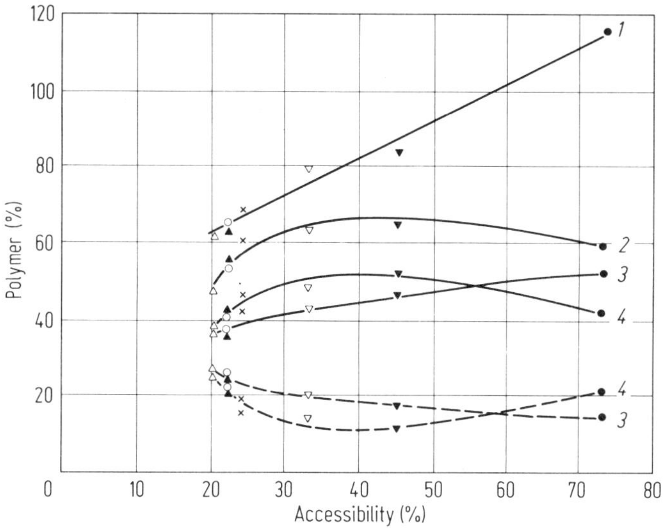

Fig. 4. Relationship between maximum graft yields and accessibility of different substrates; (———) percent PAN grafted; (– – – –) percent homopolymer formed in solution; (●) native cotton; (▼) NaOH-cotton; (▽) EDA-cotton; (×) $ZnCl_2$-cotton; (△) hydrocellulose; (▲) viscose; (○) ramie. Curves: (1–4) percent graft yields obtained by methods I–IV, respectively; $[Ce^{IV}] = 0.006$ M in 1% HNO_3; cellulose, 1 g; temperature 40 °C; reaction time 3 hr. Acrylonitrile: (*1*, *2*) 15 cm^3; (*3*, *4*) 1 cm^3. Reprinted from Ref. [6], p. 1639, by courtesy of John Wiley and Sons, Inc.

provide grafting, were carried out separately and sufficient monomer used, or when sufficient time was given for diffusion and adsorption of monomer before commencing the grafting reaction, viscose showed higher graft yields than swollen cottons. This may be seen from Figure 4. Kesting and Stannett[18] reported higher graft yields for cellophane than for cotton when both substrates were first irradiated with γ-rays from Co-60. The irradiated celluloses were than immersed in the monomer solution. Styrene, dioxane, and water formed the monomer-solvent system. The high yields of grafting obtained with viscose would also be anticipated in view of its greater swelling and accessibility when compared with cotton fibers. A similar pattern was observed when grafting was initiated by mutual irradiation with gamma rays from Co-60 (Fig. 5)[19].

Turning now to the data presented in Figure 4, in method 1, the substrate (1 g) was reacted with CAN (0.006 M in 1% HNO_3) for 3 hours at a liquor ratio of 30.1. The excess Ce^{IV} ion was then drawn off by suction. AN monomer (15 cm^3) was then introduced and grafting was conducted for 3 hours. The entire sequence of operations was carried out under nitrogen. Thus, the two reactions, i.e., initiation and grafting, were carried out in two separate steps and homopolymer formation was practically avoided. Method II was identical to method I, except that the reactions were performed in air instead of nitrogen.

In method III, the cellulose sample (1 g) was immersed for 3 hours in 30 cm^3 1% HNO_3 to which 1 cm^3 of AN was added. Then 1 cm^3 of 0.18 M CAN

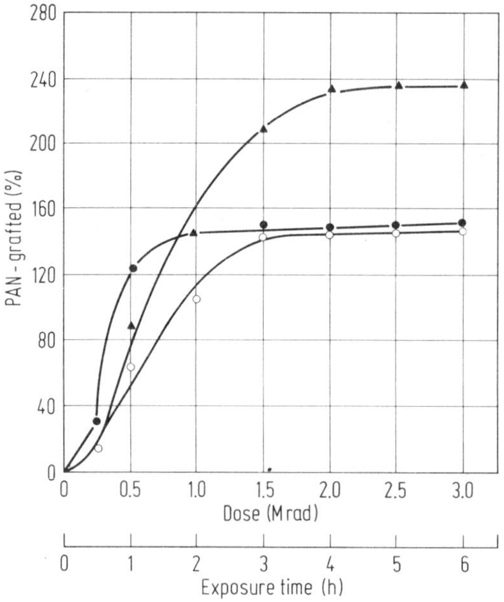

Fig. 5. Grafting of polyacrylonitrile (PAN) on different cellulosic materials using the mutual irradiation technique: (○) native cotton; (△) mercerised cotton; (●) viscose; cellulose = 0.5 g; AN = 15 cm^3, water 0.4 cm^3. Reprinted from Ref.[19], p. 1071 by courtesy of Textile Research Institute.

was added, making an initiator concentration of 0.006 M in the reaction mixture, and the reaction was allowed to proceed for a further 3 hours. In method IV, simultaneous initiation and grafting were carried out using 1 g cellulose, 1 cm^3 AN, and 30 cm^3 of 0.006 M CAN. The reaction temperature in all cases was 40 °C.

Obviously, graft yields are linearly and directly related to accessibility only when the grafting reaction is conducted under conditions which suppress homopolymerization. In method I, it is clear that a linear relationship could be obtained where there was no chance for the homopolymer to be formed. Method II was conducted under conditions identical to method I, but in the presence of air it did not show the same relation. The graft yields obtained by method I are also considerably higher than those of method II. This must be due to the presence of oxygen in the system which would accelerate the decay of free radicals produced by Ce^{IV}. The decay effect seems to be more pronounced in the case of more accessible substrates.

In methods III and IV (cf. Fig. 4), as expected, the homopolymer formed in solution during the grafting reaction decreased with increasing graft yields and vice versa. This clearly indicates that, for a given substrate, grafting and homopolymerization in solution are inversely related, both being dependent on the efficiency of the grafting reaction. In method III, the graft yields increased with increasing accessibility of the substrate. This is probably due to a greater availability of physically absorbed monomer on the more accessible substrate. Methods I and II showed higher graft yields than methods III and IV due to the significantly larger quantity of monomer used, as well as to the suppression of homopolymerization.

Here, too, it is worth noting that hydrocellulose, in spite of its high crystallinity and low reactivity, showed a substantial graft yield which was nearly equal to that of cotton, particularly when grafting was performed under nitrogen by separate initiation and grafting (Method I). On the other hand, when grafting was carried out under the same conditions but in air (Method II), hydrocellulose proved considerably less amenable to grafting than native cotton. This may be because the active sites produced by Ce^{IV} on the cellulose backbone would be expected to occur mainly on the surface of the fiber in case of hydrocellulose. These would be more susceptible to termination by atmospheric oxygen, thereby reducing the graft yield.

Etherification, esterification, or crosslinking of cellulose causes a considerable change in the physical and chemical structure of cellulose. Such changes has been found to have a significant influence on the susceptibility of cellulose toward grafting. This susceptibility depends on the nature, amount, and distribution of the substitutent groups in the modified celluloses. For example, while cellulose bearing either cyanoethyl or carboxymethyl groups showed higher graft yields, cellulose bearing both groups showed lower yields. The presence of acrylamidomethyl groups in the cellulose molecule reduces the reactivity of cellulose to grafting.

The reactivity decreases as the amount of acrylamidomethyl groups increases. The same holds true for acetylated and crosslinked celluloses. On the other hand, the presence of carbamoylethyl or cyanoethyl groups in the cellulose molecules enhances the susceptibility of cellulose to grafting, which increases with the amount of the introduced groups. Details of these studies will be given later (see Chapter 6).

2.3. Grafting Reactions

Vinyl monomers are unsaturated compounds having the general formula CH_2CHX, where X is a functional group of the type —OH, —Cl, —C≡N, etc. Their reaction with cellulose in the presence of a catalyst involves either a free radical mechanism or an ionic mechanism. Because the free radical mechanism is more usual, this will form the main part of the present discussion. However, ionic grafting has also been discussed (see Chapter 4).

Grafting of a vinyl monomer to cellulose is a typical free radical polymerization reaction which involves three distinct aspects, namely, initiation, propagation, and termination. Initiation consists of two steps. The first step is to produce free radicals on the cellulose backbone from the initiator. This is generally achieved by abstraction of a hydrogen atom from the cellulose molecule.

The second step entails the additon of a monomer molecule to the cellulose free radical, resulting in the formation of a covalent bond between the monomer and the cellulose and in the creation of a free radical on the newly formed branch. Thus, a chain is initiated. This is followed by many subsequent additions of monomer molecules to the initiated chain, thereby propagating the chain.

Termination occurs by combination, where the radicals of two growing polymer chains are coupled:

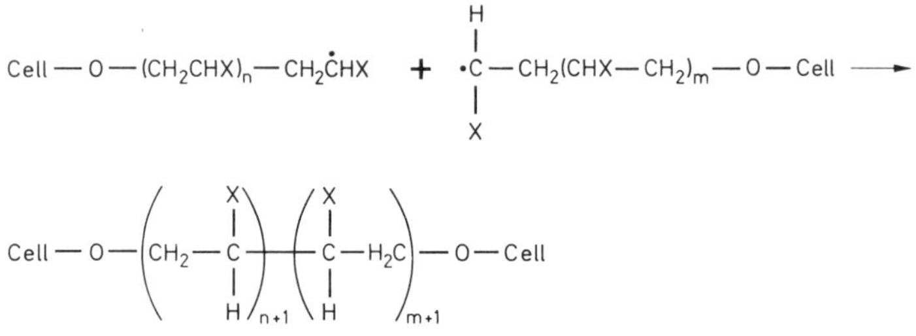

(54)

or by disproportionation where a hydrogen atom is abstracted by one chain from the other:

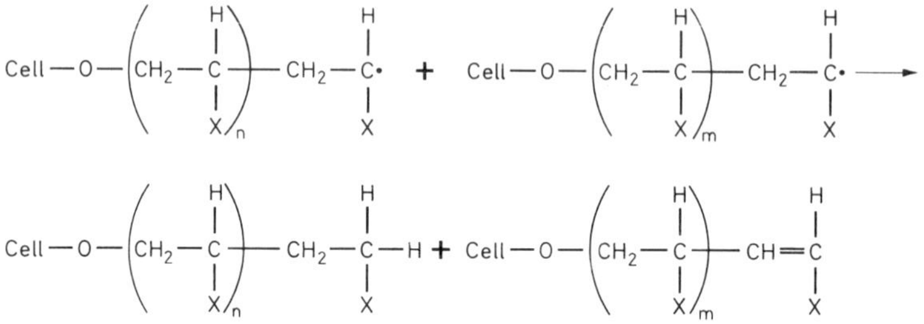

(55)

Termination may also occur by reaction with impurities, initiator, or activated monomer, or by a chain transfer process.

If, on the other hand, a monomer molecule is activated by the initiator, homopolymerization takes place.

2.4. Mechanism of Free Radical Formation on Cellulose

In one form or another, cellulose is perhaps the most abundant organic compound occuring in nature, since it forms the skeletal frame work of all high plants. The purest form of naturally occurring cellulose is obtained in cotton, where the cellulose constitutes some 90% of the fiber, but wood and bast fibers such as flax, ramie, jute, etc., also contain a high proportion of cellulose. In jute, for example, which is composed chiefly of cellulose, hemicellulose and lignin, the proportions are roughly 64% cellulose, 24% hemicellulose, and 15% lignin. For this discussion pure cellulose, or what is called α-cellulose, will be considered.

The chemical structure of cellulose is now well established and consists of long chains of β anhydro-D-glycose units combined through a hemiacetal bridge between carbons 1 and 4 (Fig. 6). The primary hydroxyl groups (at carbon atom 6) lie alternately on either side of the chain so that cellobiose forms the repeating unit. It is apparent from the structure of cellulose that one of its terminal

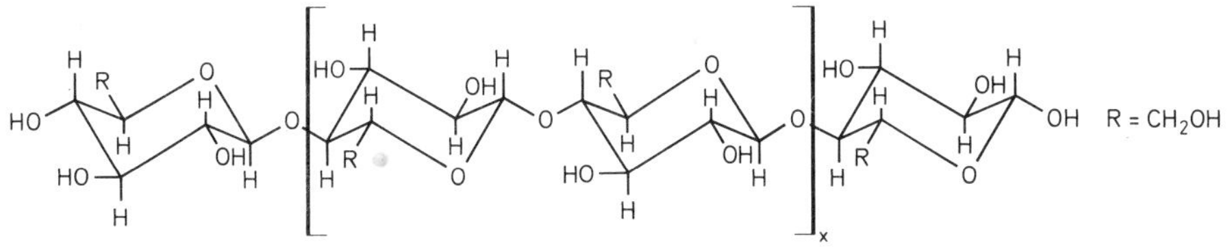

Fig. 6. Cellulose molecule.

groups is potentially aldehydic, while the other is a non-reducing hydroxyl group.

The formation of free radicals on the cellulose molecule can occur by the following possible homolytic cleavage processes: dehydrogenation, dehydroxylation, depolymerization, or cleavage of the glucopyranoside ring[20-27]. The location of the free radical sites on the cellulose molecule and within the fibrous structure will depend on the method of initiation of such sites and on the physical and chemical properties of the cellulose[28-30]. Formation of free radicals on the cellulose molecule can be achieved in a number of ways. These may conveniently be classified into three main types: chain transfer, physical activation, and chemical activation.

2.4.1. Chain Transfer

A normal free radical polymerization is carried out in the presence of cellulose or a cellulose derivative. Termination of a growing polymer chain by abstracting a hydrogen atom from the cellulose molecule gives rise to a cellulose radical which itself initiates graft copolymerization. The chain transfer process which leads to a grafting reaction can be significantly promoted by introducing groups with large chain transfer constants into the cellulose molecule (see Chapter 6).

2.4.2. Physical Activation

Free radicals capable of initiating polymerization of vinyl monomers may be introduced into the cellulose by physico-chemical or mechanical techniques. These activation processes include:

1. Use of high energy irradiation methods, such as the gamma rays from a Co-60 source. The irradiated cellulose is activated for graft formation (see Chapter 3).
2. Use of low energy irradiation methods, such as ultra-violet light rays from a mercury lamp in the presence of a photosensitizer. The latter absorbs the ultraviolet light and is activated; the active centers are then transferred to the cellulose molecule to initiate a graft chain. Increasing interest is being shown in non-sensitized systems (see Chapter 3).
3. Mechanical working of the cellulose material can produce radicals, which, in the presence of a suitable vinyl monomer, can produce graft and block copolymers. Mastication, ultrasonic wave-induced degradation, vapor swelling, and even freezing and thawing methods have been investigated using this method.

2.4.3. Chemical Activation

There are a number of strictly chemical processes which have been used to activate the cellulose through the introduction of free radicals by an independent chemical reaction. The following are typical examples of these processes:

1. Use of redox (reducing-oxidizing) systems, such as glucose hydrogen peroxide, whereby free radicals are first generated outside the cellulose molecules. These free radicals then transfer the active sites to the cellulose molecules by a charge transfer mechanism (see Chapter 4).
2. Use of the cellulose itself as the reducing component in the redox system. Thus, the active centers are directly produced on the cellulose backbone and no charge transfer mechanism is necessary to initiate cellulose graft copolymer formation. This system requires the use of such powerful oxidizing agents as cerium salts (Ce^{IV}) (see Chapter 4).
3. Initiators, such as organic peroxides or diazonium salts, are built into the cellulose molecule. The peroxide groups may be generated by treating the cellulose with an ozone-oxygen mixture, whereas the diazonium salt of cellulose can be prepared by reacting a halogenated amine with cellulose and reacting the product with nitrous acid to form the diazonium salt. The peroxide groups or diazonium salts can be decomposed into radicals situated on the cellulose backbone by heating, ultraviolet light, or by redox systems (see Chapter 4).

Graft copolymerization reactions involving cellulose and brought about by a radical mechanism using chemical initiation have certain characteristics which may be summarized as follows[31, 32]:

(i) The graft copolymerization reaction is usally carried out in a heterogeneous medium and the cellulose conversion is relatively low, not exceeding 20–30% of the weight of the initial cellulose. Even in grafting large quantities of synthetic polymer (200–500% of the initial cellulose weight), the conversion of the cellulose does not exceed 70–80%. Consequently, grafting results in a product which contains both the graft copolymer and a greater or lesser quantity of unchanged cellulose. In a number of cases, synthetic homopolymer is also formed as a result of side reactions. This mixture of products may be separated by the use of different methods into their main components. In this way the composition of the reaction products can be determined. A cellulose graft copolymer in actual use in the Soviet Union has the following composition: 30–50% cellulose, which did not participate in the graft copolymerization, and 50–70% graft copolymer with 10–30% of grafted polymer.

(ii) The molecular mass of the grafted chain is considerable and is comparable to a degree of polymerization of between 1000 and 15,000. Consequently, there is one grafted chain to 5–15 cellulose macromolecules. Depending on the conditions of carrying out the reaction, and, in particular, on the method of initiation and the degree of swelling of the cellulose, the grafted chains are localized on the surface of the cellulose fiber or on the supermolecular structure. The high molecular masses of the grafted chains are caused by the special conditions of radical polymerization in a heterogenous phase. In these conditions, the chain termination process is hindered by the decreased interaction of growing macromolecules whose mobility, because of the chemical combination with such a rigid chain polymer as cellulose, is sharply decreased. This conclusion is confirmed in that the degree of polymerization of homopolymer formed under the same conditions is 10–15 times less than the degree of polymerization of the grafted chain[32], (see also Chapter 7). It is important to note that the molecular mass of the grafted synthetic polymer in the heterogenous medium may be considerably decreased by the addition of a regulator, i.e., a chain terminator[32].

(iii) The hindered chain termination of the growing chain in a heterogeneous medium determines another characteristic of cellulose graft polymerization. The radical initiator (or component of the redox system) takes part in the chain termination, reacting not only with the macroradical but also with cellulose radicals formed in the first stage of its oxidation. Using the conditions required for carrying out graft polymerization, there take place two competitive reactions: the first is between cellulose macroradicals and the initiator, while the second is between cellulose macroradicals and the monomer. In the majority of cases the rate of the first reaction is considerably greater (5–10 times) than that of the second reaction, which leads to a general characteristic of cellulose graft copolymerization, namely, the low effectiveness of initiation. The initiator effectiveness of graft copolymerization as a rule does not exceed 10–20%[32].

(iv) The graft copolymerization reaction in a heterogeneous medium and the participation of the radical initiator in the chain termination lead to the formation of grafted chains with a narrow molecular mass distribution. The molecular mass distribution of grafted chains is considerably less broad than that of homopolymers obtained under the same reaction conditions. As a rule, for grafted polymers the ratio $\bar{M}_w/\bar{M}_n$ is not greater than 2; with graft polymerizations in aqueous emulsions, however, this ratio is lowered to 1.5–1.4[32].

(v) On carrying out the graft polymerization reaction from binary mixtures of monomers, yet another characteristic becomes evident. In this case, there is observed a synergistic effect on the graft yield for any pair of monomers. This was established earlier by Sakurada[33] during his investigation of radical grafting reactions. This is of great practical interest. For example, under certain conditions acrylonitrile can be grafted to cellulose fiber to the extent of 30%; the grafting of styrene under these conditions does not take place. However, the graft copolymerization of a mixture of acrylonitrile and styrene proceeds to 140%[32]; the grafted chain is made up of copolymers of both of the monomers used in grafting and is not a mixture of chains, each consisting of a single monomer[34].

In a more recent development, graft copolymerization has been accomplished by the "uncatalyzed" polymerization of vinyl monomers in the presence of cellulose and water and in the absence of a conventional polymerization initiator (see Section 2.8).

There are evidently a number of methods for initiating vinyl graft polymerization onto cellulose by a radical mechanism. It is of prime importance to emphasize that the most promising methods are those which meet the following requirements[32].

1. The synthesis of a graft copolymer without the simultaneous formation of homopolymer or with a minimum of this side reaction.
2. The minimum degradation of cellulose and its derivatives in the grafting process.
3. The ready availability of the initiator, and the possibility of using it while carrying out the reaction in an aqueous medium.
4. A wide field of application, that is, the possibility of performing the grafting of a large variety of monomers.
5. Low unit cost for the initiation procedure.

2.5. Charge-Transfer Mechanism of Grafting

Graft copolymerization of polar vinyl monomers onto cellulose by chemical activation and by irradiation has often been presumed to be solely the result of propagation from radical sites generated on the cellulose. However, in the case of grafting cellulose with Ce^{IV} ions and acrylamide, it has been proposed that the graft polymerization reaction is initiated on cellulose as a result of (1) an unhydrated Ce^{IV} ion-cellulose redox reaction which generates a radical site on the cellulose, which in turn propagates by the addition of uncomplexed acrylamide and, concurrently, (2) a hydrated Ce^{IV} ion-cellulose, or a hydrated Ce^{IV} ion-complexed acrylamide-cellulose reaction, which presumably generates a radical site on the cellulose and then propagates by the addition of Ce^{IV} ion-complexed acrylamide[35].

It has also been proposed that the nature of the copolymer bonding and the adsorption characteristics of monomers are related and that a charge transfer mechanism could explain the predominant type of bonding in cellulose copolymers[36]. Irradiation of cellulose produces trapped free radicals which may participate in the charge transfer bonding process. Detailed ESR evidence of radiation-induced trapped free radical formation has been included[16,37,38]. The possible relation between copolymerization and radical formation has also been considered[37,39,40]; the results strongly support a predominant free radical mechanism in grafting[39].

Nevertheless, the formation of grafted alternating copolymer from the radiation-induced polymerization of styrene-acrylonitrile and butadiene-acrylonitrile, and the radical-catalyzed polymerization of styrene-methyl methacrylate and styrene-acrylonitrile-$ZnCl_2$, as well as the uncatalyzed copolymerization of methyl methacrylate with other monomers in the presence of cellulose and water, have been suggested as indicating the participation of comonomer charge transfer complexes in the graft polymerization[33,41-51]. Consequently, it is proposed[35] that cellulose acts as a complexing agent for the activation of acceptor monomers and as a matrix for the alignment of acceptor monomer complexes (A^+ ... $^-$A-Cellulose-water) and donor-acceptor comonomer complexes (D^+ ... $^-$A-Cellulose-water).

However, the charge transfer mechanism is somewhat speculative and, until more data are available, ought to be considered unreliable.

2.6. Grafting by Chain Transfer

Growing polymer chains are capable of transferring their radical character to certain compounds called chain transfer agents. The process is called chain transfer; it terminates the polymer chain and leaves the chain transfer agent in the form of a radical that may subsequently initiate another polymerization reaction[52-63]. The ratio of the frequency of any transfer reaction to propagation is called the transfer constant, C, and is defined[64] as

$$C = k_{tr}/k_p \, ,$$

where k_{tr} and k_p are the rate constants for chain transfer and propagation, respectively.

For the synthesis of graft copolymers, chain transfer with the polymer is of great interest. Thus, during polymerization of monomer (M) by initiator (I), a growing chain radical ($P_n^{\cdot}$) is formed:

$$I \rightarrow 2R^{\cdot} \, , \tag{1}$$

$$R^{\cdot} + M \rightarrow M_1^{\cdot} \, , \tag{3}$$

$$RM^{\cdot} + nM \rightarrow P_n^{\cdot} \, . \tag{56}$$

In the presence of a polymer (A—B), where (A) is a labile atom (usually hydrogen or halogen), the growing radical may terminate by abstraction of (A) from the polymer to give rise to a free radical formation on the polymer backbone (i.e., B$^{\cdot}$). Addition of the monomer to this new macroradical produces graft copolymer:

$$P_n^{\cdot} + A{-}B \rightarrow P_nA + B^{\cdot} \, , \tag{57}$$

$$B^{\cdot} + M \rightarrow BM^{\cdot} \xrightarrow{nM} \text{graft copolymer} \, . \tag{58}$$

An alternative mechanism, leading to the same type of macroradical as illustrated in Eq. 57, is the transfer reaction between the primary radical and the polymer.

$$R^{\cdot} + A - B \rightarrow RA + B^{\cdot} \, . \tag{59}$$

For efficient grafting initiation, it is essential that the polymer to be grafted successfully competes for the primary (Eq. 59) or growing radical (Eq. 57) with the monomer (Eq. 3), the homopolymer of the monomer, the solvent, or other species present. The relative values of the various transfer constants determine the outcome of this competition. Transfer constants for polymers can be obtained from data on low-molecular-mass compounds of similar structures[65-67], assuming identical reactivity toward the attacking radical. Indeed, the validity of this assumption has been confirmed by transfer constant determinations in the presence of oligomers[68-70].

The importance of chain transfer of the polymer backbone with the primary radicals is demonstrated by the variation in grafting efficiency with the nature of the initiator, a phenomenon commonly called the *initiator effect*. Smets et al.[71] have shown that the polymerization of methyl methacrylate in the presence of polystyrene gives appreciable amounts of graft copolymer, if the reaction is initated with benzoyl peroxide. Using azobisisobutyronitrile (AIBN) or ditertiarybutyl peroxide, the degree of grafting is not significant. Their explanation of the inability of AIBN to initiate graft copolymerization is that the $(CH_3)_2\dot{C}$—CN radical from AIBN in appreciably less reactive than the $C_6H_5^{\cdot}$ and $C_6H_5COO^{\cdot}$ radicals from benzoyl peroxide because of resonance stabilization[71]. Although the $(CH_3)_2\dot{C}$—CN radical is sufficiently reactive to polymerize monomers, it does not undergo reaction with polymers. In similar experiments carried out with vinyl acetate and poly(methyl methacrylate)[71], graft copolymer formation is seen to be independent of the nature of the initiator. These results reveal that the relative reactivities toward chain transfer differ considerably for the growing radicals and the polymer backbone, as well as for the primary radicals.

An increase in the temperature may enhance the efficiency of grafting[72,73]. This finding is in agreement with the higher activation energy of the transfer reaction compared to the propagation reaction. The efficiency of grafting also increases as the concentration of initiator increases. The presence of good low molecular mass chain transfer agents, such as dodecylmercaptan, decreases considerably the efficiency of grafting[72], showing that the low molecular mass transfer agent competes with the polymer for the free radicals.

The general utility of chain-transfer reactions in obtaining graft copolymers is seriously limited in many cases by the reluctance of the backbone to undergo free radical attack. Several efforts have, therefore, been made to synthesize polymers with reactive sites characterized by high transfer constants[63,74,75]. For instance, methyl methacrylate-glycidyl methacrylate copolymers containing small amounts of epoxide groups have been synthesized[62,75]. By reacting these materials with thioglycollic acid or H_2S, thiol groups are incorporated on the polymer backbone:

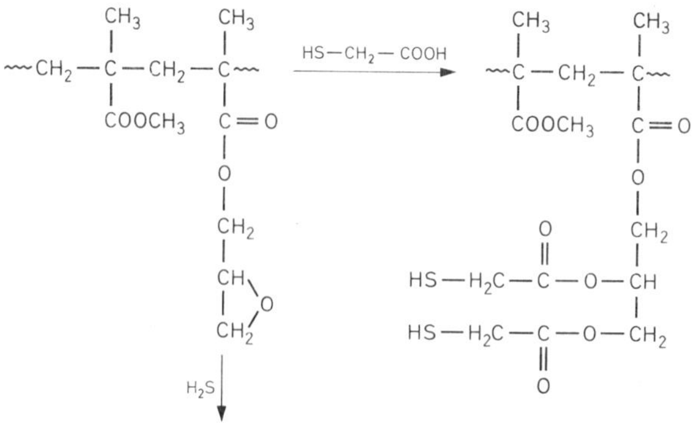

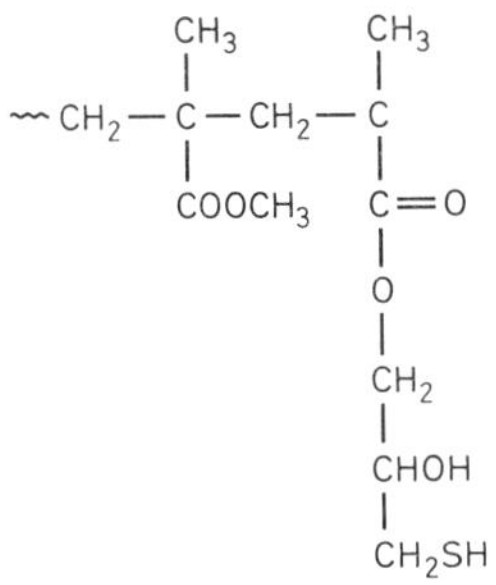

Grafting may occur by a process of chain transfer when cellulose is present during a polymerization reaction. The growing polymer chain may terminate by abstraction of a hydrogen atom from the cellulose. This leaves an unshared electron, i.e., a free radical on the cellulose chain which is capable of initiating grafting. The reactions involved are similar to those shown in Eq. (1), (3), (56), (57), and (58). Termination occurs by combination, by disproportionation, or both[76)].

The efficiency of grafting is quite low, however, as cellulose is a very poor transfer agent[77)], but it may be increased by forming a cellulose derivative with a substituent group that is more susceptible to chain transfer[78, 79)]. For example, by introducing a mercapto group, —SH, into the cellulose molecules, the chain transfer process is greatly enhanced owing to the ease with which the hydrogen atom of the mercapto group is abstracted by the growing chain radical[79)]. A simple method has been described for the introduction of the —SH group into the cellulose by reacting the latter with ethylene sulfide[79)].

The addition reaction of a cellulosic with ethylene sulfide takes place readily when cellulose acetate is first swollen in benzene. To achieve the same result with cellophane, it is necessary first to solvent-exchange the water, via alcohol, with benzene. After washing the film with benzene, styrene is added and its polymerization initiated by azobisisobutyronitrile (AIBN)[79)]. Grafting of vinyl monomers other than styrene, such as methyl methacrylate via chain transfer on mercaptoethyl cellulose, has been possible.

It has been shown that the initial rate of grafting is proportional to the square root of the AIBN concentration and roughly proportional to the concentration of the monomer. Rates of grafting were on the order of 10–15 weight percent per hour[79)].

It is of interest to point out that the same principle (i.e., grafting to a backbone containing groups susceptible to chain transfer) prompted the synthesis of a polyester from pentaerythritol dibromide and adipic acid[80)]:

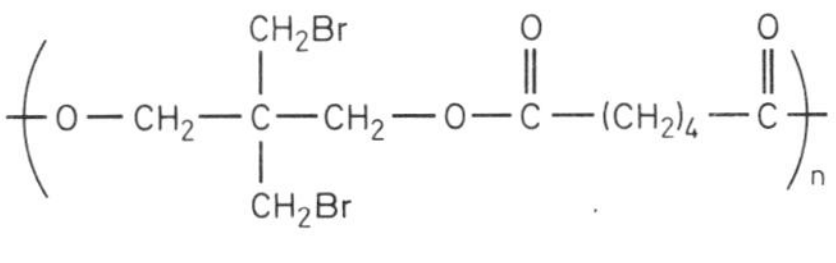

In this case, the bromomethyl groups serve as reactive sites for grafting with styrene by chain transfer.

2.7. Grafting by Mechanochemical Activation of Cellulose

When stresses are applied to macromolecules, both local segmental motions and molecular flow occur. Scissions may take place if stress of sufficient magnitude is localized in one segment of a molecule. Such scissions can lead to the formation of free radicals. Block and graft copolymers can be formed via these free radicals by subjecting polymer mixtures or polymer-nonomer mixtures to bond-breaking stresses. The various methods which have been employed are mastication, milling, ultrasonic irradiation, high frequency discharge, freezing and thawing, and vapor phase swelling.

Cold mastication of bulk polymer mixtures or polymer-monomer mixtures in an inert atmosphere is the most frequently used method for making graft and block copolymers of fully synthetic polymers using mechanochemical synthesis[81-86].

The efficiency of bond breakage is directly related to the plasticity of the system during mastication. The more plastic the mixture, the less is the chance that localized stresses will cause scission before flow of the molecules occurs. The ideal condition for efficient chain scission lies between the viscous and vitreous states. However, even glassy polymers can be sheared to yield free radicals[87]. Milling efficiencies have been studied by use of electron spin resonance measurements in various polymers[88].

Mastication of polymer-monomer mixtures has been extensively studied[89]. The influence of experimental variables on the reaction depends mainly on their effect on the plasticity of the mixture. Efficiency of copolymerization generally increases with:

1. Increase in the initial molecular mass of the polymer.
2. Decrease in temperature.
3. Decrease in the monomer (and/or solvent) concentration.
4. Increase in internal viscosity (and thus, also the resistance to shear) of the copolymer being formed.

The rate of shear and design of the masticator will also affect the efficiency, depending on the opposing effects encountered when an increase in shear rate causes an increase in the energy input to the system.

Intensive mechanical disintegration of inorganic substances, such as metals, salts, and oxides, can lead to formation of specific active centers of an ionic or free radical nature on the freshly formed surfaces[87]. These centers can react with monomers causing polymerization[90] or with polymers causing grafting[87]. It should be possible to form graft and/or block copolymers with inorganic links between some of the copolymer segments by masticating mixtures of monomers and polymers with the inorganic solids.

A general description of the mastication process may be given as follows[91]. Polymers are first swelled by monomer, generally to 15–50% monomer content, in an inert atmosphere. Equilibration often takes place up to 16 hours or more. Then the swollen mixtures are masticated under nitrogen at 15–50 °C for times generally under 1 hour. The product may be extracted with selected solvents and then fractionally precipitated or turbidimetrically titrated. A few systems have been studied in detail, e.g., poly(methyl methacrylate)-styrene and polystyrene-methyl methacrylate[91].

Swelling of lightly crosslinked polymers by solvent vapors may lead to chain rupture due to stressing of molecular entanglements at a critical solvent concentration[92]. If the solvent is a monomer, block copolymerization can result.

Copolymerization of various vinyl monomers with different polymers has been achieved by using the ultrasonic irradiation technique[93–97]. Ultrasonic irradiation of polymer solutions leads to chain rupture where cavitation occurs in the solution[98,99]. If the solution contains a monomer or another polymer, block copolymers may be formed. The efficiency of the degradation and, thus, copolymerization depends upon the efficiency of cavitation. This is a function of the gases dissolved in the solution, the vapor pressure of the solvent, the surface tension of the solution, and the minimum sonic frequency. Moreover, the presence of high molecular mass polymer is vital, since below a critical minimum molecular mass, degradation will not occur.

Block and graft copolymers of cellulose can be prepared by forming cellulose radicals by mechanical scission of the cellulose molecule, e.g., by ball milling cellulose acetate in the presence of vinyl chloride[100], by applying high shear[101], or by subjecting cellulose to swelling and freezing cycles[102]. It is important to note that the mechanochemical techniques which have been used for the preparation of block and graft copolymers are not very specific, since these methods are usually accompanied by considerable homopolymer formation.

2.8. Uncatalyzed Grafting

The grafting of cellulose and cellulose derivatives with different vinyl monomers, in methanol or ethanol in the presence of promoters such as alkyl halides, and in the absence of conventional initiators, has been studied in detail[103,104]. Such investigations were first undertaken by Japanese researchers led by Imoto[105,110].

When a vinyl monomer is heated in an aqueous suspension with cellulose at 85 °C in the absence of a conventional initiator, graft copolymerization readily occurs and the graft copolymer is accompanied by a high yield of high molecular mass homopolymer. Addition of small amounts of carbon tetrachloride to the system proved to be a good promoter of the process[104,106]. With some kinds of cellulose which are not able to initiate aqueous vinyl polymerization, the presence of carbon tetrachloride is essential[106].

Acrylates and methacrylates showed a specificity for the polymerization initiated in the system of cellulose (or silk) and water[109]. The same situation was also encountered even when carbon tetrachloride was included in the system[110]. For instance, polymerizations of methyl methacrylate, n-butyl acrylate, acrylonitrile, and styrene (2 cm^3 each) were conducted at 85 °C for 5 hours in the presence of cupra-rayon (0.2 gm), water (10 cm^3), and carbon tetrachloride (0.2 cm^3). Results of this investigation showed that acrylonitrile and styrene have no tendency to polymerize, whereas polymerization of methyl methacrylate and n-butyl acrylate resulted in a weight increase of the rayon of 490% and 303% respectively[110].

It was suggested that uncatalyzed grafting proceeds via a radical mechanism. With partially acetylated cellulose fiber, for example, some active sites were assumed to be present which would react with carbon tetrachloride to form initiating radicals[108]. This postulation was confirmed by the following observations: (a) hydroquinone was found to act as inhibitor; (b) copolymerization of methyl methacrylate and styrene (at a molar ratio of 1:0.99) with the cupra rayon-carbon tetrachloride-water system yields a copolymer consisting of 43.2% styrene and 56.8% methyl methacrylate, which is close to the calculated proportions, (50:50)[110]; and (c) grafting of vinylidene chloride, which can only polymerize by a radical mechanism, has been possible by using the methyl cellulose-methanol (or ethanol)-carbon tetrachloride systems[104].

The mechanism which has been proposed[111] for the polymerization of methyl methacrylate in the presence of cellulose (or other effective polymer) and water involves the formation of a complex between the polymeric —OH (or —CONH— in the case of silk) groups, water, and monomer as shown in Eq. 60:

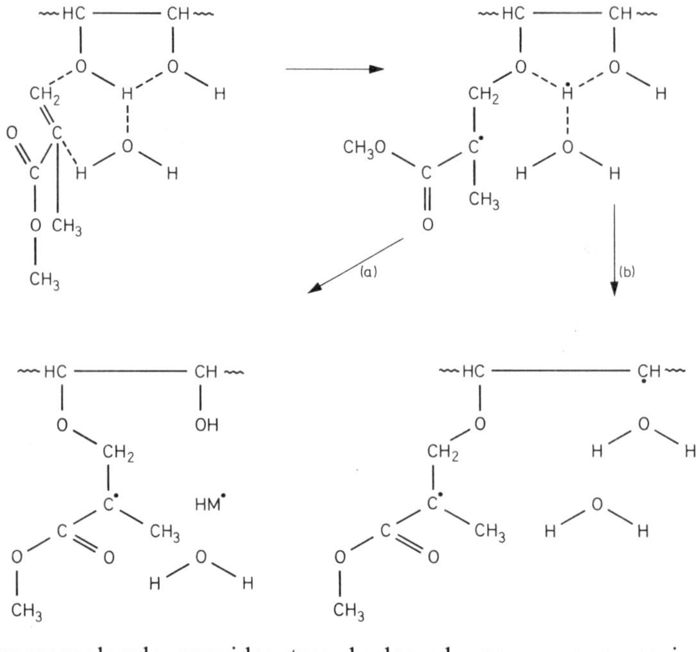

(60)

The macromolecule provides two hydroxyl groups, one serving as donor and the other as acceptor. The —OH group in the complex breaks, forming a new bond between oxygen and monomer and liberating atomic hydrogen. The hydrogen atom initiates the homopolymerization of the monomer or the graft polymerization on the polyol[35].

As already indicated, carbon tetrachloride has a dramatic effect on the cellulose-water-vinyl monomer system[105]. In the presence of a small amount of carbon tetrachloride, the conversion of monomer to polymer, after 5 hours at 85 °C, increases rapidly and then decreases as the carbon tetrachloride content is increased further. The conversion is zero in the absence of water.

Although the conversion increases in the presence of carbon tetrachloride, the grafting efficiency decreases exponentially with increasing amounts of carbon tetrachloride in the cellulose-water-methyl methacrylate system at 85 °C. The molecular mass of the homopolymer decreases with increasing carbon tetrachloride content, apparently due to conventional chain transfer.

Since the grafting efficiency in the cellulose-water-carbon tetrachloride system is lower than that in the cellulose-water system, the proposed reaction mechanism for initiation of methyl methacrylate polymerization has been modified, as shown in Eq. (61)[35].

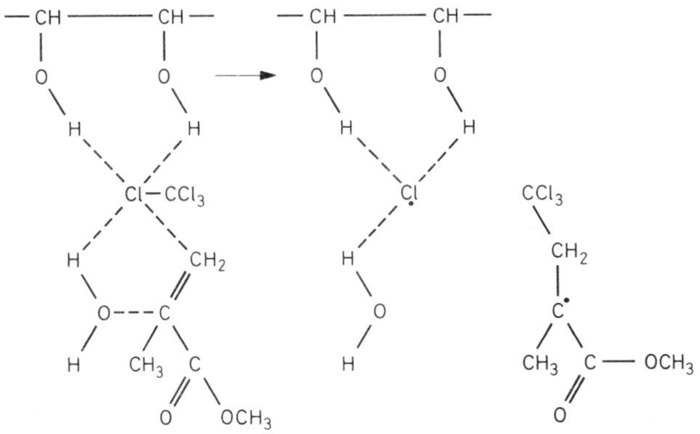

(61)

Some samples of cellulose have been found to be ineffective in initiating polymerization while others have been effective[106]. The efficiency of an inferior grade of wood pulp is greatly improved by the addition of a small amount of carbon tetrachloride, although no polymerization of methyl methacrylate occurs in the absence of either water or wood pulp. When the grade of wood pulp that is effective in the presence of carbon tetrachloride is either boiled in water, extracted with ethanol or ether, boiled in a dilute sodium hydroxide solution, or immersed in a dilute sodium borohydride solution, its effectiveness in the presence of carbon tetrachloride is somewhat reduced.

The addition of Cu^{II} ions to a mixture of methyl methacrylate and purified cotton in the presence of water greatly increases the polymer yield in the absence of carbon tetrachloride[112]. Consequently, it has been proposed that Cu^{II} ion plays an important role in polymerization in the presence of cellulose-water, and the earlier mechanism shown in Eq. (60) has been revised[35], as shown in Eq. 62 and 63.

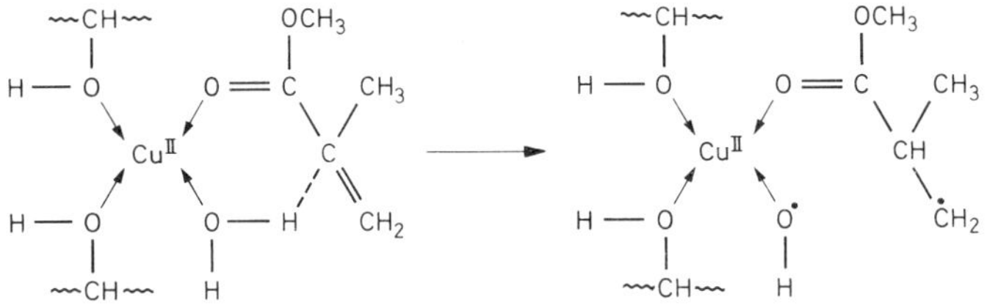

(62)

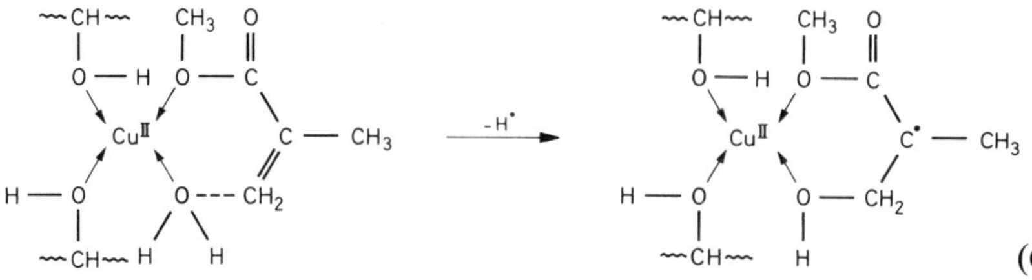

(63)

Notwithstanding the various cellulose-monomer complexes which have been proposed and the great care taken to exclude air or peroxides from the reaction mixture, the participation of oxygen in the initiation step in the uncatalyzed graft copolymerization has been considered and not completely disproved[106,112,113].

2.9. The Kinetics of Radiation-Induced Grafting onto Cellulosics

As we have seen, work has been devoted to the mechanics of grafting onto cellulose using a variety of irradiation procedures. Much more attention has been given to the effects on bulk physical properties of graft copolymerization. By contrast, the interest in studies of the kinetics of grafting has been meager, perhaps because of the complexities of such systems.

Odian and coworkers have provided the basis for an understanding of the kinetics of the grafting reactions involving polyethylene and styrene[114,115]. The principles have also been applied to nylon-styrene composites[116]. Styrene has proven to be a popular choice for fundamental study because of its low G (radical) value relative to that of cellulose, nylon, or polyethylene (See Chapter 3). The model used is partly developed from the simple kinetics of radiation-induced homopolymerization reactions (See Section 1.2.5.). This model assumes that the polymerization system has come to equilibrium before the irradiation program is started. Omichi and Araki[117] have examined the styrene — pre-irradiated poly (isobutylene oxide) system in an attempt to understand the kinetics of the grafting process, though the radiation was generated by a (2 Mev, 1 mA) resonance transformer electron accelerator at −78 °C, as described by Sekiguchi et al.[118]. Under these circumstances, grafting is non-steady state, and thus, the radical concentration [M˙] is given by[118]

$$[\dot{M}] = \int_0^t R_i \, dt - \int_0^t R_t \, dt, \tag{64}$$

while the degree of polymerization, P_n, is given by

$$P_n = \int_0^t R_p \, dt \bigg/ \left(\int_0^t R_i \, dt + \int_0^t R_{tr} \, dt - \int_0^t R_t \, dt \right) \tag{65}$$

where the symbols have the same meaning as given in 1.2.5. A combination of Eqs. (5, 9, 22, 64 and 65), together with the representation of the chain transfer reaction (through the rate of transfer) as

$$M^{\cdot} + XY \rightarrow MX + Y^{\cdot}$$

giving

$$R_{tr} = k_{tr}[M^{\cdot}][XY],$$

provides Eqs. (66) and (67) where the concentration of monomer is assumed to be constant throughout the reaction time

$$R_P \Big/ \int_0^t R_p^2 \, dt = \frac{k_p[M] \int_0^t R_i \, dt}{\int_0^t R_p^2 \, dt} - k_t/k_p[M], \tag{66}$$

$$R_P \Big/ \int_0^t R_p \, dt = (k_p[M]/P_n) - k_{tr}[XY]. \tag{67}$$

where [XY] denotes the chain transfer agent.

Omichi and Araki have shown this approach to be a reasonable indication of the non-steady state situation. Various relevant values of parameters are given in Table 1. These data indicate the presence of a viscosity or Trommsdorff effect. The model could be usefully applied to pre-irradiated cellulose grafting, thus filling a long existing gap in the general understanding of cellulose grafting when carried out under such conditions.

Fortunately, the situation with regard to simultaneous grafting from monomer solutions is now quite clearly understood from the kinetic standpoint.

Table 1. Kinetic values of the graft copolymerization of styrene on pre-irradiated P.I.B.O. [Poly (isobutylene oxide)] at 25 °C. Taken from Ref. (117) p. 186 by courtesy of John Wiley and Sons Inc.

Parameter	Suspension Irradiation	Powder Irradiation	
		Styrene-Methanol	Styrene
		(1:1)	
$\int_0^t R_i \, dt$, mol/g	2.0×10^{-7}	1.7×10^{-7}	2.1×10^{-7}
$k_p[M]$, min^{-1}	0.46×10^2	0.56×10^2	1.1×10^2
[M], mol/g PIBO	2.8×10^{-3}	4.5×10^{-3}	8.6×10^{-3}
k_p, l/mol sec	0.32	0.24	0.27
$k_{tr}[XY]$, min^{-1}	5.0×10^{-2}	1.7×10^{-2}	2.9×10^{-2}
[XY], mol/g	1.4×10^{-2}	1.4×10^{-2}	1.4×10^{-2}
k_{tr}, l/mol sec	7.1×10^{-5}	2.4×10^{-5}	4.4×10^{-5}
k_t, l/mol sec	11.3	1.0	2.3
k_p^2/k_t, l/mol sec	9.1×10^{-3}	5.8×10^{-2}	3.0×10^{-2}

Guthrie et al. have examined the grafting of styrene onto cellulose[119,120] and selected derivatives[121,122]. In their discussion of the kinetics of grafting in the dioxan-styrene-water-cellulose system, certain assumptions were made:

(i) The monomer solution is available at all the grafting sites on the cellulose.

(ii) Grafting takes place in the accessible regions of the backbone polymer. In view of the small film thickness used and the high efficiency of swelling obtained, such an assumption is reasonable.

(iii) In the expression of grafting rates and monomer concentrations, the individual volumes of cellulose, styrene, dioxan, and water are additive without contraction or expansion of volume. Though a considerable over-simplification, this assumption is thought unlikely to affect the relative validity of the various calculated terms.

(iv) The grafting reaction is not diffusion controlled. This assumption is justified, since equilibrium swelling is attained in all cases prior to the commencement of irradiation. If grafting were essentially diffusion controlled, then the method of Omichi and Araki[117] should be used. Since consistency of technique was maintained throughout and factors thought likely to influence grafting were investigated successfully, the mechanism of the grafting and homopolymerization processes can be considered. It is of interest to determine the specific rates of graft copolymerization and homopolymerization together with the k_p^2/k_t values for the various cellulose-styrene-dioxan-water systems. Data used in determination of reaction rates were restricted to the initial stages of grafting and homopolymerization.

The rate of graft polymerization and homopolymerization is given by

$$R_p = k_p \left(\frac{R_i}{2k_t}\right)^{1/2} [M] \tag{68}$$

where k_p and k_t = propagation and biomolecular-termination rate constants, respectively; $R_P \equiv R_G$ = mol of styrene grafted per liter of cellulose-styrene-dioxan-water per second, in the grafting reaction; $R_P \equiv R_H$ = mol of styrene converted to homopolymer per liter of bulk monomer solution per second, in the homopolymerization reaction; [M] is the monomer concentration (in grafting this is expressed as mol of styrene per liter of bulk solution) as see in Tables 2 and 3; and R_i = mol of free radicals produced on the cellulose polymer per liter of swollen film per second (in homopolymerization R_i is expressed as the concentration of homopolymer radicals per liter of bulk solution per second). The rate of initiation for both grafting and homopolymerization is obtained by

$$R_i = \frac{G\Phi\varrho_p}{N_A \times 100} \tag{69}$$

where G = the number of free radicals produced in the system per 100 eV of absorbed radiation energy, N_A = the Avogadro number, Φ = the radiation intensity in eV per gram of cellulose per second (= 2.26×10^{13} eV/g sec), and

Table 2. Liquid composition of the bulk medium and concentration of monomer in regenerated cellulose film for various concentrations in the bulk medium. Reprinted from Ref. (120), p. 1032 by courtesy of John Wiley and Sons, Inc.

Concentration of cellulose-dioxan-styrene-water system, g/l.				Monomer concn. in film, mol/l. of swollen film $X\ 10^2$	Styrene concn. in bulk medium mol/l.
Styrene	Dioxan	Cellulose	Water		
2.16	160	1252	3.53	2.08	0.432
6.51	202	1187	4.10	6.26	0.896
12.8	242	1125	6.50	12.3	1.73
15.7	230	1124	6.90	15.1	2.59
26.5	310	1028	11.1	25.5	3.43

Impregnation carried out at 30 °C. Styrene concentrations followed by extraction of swollen films with 98:2, dioxan/water mixtures and measurement of λ_{max} at 282 and 291 nm (123).

ϱ_p = the concentration of accessible cellulose in grams per liter of swollen film in grafting, or the number of grams of monomer solution per liter of swollen film in homopolymerization.

The results of the monomer and intensity dependence found in this study are consistent with Eq. (68), and hence this approach seems to be valid.

Rearrangement and squaring of Eq. (68) gives

$$k_p^2/k_t = 2R_p^2/R_i[M]^2 \,. \tag{70}$$

Calculated values for R_i and k_p^2/k_t for the grafting of styrene to cellulose in a dioxan-water medium are given in Table 3. Since only qualitative importance is attached to values of k_p^2/k_t, the use of this approach, which neglects termination between graft copolymer and homopolymer radicals, is justified.

Table 3. k_p^2/k_t in grafting for the bulk monomer concentrations in the grafting media. Values obtained from R_p and R_i employing a literature value (121) for the G (radical) of cellulose. Reprinted from Ref. (120) p. 1034 by courtesy of John Wiley and Sons, Inc.

Monomer concn. in bulk grafting medium, mol/l.	Monomer concn. in film, mol/l. of swollen film $\times 10^2$	R_p (mol grafted. l^{-1} of swollen film. sec^{-1})	R_i, mol $l.^{-1}$ sec^{-1} $\times 10^9$	ρ_p, g/l. of swollen film	k_p^2/k_t, l. mol^{-1} sec^{-1}	Monomer concn. in bulk, % (v/v)	Swelling at equilibrium based on initial film weight, %
0.432	2.08	4.24	1.34	1252	62.2	5	13.2
0.863	6.26	11.9	1.27	1187	57.1	10	18.0
1.73	12.3	22.4	1.20	1125	54.2	20	21.5
2.59	15.1	27.4	1.20	1124	54.8	30	23.3
3.46	25.5	46.4	1.10	1028	60.5	40	32.8

The rates of initiation, R_i, were calculated using a G (radical) value for cellulose radical production of 2.88[121]. Values of p were calculated from the known weight fractions of each of the components of a swollen film and their respective densities, measured at 30 °C. The density of regenerated cellulose film, measured by the flotation technique using benzene and carbon tetrachloride, was found to be 1.47 g/cm^3. Accepted values for k_p^2/k_t in the free-radical polymerization of pure styrene at 25 °C are of the order 1×10^{-4} l mol^{-1} sec^{-1} [124], which is smaller by 6×10^5 than the values obtained in the above-mentioned styrene-cellulose copolymerization (see Table 3). The values of Guthrie et al.[120] compare favorably with those of Odian et al., in their grafting of styrene onto nylon[116], of 20 l mol^{-1} sec^{-1}. Odian's study of grafting onto polyethylene produced values ranging from 0.12 to 8.3 l mol^{-1} sec^{-1} for k_p^2/k_t, depending on the physical nature of the polymerization medium[114].

The rate of initiation (R_i) in the radiation-induced grafting carried out by Guthrie et al.[120], together with the values of R_p, p and k_p^2/k_t, are given in Table 4.

Values for the rate of initiation, R_i, for the homopolymerization reaction were obtained through use of a composite G (radical) value. This assumes that each of the components of the monomer solution contributes to the initiation of homopolymerization to an extent dependent on its G (radical) value and its weight fraction in the mixture. The individual G values used were cellulose, 2.88[121]; water, 4.8[125]; styrene, 0.69[126]; and dioxane, 2.0[127].

Thus, denoting the weight fraction by W and letting the symbol G denote the G (radical) value, we obtain

$$\begin{aligned} G\ \text{composite} &= (G\ (\text{styrene}) \times W\ (\text{styrene})) \\ &+ (G\ (\text{water}) \times W\ (\text{water})) \\ &+ (G\ (\text{dioxan}) \times W\ (\text{dioxan}))\,. \end{aligned}$$

This calculated composite G (radical) value provides a realistic approach to the mechanism of homopolymerization. Since composite G values are used, composite values for ϱ (monomer and solvent) were calculated. Thus, ϱ (compo-

Table 4. k_p^2/k_t in homopolymerization (h) attendant during grafting, for various bulk monomer concentrations. Values obtained from rates of homopolymerization $R_p(h)$ together with values of R_i calculated from the composite G (radical) of each bulk medium. Reprinted from Ref. (120), p. 1035, by courtesy of John Wiley and Sons, Inc.

Monomer concn. in bulk, mol/l.	G (radical) composite	p (composite) g/l. $\times 10^{-3}$	$R_i(h)$, mol l.$^{-1}$ sec^{-1} $\times 10^{10}$	$R_p(h)$, mol l.$^{-1}$ sec^{-1} $\times 10^7$	k_p^2/k_t, l. mol^{-1} sec^{-1} $\times 10^4$
0.432	2.10	1.017	8.01	1.39	2.58
0.864	2.0	1.011	7.57	2.97	3.13
1.73	1.8	0.999	6.74	5.97	3.54
2.59	1.7	0.986	6.29	9.05	3.86
3.46	1.6	0.973	5.84	11.2	3.58

site) = (weight of styrene + weight of dioxan + weight of water) in g/dm^3 of the monomer solution.

The values of k_p^2/k_t obtained in this study of styrene homopolymerization are in good agreement with those of other studies and with the accepted value for the free-radical polymerization of the pure styrene. An increase of k_p^2/k_t is observed with increasing concentration of monomer due to changes in the viscosity of the monomer solution and consequent restriction of movement of the growing polymer chains.

Utilization of the known values of k_p^2/k_t and k_p for the free-radical polymerization of styrene, 1×10^{-4} l mol^{-1} sec^{-1} and 55 l mol^{-1} sec^{-1} [128], respectively, enables values for k_t to be calculated.

Rearrangement of Eq. (69) and (70) gives

$$G = N_A \times 100 \times \mathrm{R}_i/p\Phi \,, \tag{71}$$

and

$$\mathrm{R}_i = 2\mathrm{R}_p^2\, k_t/[\mathrm{M}]^2\,(k_p)^2\,. \tag{72}$$

Hence, values for G (composite) may be derived. Table (5) shows values of the composite G (radical) obtained for various monomer concentrations. It is seen that these compare favorably with those used in the experimental k_p^2/k_t values for homopolymerization.

Experimentally determined k_p^2/k_t values are in close agreement with those obtained using a literature value for k_p and a value of k_t calculated from quoted values of k_p^2/k_t and k_p. This indicates that the individual experimental values of k_p and k_t for the radiation-induced homopolymerization of styrene, as described, are probably of similar order to the literature values of k_p and k_t.

Studies by Guthrie and Haq. on grafting to hydroxyethyl cellulose film (degree of substitution = 0.4) produced the data given in Table 6 for a range initial bulk monomer concentrations[129]. Again we see evidence of the Trommsdorff type of effect, though the values of k_p^2/k_t are significantly less than those

Table 5. Calculated G (radical) values for the homopolymerization of styrene, attendant during grafting. Obtained from $R_p(h)$ together with values of R_i calculated on the basis of literature values of k_p^2/k_t. Reprinted from Ref. (120), p. 1036, by courtesy of John Wiley and Sons, Inc.

Monomer concn. in bulk, mol/l.	p (composite), g/l. of bulk solution $\times 10^{-3}$	$R_i(h)$ calculated, mol l^{-1} sec$^{-1} \times 10^{10}$	$R_p(h)$, mol $\times$ l.$^{-1}$ sec^{-1} $\times 10^7$	k_p^2/k_t, l. mol^{-1} sec^{-1} $\times 10^{-4}$ (from literature values)	G (radical), composite, calculated
0.432	1.017	7.17	1.39	2.88	1.9
0.864	1.011	8.23	2.97	2.88	2.1
1.73	0.999	8.29	5.97	2.88	2.2
2.59	0.986	8.44	9.05	2.88	2.3
3.46	0.973	7.25	11.18	2.88	2.0

Table 6. Grafting of styrene and hydroxyethyl cellulose film, various initial bulk monomer concentrations. k_p^2/k_t. Reprinted from Ref.[129] p. 73. Courtesy of I.P.C. Science and Technology Press.

Monomer concentration	p g l^{-1} of swollen film	R_i mol l^{-1} sec^{-1} $\times 10^{-9}$	$R_P = R_G$ (% Wt/wt) $\times 10^{-4}$	k_p^2/k_t l mol^{-1} sec^{-1}
0.864	1187	1.595	1.123	0.213
1.73	1125	1.511	1.684	0.166
2.59	1124	1.509	2.237	0.134
3.46	1028	1.381	3.188	0.110

Table 7. k_p^2/k_t values for carbanilated celluloses having various degrees of substitution. Taken from Ref.[121] p. 753 courtesy of Pergamon Press, Ltd.

D.S.	R_G (% wt. hr^{-1})	$[M]$ [mol (l. swollen film)$^{-1}$]	G (radical) *in vacuo*	$10^{-3} \times p$ [g. substituted cellulose (l. swollen film)$^{-1}$]	$10^5 \times R_p$ [mol styrene grafted (l. swollen film)$^{-1}$ sec^{-1}]	$10^2 \times [M]^2$	$10^9 \times R_i$ (mol l^{-1} sec^{-1})	$10^{10} \times (2R_p)^2$	k_p^2/k_t (l mol^{-1} sec^{-1})
0.0	0.74	0.123	2.90	1.122	2.24	1.51	1.20	10.0	54.2
0.02	0.51	0.142	2.45	1.120	1.52	2.02	1.03	4.62	22.3
0.27	0.45	0.874	1.50	0.761	0.92	76.4	0.43	1.68	0.512
0.52	0.30	1.110	1.10	0.590	0.48	123.0	0.24	0.49	0.159

obtained with regenerated cellulose film. This indicates the more open morphology of hydroxyethyl cellulose film relative to the unsubstituted variety.

In their work on grafting to cellulose carbanilates (degrees of substitution in the range 0.0 to 0.51), Guthrie et al. reported data whereby values of k_p^2/k_t could be calculated. The relevant details are given in Table 7. Thus, for fixed values of the styrene concentration and intensity, plots of per cent grafting against time for each degree of substitution were seen to be accurately linear, no induction periods being observed. Values of R_G ($= R_P$) were obtained from the slopes and ratio of k_p^2/k_t calculated from a knowledge of the density of the substituted celluloses, the rate of initiation, and the composite densities[130].

The procedure was as described above. Grafting was carried out in the styrene-dioxan-cellulose-water system. It is clear that the value of k_p^2/k_t falls regularly with increasing degrees of carbanilation. Since cellulose tricarbanilate is soluble in dioxan, it is not surprising that these substituted products should have a high affinity for the dioxan component of the bulk monomer system. Thus, the carbanilated substrate is much more open and, consequently, accessible to monomer/solvent than is non-substituted cellulose. The Trommsdorff type of effect becomes less pronounced as the degree of substitution increases.

This kind of kinetic analysis could be applied to a wide variety of grafting systems, provided the data are processed in the prescribed manner.

References

1. Hebeish, A.: L'Industrie Textile *991*, 415 (1970)
2. Stannett, V., Wellons, J. D., Yasuda, H.: J. Polymer Sci. *C 4*, 551 (1964)
3. Huang, R. Y-M., Immergut, B., Immergut, E. H., Rapson, W. H.: J. Polymer Sci. *A 1*, 1257 (1963)
4. Hebeish, A.: Kolorisztikai Ertesito *13*, 12 (1971)
5. Hebeish, A., Mehta, P. C.: Text. Research J. *37*, 911 (1967)
6. Hebeish, A., Mehta, P. C.: J. Appl. Polym. Sci. *12*, 1625 (1968)
7. Kulkarni, A. Y., Hebeish, A., Mehta, P. C.: Proc. 9th Tech. Conf. ATIRA, BITRA and SITRA, India 1968, P. 367
8. Kantouch, A., Hebeish, A., El-Rafie, M. H.: European Polymer J. *6*, 1575 (1970)
9. Hebeish, A., Thesis, Ph. D.: Gujarat University, India 1967
10. Geacintov, N., Stannett, V., Abrahamson, E. W.: Makromol. Chem. *36*, 52 (1959)
11. Geacintov, N., Stannett, V., Abrahamson, E. W., Hermans, J. J.: J. Appl. Polym. Sci., *3*, 54 (1960)
12. Blouin, F. A., Morris, N. J., Arthur, J. C. Jr.: Text. Research *J 36*, 309 (1966)
13. Sidykov, T. S., Shablygin, M. V., Livshitz, R. M., Rogovin, Z. A.: Vysokomol. Soedin *8*, 2035 (1967)
14. Livshitz, R. M., Sidikov, T. S., Rogovin, Z. A.: Cellulose Chem. Technol. *2*, 3 (1968)
15. Sakurada, I., Okada, T., Uchida, M., Kugo, E.: Doitai To Hoskasen *2*, 316 (1959)
16. Dilli, S., Garnett, J. L., Martin, E. C., Phuoc, D. H.: J. Polymer Sci. *C 2*, 169 (1963)
17. Huang, R. Y-M., Rapson, W. H.: J. Polymer Sci. *C 2*, 169 (1963)
18. Kesting, R. E., Stannett, V.: Makromol. Chem. *65*, 248 (1963)
19. Hebeish, A., Mehta, P. C.: Text. Research J. *38*, 1070 (1968)
20. Arthur, J. C. Jr.: Text. Research J. *28*, 204 (1958)
21. Arthur, J. C. Jr., Blouin, F. A.: Text. Research *J 34*, 733 (1964)
22. Arthur, J. C. Jr., Blouin, F. A., Demint, R. J.: Am. Dyestuff Reptr. *49*, 383 (1960)
23. Arthur, J. C. Jr., Demint, R. J.: Text. Research *J 32*, 108, (1962)
24. Blouin, F. A., Arthur, J. C. Jr.: Am. Dyestuff Reptr. *53*, 951 (1964)
25. Blouin, F. A., Arthur, J. C. Jr.: J. Chem. Eng. Data. *5*, 470 (1960)
26. Blouin, F. A., Arthur, J. C. Jr.: Text. Res. *J 28*, 198 (1958)
27. Blouin, F. A., Arthur, J. C. Jr., Orr, R. S., Ott, V. J.: Text. Res. *J 31*, 597 (1961)
28. Arthur, J. C. Jr., Baugh, P. J., Hinojosa, O.: J. Appl. Polym. Sci. *10*, 591 (1966)
29. Arthur, J. C. Jr., Hinojosa, O., Bains, M. S.: J. Appl. Polym. Sci. *12*, 1411 (1968)
30. Arthur, J. C. Jr., Mares, T., Hinojosa, O.: Text. Research *J 36*, 630 (1966)
31. Rogovin, Z. A., Livshitz, R. M.: Bull. Inst. Politech. *16*, 297 (1970)
32. Rogovin, Z. A.: J. Polymer Sci. *C 37*, 221 (1972)
33. Sakurada, I., Akada, T., Hatakayama, S., Kimura, F.: J. Polymer Sci. *C 4*, 1233 (1963)
34. Stanchenko, G. I., Livshitz, R. M., Loffe, T. A., Rogovin, Z. A.: Cellulose Chem. Technol. *3*, 367 (1969)
35. Gaylord, N.: J. Polymer Sci. *C 37*, 153 (1972)
36. Dilli, S., Garnett, J. L.: J. Polymer Sci. *A 14*, 2323 (1966)
37. Dilli, S., Ernst, I. T., Garnett, J. L.: Austral. J. Chem. *20*, 911 (1967)
38. Dilli, S., Garnett, J. L.: Nature, *198*, 984 (1963)
39. Dilli, S., Garnett, J. L.: Austral. J. Chem. *23*, 1163 (1970)
40. Stannett, V.: IUPAC Conference, Syndney, Australia 1968
41. Gaylord, N. G.: J. Polymer Sci *C 31*, 247 (1970)
42. Gaylord, N. G., Patnaik, B.: J. Polymer Sci. *B 8*, 401, 549 (1970)
43. Gaylord, N. G., Patnaik, B.: J. Polymer Sci. *B 8*, 411 (1970)
44. Gaylord, N. G., Takahashi, A.: Advan. Chem. *91*, 94 (1969)
45. Gaylord, N. G., Takahashi, A., Anand, L. C.: J. Polymer Sci. *B 9*, 97 (1971)

46. Imoto, M., Takemoto, K., Kondo, M.: Makromol. Chem. *98*, 74 (1966)
47. Nara, S., Matsuyama, K.: Kobunshu Kagaku *25*, 840 (1968)
48. Takahashi, T., Hori, Y., Sato, I.: J. Polymer Sci. *A 1*, *6*, 2091 (1968)
49. Gaylord, N. G., Antropivsova, H., Patnaik, B. K.: J. Polymer Sci. *B 9*, 387 (1971)
50. Gaylord, N. G., Kikawa, E., Takahashi, A.: J. Polymer Sci. *B 9*, 379 (1971)
51. El'garph, S. A., Konkin, A. A., Rogovin, Z. A.: Vysokomol. Soedin. *8*, 42 (1966), Polymer Sci. USSR *8*, 41 (1966)
52. Houtz, R. C., Adkins, H.: J. Amer. Chem. Soc. *55*, 1609 (1933)
53. Carlin, R. B., Shakespeare, N. E.: J. Amer. Chem. Soc. *68*, 876 (1946)
54. Carlin, R. B., Hufford, D. L.: J. Amer. Chem. Soc. *72*, 4200 (1950)
55. Roland, J. R., Richards, L. M.: J. Polymer Sci. *9*, 61 (1952)
56. Bevington, J. C., Guzman, G. M., Melville, H. W.: Nature, *170*, 1026 (1952)
57. Flory, P. J.: J. Amer. Chem. Soc. *59*, 241 (1937)
58. Angier, D. J., Watson, W. F.: J. Polymer Sci. *18*, 129 (1955)
59. Hayes, R. A.: J. Polymer Sci. *13*, 383 (1954)
60. Immergut, E. H., Mark, H.: Makromol. Chem. *18/19*, 322 (1956)
61. Fox, T. G., Gratch, S.: Ann. N.Y. Acad. Sci. *57*, 367 (1953)
62. Fox, T. G., Gluckman, M. S., Gornick, F., Graham, R. K., Gratch, S.: J. Polymer Sci. *37*, 397 (1959)
63. Jahn, E. C.: Paper presented at the Internatl. Symp. Macromol. Stockholm: 1953
64. Walling, C.: In Free Radicals in Solution. New York: Wiley 1957, p. 150
65. Basu, S., Sen, J. N., Palit, S. R.: Proc. Roy. Soc. (London) *A 202*, 485 (1950)
66. Gregg, R. A., Mayo, F. R.: J. Amer. Chem. Soc. *75*, 3530 (1953)
67. Lim, D., Wicherle, O.: J. Polymer Sci. *39*, 579 (1958)
68. Morton, M., Piirma, I.: J. Amer. Chem. Soc. *80*, 5596 (1958)
69. Okamura, S., Hayashi, K., Kitanishi, Y.: Isotopes Radiation Tokyo, *3*, 398 (1960)
70. Schulz, G. V., Henrici, G., Olivé, S.: J. Polymer Sci. *17*, 45 (1955)
71. Smets, G., Roovers, J., Van Humbeek, W.: J. Appl. Polym. Sci. *5*, 149 (1961)
72. Hayes, R. A.: J. Polymer Sci. *11*, 531 (1953)
73. Smets, G., Claesen, M.: J. Polymer Sci. *8*, 289 (1952)
74. Bamford, C. H., White, E. F. T.: Trans. Faraday Soc. *52*, 716, (1956); *54*, 268, 278 (1958)
75. Gluckman, M. S., Kampf, M. J., O'Brien, J. L., Fox, T. G., Graham, R. K.: J. Polymer Sci. *37*, 411 (1959)
76. Hermans, J. J.: Pure Appl. Chem. *5*, 147, (1962)
77. Machell, G., Richards, G. N.: J. Chem. Soc. 3308 (1961)
78. Burroughs, R. H.: Thesis, State Univ. Coll. Forestry, Syracuse, N.Y., June 1955
79. Chaudhuri, D. K. R., Hermans, J. J.: J. Polymer Sci. *51*, 373 (1961)
80. Schonfeld, E., Waltcher, I.: J. Polymer Sci. *35*, 536 (1959)
81. Kargin, V. A., Kovarskaya, B. M., Golubenkova, L. I., Akutin, M. S., Slonimskii, G. L.: Khim. Prom. 77 (1957)
82. Kargin, V. A., Plate, N. A., Dobrynina, A. S.: Kolloid Zhur. *20*, 232 (1958); through C.A., *52*, 21215 (1958)
83. Berlin, A. A., Kronman, A. G., Yanovskii, D. M., Kargin, V. A.: Yysokomol. Soedin. *2*, 1839 (1960)
84. Berlin, A. A., Makarova, T. A.: J. Gen. Chem., USSR., (English Translation) *21*, 1383 (1951)
85. Bateman, L. C.: Ind. Eng. Chem. *49*, 704 (1957)
86. Ceresa, R. J.: Polymer *1*, 72 (1960)
87. Kargin, V. A., Plate, N. A.: J. Polymer Sci. *52*, 155 (1961)
88. Butyagin, P. Yu.: Dokl. Akad. Nauk. SSSR. *140*, 145 (1961)
89. Ceresa, R. J.: Block and Graft Copdymers, Washington: Butterworths 1962
90. Kargin, V. A., Kabanov, V. A.: J. Polymer Sci. *52*, 71 (1961)
91. Angier, D. J., Ceresa, R. J., Watson, W. F.: J. Polymer Sci. *34*, 699, 1959
92. Ceresa, R. J.: Polymer *2*, 213 (1961)
93. Akutin, M. S.: Trans. Plastics Inst. (London) *28*, 216 (1960)
94. Akutin, M. S., Parlashkevich, N. Ya., Kogan, I. N., Kalinina, S. P., Menes, L. I.: Primenenie ul'braakustiki K Issled Veshchestva *10*, 47 (1960); through C.A. *56*, 13086 (1962)
95. El'Piner, L. E.: Russian Chem. Rev. (English translation) *29*, 1 (1960)

96. Henglein, A.: Makromol. Chem. *14*, 128 (1954)
97. Henglein, A.: Makromol. Chem. *18/19*, 37 (1956)
98. Grassie, N.: Chemistry of High Polymer Degradation Processes, New York: Interscience 1956
99. Jellinek, H. H. G.: Degradation of High Polymers, New York: Academic Press 1955
100. Deters, W., Huang, D.: Faserforsch. Textiltech. *14*, 183 (1963)
101. De Bataafsche, N. Y.: Petroleum Maatschappij, British Pat. 679562 1952; through C.A. *47*, 7825 (1953)
102. Ceresa, R. J.: Paper presented at the A.C.S. 138th meeting, New York, N.Y., September 1960
103. Simionescu, G. I., Ciubotariu, M.: Paper presented at the 39th Internatl. Congress of Indust. Chem. Bucharest, Romania, September 1970
104. Simionescu, G. I., Feldman, D., Ciubotariu, M.: J. Polymer Sci. C. *37*, 119 (1972)
105. Imoto, M., Takemoto, K., Sutch, M.: Makromol. Chem. *110*, 31 (1967)
106. Imoto, M., Iki, Y., Kinoshita, M., Takemoto, K.: Makromol. Chem. *122*, 287 (1969)
107. Kubushiro, K., Takemotot, K., Imoto, M.: Makromol. Chem. *120*, 139 (1968)
108. Kubushiro, K., Takemotot, K., Imoto, M.: Makromol. Chem. *138*, 1 (1970)
109. Imoto, M., Takemoto, K., Azuma, H., Kita, K., Kendo, H.: Makromol. Chem. *107*, 188 (1967)
110. Imoto, M., Ueda, K., Takemoto, K.: Makromol. Chem. *138*, 11 (1970)
111. Imoto, M., Takemoto, K., Otsuki, T.: Makromol. Chem. *104*, 244 (1967)
112. Imoto, M., Iki, Y., Kawabata, Y., Kiroshita, M.: Makromol. Chem. *140*, 281 (1970)
113. Ko, K., Sato, K.: Kobunshi, *19*, 439 (1970)
114. Odian, G., Rossi, I., Trachtenberg, E. N.: J. Polymer Sci. *42*, 575 (1960)
115. Odian, G., Rossi, I., Sobel, M., Klein, R.: J. Polymer Sci. *55*, 663 (1961)
116. Odian, G., Sobel, M., Rossi, A., Klein, R., Ackers, T.: J. Polymer Sci. *A 1*, 639 (1963)
117. Omichi, H., Araki, K.: J. Polymer. Sci., Polymer Chem. *16*, 179 (1978)
118. Sekiguchi, A., Yoshida, K., Araki, K.: Proc. 23rd Conf. High Polymers, Japan *4A05*, *4A06* (1974)
119. Guthrie, J. T., Huglin, M. B., Phillips, G. O.: J. Polymer Sci., C. *37*, 205 (1972)
120. Guthrie, J. T., Huglin, M. B., Phillips, G. O.: J. Appl. Polym. Sci. *16*, 1017 (1972)
121. Guthrie, J. T., Huglin, M. B., Phillips, G. O.: Eur. Polymer *J 8*, 747 (1972)
122. Guthrie, J. T., Haq, Z.: Polymer *15*, 133 (1974)
123. Guthrie, J. T., Huglin, M. B., Phillips, G. O.: Angew. Makromol. Chem. *13*, 199 (1970)
124. Tobolsky, A. V., Mesrobian, R. B.: Organic Peroxides. New York: Interscience 1954, P. 144
125. Hochandel, C. J., Lind, S. C.: Ann. Rev. Phys. Chem. *7*, 83 (1956)
126. Chapiro, A.: Radiation Chemistry of Polymeric Systems. New York: Wiley 1962, P. 174
127. Seitzer, W. H., Tobolsky, A. V.: J. Amer. Chem. Soc. *77*, 2687 (1955)
128. Matheson, M. S., Auer, E. E., Bevilacqua, E, B., Hart, E. J.: J. Amer. Chem. Soc. *73*, 1700, 5395 (1951)
129. Haq, Z. M. Sc.: Thesis submission, University of Leeds 1973
130. Guthrie, J. T., Ph. D. Thesis: University if Salford, U. K. 1971

3. Radiation-Induced Grafting onto Cellulosics

3.1. Introduction

Grafting by irradiation of polymers in the presence of, or followed by, treatment with monomer has been one of the most successful of all grafting methods and has been applied to an enormous number of polymer-monomer systems. Chemical bonds are ruptured in organic materials when they are irradiated with electromagnetic or particulate radiation of 100 eV or greater. Among the commonly used energy sources are Cobalt-60, electron accelerators, and nuclear reactors. Free radicals have been detected in irradiated organic systems, and if vinyl monomers are irradiated, polymerizations take place through a free radical mechanism. When organic polymers are irradiated, simultaneous cross-linking and degradation of the chains occur due to the formation of free radicals on the polymer backbone. Irradiation in the presence of oxygen, an effective free radical scavanger, produces peroxides and hydroperoxides within the polymer. In amorphous polymers which are in the vitreous state or in highly crystalline polymers, irradiation in vacuum, or even in the presence of oxygen, may lead to trapped radicals. Free radicals or peroxides formed on the polymer backbone by the radiation can be used to initiate the polymerization of different vinyl monomers, leading to graft copolymer formation.

The initiation of chemical reactions by photolysis is also well known. Selective absorption of electromagnetic radiation in the visible and ultra-violet region causes the rupture of chemical bonds and the production of free radicals. Some bonds are efficient absorbers of certain wave lengths of the incident radiation and are easily ruptured. Photolysis of polymers containing such bonds can lead to the formation of free radical sites on the polymer backbone. In the presence of a monomer, graft copolymers as well as homopolymers may be formed. When none of the bonds within a system is ruptured by this low-energy radiation, copolymerization can be otherwise promoted by the addition of photosensitizers. These are molecules which can absorb the radiation energy and transfer some of this energy to the other chemical species present. Photosensitizers may also decompose into active free radicals which can initiate the polymerization of a monomer in the presence of the polymer. In addition, these active species may abstract an atom from the polymer backbone. The ease of formation of the free radicals within any irradiated system, their concentration, stability, and location are dependent on the type of radiation, the total energy absorbed, the depth of penetration, the degree of localization and the radiation sensitivity of the material. Subsequent chemical reactions are dependent on variables such as the temperature, concentration, radical activity, and the immediate environment.

Polymer modification using radiation-induced processing offers various advantages over currently accepted processes. However, there are also certain difficulties. These are discussed in detail by Liepins et al.[1] in their review of radiation-induced processing of textiles. This review, though dealing mainly with the flammability resistance of textile polymers, gives an excellent summary of the state of the art, the types of radiation source, and their characteristics. Of all the sources so far considered, the gamma (γ) source is the only one which does not require electrical power for activation.

Two common γ sources are available Co-60 and Cs-137, though Co-60 is the one used in commercial isotope radiation plants. The reasons for this are not difficult to find. Co-60 is relatively cheap, has a useful half life of 5.25 years, is unaffected by high energy radiation from external sources, and may be used up to temperatures of 1300 K. The emission of two gamma rays per disintegration provides a total of 2.5 MeV energy per disintegration. Commercial operations require the use of multi-kilocurie sources, with continuous exposure providing high penetration levels. Several source geometries are available to satisfy particular requirements. Co-60 sources used on a commercial scale lend themselves to ready automation, are reliable and highly penetrating, and provide continuous irradiation during exposure. As stated above, these sources gradually lose strength, though in a controlled manner. This loss of strength means that the sources need replacing at approximately 2–5 year intervals. The need for automation in product delivery for irradiation is a limiting factor in the degree of flexibility in exposure doses in terms of the distance between the radiation sources and the samples being irradiated. Since the irradiation step would normally be part of an on-line system, this need not be too serious a restriction. Most commercial operations are based on a continuously moving conveyor belt system which operates in the radiation chamber.

There are several important types of electron beam radiation sources[2]. These include the Van de Graff accelerator, the resonant transformer, the linear accelerator, the Dynacote, the Dynamitron, and the insulating core transformer. In terms of radiation processing, the following features are relevant. The sources require regular maintenance, provide limited product penetration, are of constant strength, give high dose rates in the absorbing regions, require power for operation, but can be activated or deactivated as required, and are relatively inexpensive to operate.

Ultra-violet radiation-induced processing is now well established in several fields, including the curing of surface coatings and the drying of printing inks. In terms of the modification of textiles, this technique is still in its infancy.

Additional radiation sources include low-pressure plasma radiation sources and corona discharge. Low-pressure plasma radiation is used commercially to graft acrylic acid to polyester fabric and could easily be extended to provide surface modification of cellulosic textiles[3]. Corona discharge is commonly used to provide surface activation of plastic films in an attempt to improve their adhesion characteristics. This is used especially in the printing of flexible packaging films. It is somewhat doubtful if this technique has much to offer in terms of modifying the physical or chemical properties of cellulosics through graft copolymerization.

3.1.1. Radiation Units; The Concept of Dose and Dosimetry

The term 'dose' is used to denote in a quantitative manner, the radiation received by a particular substance in a radiation field. Prior to the Copenhagen and Geneva meetings of the International Commission of Radiological Units[4,5], considerable confusion existed regarding units of radiation dose. Until this time, it was common practice to express dose in terms of the roentgen, a unit of radiation flux based on the ionization produced in air. By definition, the roentgen is a unit of incident radiation dose as distinct from absorbed radiation dose. When substances of different chemical nature are irradiated in the same field, the energy dissipation in the various systems may vary greatly for the same incident radiation dose. A unit of absorbed dose, namely the RAD introduced by the above-mentioned commisions, has done much to clarify the position. As stated above, in the history of radiation-induced grafting, various terms have been used to indicate levels of exposure in a quantitative manner. In polymer processing, the three most commonly encountered units are the rad, the rep, and the roentgen. In terms of the S.I. convention, 1 rad $= 10^{-2}$ J/kg. The absorbed dose in a specified medium was, until recently, expressed as rads per second, megarads per hour, electron volts per gram per second, or ergs per gram per second, etc. For convenience, various conversions are presented here:

1 rad $= 10^{-2}$ J/kg $=$ 100 ergs/g $= 6.24 \times 10^{13}$ eV/g ,
1 rad/sec $= 10^{-2}$ W/kg ,
1 Mrad/hour $=$ 2.78 W/kg ,
1 eV/g $= 1.602 \times 10^{-16}$ W/kg ,
1 rep $=$ approximates to 0.93 rad ,
1 roentgen of $\times$ of gamma radiation is approximately equal to 0.96 rad when dealing with energy absorption in most organic polymer systems.
1 Curie (Ci) $= 3.7 \times 10^{10}$ disintegrations/sec .

Generally, one can achieve between 0.5 and 100 rads/minute for 1 Ci of Co-60, depending on the geometry of the source, the sample arrangement, and the absorbing efficiency of the sample.

The scientific literature contains many references to the design and calibration of equipment for measuring the radiation dose. For any specific application, the most suitable technique depends on the type of radiation, the dose, the radiation intensity, and also on the accuracy, simplicity, and speed of measurement required.

Techniques for determining absorbed doses can be divided into two classes: absolute methods and those involving secondary standards. Absolute methods of dosimetry include the ionization chamber, of which those defined for use with fast electron beams are described by Boag et al.[6], Laughlin et al.[7], McEhinney et al.[8], Hochanadel and Weiss[9], and by Lazo et al.[10]. There are often serious difficulties with absolute physical methods of determining radiation dose, and for many purposes, secondary standards based on a chemical change taking place within the system being irradiated have been adopted. Examples of chemical dosimeters include the oxidation of aqueous solutions of Fe^{II}sulfate to Fe^{III}sulfate[11], the reduction of aqueous solutions of Ce^{IV} sulfate[12,13], the decoloration of methylene blue in organic solvents[14,15], and the decomposition of nitrous oxide[16]. Colorimetric methods of dosimetry include, for example, the coloration of glasses and of ionic solids[17–19], the optical absorption in polystyrene and poly(methyl methacrylate)[20], dye-impregnated plastics[21], and cellophane films[22]. The most commonly used secondary dosimeter involves the oxidation of Fe^{II} ions to Fe^{III} ions and is known as the Fricke dosimeter. The standard solution consists of 1 mM of Fe^{II} sulfate in an air-saturated 0.4 M aqueous solution of sulfuric acid. The Fe^{III} ion is monitored by spectrophotometric measurement of the absorption at 304 nm. The characteristics of the Fricke system are well documented[11,23,24]. However, certain points are noteworthy. The number of oxidation events is dependent on the type of radiation used[24–26]. Also, the oxidation process depends on the presence of oxygen in the system. In using this chemical dosimeter, it is essential that the exposure be controlled so that sufficient oxygen is present at all times, if the oxidation is to proceed in a standard manner. In practice, using this dosimeter limits the total dose measureable to about 40,000 rads for gamma radiation. For polymer research, this can be a limiting factor since, unless a low intensity source is used with correspondingly long exposure times, considerable inaccuracies in dosimetry may result.

The G Value

Although the individual reactions may not be known, the sensitivity of a system to radiation-induced change may be expressed in terms of the number of changes produced by a given absorbed dose. This yield is usually expressed as the *G* value[27], which is defined as the number events of a given kind produced per 100 eV of energy absorbed. In using *G* values it is not implied that all the energy absorbed is used to produce the resultant entities. Likewise, in the definition no assumptions are made as to the mechanisms of the reaction and, for any given system, different *G* values can be given to denote each of a series of products formed in the course of a single irradiation. For example, *G* (radical) — styrene = 0.69[28]. Thus, on average, 0.69 styryl radicals are produced in the irradiation of bulk styrene for each 100 eV of gamma radiation energy absorbed.

In any detailed analysis of a radiation-induced polymerization or degradation, a knowledge of the *G* value of each of the components is invaluable. For example, the grafting of styrene onto regenerated cellulose using a solvent for the monomer and a swelling agent for the substrate requires that the *G* value of each component be known and considered. This is because the radiolysis products may have a marked effect on grafting efficiency.

3.2. Grafting by High Energy Radiation

The interaction of high-energy radiation (gamma, electron beam, etc.) with polymers, such as cellulose and cellulosics in the solid state, induces an interesting type of radiochemical reaction[29].

Polymeric molecules are more or less held in a fixed lattice, and ordinary collision mechanisms are probably largely inoperative. The molecular activation or excitation induced in the cellulosic by the absorbed radiation is not hindered by the crystalline-amorphous nature of the cellulose structure. However, the nature of the cellulose structure affects the accessability of the activated or excited sites with respect to initiation of chain reactions or to interaction with chemical reagents. Thus, for chemical modification of cellulose it is perhaps preferable to think in terms of relative accessability, since terms based on crystallinity and amorphous character do not lead to accurate descriptions of practical situations. It is likely that the half life of the activated or excited states, particularly in the more ordered regions of the cellulose, is several hours, perhaps days. This has been shown by Guthrie et al. in their examination of γ irradiated regenerated cellulose films[30,62]. Samples of regenerated cellulose were dissolved in Cadoxen [tris(ethylene diamine) cadmium dihydroxide] after exposure to known levels of γ radiation. Figure 7 shows the effect of exposure on the intrinsic viscosity of cellulose, expressed as the per cent decrease in intrinsic viscosity versus the log of the total dose[62]. Irradiations were carried out both in air and in vacuum. Vacuum-irradiated samples, on exposure to the atmosphere, undergo rapid secondary degradation. The Mark-Houwink values

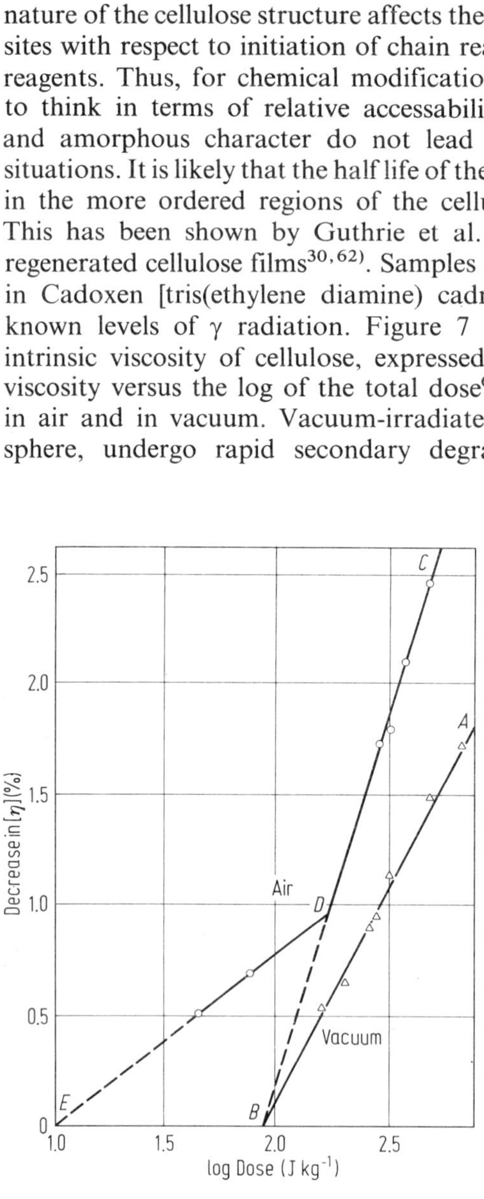

Fig. 7. Determination of the threshold dose for radiation damage on irradiation of cellulose; (○) radiation in air, (△) radiation under vacuum. Reprinted from Ref. [62], p. 1025, by courtesy of John Wiley and Sons, Inc.

for cellulose in Cadoxen at 25 °C are $K \doteq 1.8 \times 10^{-2}$ dl g^{-1} and $\nu = 0.77$ as obtained from light-scattering studies of standard samples[30a)], such that

$$[\eta] = k\bar{Z}_v^{\nu} \quad \text{and} \quad \bar{M}_v = \bar{Z}_v \times 162\,,$$

where $\bar{Z}_v$ is the viscosity-average degree of polymerization, and $\bar{M}_v$ is the viscosity-average molecular mass.

The effect of air in the cellulose degradation was shown by irradiating two series of samples, one in air, the other under vacuum. The intrinsic viscosity, $[\eta]$, for different cellulose samples was determined in the normal manner in a suspended-level dilution viscometer. Using the Mark-Houwink values quoted, the viscosity-average molecular masses were obtained. These viscosity-average molecular masses were converted to number-averages using the equation appropriate for a random distribution[30b, 30c)], (v_{13})

$$\bar{M}_v = \bar{M}_n[(\nu + 1) \cdot \Gamma(\nu + 1)]^{1/\nu}$$

where Γ denotes the gamma function. For cellulose in Cadoxen, $v = 0.77$, which gives the relationship

$$\bar{M}_v \doteq \bar{M}_n(1.894)$$

whereby the number-average molecular masses are derived.

From the variation of the number-average molecular mass with total dose, for cellulose it is possible to acquire information regarding the de-polymerization

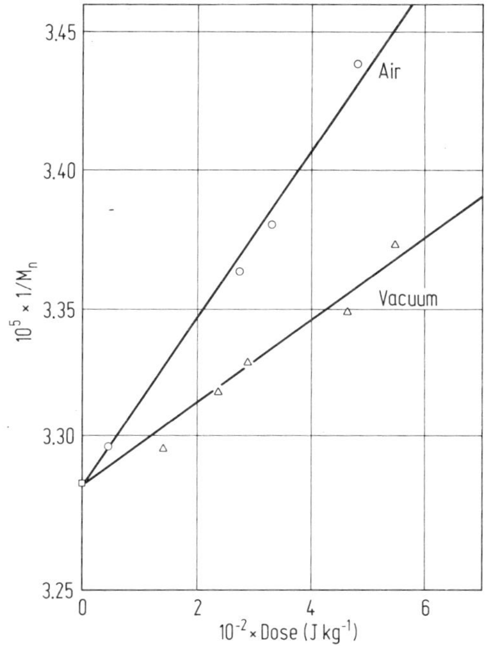

Fig. 8. Determination of G (scission) for cellulose on irradiation in (○) air and (△) vacuum. Reprinted from Ref.[62)], p. 1024, by courtesy of John Wiley and Sons, Inc.

process through the determination of G values for scission of the cellulose backbone. Thus,

$$N_A/\bar{M}_n = N_A/(\bar{M}_n)_0 + (6.242 \times 10^{17} \times G)\, r$$

where N_A = the Avogadro number, $(\bar{M}_n)_0$ = the number average molecular mass of unirradiated cellulose, $\bar{M}_n$ = the number-average molecular mass of the irradiated cellulose, r = the total dose in J/kg × 10^4, and G = the number of scissions produced in the cellulose backbone per 100 eV of absorbed radiation energy.

From this

$$1/\bar{M}_n = 1/(\bar{M}_n)_0 + (6{,}242 \times 10^{17} \times G/6.023 \times 10^{23})\, r\,.$$

Thus, $1/\bar{M}_n$, when plotted against the absorbed dose expressed in J/kg × 10^4, gives a linear plot (Fig. 8)[62]. From the slope, a value for G (scission) is obtained, since slope = $1.036 \times 10^{-6} \times G$ (scission).

For celluloses irradiated under vacuum, the value of G (scission) was found to be 13.5, whereas that in air was 24.6. This shows conclusively that oxygen enhances the radiation-induced degradation of the cellulose backbone.

Corresponding measurements made at various intervals after removal from the irradiation source gave values for G (scission) of 186 and 144 for samples irradiated in air or vacuum (Fig. 9)[62]. Electron spin resonance (e.s.r.) spectroscopic examination of samples irradiated under conditions similar to the above have demonstrated the presence of a weak, residual signal in the cellulose four years after removal of the evacuated, degassed samples from the irradiation source.

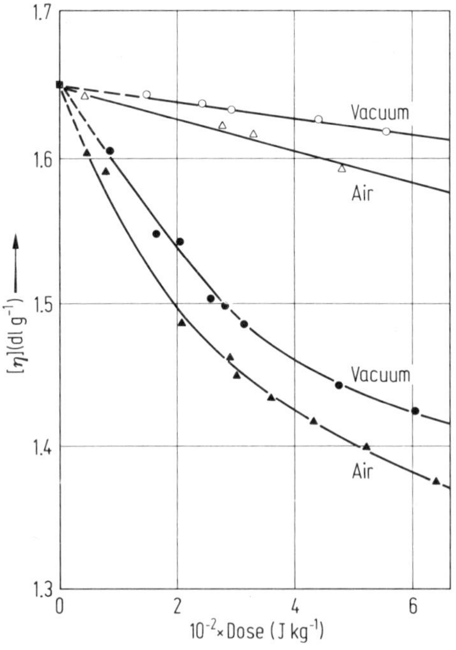

Fig. 9. After-effect degradation of irradiated cellulose: (○), (△), [η] measured immediately after irradiation; (●), (▲), [η] measured after 21 days. Reprinted from Ref.[62], p. 1026 by courtesy of John Wiley and Sons, Inc.

Such longevity of active sites belongs to those sites which are located in the inaccessible regions of cellulose. Sites located in the accessible regions are much more prone to termination or quenching by one of several possible terminating species. The half life of such species is highly dependent on their particular environment. The presence of active species in the more ordered regions (so-called trapped radicals) offers the possibility of post-irradiation reactions under many experimental conditions. The mechanism, providing relatively long-term activity, is one of energy transfer within the cellulose substrate. This will depend to a great extent on the chemical nature of each particular cellulose derivative. Hence, it is unreasonable to make generalizations regarding the suitability of particular cellulosics for post-irradiation modification. The effect of high-energy radiation on cellulose in the absence and presence of vinyl monomers will now be discussed.

3.2.1. Energy Transfer in the Irradiation of Cellulose

When carbohydrates are irradiated in the solid state by high energy radiation, energy transfer effects lead to localization of the energy within the molecule. These energy transfer effects are generally considered to be dependent on the mechanism of energy loss by the incident radiation to the chemical molecule, and initial random non-localized deposition (and subsequent dissipation) of the energy within the molecule and the rapid localization of the energy within the molecule. The localization of the energy in the molecule results in physical and chemical changes depending on the nature of the matter irradiated, such as degradation, activation of long-lived excited sites, degradation of the molecule, or depolymerization[31)]. The time required for this presumed localization process is estimated not to exceed 10^{-13} sec. Furthermore, it has been suggested that the effects of an initial non-localized deposition of high energy within matter can extend over distances ranging to 10 nm[32)].

The intermolecular and intramolecular transfers of high energy in the localization process in carbohydrates in the solid state have been shown to be dependent on the physical structure of the carbohydrate[33)], on certain complex formations of other molecules with the carbohydrates[34)], and on substituent groups on the molecule[35,36)]. The higher acid yields for irradiated α-D-glucose in the polycrystalline state as compared with the acid yields from irradiated α-D-glucose in syrup or freeze-dried states were interpreted to indicate a mechanism of energy transfer which was facilitated by the crystalline state[33)]. On irradiation, complexes of Schardinger dextrins with *p*-nitrophenol gave the characteristic electron spin resonance spectrum for irradiated *p*-nitrophenol. The absence of the characteristic electron spin resonance spectrum for the irradiated dextrin was interpreted to indicate a mechanism of energy transfer from the dextrin to the *p*-nitrophenol[34)]. The radiation protection of α-D-glucose involved relatively large amounts of energy. From the dimensions of α-D-glucose crystals and the dimensions of β-cyclic dextrin, it is obvious that intermolecular energy transfer occurs over some distance. Perhaps, in the case of freeze-dried crystals of α-D-glucose about 1 μm in diameter, intermolecular energy transfer occurs over distances greater than 10 nm, the diameter of a spur of nonlocalized energy deposition[32].

In his excellent review of radiation-induced grafting onto cellulosic substrates, Moore gives comprehensive coverage to the action of ionizing radiation on cellulose[37]. He points out the established fact that the chemical and physical effects of radiation on cellulosics are dependent on the type of radiation, the dose rate and total dose, and the presence or absence of oxygen and moisture. Among the major changes are chain 'scission'[38–41], carboxyl formation; and carbonyl formation in the ratio 1:1:20[42]. The solubility of cellulose in water and sodium hydroxide increases with increasing doses, and evidence is available to show that peroxides are formed on irradiation of cellulose in air[43]. Charlesby has shown that the accessible and inaccessible regions in cellulose are equally subject to main chain cleavage, although those celluloses showing the greater level of accessibility (e.g., rayon) are more easily degraded than the less accessible forms of cellulose (e.g., native cotton)[44]. Although main chain cleavage is a major event in radiation-induced damage, some degradation of the glucosidic rings also takes place since hydrolysis of strongly irradiated cellulose yields xylose, arabinose, and glucoronic acid, as well as glucose[45].

Since cellulose and its derivatives may be irradiated for purposes unrelated to grafting reactions (examples include sterilization of packaged pharmaceuticals and the curing of coatings containing cellulose derivatives), conditions associated with the protection of celluloses against radiation-induced damage deserve attention.

3.2.2. The Radiation Protection of Cellulose

Radiation protection of the glycosidic linkages in the glucosides and in cellulose by the introduction of aromatic groups into the molecules was interpreted to indicate that the introduction of aromatic groups into those carbohydrates affected energy transfer, particularly localization of the energy. Hence, it was suggested that intramolecular energy transfer could occur from the aromatic groups through linkage to the carbohydrate molecule, before localization of the energy, to cause chemical change. Alternatively, energy deposited within the carbohydrate molecule could be rapidly localized, possibly through transitions between electronic levels, in an essentially inert group, such as an aromatic group, from which it could be dissipated[37].

Grafting reactions and chemical substitution have provided considerable information concerning the processes of intramolecular energy transfer of high energy in carbohydrates[35,36]. Such information is important since the concept offers a means of protecting cellulose against otherwise inevitable damage when it is irradiated. However, such protection is achieved only at the expense of the number of activated sites which are available for any subsequent radiation-induced process such as graft copolymerization. As such, radiation protection is an important factor in any commercial application.

Arthur et al.[46–49] have investigated the protection against radiation given to cellulose through introduction of a wide variety of aromatic groups. These include phenylcarbamoyl, benzyl, naphthoyl, cinnamoyl, benzyl, benzydryl, and trityl groups. Only the benzyl group failed to give some measure of protection. The extent of substitution required to impart a standard level of protection was found to depend on the nature of the group, the type of linkage to the cellulose backbone, and the stability of such a linkage. It was found that where the absorbed radiation cleaved the aromatic group from the main chain, no protection against radiation effects took place. It has been shown that a modification of the aromatic group, thus bringing about a reduction in the effective number of available π electrons, results in a corresponding decrease in the level of protection given to the cellulose. Studies by Chidambareswaran et al.[50], Guthrie et al.[51], and Singh et al.[52] have confirmed this protective nature of aromatic groups toward cellulose. Chidambareswaran et al. compared benzoylation of cellulose with allylation in terms of their effect on radiation protection and found that, while allyation influenced both the nature and yield of free radicals of the cellulose as seen through e.s.r., benzoylation affected only the yield.

Guthrie et al. have examined the carbanilated cellulose system in some detail[51]. Carbanilate groups are effective in protecting cellulose from radiation-induced damage, even at relatively moderate degrees of substitution (D.S.), as seen using e.s.r. spectrocopy. Irradiation of carbanilated cellulose in vacuo produces the cyclohexadienyl radical. No cyclohexadienyl radicals were observed with samples irradiated in air since they nad been quenched by atmospheric oxygen and moisture. Ultimately, the e.s.r. signal degenerates to one similar to that provided by cellulose which has been irradiated in air, though of much reduced intensity. Typical *G* (radical) values, obtained from controlled radical yield

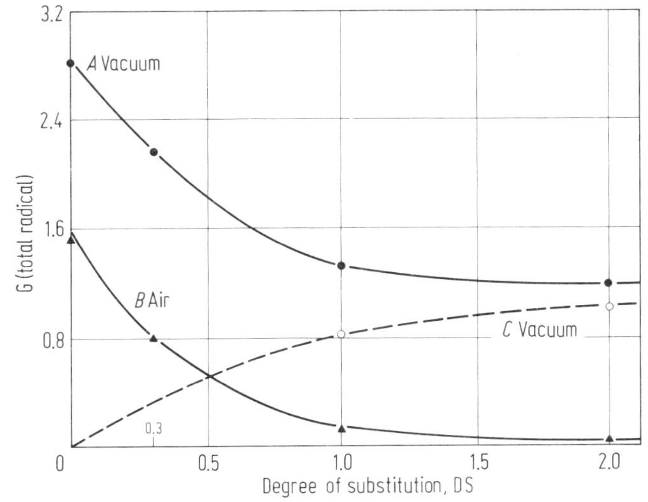

Fig. 10. Change in *G* (*x* radical) value with increasing substitution. *A* (●), $x = G$ (total radical) in vacuum, i.e., inaccessible + accessible + cyclohexadienyl radicals. *B* (▲), $X = G$ (total radical) in air, i.e., inaccessible and cyclohexadienyl radicals. c: (○), $x = G$ (cyclohexadienyl radicals) in vacuum. Reprinted from Ref.[51], p. 753, by courtesy of Pergamon Press Ltd.

studies, are as follows: unsubstituted cellulose (D.S. = 0.0) 2.86; carbanilated cellulose (D.S. = 0.3), 2.20; carbanilated cellulose (D.S. = 1.0), 1.33; carbanilated cellulose (D.S. = 2.0), 1.19, (Fig. 10). In the spectra of carbanilated cellulose (shown in Fig. 11), two radical species are indicated, those in the accessible and inaccessible regions of cellulose and the cyclohexadienyl radicals. Figure 12 gives an indication of the decay of the cyclohexadienyl species after irradiation under vacuum.

The lifetime of the species producing the nine line signal was found to be approximately 5 hr, after which a spectrum similar to that of an air-irradiated, substituted cellulose was obtained, but of reduced intensity. It is noteworthy that the cyclohexadienyl radical was not observed with the carbanilated cellulose that had been irradiated in air.

Quenching of radical activity in the inaccessible regions is best explained in terms of transfer of energy to the aromatic group from the cellulose backbone, since the substitution reaction can only occur in the accessible regions.

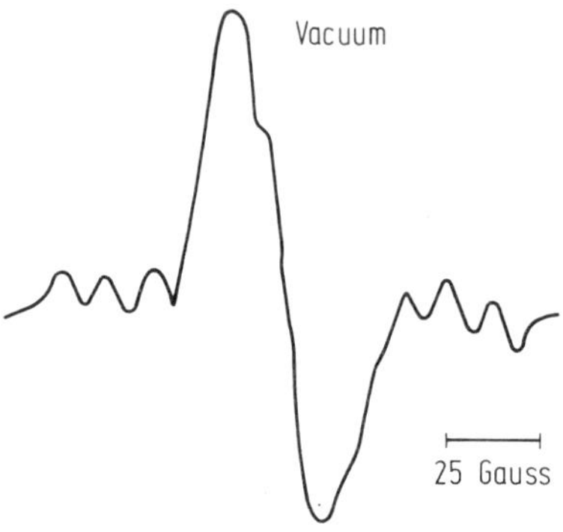

Fig. 11. E.s.r. spectrum of a typical cellulose carbanilate (D.S. = 2.0), irradiated under vacuum. Reprinted from Ref.[51)], p. 750, by courtesy of Pergamon Press, Ltd.

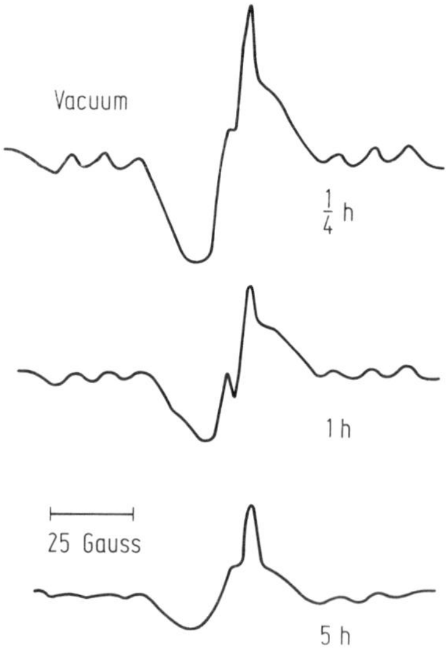

Fig. 12. Lifetime of the cyclohexadienyl radical from carbanilated cellulose, irradiated under vacuum as detected by e.s.r. Reprinted from Ref.[51)], p. 751, by courtesy of Pergamon Press, Ltd.

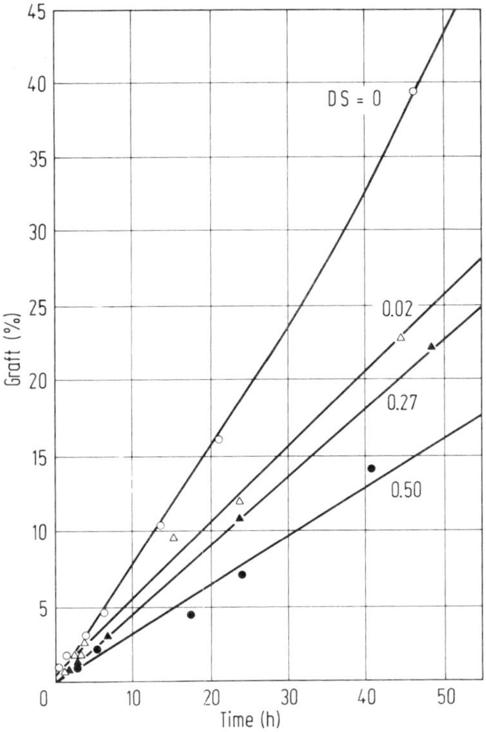

Fig. 13. Variation in the extent of grafting with degree of substitution of carbanilated celluloses, D.S.: (○), D.S. = 0; (△), D.S. = 0.2; (▲), D.S. = 0.27; and (●), D.S. = 0.50. Reprinted from Ref.[51], p. 750, by courtesy of Pergamon Press, Ltd.

This should provide a means of protecting cellulose against damage without altering the nature of the cellulose too drastically. In this way, grafting to reasonable levels should still be possible. Justification for this statement is found in Figure 13, which shows the changes in grafting behavior, for the grafting of styrene onto cellulose and carbanilated cellulose, with increasing substitution of the parent polymer. The decrease in grafting is marginal when compared to the protection against damage which is provided by carbanilation.

Graft polymerization reactions may be used to impart stability to radiation-induced damage in certain instances. In an e.s.r. study of cellulose acetate-styrene grafting, Campbell et al. observed a slow decay in the concentration of trapped radicals after the addition of monomer to the system[53]. Such a reduction in radical concentration may be explained through energy transfer to the grafted areas of the copolymer and ultimate dissipation of the energy as heat. Since grafting occurs only in the accessible regions, transfer of energy from the inaccessible to the accessible regions is indicated. Evidence supporting the above view has been accumulating[54–56]. More tangible evidence of the effectiveness of aromatic groups in protecting cellulose against radiation-induced damage is seen in the retention and frequent improvement of the mechanical properties and the corresponding reduction in main chain cleavage. Radiation doses required to start deterioration of celluloses containing a low degree of substitution by aromatic groups are usually at least ten times those required for sterilization (of unsubstituted cellulose) and thirty times those needed to induce grafting.

In proposing that protection be used on a routine basis, we must be aware of changes which may take place in the behavioral characteristics of the substrate and the uses to which the cellulose will ultimately be put. As an example, we have problems which arise when protected celluloses are expected to withstand high temperature processing, even on a molecular, localized scale. Under such conditions, cleavage of these protective groups may occur, rendering the cellulosic useless for a particular application.

3.2.3. The Radiation Chemistry of Cellulose

The penetrating power of high energy radiation, particularly gamma and X-rays, is not influenced by the order-disorder character of cellulose and its derivatives[47)]. However, secondary adjustments and rearrangements are very much dependent on the physical structure of the irradiated material. Electronic equilibrium must be attained between the radicals and excited molecules, whose freedom of movement is restricted by their solid state environment and by other activated species which are in close proximity. Such an equilibrium may be achieved by intramolecular rearrangement, by energy transfer, or by interaction with any mobile molecules or activated species present, or it may be produced during irradiation.

The nature of the structure of cellulose affects the accessibility of the activated or excited sites to chemical reagents and to initiation of chain reactions. Cotton cellulose is itself a unique polymeric product. Cellulose consists of long linear chains which pass through both highly ordered and highly disordered regions. The ordered areas of the cellulose are almost impenetrable to most gaseous and liquid systems. Temperature elevation does not produce chain mobility. Even the less ordered portions of the structure are relatively fixed. These properties of cellulose are the result of linearity of the chain, of the steric requirements of the monomer units, and especially of the strong hydrogen bonds formed between hydroxyl groups of neighboring chains that are in proper alignment.

Cellulose is made up of repeating anhydroglucose units, which contain carbon, oxygen, and hydrogen atoms. Each anhydroglucose unit has an equal probability of being affected by the radiation, and each carbon and oxygen atom of this unit has at least a similar probability of being ionized. However, the chemical adjustment and rearrangements which take place after initial ionization and excitation are very much dependent on the chemical structure of the irradiated material. That is, the nature of the changes produced in the cellulose by high energy radiation is dependent upon the particular carbon atom ionized or the particular carbon atom to which the ionized atom is bound, the chemical nature (primary hydroxyl, secondary hydroxyl, potential aldehyde) and position of this carbon atom in the basic glucose unit, and the atmosphere under which the cellulose is irradiated or the atmosphere to which it is thereafter exposed. Thus, in order to determine truly definite mechanisms for the radiation chemistry of cellulose, both the chemical and physical properties of the cellulose after irradiation must be evaluated.

3.2.3.1. Molecular and Structural Properties of Irradiated Cellulose

Opinions vary as to the sensitivity of cellulosics to high energy ionizing radiation and also to lower energy exciting ultra-violet energy. However, there is little doubt that some forms of cellulose are more prone to degradation than are others. Here we view the problem through changes observed in the physical properties, which can be interpreted in terms of the sensitivity and structure of particular cellulosics. Several factors contribute to the overall deterioration, including the presence of moisture, oxygen, additives, sensitizers, protecting groups, the dose rate and total dose, and the physical form of the cellulosic (powder, film, pulp, etc.). It is generally accepted that the dose levels needed to encourage grafting are significantly below those likely to cause extensive damage to the cellulosic. However, such a generalization may be somewhat misleading. Factors of importance to any detailed study of the changes in the molecular and structural properties of cellulose and its derivatives are the reaction conditions, substrate composition and morphology, previous chemical and physical modifications, sample purity, and the presence of trapped radicals.

When cellulose acetate is exposed to X-rays for various periods to time up to 190 hours, the well-defined rings observed in X-ray diffraction photographs of the original sample show an increasing diffuseness for irradiated samples as the exposure time is increased. Based on the diffuseness of the X-ray diffraction lines of the irradiated samples and on phenol delustering tests, it has been suggested that X-ray irradiation decreases the crystallinity of the samples[57]. Studies of the effect of high energy cathode rays on wood, wood pulp, and cotton linters have shown that the irradiated cellulose has a higher rate of hydrolysis and a greater quantity of easily hydrolyzed material than the unirradiated sample[36]. This would also seem to indicate that irradiation produces a reduction in the level of order, resulting from a disruption of the ordered regions of cellulose. However, the order — disorder nature of cotton cellulose apparently is not changed by gamma radiation. The infrared absorption spectrum of irradiated cellulose retains all the absorption bands of the purified cotton control

Table 8. Some chemical and physical properties of purified cellulose irradiated with gamma rays in oxygen atmosphere. Adopted from Ref.[61], p. 384, by courtesy of A. Howes Publication

Dosage (roentgens)	Zero	1×10^5	1×10^6	5×10^6	1×10^7	5×10^7	1×10^8
Degree of polymerization	4400	3100	800	290	180	72	39
Carbonyl Groups (mmol/g)	0.00	0.01	0.06	0.29	0.58	1.92	3.04
Carboxyl Groups (mmol/g)	0.002	0.004	0.007	0.022	0.039	0.107	0.155
Water Solubility (%)	0.1	0.0	0.0	0.2	0.4	4.5	9.7
Alkali Solubility (%)	0.8	1.5	4.0	15.8	30.1	61.4	74.4
Pressley Index (lb/mg)	7.61	7.78	7.12	5.86	4.31	—	—

without a lessening of intensity. It was also shown that a loss in crystallinity of cellulose results in a broadening and smoothing-out of the series of sharp bands given by cellulose in the 7.0 through 9.5 μm region. The irradiated cellulose did not show this behavior. Further radial X-ray spectrometer tracings of the irradiated samples indicate that irradiation does not decrease the level of apparent order in the cellulose. The substantial increase in the rate of acid hydrolysis and the amount of iodine sorption of irradiated cellulose, as compared to the control, could be due to the chemically altered structure rather than to an alteration of the crystalline structure of cellulose[58-60].

The major molecular changes which occurred on irradiation of cellulose with high energy radiation, such as gamma rays from Cobalt-60, thermal neutrons, etc., have been shown to be depolymerization and formation of acid and reducing groups.[61]. The ratio of the number of chain changes to the number of acid groups formed to the number of reducing groups formed was 1:0.5:19. Over a dosage ranging from 10^5 to 10^8 roentgens, these molecular properties of the cellulose molecule (initial degree of polymerization 4,400 or 700,000 g/mol weight average molecular mass) were related to dosage, as follows,

$$\ln P = K' \ln Nn + K'' \qquad (64)$$

where P = molecular property, Nn = dosage in roentgens, K' and K'' = constants. P is expressed as degree of polymerization, or in mols of acid groups, or reducing groups per unit mass of cellulose.

Guthrie et al.[62] have shown that regenerated cellulose films, on irradiation in a 98/2 (dioxan:water) mixture have a very low threshold for the commencement of radiation-induced damage. Irradiation in air or under vacuum gives a threshold for damage, assessed viscometrically, of less than 1 krad (Fig. 7). The role of the liquid medium in this degradation is not clear. However, the radiolysis products of both the components of the liquid medium are potentially good radical initiating species and would tend to enhance the prevailing degradation processes. It is unlikely that this low radiation dose has any influence on the bulk properties of cellulose or its derivatives. The threshold for notable changes in mechanical or tensile properties is thought to be considerably in excess of 1 krad. In a relatively early study, it was noted that the fibrous structure of cellulose disintegrated at dosages higher than five to 10 million roentgens, with a resultant loss of textile properties. Tables 8 and 9 show the properties of purified cellulose irradiated with gamma rays under a variety of conditions[61]. Similar results were obtained when irradiation was performed in a "pure" source of thermal neutrons in the dosage range of 5×10^{14} through 1×10^{17} neutrons[63].

At lower radiation dosages, where cotton cellulose retains its fibrous properties, small increases in tensile strength have been reported by some investigators, with decreases being observed at higher incident doses[59,64,65]. Also, it has been reported that when water-soluble derivatives of cellulose in dilute aqueous solution were irradiated, a gel formation was noted[46-48]. These changes could indicate increased intermolecular bonding. However, decreased intermolecular bonding on irradiation of cotton cellulose was determined from the viscous flow of dilute solutions of irradiated cotton. The enthalpy of activation of viscous flow of dilute

Table 9. Some chemical and physical properties of purified cellulose irradiated with gamma rays in nitrogen atmosphere. Adapted from Ref.[61], p. 384 by courtesy of A. Howes Publication

Dosage (roentgens)	0.0	1×10^5	1×10^6	5×10^6	1×10^7	5×10^7	1×10^8
Degree of polymerization	4400	3100	880	320	210	79	56
Carbonyl Groups (mmol/g)	0.00	0.01	0.05	0.22	0.36	1.30	2.66
Carboxyl Groups (mmol/g)	0.002	0.002	0.005	0.015	0.023	0.070	0.139
Water Solubility (%)	0.1	0.0	0.0	0.1	0.5	3.3	10.4
Alkali Solubility (%)	0.8	1.0	4.9	15.9	24.1	57.8	71.1
Pressley Index (lb/mg)	7.61	7.80	7.11	5.79	4.81	—	—

solutions of purified or irradiated cottons was relatively constant. In the case of irradiated cottons, the lower negative value for the entropy of activation of viscous flow of dilute solutions indicated that less alignment of the irradiated cellulose molecules was required for viscous flow than in the case of purified cellulose molecules. Consequently, higher values for the free energy of activation of viscous flow of dilute solutions of purified cellulose were obtained than in the case of irradiated cellulose. This decreased bonding in irradiated cellulose is probably due to the decrease in the number of interacting anhydroglucose units in the molecule[69].

The moisture regain of cellulose is not greatly changed by irradiation. Changes in moisture regain are usually interpreted as reflecting changes in the submicroscopic structure of the cellulose if the treatment has not altered the chemical composition of the cellulose. The small changes in moisture regain values on irradiation of cellulose would suggest that there is no marked alteration of the submicroscopic structure of the fibers, and that the observed chemical changes do not markedly affect the regain values[59].

Changes in the fibrillar organization of cotton fibers after exposure to gamma, high-energy electron, and thermal neutron radiations in various ambient atmospheres were evaluated by the techniques of optical and electron microscopy[70]. Radiation exposures evaluated were: gamma rays, 10^5 to 10^8 r. (roentgens); 2 Mev electrons, 5×10^4 to 10^8 rep. (roentgens equivalent physical); integrated thermal neutron fluxes, 5×10^{14} to 10^{17} nvt (thermal neutron/cm^2/sec. x exposure time in seconds). Figure 14–17 show the progressive decrease in the size of fibrillar fragments of gamma-irradiated fibers with increasing dose. Distinctive changes in the fibrillation pattern of either unpurified or purified cotton did not occur at levels of 10^5 and 10^6 r., but fibers receiving up to 10^7 r. formed shorter fragments, and at 10^8 r. some of the sheets of fibrils broke into short fragments and particles. The fragmentation pattern of the electron-irradiated fibers differed from unexposed fibers only at dose levels of 10^8 rep., but the same type of change as seen in gamma-irradiated fibers was found. The only changes noted in the pattern of neutron-irradiated samples were the formation of spicules, or loose ends of fibrils within the sheets of fibrils, and some evidence of

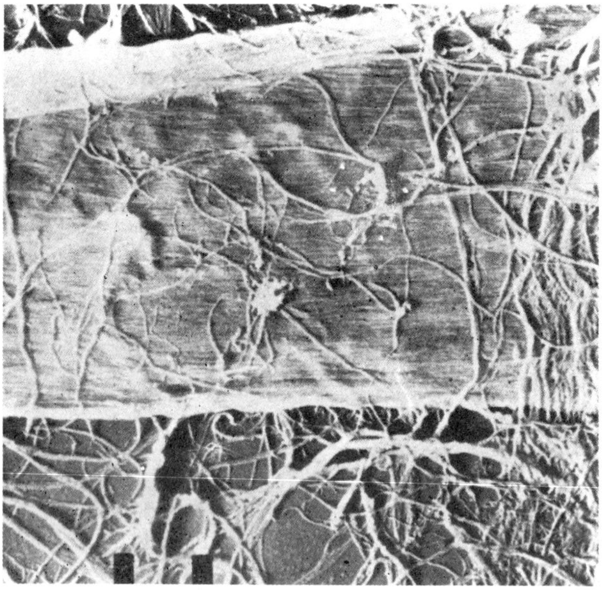

Fig. 14. Electron micrograph of fragment of unexposed purified cotton. Reprinted from Ref.[70], p. 513, by courtesy of Textile Research Institute.

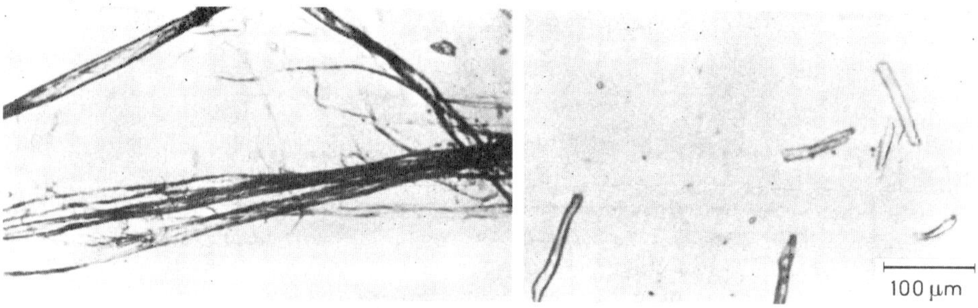

a b

Fig. 15a, b. Photomicrographs of fragments of gamma-irradiated purified cotton **a** unexposed, **b** 10^8 r. Reprinted from Ref.[70], p. 513, by courtesy of Textile Research Institute.

shortening of the individual fibrils. These indications of chain cleavage were not truly distinctive except in samples exposed to 10^{17} nvt (Fig. 16b).

The effects of larger doses of gamma radiation on purified cotton are illustrated in Figure 17. Although excessive chain scission must have occurred to produce the small particles seen in most of the photographs, the predominant change in pattern is in the formation of the short chunks of fibrils which appear fused and brittle. Most of the microfibrillate texture has been lost, and the smooth appearance suggests that interfibrillar fusion has occurred[45].

Scanning electron microscopy was used by Guthrie and Percival in their examination of the effects of high energy radiation on the fibrous character

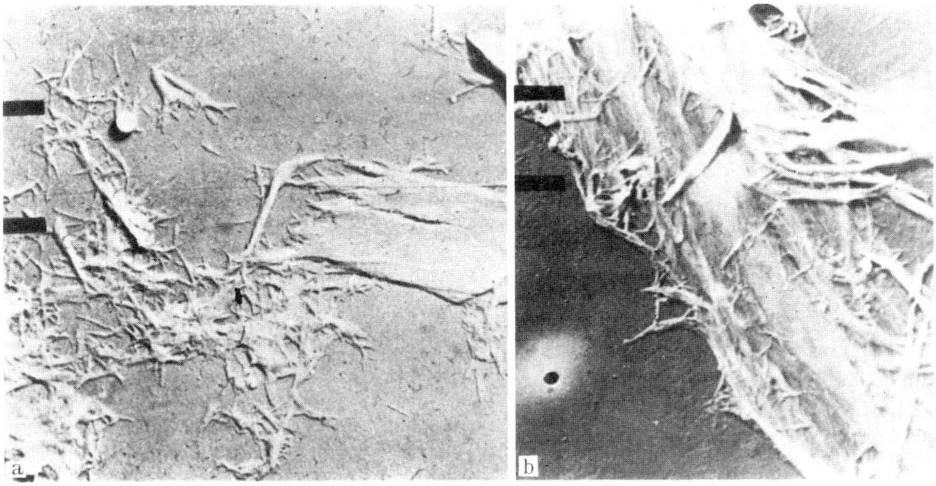

Fig. 16a, b. Electron micrographs of fragments of irradiated purified cotton (compare with fig. 14). **a** Gamma-irradiated, 10^8 r, N_2 atmosphere; **b** neutron-irradiated, 10^{17} nvt. in air. Reprinted from Ref.[70], p. 514, by courtesy of Textile Research Institute.

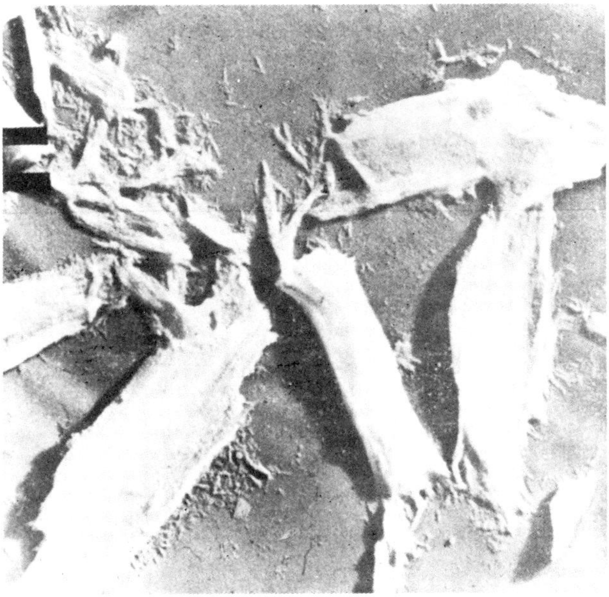

Fig. 17. Electron micrograph showing typical appearance of fragments of gamma-irradiated cotton, 4×10^8 r, in air. Reprinted from Ref.[70], p. 514, by courtesy of Textile Research Institute.

of regenerated celluloses[71]. Samples of both irradiated and unirradiated cellulose film were converted to the carbanilate and the resulting derivatives examined. Damage was seen through the breakdown of the microfibrils and the formation of a layer-like structure. Thermal analysis of the same samples demonstrates the complexity of the consequences of irradiation of cellulosics. Unirradiated

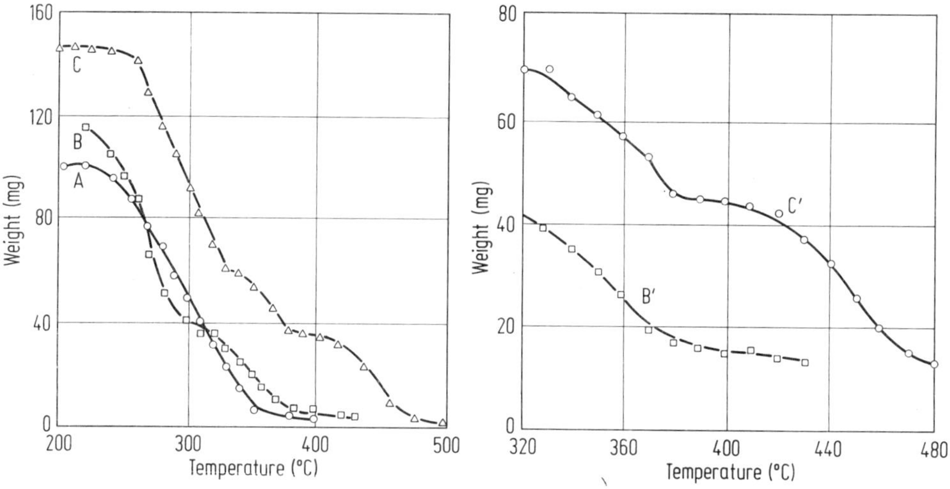

Fig. 18. Differential thermograms of *A*, unirradiated cellulose tricarbanilate (CTC); *B*, irradiated CTC; *C*, the C.T.C. — polystyrene copolymer. Inset (*B*′, *C*′) expansion of *B* and *C* at elevated temperatures. Reprinted from Ref.[71], p. 532, by courtesy of I.P.C., Science and Technology Press, Ltd.

cellulose, on conversion to the carbanilate, shows a reasonably straightforward thermal profile. This observation is supported by thermogravimetric analysis which carries a single weight loss versus temperature increase profile. Irradiated cellulose converted to the carbanilate shows complex thermal properties. The details are shown in figure 18. The thermograms shown in Figure 18 relate to unirradiated cellulose converted to the tricarbanilate (A), irradiated cellulose extracted from the apparently grafted copolymer and subsequently converted to the tricarbanilate (B), and the cellulose — PS copolymer whose cellulose component has been converted to the tricarbanilate (C). Thus, we have unirradiated CTC (A), irradiated CTC (B), and the CTC-PS, (PS denotes polystyrene), copolymer (C).

From C, the composition of the copolymer was calculated at 76% cellulose tricarbanilate and 24% polystyrene. This result is in reasonable agreement with those obtained by other methods of analysis: elemental analysis (70.6%: 29.4%), differential refractometry (70%:30%), and gravimetric procedure (67.6%:32.4%). Hence, we have another satisfactory means of obtaining the composition of cellulose copolymers. The observed differences between the thermograms A and B and the similarities between B and C in the 320°–420 °C temperature range give an indication that high energy gamma radiation causes pronounced changes in the reactivity of cellulose which are maintained in subsequent carbanilation. Thus, the CTC-PS copolymer has similarities with the irradiated CTC, though it possesses the added stability afforded by the grafted polystyrene. This topic is covered more fully in Chapter 7.

3.2.3.2. Chemical Properties of Irradiated Cellulose

The changes in the chemical structure of cellulose brought about by irradiation have been extensively studied[38, 72–79]. For this purpose, analyses were made of the volatile products, residual cellulosic products, and products obtained by fractionation of the residues.

The gases, evolved or absorbed by cotton cellulose in atmospheres of nitrogen, oxygen, and air, were determined. Absolute changes in pressure measured at constant volume and constant temperature, during irradiation of cotton cellulose in sealed containers, were determined. The initial decrease in pressure of the oxygen atmosphere indicated that oxygen was reacting with the activated cellulose at a faster rate than gases were being evolved. The initial decrease and subsequent increase in pressure in the air atmosphere indicated reaction of oxygen with the activated cellulose and evolution of gases[61)].

Mass spectrometry indicated complete reaction of oxygen present in the air atmosphere and almost complete reaction of the oxygen atmosphere with activated cellulose at the massive doses investigated. Mass spectrometry also indicated that the principal gases evolved were hydrogen, carbon monoxide, and carbon dioxide. Infra-red spectroscopy confirmed the presence of carbon monoxide, carbon dioxide, and water vapor. After irradiation of cotton cellulose in an oxygen atmosphere to a dosage of 10^8 roentgens, the composition of the gaseous atmosphere was carbon monoxide (60%), carbon dioxide (20%), hydrogen (9%), and oxygen (11%). The magnitude of the amount of oxygen consumed indicated that it was an important effect in the irradiation of cellulose[61)].

Irradiation of basswood with high velocity electrons altered its structure in such a manner that some of the insoluble carbohydrate become available to rumen bacteria. Maximum availability occured at a dose of 10^8 equivalent roentgens, and at this point the wood becames hydroscopic and friable. Formation of reducing sugars in the soluble products was also estimated[76)]. Exposure of wood, wood pulp, and cotton linters to cathode rays at a dosage of 10^8 equivalent roentgens caused a substantial destruction of the carbohydrates, as measured by quantitative determination of the potential sugar content, before and after irradiation. The percentage of the carbohydrate destroyed upon irradiation at this dosage, based upon the differences between these two values, was found to be 14% for cotton linters, 17% for wood pulp, and 9% for wood. On the other hand, at a dose of 5×10^8 equivalent roentgens, the cellulose was so degraded that it was found to be completely water soluble[31)].

Irradiation of cotton cellulose with beta rays was found to increase the water solubility slightly. The water-soluble portion exhibited an increased acidity. The irradiated samples reduced Fehling's solution and showed a greater affinity for methylene blue (basic dye) than did the controls. The irradiated yarns showed a decreasing affinity on pretreating for 30 minutes with a boiling sodium hydroxide solution. The water-extracted, irradiated samples were found to reduce KI, but this property was not exhibited by the water extract. It was then concluded from these results that irradiation produced an oxycellulose of the acidic type in which some cellulose peroxides were present[64)].

With increasing radiation doses, the solubility of degradation products in water and in dilute sodium hydroxide increased. Cellulose, after receiving a dose of 10^8 roentgens, was about 10 percent soluble in water and about 70 percent soluble in one percent sodium hydroxide solution. The high solubility of the irradiated cellulose in sodium hydroxide solution, as compared to solubility in water, may be due to the formation of a great number of reducing groups. In basic solution, the interaction of sodium hydroxide with irradiated cellulose may lead to further chain cleavage and, consequently, to a marked increase in solubility in base as compared to that in water (Tables 8 and 9)[61].

Using radioactive tracer cations, it was demonstrated that irradiated cotton cellulose had base-exchange properties. Salt bonds were probably the major linkages involved, since any cation could be replaced with another cation by varying their relative concentrations[61, 77].

The distribution of acid and reducing groups in different fractions of the irradiated cotton cellulose was determined. The number of acid groups increased as the molecular mass of the fractions decreased, probably indicating a chain end effect. At a massive dosage of 10^8 roentgens, about one acid group was formed for each two chain cleavages. The infra-red data on the various fractions indicated that the band at 5.75 μm (—C=O) showed a distribution among the fractions of irradiated cellulose similar to that obtained by acid group titration. Considering the large number of reducing groups formed, the relatively weak intensity of the band at 5.75 μm indicated that the reducing groups were probably combined with water to form hydrates, or that they internally formed hemiacetals. Apparently, the band at 5.75 μm indicated the >C=O of the acid group[59, 78].

Paper chromatographic data established the presence of glucose, cellobiose, and a homologous series of cellulodextrins. The alcohol-soluble fraction of the irradiated cotton cellulose exhibited about 17 components. One of the major acid monosaccharide units prodced was 2-keto-gluconic acid. The other major low molecular mass acid components were at least partially identified as the lactones of 2-ketogluconic acid and 2-ketocellobionic acid. Ultraviolet spectral data also indicated the presence of a chromophore group of the keto acid, 2-ketogluconic acid. The group was present in the low molecular mass fraction and in the intermediate fraction in approximately the same ratio as was found by acid titration of these same fractions[47].

Arthur et al.[72] have examined in considerable detail the oxidative reactions of cellulose initiated by free radicals. The formation of free radicals through exposure to high energy radiation increases the reactivity of cellulose mainly through dehydrogenation and the occurrence of oxidative depolymerization reactions. Evidence of secondary reaction has been supplied by Panchenkov et al.[73] from their work on the mechanical properties of cellulose and its esters. A significant reduction in the breaking strength and elongation at rupture was observed to occur on irradiation in the absence of a monomer. The rate of decomposition was found to be rapid at high dose rates. It appears that the degradation effects are markedly enhanced by the presence of air and moisture. Imamura and Ueno have provided insight into the influence of radiation exposure on dissolving pulps[79–82]. It has been shown that both gamma and X-radiation produce the same chemical and physical effects. The rate of main chain cleavage was found to be directly proportional to the dose rate and to be markedly enhanced by elevation of the temperature. Potential industrial processing is suggested by this work, especially in aspects of cellulose solution and derivativization. Here the molecular mass and the molecular mass distribution are important properties. Further investigation has shown that xanthation of irradiated wood pulp, having an average degree of polymerization of approximately 400, provides, on coagulation and regeneration in the standard manner, films of superior properties to those derived from oxidation-induced degradation[83)].

The effect of gamma-radiation on the low temperature, dilute acid hydrolysis of cellulose has been investigated by Kunz et al.[84)], who obtained greater yields of glucose with irradiated samples than with the unirradiated ones. This work indicates that low molecular mass products of crude cellulosic materials are obtainable in resonable yields through extensive preliminary irradiation, provided the cellulosic substrate does not have the means to undergo extensive delocalization of the incident energy.

3.2.3.3. Effect of Storage on the Properties of Irradiated Cellulose

Irradiation of cellulose with high energy radiation yields products containing relatively stable free radicals, easily shown by electron spin resonance spectroscopy[38, 85-88)]. Differences in the crystallinity of celluloses seem to have no obvious effect on the yield of free radicals, though differences can be seen in the nature of the e.s.r. spectra. These depend greatly on the conditions of preparation and measurement. Hence, the type of cellulose used, its physical form, chemical composition, the total radiation dose, the gaseous environment, the presence of moisture, and the temperature all have an effect on the nature of the signal as well as on the radical yield. Generally, the e.s.r. spectra of gamma-irradiated cellulose and its derivatives are asymmetric, though the level of symmetry is significantly greater for pure cellulose than for its derivatives or for regenerated cellulose[87, 88)]. Benzoylated cotton has been shown to provide an e.s.r. spectrum similar to that of unmodified cellulose. This is in marked contrast to carbanilated cellulose, which exhibits a marked tendency toward the generation of the cyclohexadienyl radical. This indicates the enhanced protection given by the carbanilate group relative to the benzoyl group. However, the radical yield, as seen in the spectra of benzoylated cellulose and carbanilated cellulose, is markedly reduced relative to unmodified cellulose[38, 39, 51)].

A knowledge of the method of production, nature, and half-lives of the free radicals on the cellulose backbone would be of particular interest for a number of reasons. First, it would assist in further evaluating the reactions involved in the radiation chemistry of cellulose, particularly with respect to the initial degradation process. Secondly, using high energy radiation, post-irradiation degradation reactions could be a disadvantage in the development of commercially useful products. Thirdly, the presence of a long-lived radical could, on the other hand, have a practival advantage in graft polymerization reactions, since a wider range of experimental conditions could be used.

The intrinsic viscosity of irradiated wood cellulose (ca. 0.3% moisture, ca. 0.5×10^6 roentgens, initial intrinsic viscosity of unirradiated cellulose ca. 4) has been found to decrease for post-irradiation periods up to 30 days to a value of about 2. This post-irradiation decrease in intrinsic viscosity accounted for considerably more than half of the total decrease observed. Post-irradiation effects in low molecular mass cellulose (filter paper, initial intrinsic viscosity of unirradiated cellulose ca. 4.6, degree of polymerization ca. 880) were similar[89, 90].

The effect of Co-60 irradiation, fission-products irradiation, and electron irradiation on the post-irradiation stability of irradiated cotton cellulose, as measured by changes in the tensile strength, specific viscosity, and reducing and acid group content are shown in Tables 10, 11 and 12. Apparently, there is not a great effect on the stability of the irradiated, purified cellulose when related to type of radiation, dose rate, or atmosphere in which the irradiation occurred. However, at dosages which are normally used to sterilize cotton products (2.3×10^6 roentgents or higher), and for post-irradiation times of one year or less, the decreases in the Pressley index and specific viscosity of the irradiated cellulose were small for both electron and cobalt-60 irradiation. For fission-products irradiation and for post-irradiation of more than 3 years at 0.1×10^6 roentgens, the specific viscosity of the irradiated cellulose decreased measurably with time.

The radiation-induced depolymerization of rayon cellulose, in comparison to cotton cellulose, in the presence and absence of air has been studied[92]. The presence of air has only a slight effect on the depolymerization of cellulose. This was observed with rayon and cotton. However, the degree of polymerization of cotton drops to about half, whereas the degree of polymerization of rayon

Table 10. The effects of cobalt-60 irradiation on some properties of cotton cellulose*. Adapted from Ref.[91], p. 733, by courtesy of Textile Research Institute

Post-irradiation time, days	Pressley Index lb/mg		Specific viscosity	
	Irradiation atmosphere			
	Air	Nitrogen	Air	Nitrogen
0	7.5	6.9	2.7	2.9
1	7.0	7.0	2.7	—
2	—	—	—	2.6
6	—	—	2.6	—
7	7.1	6.8	—	—
8	—	—	—	2.3
14	—	6.8	2.5	2.3
21	6.9	—	—	—
27	—	—	2.5	—
42	6.9	—	—	—
50	—	6.7	—	—
77	—	6.8	2.4	—
104	6.9	—	—	—
121	—	—	—	—
133	—	6.3	—	—
152	—	6.2	—	—
364	6.9	—	—	—
378	—	6.7	—	—

* Dose rate: 0.44×10^6 r/hr; dosage: 2.3×10^6 r; samples irradiated in air contained about 6.8% moisture, and those in nitrogen contained about 0.5% moisture. Irradiated samples stored in air at 21 °C and 65% R.H.

Table 11. The effects of fission products irradiation on some properties of cotton cellulose*. Adapted from Ref.[91)], p. 734, by courtesy of Textile Research Institute

Property and Post-irradiation time	Dosage, r		
	0.1×10^6	1×10^6	5×10^6
Oxygen irradiation atmosphere			
Pressley index, lb/mg			
3– 4 months	7.8	7.1	5.9
40–41 months	7.2	6.7	5.2
Specific viscosity			
3– 4 months	26	3.9	0.98
40–41 months	17	2.9	0.82
Reducing groups, mmol/g			
3– 4 months	0.01	0,06	0.29
40–41 months	0.02	0.09	0.29
Acid groups, mmol/g			
3– 4 months	0.004	0.007	0.022
40–41 months	0.003	0.006	0.018
Nitrogen irradiation atmosphere			
Pressley index, lb/mg			
3– 4 months	7.8	7.1	5.8
40–41 months	7.5	6.7	5.4
Specific viscosity			
3– 4 months	27	4.4	1.1
40–41 months	19	3.6	0.95
Reducing groups, mmol/g			
3– 4 months	0.01	0.05	0.22
40–41 months	0.01	0.07	0.26
Acid groups, mmol/g			
3– 4 months	0.002	0.005	0.015
40–41 months	0.004	0.007	0.017

* Dose rate: 3 to 6×10^6 r/hr; samples contained about 6.8% moisture.

drops to three-quarters of the initial value after irradiation of 1 Mrad and storing for 2 months. It has also been reported that the number of scissions of chemical bonds of cellulose molecule is not only independent of the radiation dose, but also independent of the degree of polymerization and of the microstructure of the initial cellulosic fibre.

3.2.3.4. Location of the Free Radicals in the Cellulose Structure after Irradiation

Evidence from various sources has shown that the radicals formed on cellulosics by radiation are located in both the accessible and inaccessible regions. In terms of secondary reactions, however, it is those radicals which are present in the accessible regions which are the more significant. The main sources of support for this view are (i) the dramatic effect of water on grafting reactions, (ii) localization of the grafted polymer in the accessible regions, (iii) conductivity measurements, (iv) dielectric constants, and (v) e.s.r. studies.

Table 12. The effects of electron irradiation on some properties of cotton cellulose*. Adapted from Ref.[91], p. 734, by courtesy of Textile Research Institute

Post-irradiation time, days	Pressley index, lb/mg	Specific viscosity
Dosage: 2.3×10^6 r		
14	7.3	2.1
21	6.7	2.1
28	7.1	2.1
42	7.0	2.1
70	7.0	2.0
133	6.7	2.0
197	6.6	2.0
Dosage: 4.9×10^6 r		
14	6.4	1.1
21	6.0	1.1
28	6.5	1.1
42	6.3	1.1
70	6.1	1.1
133	6.1	1.1
197	5.9	1.0

* 7-MeV linear accelerator for electrons; dose rate: about $280 \times \times 10^6$ r/hr; samples contained about 6.8% moisture and were irradiated in air.
Reducing groups (mmol/g) and acid groups (mmol/g) at 2.3×10^6 r were respectively 14 days, 0.11 and 0.008; 197 days, 0.12 and 0.010. At 4.9×10^6 r, the values were 14 days, 0.22 and 0.016; 197 days, 0.26 and 0.016

It is known that cellulose can absorb large amounts of water vapor on its internal surface (8% (w/w) of water under atmospheric conditions and at room temperature. Adsorption is ascribed to the hydrogen bonding of water molecules by the available cellulose hydroxyl groups in the accessible regions. In addition, there is the possibility of a small contribution from the fringes of the more ordered zones[93]. The highly ordered regions are not penetrated by water vapor[94-96]. From infra-red data, a direct relationship has been found between the extent of water adsorbtion and the size of the non-ordered fraction. Hence, only those cellulose hydroxyl groups which are not hydrogen-bonded in a regular, ordered manner are available as adsorption sites for water molecules. As has already been pointed out, grafting occurs exclusively in the accessible regions of cellulose. The presence of water also has a dramatic effect on irradiation grafting. The influence of water on trapped radical formation in irradiated celluloses is shown in Table 13. The conductivity data (Tables 14 and 15) are of a significantly greater magnitude for non-dry celluloses, irradiated under three different conditions, namely, in air, under vacuum, and in potassium iodide solution. This observation holds true, even on increasing the total absorbed dose level[97]. However, depending on the particular monomer, the copolymerization medium, and the nature of the substrate, there arises the

Table 13. ESR spin concentrations for irradiated celluloses. Reprinted from Ref.[97], p. 108, by courtesy of John Wiley and Sons, Inc.

Irradiation Conditions	'Non-Dry'		'Dry'	
	10^6 rads	10^8 rads	10^6 rads	10^8 rads
Air	5.14×10^{15}	4.53×10^{15}	5.23×10^{15}	4.75×10^{16}
Vacuum	8.87×10^{15}	8.18×10^{16}	9.02×10^{15}	8.73×10^{16}
KI solution	4.63×10^{15}	9.99×10^{15}	4.31×10^{15}	1.18×10^{16}

Table 14. Conductivities of cellulose samples (ND).[a] Reprinted from Ref.[97], p. 109, by courtesy of John Wiley and Sons, Inc.

Irradiation Dose	Conductivity (ohm^{-1} cm^{-1})		
	Irradiated in Air	Irradiated under Vacuum	Irradiated in KI Solution
10^4 rads	3.17×10^{-16}	3.00×10^{-16}	2.30×10^{-16}
10^6 rads	6.93×10^{-16}	7.10×10^{-16}	6.00×10^{-16}
10^8 rads	2.24×10^{-14}	2.60×10^{-14}	8.87×10^{-15}
Cellulose	2.48×10^{-16}		

[a] Measured with gold guard electrode.
ND = non-dry

Table 15. Conductivities of cellulose samples (D).[a] Reprinted from Ref.[97], p. 109, by courtesy of John Wiley and Sons, Inc.

Irradiation Dose	Conductivity (ohm^{-1} cm^{-1})		
	Irradiated in Air	Irradiated under Vacuum	Irradiated in KI Solution
10^4 rads	2.29×10^{-16}	2.07×10^{-16}	2.40×10^{-16}
10^6 rads	2.39×10^{-16}	2.11×10^{-16}	2.28×10^{-16}
10^8 rads	5.12×10^{-16}	5.08×10^{-16}	3.46×10^{-16}
Cellulose	2.17×10^{-16}		

[a] Measured with gold guard electrode.
D = Dry

possibility that grafting may enhance the availability of accessible sites by opening up the internal structure of the substrate. This allows access to otherwise inaccessible zones. For most systems, however, grafting in such zones will always be a minor component of total grafting.

The conductivity data have been taken as further evidence that radiation grafting in cellulose occurs exclusively in the accessible regions. Furthermore, conductivities for non-dry samples, irradiated in air and under vacuum, were similar for a particular radiation dose, but the conductivities of samples irradiated in KI solution were lower. The decrease in values for the KI samples are in agreement with the e.s.r. measurements of unpaired spin concentrations. Thus, it does appear that the conductivity may be determined, at least in part, by the concentrations of alkoxy radicals, since potassium iodide inhibits the formation of alkoxy radicals[98], resulting in a lower number of unpaired spins as found by e.s.r.[97].

That grafting occurs essentially in the accessible regions of cellulose has been indicated by dielectric constant measurements[97]. The data of Table 16 reveal that the higher dielectric constants for "nondry" celluloses, under all irradiation conditions, are consistent with Rosenberg's theory[99] in which the dependence of dielectric constant on adsorbed water is represented by the equation

$$K' = K/(1 - Kxm) ,$$

whre K is the static dielectric constant, K' is the effective dielectric constant, m is the amount of adsorbate, and x is a constant.

Additionally, it has been found that variations in dielectric constant of various celluloses are in agreement[100-101] with X-ray determinations of the degree of order in similar celluloses; thus, an increase in dielectric constant is associated with decreasing order. In Table 16, though slight differences were found between values in the non-dry and dry series, the dielectric constants within each series were essentially the same. It follows from this that the degree of crystallinity in cellulose is not markedly affected by radiation up to 10^8 rads. This conclusion is consistent with X-ray and infra-red measurements[59,61,102].

E.s.r. spectra generated by predried irradiated cellulose I and predried irradiated cellulose II showed only small differences in the trapped radical spectra[103]. Most of the free radicals formed in the irradiated celluloses are apparently trapped in the amorphous, or less ordered, regions of the cellulosic structure[104]. About 70% of the radicals formed in irradiated cellulose I and about 90% of the radicals

Table 16. Dielectric constants for cellulose and irradiated cellulose samples. Reprinted from Ref.[97], p. 111, by courtesy of John Wiley and Sons, Inc.

Sample	Dielectric Constant	
	'Non-dry'	'Dry'
Cellulose	2.21	2.00
10^4 rads (air)	2.19	1.99
10^4 rads (vacuum)	2.20	1.97
10^4 rads (KI)	2.17	1.99
10^6 rads (air)	2.15	1.95
10^6 rads (vacuum)	2.16	1.98
10^6 rads (KI)	2.16	1.96
10^8 rads (air)	2.13	1.94
10^8 rads (vacuum)	2.15	1.93
10^8 rads (KI)	2.14	1.95

formed in cellulose II are scavenged by contacting the sample with water. The spectra of the free radicals scavenged by water are almost indentical for both irradiated celluloses. Contact of the irradiated celluloses with ammonia (liquid) scavenged most of the trapped free radicals[104].

It is accepted that e.s.r. investigations are closely dependent on the conditions of measurement and sample environment. Hence, the relatively simple picture given above should be examined in more detail, expecially with regard to radical type and location.

It has recently been shown[105] that the extent of copolymerization of cellulose I and II with ethyl acrylate from methanol-water solution by γ-radiation was directly related to the initial concentration of free radicals in the irradiated cellulose. However, a fraction of the radicals, about 10% of the maximum concentration in cellulose obtained at a dosage of 5.2×10^{19} eV/g, was apparently in the highly ordered regions of the cellulosic structure. These radicals were not scavenged and did not initiate copolymerization reactions.

3.2.3.5. Mechanisms of Radiation-Induced Reactions in Cellulose

Studies on the e.s.r. of irradiated crystalline sugars related to cellulose suggested that, on the basis of the chemical structure of glucose, the most reasonable positions for formation of stable radicals would be the C_1 and C_5 positions. The spectra obtained on irradiated glucose three months after irradiation were interpreted to indicate the presence of two radical species, one of which was attributed to a radical at the C_1 position and the other to a radical at either the C_5 or C_6 position[106]. However, other studies on the irradiation of crystalline sugars suggested involvement of the C_1, C_2, and C_6 positions in stable free radical formation[107–110].

In proposing mechanisms of interaction of high energy radiation with cellulose, the interaction of radiation with each carbon atom in the anhydroglucose unit of the cellulose molecule will be considered independently. For example, the mechanism involved in the formation of a reducing group at the C_1 position would not be that involved in the formation of the same group at the C_6 position. The C_1 position is potentially an aldehyde group, while the C_6 position is a primary hydroxyl group. The proposed mechanism will be concerned with the method of activation of each particular carbon atom, each atom being activated by the ionization of the carbon atom itself or by the ionization of one of the atoms or groups bound to it. Since the energy of incident radiation would be sufficient to activate any carbon, oxygen or hydrogen atom with which it interacts, the activity reactions should be chemically non-specific and random in nature. The secondary reactions, initiated by the activation, will probably be chemically specific in nature[61].

At massive dosages, when the radiation interacts with the C_1 carbon, chain cleavage results with the formation of 2-ketogluconic acid on the reducing end of the chain utilizing one mol of oxygen in the formation of the acid and leaving an unaltered glucose unit on the non-reducing end of the other chain. Interaction with the C_4 carbon results in a chain cleavage, with the reducing end of the chain being liberated as an unaltered glucose unit and the production of a ketone group in the C_4 position of the nonreducing end of the other chain, using half a mol of oxygen. On the other hand, interaction of radiation with the C_2, C_3, C_5, and C_6 carbons results in reducing group formation with evolution of hydrogen and without chain cleavage. The distribution of reducing groups in the various fractions of the irradiated cellulose indicated that these groups were equally distributed along the cellulosic chain, suggesting involvement of the C_2, C_3, C_5, and C_6 carbons in the interaction[47)].

When available, considerable quantities of oxygen were consumed[47)]. The oxygen consumed in the reaction could go into formation of acid groups or into formation of carbon dioxide, but this would account for only a small fraction of the oxygen. Peroxide groups could be formed in the cellulose, consuming oxygen, or oxygen could react with hydrogen, liberated in the formation of reducing groups, to form water. Apparently, the predominant reaction on irradiation of cellulose is dehydrogenation with the production of reducing groups, which could be either ketones or aldehydes. It is interesting to point out here that the carbon monoxide and carbon dioxide evolved in the irradiation are considered to be secondary degradation products formed by decomposition of aldehydes and/or ketones in the case of carbon monoxide and in the case of carbon dioxide, by decarboxylation of carboxyl groups. Irradiation produced a much greater quantity of reducing groups than acid groups; thus, a greater yield of carbon monoxide than carbon dioxide is formed.

A comparison of the e.s.r. spectra of cellulose I and II irradiated in air, nitrogen, and oxygen atmospheres is shown in Table 17[111)]. Irradiated cellulose I had a three-line spectrum in each case. Equal interaction of two hydrogen nuclear spins with the free radical would be required to give the three-line spectrum. Due to the

Table 17. A comparison of the e.s.r. spectra of gamma-irradiated cellulose I and II. Adapted from Ref.[111)], p. 634, by courtesy of Textile Research Institute

Property	Atmospheric Conditions of Irradiation		
	Air*	Nitrogen**	Oxygen**
Spectrum, lines			
Cellulose I	3	3	3
Cellulose II	5	5	5
Line width, gauss Cellulose I			
major peak	13	19	18
minor peak	46	(48)***	(48)***
Cellulose II			
major peak	11	21	20
Nearest to			
major peak	38	(38)***	(38)***
Other peak	56	(57)***	(57)***
Relative Signal Strength			
Cellulose I	56:100:56	100 (major)	67 (major)
Cellulose II	39:63:75:63:39	49 (major)	48 (major)

* Radiation dosage 2.1×10^{20} eV/g
** Radiation Dosage 1.0×10^{19} eV/g
*** Estimated from one-half of spectra

delocalizing energy available from the oxygen atoms, the formation of radicals at C_5, C_1 or C_4 by dehydrogenation would be possibilities. Dehydrogenation at C_5 would give the radical

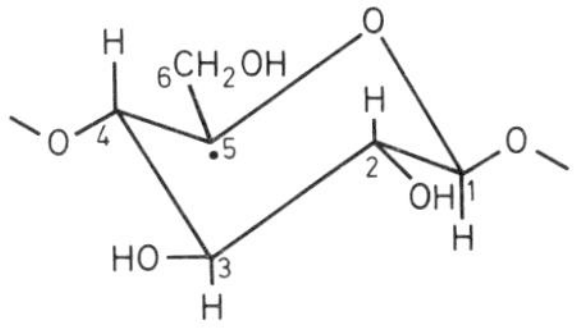

which, due to interaction with the equivalent hydrogen atoms on C_6, would give a three-line spectrum in the ratio of about 1:2:1. This is approximately the experimentally determined ratio of the relative signal strength of the three lines in the spectrum obtained (56:100:56), 1:1.8:1.

Two other possible radicals which would give a three-line spectrum would be:

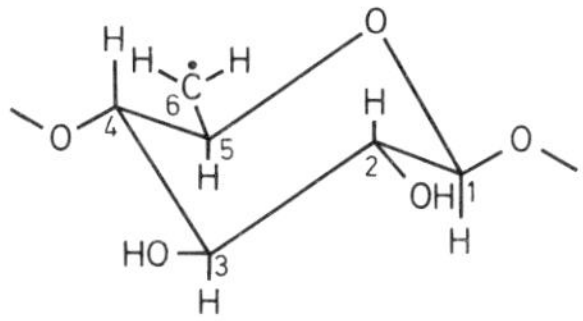

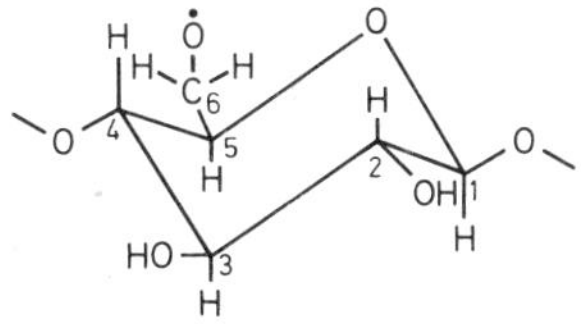

The H or OH on C_6 would be cleaved by gamma radiation. However, due to the lower delocalizing energy available at C_6, compared with that at C_5, after intermolecular energy transfer, the formation of a radical at C_5 would be more likely to occur in cellulose I.

The e.s.r. spectrum of irradiated cellulose II consisted of five lines. This could occur by the formation of two free radicals which individually yield three lines. The principal lines of the spectra of these radicals would occur at about the same value, and the less intense lines, due to equivalent hydrogen interactions of each spectrum, would occur at different line widths. One radical would likely form at C_5, as described for cellulose I, and the second radical at C_6 on the same anhydroglucose unit cell by either H atom or OH group cleavage. The formation of this second radical in cellulose II, compared with one radical in cellulose I, is further evidence of difference in the internal forces in the unit cell on changing the lattice type from I to II. The formation of the free radical at C_6 in cellulose II would indicate an increase in delocalizing energy at C_6, probably resulting from rearrangement in hydrogen bonding at C_6 in the unit cell of cellulose II. In an e.s.r. study of the post-irradiation reactions of cellulose and acrylonitrile, Baugh et al.[74] followed the effects of atmosphere, moisture content, and solutions of acrylonitrile on the yield and stability of free radical sites in irradiated cellulose. Evidence is given supporting the formation of long-lived radicals within the molecular lattice. In addition, short-lived radicals are formed from chain cleavage, giving singlet spectra which were readily accessible to interaction with water. However, other free-radical sites were formed in the inaccessible areas. These were unaffected by exposure of the irradiated cellulose to moisture or aqueous solutions, even after contact times as long as three days. The suggestion has been made that the long-lived radical sites in cellulose I (containing regain moisture) result from dehydrogenation at C_5 of the anhydroglucose unit. The long-lived radical sites in cellulose II (containing regain moisture) are thought to arise from dehydrogenation at C_5, together with dehydrogenation of the —OH group or dehydroxylation at C_6. Baugh et al. also observed that contact of irradiated cellulose with solutions of acrylonitrile in 75% aqueous $ZnCl_2$ produced a decrease in the signal strength with an increase in the contact time. A residual signal providing an ill-defined spectrum was assigned to the inaccessible radicals located within the cellulose matrix.

Other observations regarding the location of radical sites on the anhydroglucose unit have been made. Together with radicals produced as a consequence of main chain cleavage, such sites have been the subject of considerable interest.

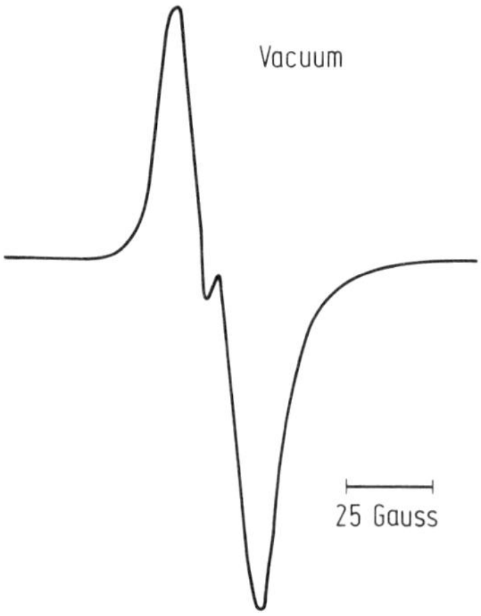

Fig. 19. E.s.r. spectrum of cellulose, irradiated under vacuum. Reprinted from Ref.[30], p. 214, by courtesy of John Wiley and Sons, Inc.

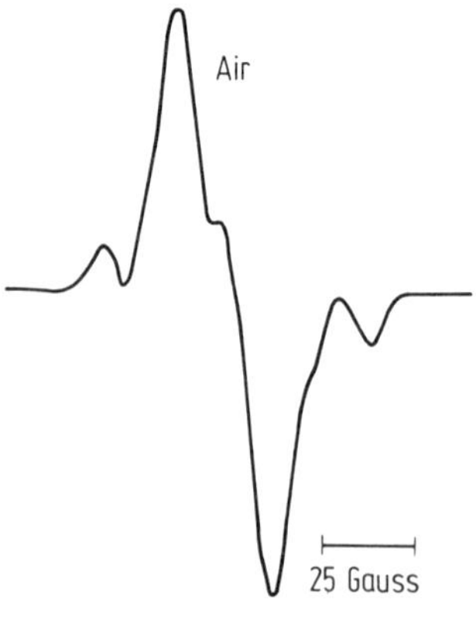

Fig. 20. E.s.r. spectrum of cellulose, irradiated in air. Reprinted from Ref.[30], p. 214, by courtesy of John Wiley and Sons, Inc.

Figures 19 and 20 represent data obtained by Guthrie et al. in their examination of regenerated cellulose films, irradiated under high vacuum and air respectively, using e.s.r. spectroscopy[35]. Samples irradiated under vacuum gave rise to an asymmetric doublet, while those irradiated in air exhibit a four-line signal. On exposure of vacuum-irradiated cellulose to the atmosphere, a spectral change was observed indicating a gradual conversion to the four-line signal previously referred to (Fig. 20). The doublet obtained with vacuum-irradiated cellulose may be satisfactorily explained by the formation of a radical at C_1 of the anhydroglucose unit through hydrogen abstraction:

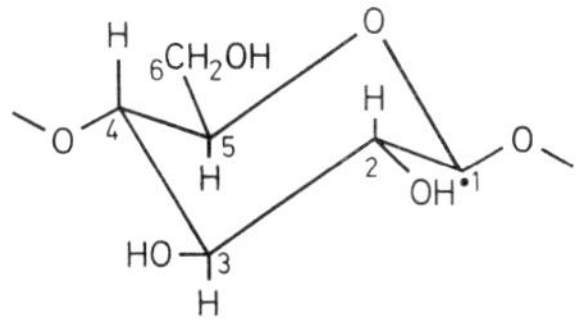

The doublet arises through interaction between the unpaired electron on C_1 and the proton bonded to C_2. Deformation in the anhydroglucose unit, relative to the undamaged glucose molecule[112], is taken as supporting the occurrence of coupling between the unpaired electron on C_1 and the proton on C_2. The formation of a quartet on irradiating cellulose in air or on exposing vacuum-irradiated samples to the atmosphere does not concur with most e.s.r. investigations of irradiated cellulose II. One explanation involves scission of the main chain to produce two species, a singlet and a quartet. However, the overall spectrum emerges as a quartet because of overlapping of the singlet. A radical on the bridging oxygen atom would give rise to a singlet spectrum, while a radical at C_1, due to scission of the C_1—O bond, would lead to a quartet being observed. The overall spectrum emerges as a quartet. The change from a doublet to a quartet signal indicates the possibility that main chain cleavage is closely associated with secondary reactions arising from contact of the primary radical sites with air and moisture. It would appear that such a scission is not a significant factor in the irradiation of cellulose under vacuum until exposure to external agencies takes place. From this work, Guthrie et al.[30] obtained a G (radical) value of 2.9 for vacuum-irradiated cellulose, which is in good agreement with that of other workers[83]. However, their value of 1.6, observed for cellulose irradiated in air, is high when compared to other observations of similar systems, namely 0.52[85] and 1.2[31,44,59,74,113], but using cellulose I.

Dilli et al.[98] have carried out a comprehensive program of cellulose irradiation and have arrived at various important conclusions from a qualitative examination of the e.s.r. spectra they obtained. Thus, irradiation in air favors alkoxyl radical formation (relative to triplet species) to a greater extent than does irradiation in vacuum. Also, decay appears to involve the preferential disappearance of the 'triplet' relative to the singlet species. Dilli et al.[98] concluded from their work that substantially the same types of radiolytic processes operate in air as in vacuum, but at different rates. In addition, they observed that water accelerates the relative decay of the triplet, while potassium iodide inhibits alkoxyl radical formation. Such observations are consistent with the general picture formed regarding primary and seondary reactions of irradiated celluloses; the main impression is that of the complex nature of the system.

Various groups have attempted kinetic analyses of the reaction profiles obtained from studies of radical yields and decay. Kesting and Stannett[33)] reported that the radical yield build-up in cellulose in the range 0 to 4.5 Mrad could be described by a steady-rate equation of type

$$a^{-1} \tanh^{-1} (B/a) = t + c\,,$$

where B is the concentration of radicals which decay to non-radical products; a and c are constants with $a = k_1/k_2$; k_1 is the rate constant for the formation of radicals, B, at a given radiation intensity; and k_2 is the rate of their decay. This decay is usually assumed to be second order in character. Experimental data[98)] provided by Dilli et al. are consistent with this steady state expression if a more complex structure is taken of the model for the cellulose used in their experiments. Such a model must make allowances for the existence of long-lived radical species.

In concluding this section on radical type, yield, and location, we must add a cautionary note regarding the concept of G (radical) values. It is possible to make comparisons between G (radical) values from e.s.r. measurements and G (chemical products) as derived from monitoring secondary chemical reactions which arise from exposure to radiation. Such secondary reactions include chain scission and graft copolymerization. It is generally accepted that G (radicals) is lower by an order of magnitude than G (chemical products) in cellulose irradiation, especially if attention is focused on chain scission events. However, it is misleading to seek quantitative relationships between these two types of G value, since chemical yields are determined by methods which convert some of the radicals into measurable chemical products. Nonetheless, as will be seen later, each type of G value is useful in its own right.

3.2.4. High Energy Radiation-Induced Grafting of Cellulosics

Gamma radiation, X-rays, alpha particles, and protons have been applied to a very large number of polymer-monomer systems for a wide variety of reasons[112–146)]. These cited references give a reasonable indication as to the manner in which grafting technology has developed. Early applications to cellulose were not very successful because, although appreciable grafting was achieved in certain instances, this occurred only at the cost of excessive degradation which led to deterioration in the physical properties of the cellulosic substrate.

As described earlier, such degradation can be reduced by partial radiation protection and by ensuring that the cellulose is in a highly swollen state (more accessible to monomer) before commencement of irradiation. This second point ensures that the conditions most conducive to grafting are achieved. Again, as more information became available regarding effective radical yields, it became possible to predict that grafting, rather than involuntary homopolymerization, would be the prevalent process. Hence, the more efficient processes are those in which the G (radical) value of the substrate is considerably in excess of that of the monomer. One example of such a system is provided by the cellulose-styrene composition which has the G (radical) values of 2.88 and 0.69 respectively.

With all types of high energy irradiation, the primary event consists of the formation of ions resulting from the scission of C—C or C—H bonds belonging to either the cellulose, the monomer, or the solvent. The ions are rapidly converted into free radicals and in nearly every known case of irradiation polymerization or irradiation grafting, a radical mechanism, rather than an ionic mechanism, accounts for the initiation and growth steps[147)]. Thus, the reactions of cellulose with vinyl monomers are apparently initiated at free radical sites, which are formed on the cellulose by interaction and localization of the energy of the ionizing radiation.

Judging from what is known of the effect, discussed earlier, of high energy irradiation on cellulose it is possible to assume that hydrogen atoms are abstracted from the cellulose. If this takes place in the presence of the monomer, grafting can occur immediately. Although direct hydrogen abstraction is a likely process, it must still be considered a hypothesis, and it is almost certain that, in the presence of water, hydroxyl radicals play a prominent part[148)]. Leavitt[66)] has observed crosslinking of cellulose derivatives in aqueous solutions when exposed to gamma irradiation and formulated the hypothesis that the first step in this process is the production of the ˙OH radicals from water. It is believed that these ˙OH radicals subsequently abstract hydrogen atoms from the cellulosic macromolecules.

If significant graft polymerization of monomer or polymers on to cellulose is to be initiated by high energy radiation and, at the same time, the relative high molecular mass and fibrous structure of the cellulose are to be retained, several important factors are worthy of consideration. These include the radiochemical yields of cellulose reactions, the radiochemical yields of graft copolymerization reactions, and those experimental conditions needed to increase the accessibility of activated sites in the cellulose to monomers.

3.2.4.1. Radiochemical Yields of Irradiated Cellulose

The radiochemical yields of the major molecular changes in cotton cellulose in an oxygen atmosphere are about 14 electron volts of energy per chain cleavage, 1.4 per reducing group formed, and 12 per acid group formed. These relatively high values indicate that, at least in the case of cleavage and acid groups, chain reactions were not initiated in the cellulose molecule by radiation. Irradiated cellulose depolymerized on standing and had a small, but measurable, increase in acid groups; however, these changes did not significantly increase the radiochemical yields. In a nitrogen atmosphere, the energy required per molecular change was slightly higher than in an oxygen atmosphere[61)].

At dosages of 5.2×10^{19} electron volts per gram of cellulose (about 10^6 roentgens of Co-60 gamma radiation) or less, little or no visible change in the fibrous structure of cellulose occurred. The number of chain cleavages per molecule was less than five reducing the average molecular mass from about 700,000 to 130,000. At this molecular mass the physical properties of fibrous cotton cellulose were only slightly changed, as compared with those of unirradiated cellulose[54)].

3.2.4.2. Radiochemical Yield of Irradiation Grafting

Some of the most important factors influencing the radiochemical yield of graft polymerization reactions of various monomers onto cellulose and its derivatives are (i) the concentration of monomer in the treating solution; (ii) the solvent; (iii) the ratio of the monomer concentration to cellulose concentration; (iv) the ratio of the monomer concentration to the solvent concentration; (v) prior chemical modification of the cellulose, and (vi) the absence of oxygen and other quenchers, particularly in post-irradiation reactions.

These factors have led to the utilization of several different approaches for developing practical techniques in grafting reactions. The indirect effects of the solvent used to disperse the monomer (and the monomer concentration), on the interaction of acrylonitrile or styrene with irradiated cellulose, have been clearly demonstrated[126, 136, 146]. For example, the radiochemical yield of acrylonitrile grafted to cellulose (expressed by the authors as molecules of monomer per hundred electron volts) was 82 using a medium containing 65% (w/w) of zinc chloride, 1400 using 75% (w/w), and 8100 using 81% (w/w) of zinc chloride. The yield of styrene grafted to cellulose was zero using pure monomer. However, for mixtures of styrene and methanol the radical yield values were 99-1, 440; 98-2, 5100; and 70-30, 4300. The indirect effects of the solvent on the radiochemical yield were probably due for the most part, to the interaction of the solvent with both the monomer and the activated cellulose through increasing

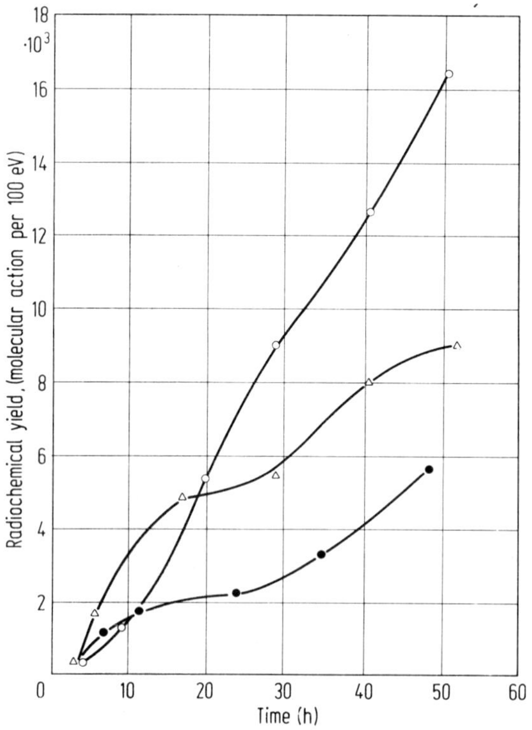

Fig. 21. Effect of dosage on the radiochemical yield of acrylonitrile-cellulose copolymer (post-irradiation procedures; about 25 °C; 32% acrylonitrile in 80% $ZnCl_2$); (○), 5.2×10^{19} e.V./g cellulose; (△), 12×10^{19} e.V./g; (●) 32×10^{19} e.V./g. Reprinted from Ref. [136], p. 2820, by courtesy of John Wiley and Sons, Inc.

the accessibility of the activated sites to the monomer. In addition, account must be taken of the radiolysis products obtained from irradiating methanol, since these have the potential of acting as secondary initiators. With the methanol-styrene system, the increased activation of the styrene is important since it leads to involuntary homopolymerization, a process which is wasteful of monomer. Such homopolymerization provides problems in subsequent graft copolymer characterization (Chapter 7).

Irradiation of a solution of styrene-methanol, in which cotton cellulose is immersed, yields an initial rate of polymerization which is about three times the rate obtained in the absence of cellulose. Apparently, the presence of radiation-activated cellulose increases the rate of effective activation of styrene and, consequently, the rate of styrene homopolymerization. This has been termed an indirect effect of cellulose[136)]. Prior chemical modification of cellulose by cyanoethylation also increases the radiochemical yield of the reaction between acrylonitrile and cellulose. At a degree of cyanoethylation of zero, the yield was 8100; at 0.3, 12,000; at 0.7, 14,000; at 1.0, 18,000; and at 1.3 it was 30,000[136)].

As shown in Figure 21, irradiation of cotton cellulose under nitrogen and post-irradiation reaction of the activated cellulose with solutions of acrylonitrile in the absence of oxygen gave graft polymers of acrylonitrile and cellulose with high

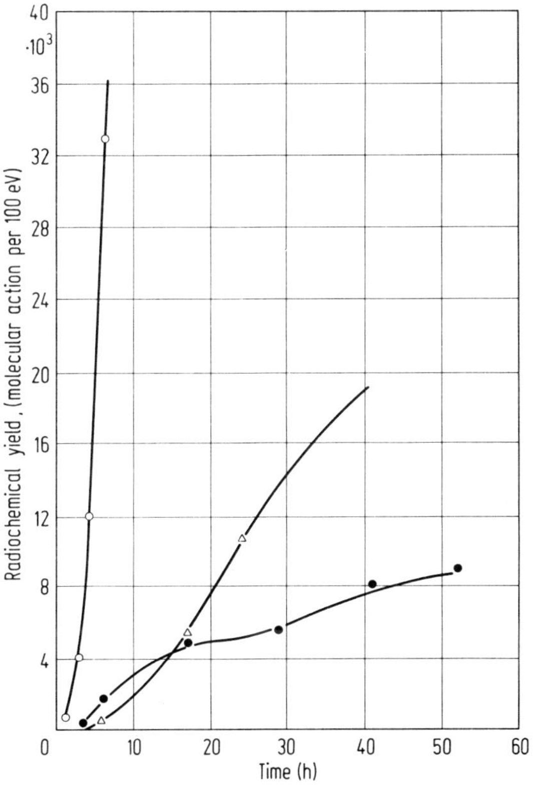

Fig. 22. Effect of composition of solvent-monomer mixture on radiochemical yield on acrylonitrile-cellulose copolymer (post-irradiation procedure; about 25 °C; cellulose-irradiated dry in nitrogen to 12×10^{19} e.V./g: (○) 16% acrylonitrile in 75% $ZnCl_2$; (△) 16% acrylonitrile in 80% $ZnCl_2$; (●) 32% acrylonitrile in 80% $ZnCl_2$. Reprinted from Ref.[136)], p. 2821, by courtesy of John Wiley and Sons, Inc.

radiochemical yields[136]. The composition of the solvent-monomer mixture is an important factor in achieving a high radiochemical yield[136]. Figure 22 shows that high radiochemical yields of graft copolymers were obtained when aqueous $ZnCl_2$ was used as the solvent. Water and N,N′-dimethylformamide as solvents gave low yields. This, then, sets the scene for examining the techniques used in radiation-induced grafting reactions[137].

3.2.4.3. Techniques of Radiation-Induced Grafting

Generally, the experimental procedures designed to give maximum yields in graft copolymerization reactions can be classified into two types:
A. Mutual Irradiation Grafting. Here the monomer and the substrate are irradiated together. This technique, usually referred to as the simultaneous technique, has the apparent advantage of making use of both the short-lived and long-lived active sites. However, as mentioned before, the formation of homopolymer may cause problems.
B. Post-Irradiation Grafting. This term is used to define the approach in which the substrate is irradiated first and exposed to the monomer at a later stage. The time lapse between irradiation and exposure is one of the important features of this technique, which has the advantage of minimizing homopolymer formation. However, because of the time lapse, the long-lived radical sites are a major factor in the initiation of graft polymerization.

Obviously, the radiolyses of all components of the grafting system i.e., cellulose, monomer and solvent, are factors in initiating graft polymerization when the simultaneous procedure is used. In both procedures, the initiation of grafting by activated sites trapped in the cellulose molecule is important. It follows from this that, regardless of the technique employed, the accessibility of the trapped activated sites to the vinyl monomer can determine the radiochemical yields of given reactions. The use of selected solvents for the monomers which can penetrate the more ordered regions of cellulose would also be advantageous in increasing radiochemical yields, as was indicated earlier. We shall now deal with each class in some detail.

A. Mutual Irradiation Grafting

When cellulose is immersed directly in the monomer and then exposed to high energy radiation, such as high energy electrons or gamma radiation from a cobalt 60 source, only a small amount of graft copolymerization usually takes place. This is because cellulose is a highly hydrogen-bonded polymer, and most vinyl and acrylic monomers are unable to penetrate beyond the surface of the cellulose substrate. Obviously, the significance of this problem varies with modified celluloses in a way which depends on the particular modification, as well as the monomer used. There are two ways to make cellulose accessible for grafting: use of a swelling agent for the cellulose in the reaction mixture, and introduction of the monomer throughout the substrate via the inclusion technique. The former approach is that adopted by most workers[114,216] irrespective of the substrate form, the monomer used, or the physical modification required from the grafting program. In the latter approach, the cellulose is first swollen in water or a similarly good swelling agent. The water is then replaced by methanol; the methanol is exchanged with a liquid having similar characteristics to the monomer, e.g., benzene if styrene is to be used, or pyridine for vinyl pyridine; and this liquid is finally replaced by the monomer[217]. In this way the monomer is introduced throughout the substrate at a high internal concentration and made available for grafting as soon as the cellulose is activated. Although this approach is less wasteful of the monomer than that involving a swelling agent directly, the multiple stages make the process cumbersome and inconvenient. However, in certain instances, e.g., timber modification or pad-batch textile processing, the inclusion technique is viable.

Role of Solvents in Mutual Irradiation Grafting

A number of investigators have shown that irradiation of cellulose, together with monomer and solvent, brings about a substantially increased grafting level when compared with irradiation of cellulose and monomer alone[97,114,213]. Okamura et al.[114] irradiated cellulose, that was preswollen in 5–10% formamide in methanol solution, with gamma rays in the presence of a methanol solution of styrene. Polymerization on the fiber reached several hundred percent. The concentration of formamide in methanol affected the amount of grafting (on viscose rayon). The maximum weight increase of the fiber was found at 70% styrene concentration in methanol.

Sakurada et al.[115–119] employed gamma irradiation in inducing graft copolymerization of vinyl monomers onto poly(vinyl alcohol) and cellulose. When the dry substrate was suspended in a large excess of the monomer, no grafting occurred and only homopolymer was formed. Grafting proceeded smoothly only when the substrate contained more than 5% water. They attributed this to the enhanced diffusion of monomer into the polymer substrate and also postulated the increased formation of polymer radicals by secondary abstraction by $H^{\cdot}$ or $OH^{\cdot}$ radicals produced in the radiolysis of water.

A Canadian patent by Hendricks[125] describes the effect of water and other aqueous swelling agents in the direct radiation grafting of vinyl monomers onto cellulose. The patent refers to a substantial improvement in grafting brought about by swelling the substrate in various liquids. Chapiro and Stannett[120] report that grafting of poly(vinyl alcohol), cellophane, and cotton with vinyl monomers by irradiation in the absence of any swelling agents leads only to surface grafting, whereas considerable depth of grafting is achieved in the presence of water. Kenaga et al.[145] impregnated wood cross-sections with water-saturated dioxane solutions of styrene and irradiated with cobalt-60 gamma rays; they noted that the percent of polymer retention decreased with increasing dose rate, thereby showing the importance of monomer diffusion to free radical sites in determining the rate of polymerization.

Huang and Rapson[144] studied the effect of different swelling agents, both miscible and immiscible, on the direct radiation graft copolymerization of styrene onto cotton cellulose and rayon by gamma-ray irradiation from Cobalt-60. They stated that direct radiation grafting onto cellulose is a heterogeneous reaction in which the diffusion of the styrene monomer into the cellulose plays a key role. No grafting was obtained when unconditioned cellulose was irradiated in bulk styrene. Grafting took place readily, however, when cellulose was first preswollen with a polar swelling agent such as water, formamide, formic acid, or dimethyl sulfoxide, and then placed in bulk styrene, together with methanol, ethanol, or acetic acid, and irradiated. The same holds true when grafting is carried out in solutions containing swelling agents. This may be observed in Figure 23.

The enhanced grafting obtained by preswelling the cellulose, or by carrying out the grafting in solutions containing swelling agents, has been interpreted in terms of "a plasticizing effect". This effect is believed to be caused by the breaking of hydrogen bonds with water or other swelling agents, thus providing greater segmental mobility of the cellulose molecules and facilitating the diffusion of the monomer. Furthermore the fact that anhydrous swelling agents caused styrene to graft onto cellulose indicates that swelling of cellulose is more important than increased formation of cellulose radicals by secondary abstraction with ˙H or ˙OH radicals produced in the radiolysis of water, as postulated by Sakurada et al.[115–119].

The effect of increasing radiation dose on the grafting of rayon with styrene was also examined by Huang and Rapson[144]. This is shown in Figure 24. Grafting increase linearly with radiation does until approximately 100% grafting and then levels off. Leveling off of grafting could be traced back to the fact that at higher levels of grafting, the reaction becomes diffusion-controlled. The efficiency of grafting is shown in the same figure and is seen to decrease with increasing radiation dose, a phenomenon which is similar to that reported by Kenaga et al.[145].

Another comprehensive study of the effect of various swelling agents on grafting cellulose with a vinyl monomer was reported by Majumdar and Rapson[141].

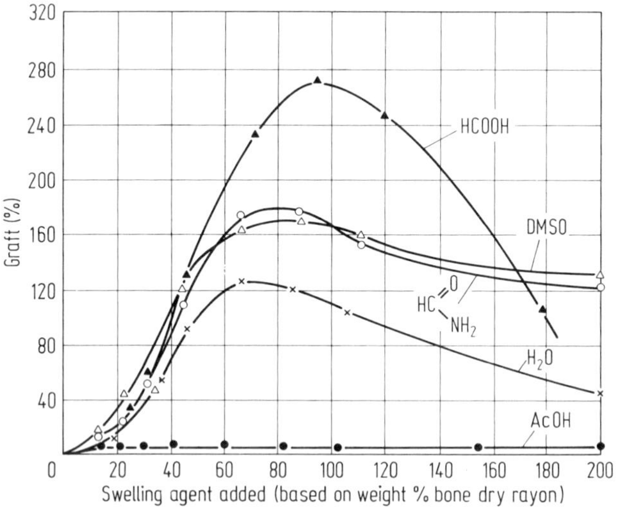

Fig. 23. Effect of nonaqueous swelling agents on the direct radiation grafting of styrene onto rayon. Total dose 4 Mrads; temperature 37 °C. Reprinted from Ref.[144], p. 176, by courtesy of John Wiley and Sons, Inc.

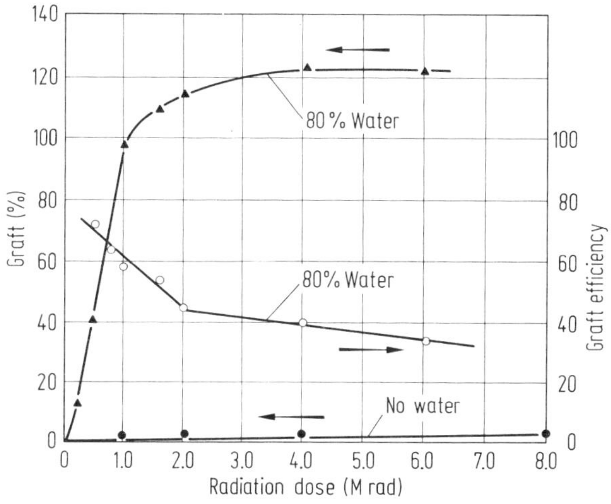

Fig. 24. Effect of radiation dose on grafting and grafting efficiency in the direct radiation grafting of styrene on rayon: (▲) grafting on rayon swollen with 80% water; (●) grafting of rayon without preswelling with water; (○) grafting efficiency of rayon swollen with 80% water. Reprinted from Ref.[144], p. 176, by courtesy of John Wiley and Sons, Inc.

They grafted polystyrene onto bleached jute yarn by irradiating jute-styrene mixtures with gamma rays from cobalt-60 under a variety of conditions. Some of the results obtained are given in Figures 25 and 26. It is obvious that no grafting occurred unless a swelling agent for jute was added, such as water, methanol, ethanol, formic acid, acetic acid, dimethyl-formamide, or dimethylsulfoxide.

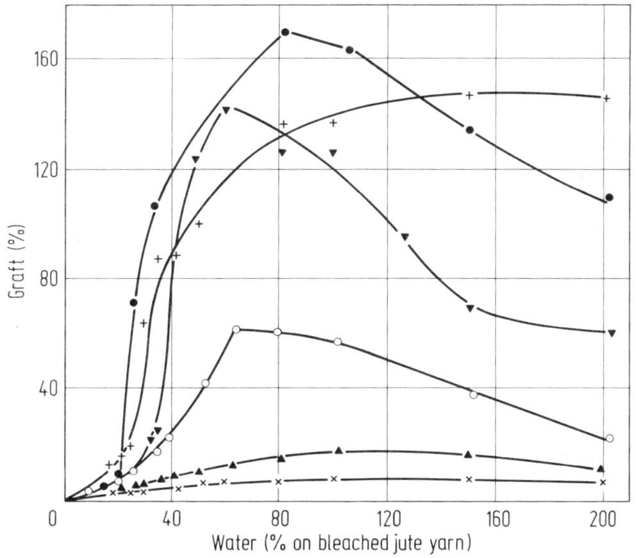

Fig. 25. Effect of water on grafting of styrene onto jute yarn at different dose levels: 10 cm^3 styrene, 0.5 g jute. Symbols: (×), 0.5 MR (megarad) water added to bleached jute yarn, (▲), 1.0 MR, water added to bleached jute yarn; (○), 2.0 MR, water added to bleached jute yarn; (●), 4.0 MR, water added to bleached jute yarn; (-+-), 4.0 MR, bleached jute yarn, water added to styrene; (▼), 4.0 MR water added to unbleached jute yarn. Reprinted from Ref.[141], p. 1011 by courtesy of Textile Research Institute.

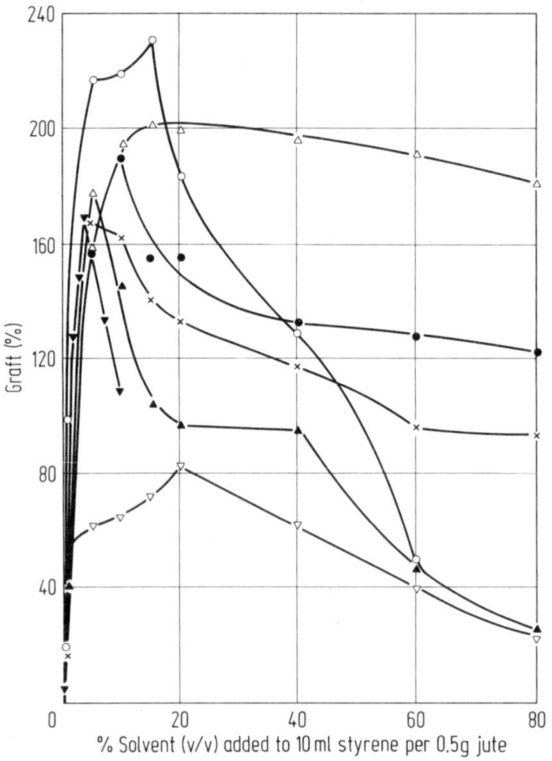

Fig. 26. Effect of cellulose swelling solvents on the grafting of styrene onto bleached jute yarn. Dose, 4 MR, moisture, 6%. Symbols: (▼, water; (●), methanol; (△), ethanol; (×), acetic acid; (▽), formic acid; (▲), dimethyl sulfoxide; (○), dimethylformamide. Reprinted from Ref.[141], p. 1013, by courtesy of Textile Research Institute.

Increasing amounts of water to jute gradually increased the amount of polystyrene added on, up to a maximum at 80–100% water on jute (Fig. 25). More water decreased the amount of grafting; this was probably due to saturation of the jute by excess liquid water and formation of a barrier to diffusion of styrene.

With other swelling agents, grafting increased as the amount of solvent was increased up to a certain point, and then it decreased (Fig. 26). The increase may be attributed partly to the radiolysis of the solvent, partly to the enhanced swelling of cellulose, and partly to the increased compatibility of the hydrophobic styrene and the hydrophilic cellulose. The decrease with larger amounts may be due to the increasing rate of termination of polymerization, producing smaller polystyrene branches, or to dilution of styrene with solvent. The effect of water on the grafting of chemically modified cellulose depends on the particular type of substitution, as well as the degree of substitution involved. Hence, the role of water as a swelling agent is reversed when grafting onto cellulose tricarbanilate is attempted. In the absence of water, little or no grafting occurs in the styrene-rayon-methanol system and in systems in which the methanol is replaced by ethanol or acetic acid. However, these solvents show efficient grafting on cotton even in the absence of water. Sakurada and Sakaguchi have shown that a 60 percent water content in cotton is the optimum for grafting with styrene[214]. DasGupta[209,211,215] has reported a 10- to 12-fold increase in the rate of grafting of styrene on cotton when water is added to the grafting solutions in amounts of up to 20 percent. Water has been found to be essential for the grafting of methyl methacrylate to filter paper from acetone solution[216] and to newsprint[209], although a 20 percent graft of vinyl acetate to newsprint has been achieved in the absence of water.

Dilli et al.[97] have shown that the structure of the solvent has a strong effect on grafting of vinyl monomers by irradiation. When a suitable solvent (e.g., a solvent which swells cellulose, is miscible in all propertions with the monomer, and can form radicals from the primary radiation), was used at a correct dose and dose rate, an accelerated grafting reaction (Trommsdorff effect) was observed. Indeed, this gel effect is valuable for an understanding of the fundamental mechanism of the reaction, and also for possible commercial utilization of the process. The peak in the Trommsdorff effect can be significantly enhanced or decreased by the use of certain organic additives[150]. Attention has been paid to the cellulose-styrene-methanol system using dose rates in the 0.045 to 0.77 Mrad/hr range in a Cobalt-60 facility. The irradiations were carried out in air[150]. The additivies used by Dilli and Garnett included glycol, acetic acid, ethyl acetate, chloroform, acetone, aniline, benzaldehyde, hydroquinone, thiourea, thiophen, pyridine, benzene, naphthalene, pyrene, phenanthrene, durene, hexamethylbenzene, stilbene, acridine, fluorobenzene, chlorobenzene, and bromobenzene. It has been shown that a low concentration (10^{-2} M) of the non-aromatic additives results in a reduction in grafting yields; while a higher concentration (10^{-1} M) gives sensitization as a common feature. Radical scavengers such as hydroquinone and thiourea suppress grafting, provided their concentration is sufficiently high. Of the aromatic additives, only anthracene and acridine show any influence on the grafting reaction. Studies of this kind are important since certain commercial outlets for radiation grafting will involve the use of additives. An example is provided by grafting reactions in whole timber, which must be efficient even in the presence of lignins and other impurities.

Dilli and Garnett have carried out a study of the grafting of styrene, methyl methacrylate, vinyl acetate and 2- and 4-vinyl pyridine to cellulose in solvents of the non-wetting type[158]. Generally poor grafting levels were achieved, irrespective of variables such as the dose rate, the total dose and the monomer concentration. The solvents for the monomer used in this study include benzene, cyclohexane, hexane, ethyl acetate, n-butanol, n-hexanol, and n-octanol. Grafting was carried out in air and under vacuum. An oxygen effect on the grafting reactions was found with the hydrocarbon solvents and the alcohol solvents in that much lower, or negligible, grafting was observed as compared to grafting carried out under vacuum. However, it must be pointed out that even the grafting levels obtained under vacuum were much less than those acquired using wetting solvents under otherwise identical conditions. The influence of benzene on grafting in solutions of styrene in methanol has been shown to be one of reducing the level of grafting over a range of different monomer concentrations. This is not really surprising, since it has been shown that benzene can efficiently scavenge methanol fragments which are formed during the radiolysis of benzene-methanol solutions[218,219]. Thus, we can see that the role of the wetting solvent is not restricted to that of providing enhanced access to the active sites in the substrate. Dilli and Garnett[158] have interpreted the part played by non-wetting solvents in terms of radical scavenging and, in particular, by a charge transfer theory which will be discussed later.

As an extension to their study of the influence of solvent type on grafting, Garnett et al. have examined the contribution made by mineral acids during grafting[167, 168]. It has been decisively shown that in the absence of air, at both high a low dose rates, the presence of certain mineral acids enhances the grafting of styrene in methanol to cellulose when the mutual irradiation technique is used. Sulfuric acid is the most effective of the mineral acids in this enhancement (Tables 18 and 19). In the presence of air, enhancement of grafting is observed only at high dose rates. The use of mineral acids induces a Trommsdorff peak in the grafting system and may also enhance any tendency toward Trommsdorff behavior, which is prevalent with the system in the absence of acid (Table 20). These enhanced acid effects in radiation grafting are attributed to an increase in the hydrogen atom yields. For the actual specific mechanism of the acid

Table 18. Radiation — induced grafting of styrene in methanol, to cellulose, in the presence of mineral acids. Reprinted from Ref.[168], p. 1460, by courtesy of the Commonwealth Scientific and Industrial Research Organization (C.S.I.R.O.), Melbourne, Australia

Styrene % by Volume	Graft (%) in presence of				
	No acid	H_2SO_4 (1 M)	HCl (2 M)	HNO_3 (2 M)	ACOH (2 M)
10	6.2	27.8	26.5	1.0	5.8
20	30.3	82.7	— (A)	7.5	23.2
30	41.3	33.7	18.3 (A)	9.1	38.8
40	—	19.7	19.1 (A)	11.2	36.8
50	38.6	—	—	—	—
60	—	25.5 (A)	15.1 (A)	20.6	42.5
80	56.1	20.6 (A)	8.6 (A)	8.0	42.2 (A)

(A) signifies that phase separation occurred. Dose rate 26.4 krad/hr. Total dose —0.20 Mrad. Irradiations in evacuated vessels.

Table 19. Effect of molarity of sulfuric acid on radiation grafting of styrene in methanol to cellulose at constant dose rate. Reprinted from Ref.[168], p. 1460, by courtesy of the Commonwealth Scientific and Industrial Research Organization (C.S.I.R.O.), Melbourne, Australia

Styrene % in Methanol	Graft (%) at Sulfuric Acid Concentrations of (M)							
	0	0.0011	0.0054	0.011	0.054	0.11	0.54	1.1
10	1.2	5.4	15.1	7.5	8.8	11.0	19.4	21.2
20	8.9	22.6	28.6	28.7	38.4	48.2	58.5	74.7
30	13.9	41.0	lost	44.7	47.5	49.6	39.3	27.0
40	16.3	27.5	37.1	17.8	34.0	30.2	20.2	16.1
60	29.8	31.2	28.0	29.1	26.8	23.0	16.5	15.1
80	34.1	24.8	21.7	26.5	19.3	21.3	12.4	10.1

Irradiation in evacuated vessels. Dose rate 27.3 krad/hr to a total dose of 200 krad.

Table 20. Effect of mineral acid on the Trommsdorff effect in radiation-induced grafting of styrene in methanol to cellulose. Reprinted from Ref.[168], p. 1464, by courtesy of the Commonwealth Scientific and Industrial Research Organization (C.S.I.R.O.). Melbourne, Australia

Styrene % by volume	Graft (%) at Sulfuric Acid Concentrations (M) of				
	0.0	0.00009	0.0009	0.009	0.07
15	35.1	38.4	31.5	50.8	78.7
20	61.9	69.3	65.6	87.2	109.8
25	99.3	112.5	110.8	123.3	162.7
30	104.5	113.9	98.8	109.6	(A)
35	100.9	102.8	80.0	90.7	124.0
45	80.5	52.6	42.0	48.4	82.5

(A) denotes that a very high level of graft conversion was achieved.
Dose rate 6.77 krad/hr. Irradiation time 17 hrs. Irradiations in deoxygenated solutions under nitrogen.

effect on grafting, the data are consistent with Dilli and Garnett's model involving charge transfer complexes (Section 2.4.4). However, the possibility of the involvement of other mechanisms, including energy transfer and ionic processes, should also be considered.

Of greater importance, perhaps, than the choice of particular monomer-solvent media is the actual ratio of the monomer to its solvent. Of the monomers which have been grafted to cellulose, styrene and methyl methacrylate are the most easily grafted. Butadiene, acrylic acid, acrylic esters, vinyl pyridines, and vinyl acetate have also been grafted, usually from methanol solutions.

Binary and tertiary mixtures of styrene, butadiene, and acrylonitrile have produced grafting of 10,000 percent at high graft rates[210, 212, 214, 220–222]. Methanol is widely accepted as being the most useful solvent in grafting to cellulose. The other solvents vary in usefulness according to the conditions, but an approximate order may be given as formamide, ethanol, dimethylformamide, dimethyl sulfoxide, acetone, acetic acid, dioxan, n-propanol, n-butanol, n-amyl alcohol, and benzene. Recent developments in non-aqueous solvents for cellulose may improve the rating of dimethylformamide and dimethylsulfoxide in the near future, especially if the current aspirations are realized. Listings are available for optimum ratios of a range of monomer-solvent compositions and are given in the excellent review provided by Moore[37].

Arthur and Blouin[136] reported that in the mutual irradiation procedure, grafting is probably initiated in part by radiation-activated sites on the surface of the cellulosic polymer; however, the importance of the accessibility of the activated sites to the initiation of graft polymerization is shown by the effects of solvent, the concentration of the solvent-monomer mixture, and the relative amount of cellulose to the solvent-monomer mixture, on the radiochemical yields. Typical examples of the results obtained are given in Tables 21, 22, and 23[136]. The radiochemical yield of styrene grafted to cellulose was zero when

Table 21. Effect of composition of solvent-monomer mixture on the radiochemical yields of graft copolymers of cellulose and styrene*. Adapted from Ref.[136], p. 2818, by courtesy of John Wiley and Sons, Inc.

Composition		Radiochemical yield, molecular actions/100 e.V.	
Styrene %	Solvent %	Methanol	N,N'-Dimethylformamide
100	0	0	0
99	1	440	220
98	2	5100	1000
97	3	5000	3300
96	4	4600	3400
95	5	—	3600
94	6	—	4000
92	8	4000	—
90	10	—	4200
85	15	4300	4700
80	20	—	4800
75	25	4800	5000
70	30	4300	5000
67	33	4800	4300
60	40	4600	4200
50	50	4500	3300
40	60	4500	2200
33	67	4000	1400
30	70	—	1200
25	75	3300	—
20	80	2200	440
15	85	1800	220
0	100	0	0

* Simultaneous procedure; radiation dosage: 5.2×10^{19} e.V./g cellulose; monomer-solvent/cellulose = 10/1; temperature, about 25 °C.

Table 22. Effect of composition of solvent-monomer mixture on the radiochemical yields of graft copolymers of cellulose and acrylonitrile*. Adapted from Ref.[136], p. 2818, by courtesy of John Wiley and Sons, Inc.

Solvent (aqueous $ZnCl_2$) %	Monomer concentration %	Radiochemical yield, molecular actions/100 e.V.
65	32	80
75	20	1100
75	32	1400
75	40	1800
80	32	8100

* Simultaneous procedure; radiation dosage; 4.2×10^{19} e.V./g cellulose; solution/cellulose = 25/1 overnight at about 25 °C, then padded to 2–4/1, prior to irradiation at about 25 °C.

Table 23. Effect of solvent monomer mixture of cellulose on the radiochemical yields of graft copolymers of cellulose and styrene*. Reprinted from Ref.[136], p. 2819, by courtesy of John Wiley and Sons, Inc.

Monomer-Solvent/ Cellulose, g/g	Radiochemical yields, molecular actions/100 e.V.
1	0
2	430
3	1200
5	4100
10	4300
20	4500
50	4600
100	4600

* Simultaneous procedure; radiation dosage = 5.2×10^{19} e.V./g cellulose; monomer-solvent = 70% styrene — 30 % methanol; temperature about 25 °C.

a pure monomer was used but significantly increased in the presence of a solvent. The indirect effects of solvents on radiochemical yields were probably due primarily to the interaction of solvent with both the monomer and the activated cellulose leading to an increase in the accessibility of activated sites to monomers. For instance, more free radicals were produced in methanol than in styrene, and the interaction of activated methanol with styrene may have increased the effectiveness of styrene, as shown in Table 21. Demint et al.[146] studied the radiation-induced interaction of styrene with cotton cellulose. With radiation excited cellulose, methanol and N,N′ dimethylformamide increased the efficiency of activation and interaction of styrene with fibrous cotton celluloses. The importance of these phenomena in the interaction, during high energy irradiation of acrylonitrile in aqueous zinc chloride with cotton cellulose, was also reported[123,128,129,137]

Inclusion Technique: Instead of using solvents, Huang et al.[217] brought styrene into close contact with cellulose by extension of the "inclusion" technique of Staudinger[224] and Krässig[225]. A 5 g. portion of cotton cellulose of known dry weight was treated in a Hydrapulper with distilled water for 10 minutes and then allowed to swell for 2 hours. The water was removed under suction and gradually exchanged for methanol. The methanol was exchanged five or six times, 100 cm^3 of methanol being used each time, and the time between exchanges was 2 hours. The methanol was exchanged twice with benzene by adding 100 cm^3 of benzene and allowed to stand 6 hours. The benzene was, in turn, finally exchanged twice with styrene. The included samples were placed in test tubes, stoppered in air, flushed with nitrogen, or sealed in vacuum prior to irradiation. Some of the results of the gamma-ray-induced graft copolymerization of styrene onto cotton cellulose using this technique are presented in Tables 24 and 25[217]. As can be seen, when styrene was not "included" in the cellulose, very little grafting occurred, as compared to the considerable amount of grafting

Table 24. Styrene-cellulose graft copolymers formed by gamma-irradiation in vacuum. Reprinted from Ref.[217], p. 1263, by courtesy of John Wiley and Sons, Inc.

Graft copolymer sample no.	Radiation dose, $\times 10^{-6}$ r	Grafting %	Grafting Efficiency
4-0	0	1.7	—
4-1	0.25	7.7	2.84
4-2	0.50	14.3	1.91
4-3	1.00	50.0	1.72
4-4	2.00	66.4	1.53
4-5	4.00	51.4	0.99
4-7*	1.00	177.0	
4-8*	1.00	169.2	
4-9*	2.00	158.3	
4-10*	2.00	166.7	

* Post irradiation heat treatment, 100 °C, 48 hr.

Table 25. Styrene-cellulose graft copolymers formed by gamma-irradiation in air. Adapted from Ref.[217], p. 1263, by courtesy of John Wiley and Sons, Inc.

Radiation dose $\times 10^{-6}$, r	Grafting %	Grafting efficiency	Note
1.00	27.8	1.59	Styrene included in cellulose
2.00	82.9	1.39	
4.00	111.2	1.08	
1.00	26.4	1.03	Styrene included additional styrene
2.00	71.6	1.00	added
4.00	124.9	1.10	
1.00	107.4	—	styrene included in cellulose and post-
1.00	110.2	—	irradiation heat treatment 100 °C, 48 hr.
2.00	197.0	—	
0.50	4.9	—	Styrene not included
1.00	4.5	—	
2.00	3.3	—	
4.00	4.6	—	
6.00	4.2	—	

obtained after the inclusion treatment of the cellulose irradiated in vacuum or in air. The presence of air appears to enhance, rather than inhibit, the grafting reaction at radiation doses above 2×10^6 roentgens. In the low reaction dose range, grafting is more effective in vacuum. Oxygen seems to play a role in the intermediate radical formation at low dosage ranges.

The Effect of Temperature on Grafting

In many studies, little effort is made to control the temperature during graft copolymerization through radiation induction. This is surprising, especially in view of the significant effect which temperature can have on grafting. Highly accessible cellulose usually requires higher grafting temperatures than are normally used with cotton. Dilli and Garnett have examined the styrene-cellulose-methanol system in great detail throughout the range 40° to —196 °C (Tables 26 and 27). The aim of this study was an observation of possible ionic contributions to the grafting reaction mechanism. The study was extended to include a series of substituted styrenes with both electron-donating and electron-withdrawing substituents[157]. In total, seventeen monomers were used, including o-, m-, and p-mono halogenated styrenes, a number of alkyl.styrenes, and p-methoxystyrene. We shall now consider some of the aspects of temperature dependence of grafting reactions involving cellulose as the trunk polymer. For the styrene-cellulose-methanol system (grafting on Whatman No I filter paper), the apparent overall activation energy for the grafting process in 80 percent and 60 percent monomer solutions was obtained using the data plotted in Figure 27 and estimated at 22 and 24 kJ mol^{-1} respectively, for the temperature range varying from 40 °C to 1 °C, which agrees favorably with the value reported by Miyama[226-227] and those found for the liquid phase polymerization of styrene[228-229]. Dilli and

Table 26. Effect of temperature on grafting yields for styrene on cellulose in methanol. Reprinted from Ref.[157], p. 1165, by courtesy of the Commonwealth Scientific and Industrial Research Organization (C.S.I.R.O.), Melbourne, Australia

Temperature	Grafting (%) at Monomer Concentration (% *V/V*)			
°C	20	40	60	80
1	1.5	6.2	10.8	13.3
11	2.5	11.1	17.9	22.4
23	8.4	23.5	30.6	35.3
31	14.9	29.7	34.6	42.5
40	21.1	34.2	38.5	43.1

Irradiation for 213 minutes at 0.145 m rad/hr using styrene-methanol mixtures. Extraction was in benzene (100 hr) and chloroform (48 hr) to constant weight. The figures represent mean values of *S* determinations.

Table 27. Low temperature grafting of styrene to cellulose in methanol. Reprinted from Ref.[157], p. 1166, by courtesy of the Commonwealth Scientific and Industrial Research Organization (CSIRO), Melbourne, Australia

Reaction Temp. (°C)	Grafting (%) at Styrene Concentration (%)			
	20	40	60	80
25	15	32	33	48
0	2	7	12	19
— 78	2	2	5	23
—196	2	2	2	2

Irradiation in air at 0.045 M rad/hr for 4 hours. Figures represent the mean values of 4 determinations.

Garnett[157] suggest that, for such results, the kinetics can be expressed by simple kinetic schemes in which the overall activation energy, E, is equated with the difference between the activation energies for propagation (E_p) and termination (E_t); thus, $E = E_p - {}^1/_2\, E_t$. This activation energy (E) is approximately, 20–40 kJ mol^{-1}, assuming the energy, E_i, associated with initiation by γ radiation, to be zero. Further dilution of the monomer solution leads to an increase in the overall activation energy. At temperatures below −78 °C, grafting is much reduced, though there is evidence that an ionic mechanism operates at these temperatures. In other examples in which the temperature has been controlled, grafting is increased markedly as the temperature is raised to 60 or 70 °C[181, 187, 201–203].

Types of Copolymers Formed by High Energy Radiation Grafting

Possible products from the irradiation of cellulose can be presented as follows: —

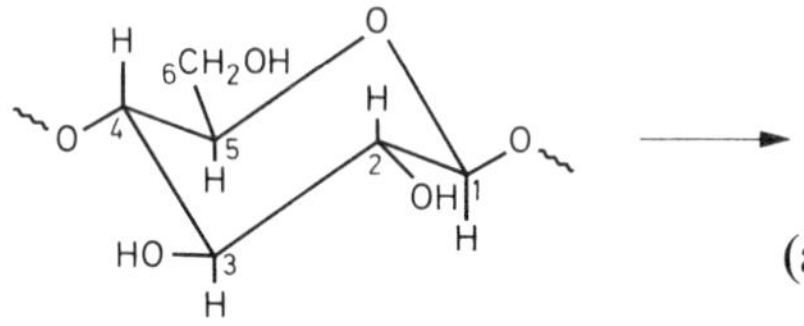

(a) 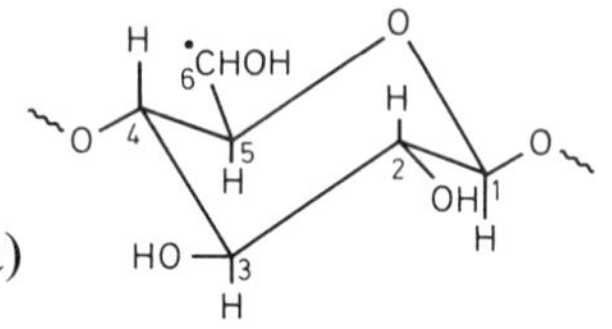 or

(b) 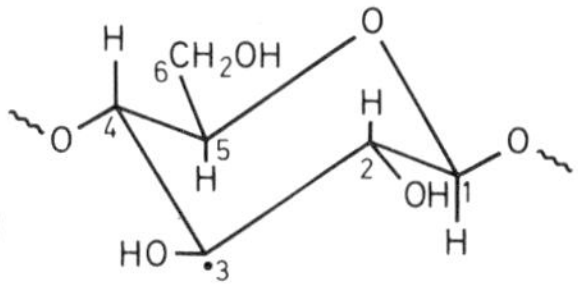 or (c) 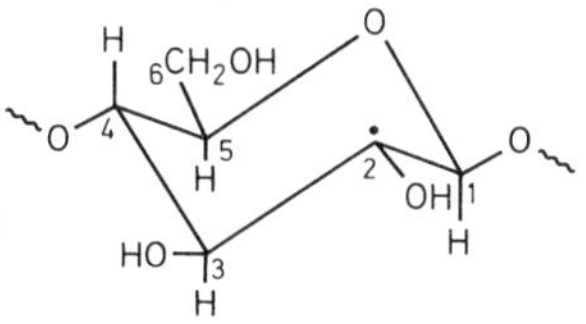 or

(d) 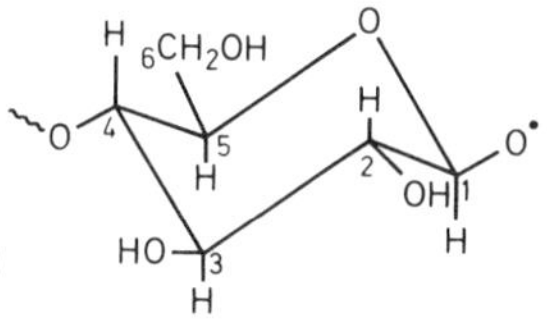 or (e) 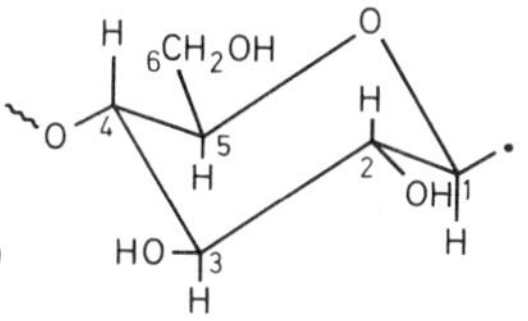

More recent evidence suggests the following as possible products[62, 74, 89]:

(f) 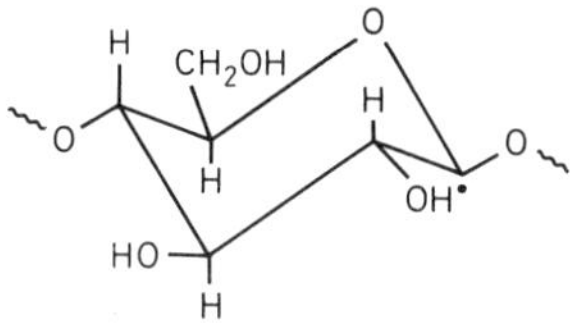 or (g) 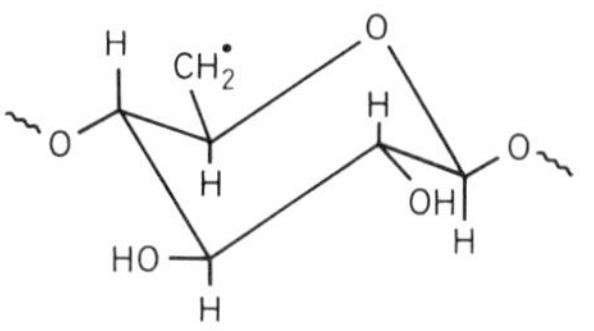 or

(h) 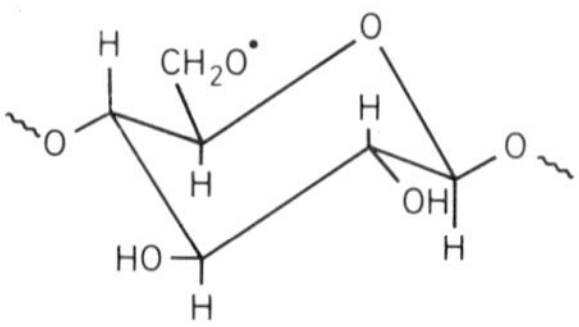

In the presence of a monomer, chain radicals such as (a), (b), (c), (f), (g), and (h) will lead to the formation of graft copolymers, whereas (d) and (e)

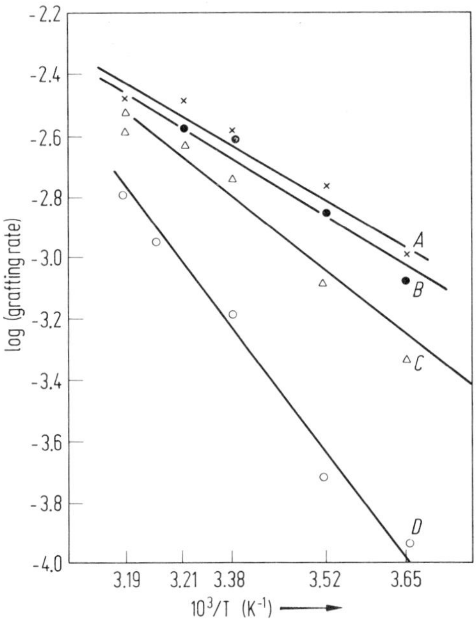

Fig. 27. The effect of temperature on grafting yields in grafting to cellulose. Arrhenius plots, *A*, *B*, *C*, and *D* are for solutions 80, 60, 40, and 20% styrene in methanol respectively. Grafting rate initially obtained in $\% \ s^{-1}$. Reprinted from Ref.[157], p. 1165, by courtesy of the Commonwealth Scientific and Industrial Research Organization (C.S.I.R.O.), Melbourne, Australia.

will give rise to block copolymers. Actually, there arises the possibility of several different types of radicals being present in a given chain. The lifetime of such species will be very dependent on the particular environment in which they are generated. Although the available evidence suggests that sites are formed on the anhydroglucose unit in preference to inter-unit positions, it is known that in many examples both types of reaction are found to take place. In addition, there will be radicals formed by radiolysis of the monomer and, if there is a solvent, by radiolysis of the solvent. These monomers and/or solvent radicals lead to formation of homopolymer. Use of mutual irradiation techniques, therefore, always yields a mixture of graft copolymer and homopolymer. The ratio of graft copolymer to homopolymer will be determined by the relative rates of grafting and homopolymerization. To achieve maximum grafting, a monomer should be selected which requires very large radiation doses in order to polymerize, i.e., a monomer with a very low value of *G*, where *G* is the number of molecules converted per 100 electron volts of energy absorbed[139,140,234,235]. Fortunately, however, initiation of free radicals on the cellulose backbone is more likely to occur than initiation of free radicals on the monomer, since the cellulose molecule interacts more readily with high energy radiation, particularly through dehydrogenation[47,61], than the vinyl monomer. Also, the activated sites on the cellulose molecule will probably tend to be longer-lived, particularly in crystalline regions, than the activated sites on the vinyl monomer.

B. Post-irradiation Grafting

This technique consists of the irradiation of cellulose in the absence of the monomer, followed by its immersion in a solution of the monomer. Pre-irradiation may be carried out in the presence or absence of air. In the presence of air, the irradiated cellulose contains peroxy-carbonyl and carboxyl groups, as well as trapped radicals. The latter are formed by breaking C—C, C—H, or C—O bonds in the cellulose and are long-lived because of their inability to come sufficiently close to another radical for a reaction to occur.

In a number of patents, is has been stated that irradiated cellulose acetate, and possibly irradiated cellulose, can initiate post-irradiation graft polymerization reactions[236–239].

Kobayashi has described the grafting of rayon with styrene by pre-irradiating the rayon in the presence of hydrogen peroxide solution, water or air, and subsequently heating it with styrene[122]. More grafting occurred after irradiation in hydrogen peroxide solution than after irradiation in either air or oxygen. The graft yield increased with increasing concentration of hydrogen peroxide. There was a non-linear correlation between the number of active sites and the extent of grafting. Both of these increased with radiation dose and decreased if the pre-irradiated samples were heated before exposure to the monomer, i.e., styrene.

Chapiro[240, 241] reported that peroxides are formed when a polymer is irradiated in air. These peroxides are subsequently decomposed at an elevated temperature, inducing graft copolymerization in the presence of vinyl monomers. Ballantine et al.[242] have reported that free radicals are formed when a polymer is irradiated in vacuum; these are trapped in the polymer and subsequently behave as graft copolymerization initiators. Even in the presence of air, some radicals result.

Shinohara and Tomioka[243] studied graft copolymerization by irradiation of polyethylene in air and reported that free radicals, as well as peroxides, are formed. This is in apparent contrast to the findings of Chapiro[240], who could not detect any free radicals on irradiation of polyethylene in air. The disagreement seems to be due to differences in irradiation conditions. The free radicals are capable of inducing graft copolymerization at low temperature, where peroxide grafting is very slow. The trapped free radicals disappear on standing in air, probably by combining with oxygen. In styrene grafting, copolymerization is induced by both free radicals and peroxides when the temperature is elevated.

Sakurada et al.[244] carried out experiments on the grafting of styrene onto cellulose fibers by post-irradiation techniques where the irradiation was performed by electron beams, in air, at atmospheric pressure and 65% relative humidity, and the grafting at 50 °C. They reported the following:

(1) The activity for the grafting of the irradiated cellulose decreases only a little when it is kept for a long time (10 days), in vacuum at room temperature, or in air at a temperature as low as −78 °C.
(2) The activity decreases remarkably when the irradiated cellulose comes into contact with water or humid air at room temperature.
(3) The activity decreases on heating the irradiated cellulose above room temperature. The higher the temperature, the greater the rate of loss of the activity.

From the above observations, Sakurada et al. came to the conclusion that the reaction might be regarded as being initiated not by peroxides, as often believed, but rather by trapped radicals[215–216].

Studies on grafting onto poly(vinyl alcohol) fibers were also carried out with similar results to those obtained with cellulose[92]. It was observed by a quantitative measurement of the oxygen consumption that an appreciable amount of oxygen was taken up by the poly(vinyl alcohol) during irradiation in the air. It is therefore probable that some peroxides are formed during pre-irradiation, but the presence of air or the amount of combined oxygen on poly(vinyl alcohol) had practically no effect on the rate and extent of the grafting. These results may be regarded as indicating very strong support for the argument that trapped radicals, and not peroxides, play the more important role in grafting onto poly(vinyl alcohol) by the post-irradiation technique.

Post-irradiation grafting onto cellulose and polyethylene has been shown[92)] to provide further support for the argument that grafting by this technique is initiated not by peroxides, but by trapped radicals independent of the presence or absence of air during pre-irradiation. The main difference between polyethylene and cellulose may be traced back to two factors, i.e., hygroscopicity and the glass transition temperatures. Polyethylene is not hygroscopic, whereas cellulose is. The glass transition temperature of dry cellulose is above room temperature; therefore, radicals formed are stable as long as they are not heated or in contact with water. On the other hand, the glass transition temperature of polyethylene is below room temperature, so that the trapped radicals are not stable and oxygen can easily penetrate into polyethylene during irradiation to form peroxides[92)].

When cotton, regular viscose rayon, and polynosics were pre-irradiated with gamma rays from Co-60 and then grafted (under different conditions) with styrene at 50 °C, no difference in the percent graft between vacuum and dry-air irradiated fibers was observed. This was the case with the three cellulosic materials. On the other hand, the presence of water has a dramatic effect on the grafting. With respect to cotton, the percent graft drops to ca. $^1/_3$ of the value which was obtained in vacuum or dry air irradiation, whereas in the case of regular viscose rayon and polynosics, the presence of water exerts a more pronounced effect so that the wet-air irradiated fibers have practically no activity for grafting[92)]. As previously shown[244)], dry irradiated fibers lose their activity for grafting when they are brought into contact with water. These phenomena cannot be explained if the active species are peroxides.

It is known that the glass transition temperature of cellulose is lowered by the presence of water, dropping below room temperature[248)]. If it is assumed that radicals are initiators for the graft polymerization, the above-mentioned phenomena may be explained quite simply, because above the glass transition temperature, radicals formed on cellulose molecules disappear by mutual reaction or by reaction with water molecules.

Since the reaction temperature of 50 °C is not a suitable temperature to decompose peroxides and initiate polymerization, grafting onto polynosics was performed at 80 °C and 117 °C. Results of these experiments showed that vacuum or dry-air-irradiated fibers had much greater activity for grafting then wet, air-irradiated fibers. It was therefore concluded that radicals formed by irradiation play the most important role in the grafting[249)].

When the grafting is initiated by peroxides, the influence of temperature on percent graft is completely different. For instance, reported experimental results of grafting acrylonitrile on polyethylene film revealed that only vacuum-irradiated film acquired a tendency to be grafted at 50 °C. On the other hand, films irradiated in the presence of air exhibited almost no tendency to be grafted at this temperature. It follows from this that the activity of vacuum-irradiated film is due to free radicals. At 80 °C, the air-irradiated and wet-irradiated films begin to show activity towards grafting, whereas the activity of the vacuum-irradiated film remains almost unchanged. When the reaction temperature was elevated to 115 °C, the activity of films irradiated in the presence of air was enormous; this is unequivocally initiated by decomposition of peroxides[72)].

Kesting and Stannett[246)] measured the build-up and decay of free radicals in the process of gamma post-irradiation grafting of styrene onto cotton cellulose from electron spin resonance measurements, as well as the decrease in grafting after storing the irradiated materials before immersion in monomer. Their results indicate that the grafting is initiated by trapped radicals. They also obtained some evidence for the occurrence of peroxide-initiated grafting at higher polymerization temperatures. These authors suggest that the peroxides might have arisen from the reaction between cellulose and water rather than oxygen. During irradiation, the small amount of water present tended to increase the amount of polymerization. The greater yields are presumably due to the greater accessibility of the moist samples to the monomer, particularly in the early stages of the reaction. The effects of water confirm similar results with cellulose and poly(vinyl alcohol)[120)].

Majumdar and Rapson[250)] reported that where both styrene and bleached jute containing 80% water were separately irradiated and subsequently heated together, grafting of polystyrene onto the bleached jute increased steadily with time, and that the percent grafting was greater than when the components were not pre-irradiated or when only the bleached jute was pre-irradiated. When styrene alone was irradiated, grafting on subsequent heating was substantial and even greater than that of simultaneous grafting[233)]. This great increase in grafting in these two cases may be due to the formation of peroxides in the air-saturated styrene by irradiation in the presence of oxygen; such peroxides subsequently initiate grafting on heating. This explanation is plausible, since styrene dissolves substantially more oxygen from air than does water.

Blouin and Arthur[134)] studied the post-irradiation reactions of the irradiated cellulose with acrylonitrile under a variety of conditions. Variables studied include atmosphere of irradiation, temperature of grafting, nature of the swelling agent, irradiation dose, and time of grafting reactions. Irradiation was performed by using high-energy gamma radiation from Co-60 at selected dosages up to 6×10^8 roentgens. Some of the results obtained are given in Table 28 and Figure 28.

It was observed that treatment of irradiated cellulose with acrylonitrile alone gave extemely low graft yields at the higher dosages studied, even at high temperatures. Treatment of the irradiated cellulose with 5% aqueous acrylonitrile gave somewhat higher yields, particularly at elevated temperatures. No significant differences in yield were observed with this treatment in the range of 60 to 100 °C. Somewhat higher graft yields were obtained on treatment of the irradiated cellulose with 32% acrylonitrile in N,N′-dimethylformamide, both at 23° and 80 °C, than with the other two systems. It was also observed that maximum graft yields were attained after 3 to 7 hours and no additional

Table 28. Post-irradiation graft polymerization reaction of irradiated cotton with acrylonitrile at various temperatures*. Reprinted from Ref.[134], p. 731, by courtesy of Textile Research Institute

Dosage roentgens	Reaction Conditions**		Grafted Polymer, %
	Monomer and solvent	Temp., °C	
2.3×10^6	5% Acrylonitrile in H_2O	80	9.3
	32% Acrylonitrile in DMF	80	20.4
6.1×10^6	Acrylonitrile	23	0.4
		80	3.5
	5% Acrylonitrile in H_2O	23	5.1
		60	15.1
		80	13.7
		100	12.8
	32% Acrylonitrile in DMF	23	15.5
		80	26.9

* Irradiated, vac.-dry, in N_2
** 22 hr.

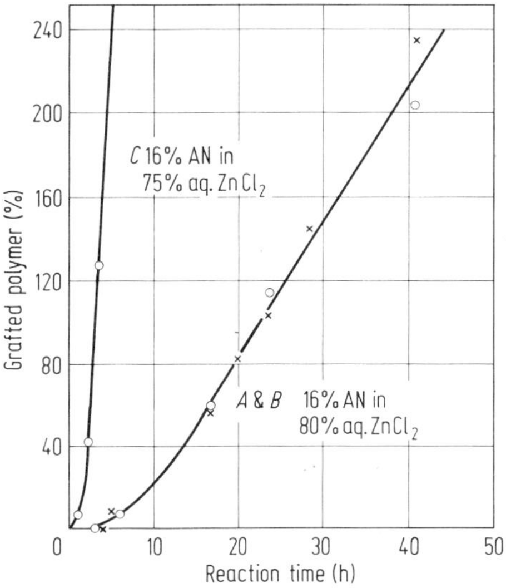

Fig. 28. Post-irradiation reaction of irradiated cotton (2.3×10^6 roentgens) with acrylonitrile (16%) in aqueous $ZnCl_2$ (75% and 80%) at about 23 °C. Irradiation conditions; *A* and *C* vac.-dry, in N_2 and *B* air-dry, in air (marked *X*). Reprinted from Ref.[134], p. 732, by courtesy of the Textile Research Institute.

increases occurred on prolonged heating. The authors pointed out that if cellulose irradiated in an air-dry state to a dosage of 6.1×10^6 roentgens was treated with 5% aqueous acrylonitrile at 23° or 80 °C, no significant polymerization occurred[134].

Higher graft yields could be achieved by irradiating cellulose and subsequently treating it in acrylonitrile dissolved in aqueous zinc chloride solution, as can be seen from Figure 28. It was then concluded that post-irradiation graft copolymerization of acrylonitrile onto cotton cellulose was initiated by the long-lived free radicals produced in the cotton by the radiation. Irradiation in an inert amosphere (absence of moisture and oxygen) produced free radicals in the accessible and inaccessible (crystalline) regions of the cellulose structure. Free radicals in the accessible regions of the cellulose structure initiated low yields of graft polymerization of acrylonitrile onto cotton in various non-swelling solution media. High graft yields were attained even at low dosages when the treating solution was able to penetrate the crystalline regions of the cellulose structure[135].

Blouin et al.[135] irradiated dry cotton in a nitrogen atmosphere with gamma ray (Co-60 source) to a dosage of 10^6 roentgens. The irradiated cellulose was then treated with acrylonitrile, methyl methacrylate, or vinyl acetate in 70, 75, or 80% aqueous zinc chloride. The rates of graft polymerization of these vinyl monomers were studied, and the radiochemical yields calculated from these conditions and based on the maximum graft polymer content, are given. Some of these results are shown in Figure 29 and Table 29. Obviously, the vinyl acetate reacted with irradiated cotton cellulose at a slower rate and to a lower extent than the other monomers studied. However, the radiochemical yields were good. The most interesting feature of this study is that a wide range of graft polymer contents could be attained at dosages low enough to avoid serious radiation damage.

The post-irradiation technique has the advantage that the grafting reaction may be carried out at different times and locations than the irradiation operation. Furthermore the same pre-irradiated substrate may be subjected to a variety of different monomers, depending on the end use desired under certain conditions. It is also possible to prepare graft copolymers uncontaminated by homopolymer. The post-irradiation technique leads to high graft yields and longer branches[251]. This may be seen from the data given in Table 30. However, it is more likely to cause cellulose degradation than the mutual irradiation technique.

Imrisov and Maryska[252,253] employed i.r. spectroscopy in an attempt to quantify the extent of grafting in a more realistic manner than is apparent with gravimetric methods. Samples of graft copolymer (cellulose or derivatives copoly-

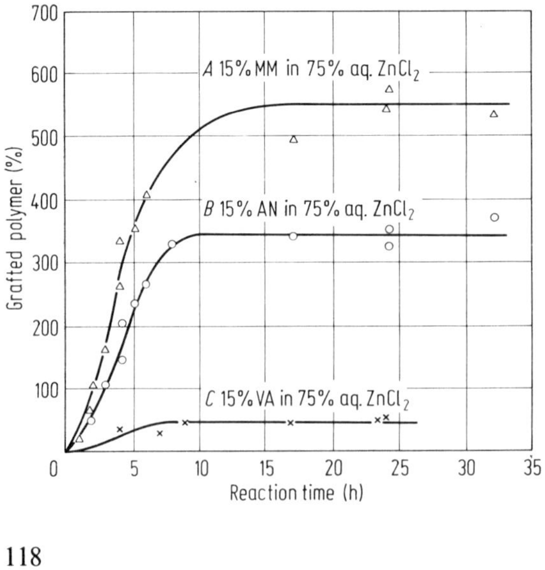

Fig. 29. Post-irradiated grafting of *A* methyl methacrylate, *B* acrylonitrile and *C* vinyl acetate in 75% aqueous $ZnCl_2$ at about 23 °C. Irradiation conditions; dry and in nitrogen, 10^6 R. Reprinted from Ref.[135], p. 311, by courtesy of Textile Research Institute.

Table 29. Radiochemical yields of post-irradiation graft polymerization reactions of vinyl monomers. With irradiated cotton cellulose*. Reprinted from Ref.[135], p. 311, by courtesy of Textile Research Institute

Monomer	Aqueous $ZnCl_2$ %	Molecular actions/ 100 e.V.
Vinyl acetate	70	6,400
	75	6,700
Acrylonitrile	70	43,000
	75	75,000
	80	68,000
Methyl methacrylate	75	63,000
	80	65,000

* Purified cotton fibers irradiated dry and in nitrogen to a dosage of 10^6 r and after irradiation treated with 15% monomer in 70, 75 or 80% aqueous $ZnCl_2$ at about 23 °C.

Table 30. Mutual versus pre-irradiation grafting of styrene to cellulose*. Adapted from Ref.[251], p. 249, by courtesy of Huthig & Wepf Verlag, Basel

Substrate	Dose MR	% Graft mutual	Mol. Wt. (visc.) · 10^{-6} Side Chains (mutual)	% Graft Pre-Irrad.	Mol. Wt. (visc.) · 10^{-6} Side Chains (Pre-irrad.)
Cellophane	0.32	69.6	1.42	64.7	1.99
	0.97	71.8	1.08	84.0	1.67
	1.62	72.6	0.97	90.0	1.52
	2.92	76.0	0.90	96.0	1.30
	3.89	77.4	0.92	98.7	1.28
Cotton	0.32	19.6	1.07	22.6	3.02
	0.65	22.7	0.90	45.6	2.89
	0.97	25.9	0.67	55.2	2.78
	1.30	30.1	0.60	63.0	2.45
	3.24	47.5	0.31	83.4	2.26

* Dose rate 3.24 · 10^5 R per hr, time of contact between monomer and pre-irradiated cellulose, 72 hrs at 25 °C; contact time between monomer and cellulose, 3 days at 25 °C before irradiation and one day after irradiation in the mutual method.

merized with acrylonitrile), were prepared for analysis using the KBr, pressed disc method. The absorption band at 2249 cm^{-1} was monitored. For a similar program using cellulose-co-styrene, the band at 700 cm^{-1} was used to determine the extent of grafting. Although quantitative infra-red spectroscopic analysis of cellulose cannot be applied generally, there are instances when it may prove useful[149]. Nakamura et al. have used electron spin resonance spectroscopy in their study of post-irradiation grafting[254–256]. E.s.r. was used in an assessment of radical accessability during the grafting of ethyl acrylate to cotton cellulose. Water was seen to accelerate the rate of grafting, and maximum grafting was achieved at phase boundary regions, presumably through the operation of a Trommsdorff type of effect. Grafting in acetone was found to be much less efficient than grafting in a mixture of methanol and water. With acetone/water, no Trommsdorff effect was observed. In acetone/water system, the e.s.r. spectra of the scavenged free radicals appeared as singlets. This supports radical formation at C_1 or C_4 or the anhydroglucose unit, as found by Guthrie et al.[89]. The lack of a Trommsdorff effect with acetone/water as the solvent medium arises from the much greater solubility of poly(ethyl acrylate) in acetone/water than in methanol/water mixtures. As may be expected, decreased grafting rates were observed with an increase in the temperature and time of irradiation storage. A corresponding decrease in the moisture regain of the irradiated cellulose was also observed. Oxygen atmospheres caused a greater decrease in grafting than did exposure in nitrogen atmospheres[256]. For these decreases, it was concluded that increased intermolecular bonding occurred during radiation storage in both nitrogen and oxygen. From the study of the temperature dependence of grafting in the range 25–60°, it was concluded that there was no evidence for the formation or decomposition of cellulose peroxides during these reactions, and that formation of graft copolymer depended primarily on the concentration of free radicals in the irradiated cellulose at the time of copolymerization. These observations are somewhat contrary to those observed in grafting to other substrates, e.g., polyethylene, in which it has been reported that the thermal decomposition of peroxide groups in the polyethylene leads to enhanced post-irradiation grafting[229,240,242]. A mechanism for this post-irradiation-induced intermolecular bonding has been proposed by Nakamura et al. This mechanism can be seriously considered for post-irradiation grafting to the various forms of cellulose, but it cannot be entertained for derivatives because of effects such as steric interference and energy transfer reactions. The suggested mechanism is as follows:

$$\text{Cell—H} \xrightarrow{\gamma} \text{Cell}^{\cdot} + \text{H}^{\cdot} \quad (73)$$

$$\text{Cell}^{\cdot} + \text{Cell}^{\cdot} \xrightarrow{\Delta} \text{Cell—Cell} \quad (74)$$

$$\text{Cell}^{\cdot} + O_2 \rightarrow \text{Cell—O—O}^{\cdot} \quad (75)$$

$$\text{Cell}^{\cdot} + \text{Cell—O—O}^{\cdot} \xrightarrow{\Delta} \text{Cell—O—O—Cell} \quad (76)$$

$$\text{Cell—O—O}^{\cdot} + \text{Cell—H} \xrightarrow{\Delta} \text{Cell—O—OH} + \text{Cell}^{\cdot} \quad (77)$$

$$\text{Cell}^{\cdot}\text{—nM} \rightarrow \text{Cell—(M)}_n^{\cdot} \quad (78)$$

$$\text{Cell—O—OH} \xrightarrow{\Delta} \text{Cell—O}^{\cdot} + {}^{\cdot}\text{OH} \quad (79)$$

$$\text{Cell—O}^{\cdot} + \text{nM} \rightarrow \text{Cell—O—(M)}_n^{\cdot} \quad (80)$$

$$\text{nM} + {}^{\cdot}\text{OH} \rightarrow \text{(M)}_n^{\cdot} + \text{HOH} \quad (81)$$

where Cell—H represents the cellulose molecule; Cell$^{\cdot}$, the cellulose radical; M, the monomer; and M$^{\cdot}$, the monomer radical. The formation of trapped radicals in cellulose probably results from dehydrogenation (Eq. 73). In nitrogen, heating of irradiated cellulose allows increased mobility of the free radical sites, particularly in the more ordered regions, leading to increased intermolecular bonding (Eq. 74). In view of the size of individual chains, depending on their excluded volume, there also arises the possibility of intramolecular crosslinking, though this would be a minor event relative to intermolecular crosslinking. Intramolecular crosslinking does not give a fully satisfactory explanation for the reduced moisture regain.

Heating of irradiated cellulose in oxygen provides the possibility of reactions (74) to (77) occurring, particularly in less ordered regions. Such reactions could lead to the formation of intermolecular bonds and an ultimate reduction in the number of sites otherwise available for grafting. Grafting polymerization is possible, as shown by Eqs. (78–80). Homopolymerization becomes a reality if Eq. (81) is followed. Again, it should be stressed that his sequence has general application.

Hinojosa and Arthur[257] have examined the significance of the propagating radical in the copolymerization of methacrylic acid with gamma-irradiated cellulose. In an e.s.r. spectroscopic investigation of irradiated poly(methacrylic acid), Abraham et al.[258] concluded that the propagating radical was

$$-CH_2-\dot{C}(CH_3)-COOH$$

This conclusion is supported by Hinojosa and Arthur, who used the cellulose-poly (methacrylic acid) radical in copolymerization reactions through immersion in solutions of other monomers such as acrylonitrile or acrylic acid. This approach provides useful insight into ways of preparing grafted block copolymers of cellulose[255].

Post-irradiation grafting onto cellulose using a variety of monomers has been attempted. Points which have been investigated include those outlined above. Additional aspects covered are grafting from binary mixtures of monomers[259], one of which was acrylonitrile; the temperature dependence of grafting styrene under both dry and aqueous conditions in the range 50°–117 °C[260]; the effect of dimethylformamide on the grafting of methyl vinyl ketone from a methanol/water medium[261]; the effect of grafted acrolein on the crosslinking tendencies of cellulose[262]; an apparent activation energy of 640 kJ/mol in the grafting of methyl methacrylate in the temperature range from room temperature to 65 °C[263]; and comonomer distribution in the grafting of butyl methacrylate onto cotton fiber[264]. Also covered are the characteristics of vapor phase grafting using vinyl fluoride and vinylidine fluoride as the comonomers, in which it was found that water vapor markedly enhanced the level of grafted add-on[265–266], and the grafting of butyl methacrylate and acrylonitrile, singly and in mixtures, with a view to control of structural modifications and changes in the dyeing behavior of the cellulose[267].

Post-irradiation grafting techniques have also been applied to modified celluloses. Here we include rayon[268–270], cellulose acetate[166,269,271,272], ethyl cellulose[273], nitrocellulose[274], and cellulose carbanilates[51].

3.2.4.4. Mechanisms of Irradiation Grafting

It has been reported that a reasonable proportion of the radiation-induced radicals in cellulose have lifetimes of years and are generally inaccessible to molecules in non-swelling media[85,98,275], since the radicals effectively become a characteristic of the bulk system. Thus, in terms of the grafting mechanism, two general types of copolymer bonding may be considered. First, those radicals which are accessible may form conventional bonds (grafts) with the monomer, presumably σ-bonds. Other types of bonding in copolymerization are also possible, though their significance will vary depending on the substrate (substituted or unsubstituted celluloses). Most mechanistic studies have been applied to the more simple cellulosics. The reasons for this are not difficult to understand, since they are mainly associated with the demand for new materials. Thus, little effort has been expended on mechanistic studies of their formation. A different mechanism for bonding in graft copolymerization to cellulose is that the so-called "inaccessible" radical sites may contribute to the bonding if charge transfer interactions are involved[276]. The proposed types of charge-transfer species[97] are discussed below.

a) General Charge-Transfer Complex Formation

Using styrene and irradiated cellulose as the simplest example, the π-electrons of styrene may be delocalized to form a stable charge-transfer bond with the free valencies of irradiated cellulose. Analogous bonding is known in inorganic complexes[277,278] and conventional aromatic substitution reactions. Thus, as the first step in grafting, styrene is complexed in a flat position (for maximum overlap) on the surface of the irradiated cellulose (Eq. 82), with $P^{\cdot}$ denoting a radical either on the surface or within the cellulose, whether the radical indicated by $P^{\cdot}$ is at the end of a growing chain or otherwise:

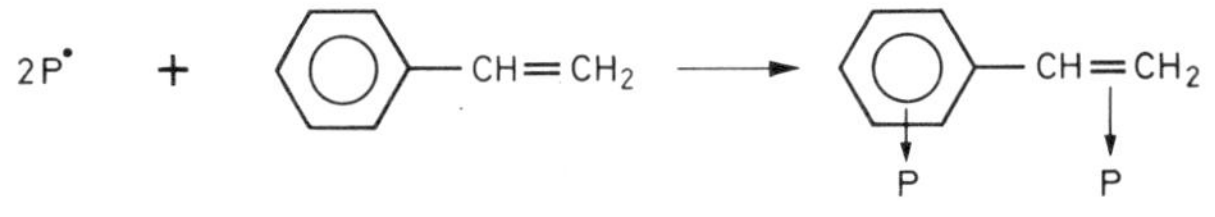

(82).

(1) Grafting at the Surface: Two reactions from Eq. (82) are now possible, involving radicals at the surface or in the bulk. Because of ease of accessibility of radical sites at the surface of the cellulose, further (π-σ reaction) may occur to give stable σ-bonded species, similar to inorganic (π-σ) conversion[278,279], Eqs. (83) and (84),

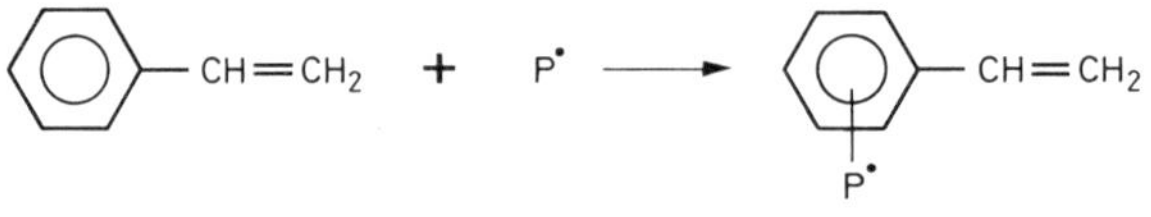

(83)

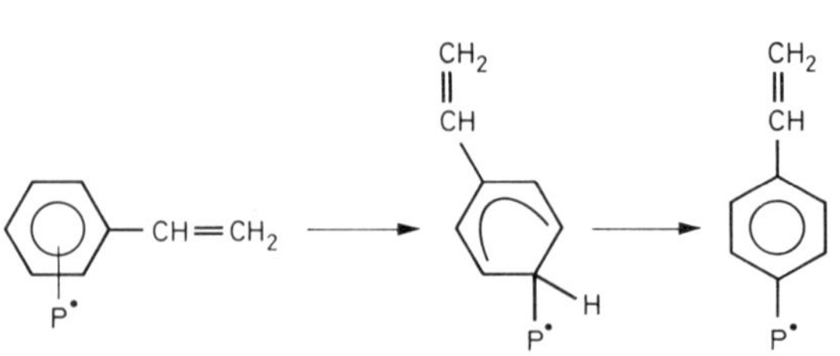

(84)

Analogous (σ-π) conversions can be shown for σ-bonding with the side chain (Eq. 85 and 86)

(85)

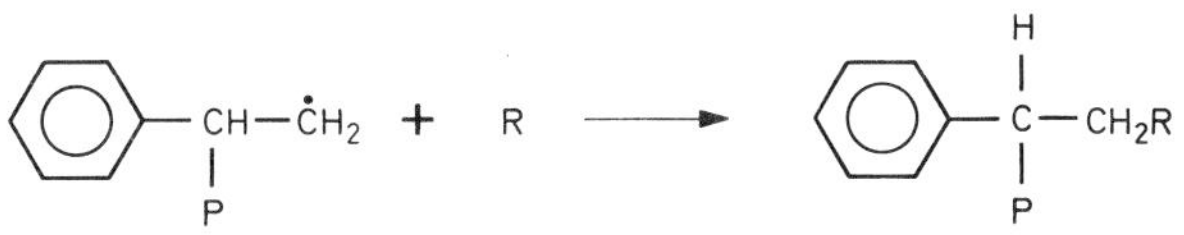

(86)

If R is a monomer, grafting will occur.

(ii) Grafting in Bulk: Here the mobility of polystyrene chains and radicals is impeded and the (π-σ) conversions are restricted; however, the long polystyrene chains can still be grafted by a series of charge-transfer bonds (Eq. 87). Such bonding would be extremely strong and would tend to keep homopolymer locked between the chains as "graft".

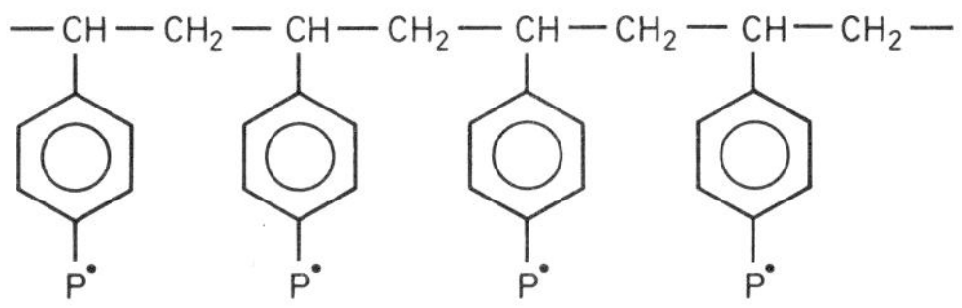

(87)

The charge-transfer mechanism can also be used to interpret bonding involving radiation-induced grafts between cellulose and monomers other than styrene, e.g., acrylonitrile, methyl methacrylate, etc. With certain of these monomers, the functional group could act as the donor for charge-transfer bonding in the bulk [Eq. (88)].

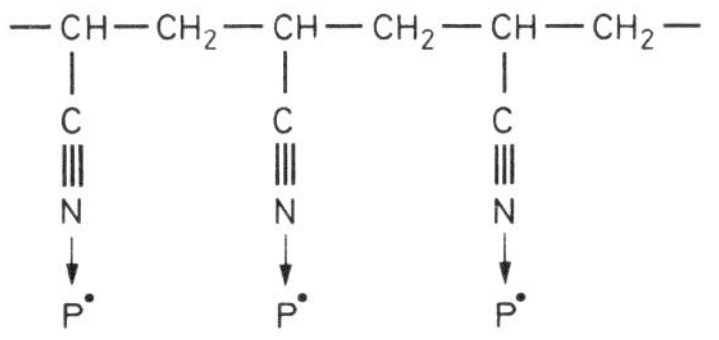

(88)

Evidence consistent with this type of bonding in Eq. (88) is provided by the effect of ammonia on the e.s.r. signal of irradiated cellulose[135]. Ammonia completely removes the signal. Part of this effect could be attributed to delocalization of the nitrogen lone-pair, as in Eq. (89).

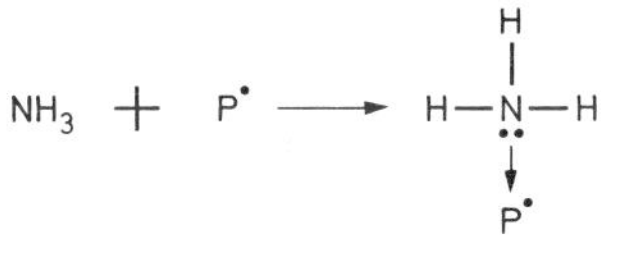

(89)

Futhermore, compounds containing NH_2 or heterocyclic nitrogen groups generally appear both to inhibit the grafting of styrene and also to reduce the intensity of the Trommsdorff peak[134]. This is obviously due to strong adsorption of the inhibitor, presumably bonding by charge-transfer forces similar to those depicted in Eq. (89).

Grafting with monomers other than styrene

Dilli and Garnett have devoted much attention to the grafting of substituted styrenes onto cellulose[157]. There is little published work on this topic, yet it provides an excellent system for obtaining an understanding some of the mechanisms of grafting because of the ease of substituting into the benzene ring one or more electronegative or electropositive groups. The reactivity of styrenes in free radical polymerizations has been shown to be as follows: styrene with electron-donating groups < styrene < styrene with electron-withdrawing groups[280]. It may be noted that a head-to-tail addition of monomers to the growing radicals will be encouraged because of the greater resonance stabilization possible in the radicals so formed.

Dilli and Garnett[157] have verified this classification through examination of various substituted styrenes. Table 31 contains the monomers used and the

Table 31. Grafting with substituted styrenes in methanol. Reprinted from Ref.[157], p. 1169, by courtesy of the Commonwealth Scientific and Industrial Research Organization (C.S.I.R.O.). Melbourne, Australia

Monomer	Grafting Yields (%)		Hammett Subst. Const.
	A	B	
styrene	30	16	—0.268
p-methoxystyrene	0		
o-methylstyrene	110		—0.170
p-methylstyrene	6		—0.151
p-ethylstyrene	2		
p-n-butylstyrene	2		
2,4-dimethylstyrene	10		
2,5-dimethylstyrene	19		
2,4,6-trimethylstyrene	0		
o-chlorostyrene	74	23	
m-chlorostyrene	a	93	+0.373
p-chlorostyrene	a	90	+0.227
o-bromostyrene	27		
m-bromostyrene	a, b	a	+0.391
p-bromostyrene	c		+0.232
m-fluorostyrene	a	44	+0.337
p-fluorostyrene	a	56	+0.072
2-vinyl pyridine (d)	3		
4-vinyl pyridine (d)	0		

Notes a) Reaction was too great to measure under the specified conditions.
b) After repeated extraction in boiling chloroform (ten extractions) the original mass of cellulose was recovered, i.e., there was no grafting. Infrared spectrum confirmed the nature of the residue.
c) Extraction of the homopolymer-incomplete with this monomer.
d) Although not styrenes, these monomers are included for comparison.

extents of grafting achieved. Where available, the Hammett substitution constants are given. As can be seen from Table 31, the results of grafting for the substituted styrenes fit reasonably into the approximate grading. Low grafting yields are obtained for those monomers containing electron-donating substituents, whereas extensive grafting occurs with monomers in which electron withdrawing groups were substituted into the benzene ring. The grafting yields obtained by Dilli and Garnett for the p-alkylstyrenes and p-methoxystyrene were consistent and appreciably smaller than for styrene. The decrease arises as a consequence of deactivation resulting from the combined conjugative (+M) and inductive (+I) effects in these compounds. It follows that the tautomeric component contributing to deactivation can be eliminated by metasubstitution of these groups in styrene; enhanced grafting yields should then be observed.

o-Methylstyrene provides us with a departure from the trend among the electron-donating substituents. Examination of molecular models indicates that steric hindrance to conjugation is significant in this monomer. Such hindrance is even more severe with 2,6-dimethylstyrene, and no grafting occurs with 2,4,6-trimethylstyrene.

Dilli and Garnett have provided a comprehensive discussion of these observations[157]. With 2,4,6-trimethylstyrene, the deactivation due to the three methyl groups is large. The stability of the radical will also be considerably reduced as a result of the reduced degree of conjugation imposed by the two ortho-substituents, forcing the vinyl group out of the plane of the benzene ring. In this respect, the stability of the radical is important for singly substituted o-methylstyrene, since a very stable π-allylic system is produced only with this compound; this could be responsible for the exceptional tendencies toward grafting which are shown by the monomer (110% cf styrene, 30% and p-methylstyrene, 6%, under identical conditions). Thus, we have

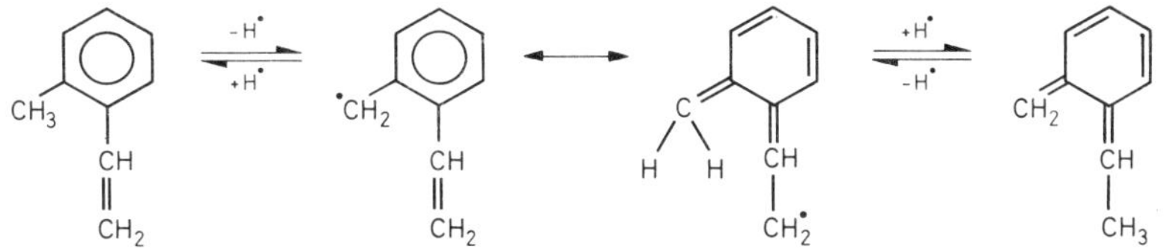

Some of this reactive character is possessed by 2,4-dimethylstyrene and 2,6-dimethylstyrene. In 2,4-dimethylstyrene, the methyl group at the 4-position can transfer charge to the vinyl double bond by hyperconjugation and reinforce the inductive effect (+I) of the methyl group, leading to an overall decrease in grafting tendencies. When the methyl group is in the 5-position, as in 2,5-dimethylstyrene, the inductive effect operates alone since no tautomerism is possible. This explains the higher grafting yield as seen with 2,5-dimethylstyrene.

For substituted styrenes containing electron withdrawing groups, the grafting results are generally consistent with the theory outlined above. The major exceptions are m-bromostyrene and p-bromostyrene. With m-bromostyrene, no grafting was observed. This result is ascribed to the inductive effect (—I) and is contrary to what might normally be expected, considering the electronegativities of the halogens. The picture presented by p-bromostyrene is complex owing to its strong tendency toward crosslinking on polymerization.

For the remaining monomers in Table 31, with the exception of meta-substituted compounds, inductive and tautomeric effects operate. A consideration of the opposing electronic effects suggests that the expected order of reactivity would be greater for ortho- and meta- than for para- substituted monomers in a series. The comparison of ortho and meta derivatives is more complex but will be in the order meta > ortho, provided that the tautomeric and inductive effects for the ortho position are equal to, or less, than the inductive effect of the meta-substituent.

With these guidelines in mind, let us examine the halogen-substituted monomers in Table 31[157]. The results show a deviation away from the predicted order, which indicates that grafting efficiency ought to proceed F > Cl > Br. The practical order is Cl > F > Br. Simple tautomeric and inductive effects are not adequate alone in providing an explanation for these results. The grafting yields are given added significance by the Hammett σ constants for the relevant meta- and para-substituents. The constants show the directional influence of the substituent in the transfer of charge to and from the ring and clarify the interpretation of, for example, the result obtained with m-bromostyrene. The relative magnitude of σ does not explain the high grafting yields obtained with para-fluorostyrene. However, arguments based on structural considerations should not be overemphasized, and due account must be taken of the particular conditions under which a study is carried out. This is especially true in grafting reactions, particularly where cellulose is concerned. The picture with regard to cellulose derivatives is even more in need of examination than is that of grafting to cellulose itself. This is one area of investigation which could be usefully expanded.

b) Radical Scavenging in Copolymerization

The predominant participation of free radical processes in the grafting reaction has been shown by the effects of hydroquinone and thiourea on grafting yields[97]. In terms of radiation chemistry, grafting of styrene in methanol may be considered[160] as the radiolysis of a binary mixture in the presence of a heterogeneous additive (cellulose). In radiation chemical studies of binary mixtures[144, 281, 282], aromatics such as benzene and styrene were shown to scavenge radiolytically-produced free radicals by both addition and substitution processes (Eq. 90–92).

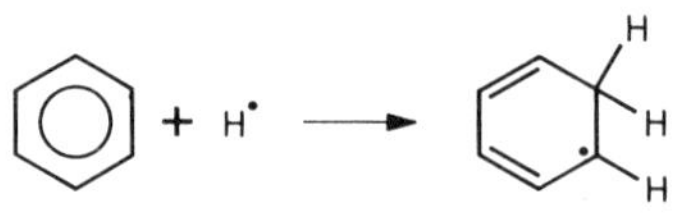

(90)

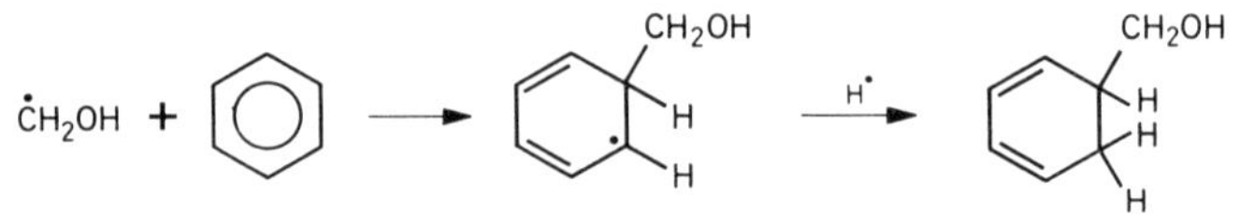

(91)

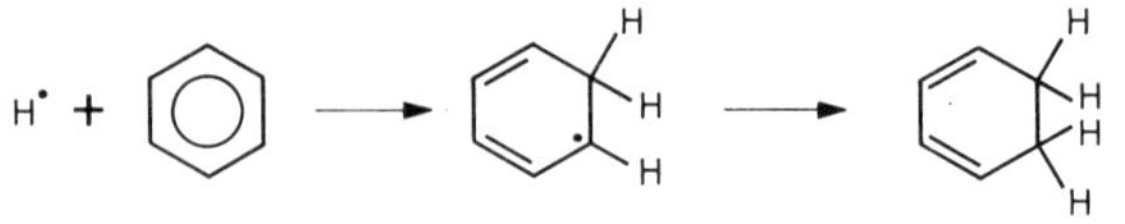

(92)

Thus, at low monomer concentration (0–10%) the solvent will be in large excess and the behavior of the system will depend essentially on the radiation chemistry of methanol. Under these conditions, styrene will scavenge solvent radicals predominantly by an addition mechanism (Eq. 91), yielding essentially homopolymer at the expense of graft. Similarly, at high monomer concentrations (70%), the relative amounts of methanol and styrene are not sufficient to prevent styrene from scavenging other styrene radiolytic fragments yielding extensive homopolymerization again at the expense of grafting. In the 30–70% region, a compromise between the two extremes occurs and the most favorable grafting conditions are obtained.[97]

c) Energy Transfer and Electron Scavenging Processes

The sensitization effect of napthalene at very low concentrations (10^{-4} M) and low dose rates suggests that energy transfer and/or electron scavenging processes are taking place in these solutions[97]. At low solute concentrations, naphthalene can scavenge solvated electrons to give the naphthalene negative ion (Eq. 93).

$$C_{10}H_8 + e^{\ominus}_{(Solv.)} \rightarrow C_{10}H_8^{\ominus} \; . \tag{93}$$

These solvated electrons participate actively in the fundamental radiolysis pathway of simple compounds, and they obviously will also contribute to radiation grafting processes. Naphthalene also scavenges hydrogen atoms, particularly in methanol, but this process does not become appreciable unless the concentration of naphthalene exceeds 10^{-3} M. Thus, in the concentration range (10^{-4} M) where naphthalene sensitizes the reaction, the mechanism of the enhanced grafting effect may well be due to energy transfer or electron scavenging.

These discussions make it evident that the charge-transfer theory can be used to explain the graft copolymerization of vinyl monomers onto cellulose by high energy radiation. It may also apply to grafting induced by ultraviolet radiation and Ce^{IV} ions.

3.3. Grafting by low Energy Radiation in the Presence of Sensitizers

Low energy photons may initiate polymerization of a vinyl or acrylic monomer if the irradiation is carried out in the presence of a sensitizer. Such a sensitizer must become active on exposure to the particular wevelength range of the incident radiation[283–285]. In addition, the lifetime of the activated species must enable secondary reactions, such as initiation of polymerization, to take place. The efficiency of sensitization is often enhanced through use of a synergist, though this is not always necessary. Certain monomers can be activated without the need for a sensitizer. However, even here, the use of a sensitizer/synergist combination can dramatically improve the efficiency of a polymerization process.

Low-energy, radiation-induced grafting involves the use of ultraviolet or visible light supplied by a suitable source. The energy is used to cause excitation of the sensitizer, causing the generation of radical species which may then attack the substrate[283]. Generation of active radical sites on the substrate is followed by grafting. Thus, the possibility of post-irradiation and simultaneous irradiation processes arises as with high energy, radiation-induced grafting. However, there are major mechanistic differences. Another feature in common with high energy processing is that of substrate damage. With low energy sources, this is termed photo-tendering. Thus, during the grafting process, deterioration of the substrate may occur either by localization of incident energy or by direct attack at main chain links by activated species.

Harries et al.[286] have examined the value of ultraviolet light as an initiator of free radical chain reactions in cellulosic substrates. Four classes are suggested: graft copolymerization reactions of vinyl monomers to modify the cellulosic substrate macromolecularly; crosslinking reactions within the substrate; homopolymerization within the substrate to provide occluded composite systems and oxidative depolymerization reactions. Modifications of the properties of cellulose in the solid state by photo-initiated reactions have been reported. These include oxidative depolymerization[287–295], crosslinking[296–300], and graft and homopolymerization[301–315]. Homopolymer formation on or within the cellulosic substrate usually imparts nondurable properties while increasing the general stiffness, especially with textiles. Oxidative depolymerization leads to a loss in tensile properties and hence will be discussed in some detail here.

3.3.1. Photo-tendering in Ultraviolet-Light-Sensitized Grafting

The mechanism of photo tendering by sensitizers has received considerable attention. Some of this attention has been aimed at cellulosics[316,317], using low molecular weight compounds as models[318]. It is generally accepted that photo tendering action is due to radical reactions. However, some controversy has arisen over the mechanism of photo-tendering. Since some photo-tendering during ultraviolet-light-induced processing is almost inevitable unless precautions are taken, we shall consider the point here.

Light of wavelengths less than 340 nm cannot induce degradation of cellulose. Hence, u-v absorbers are used as photo initiators. However, certain dyes and compounds, such as zinc oxide and titanium dioxide, are capable of absorbing light in the near ultra-violet or visible part of the spectrum and use this energy to induce the degradation of cellulose. One may therefore consider photo-tendering a common phenomenon in many composite systems. The more general problem of photosensitized processes in polymer chemistry has already been exhaustively treated[319] and will not be covered further.

The photoinitiated tendering of cellulose by vat dyes has been recognized for many years[320–329]. The anthraquinone vat dye (AQ) is photosensitized by exposure to ultraviolet or visible light forming an excited dye molecule. This excited dye molecule may then proceed to create radical sites in the parent cellulose substrate containing the (AQ) dye and provide a route to graft copolymerization. The mechanisms which are involved in the absence of, as well as in the presence of, air appear to be[328]:

(i) Absence of oxygen

$$\mathrm{AQ^{a} + h\nu \longrightarrow AQ^{*}\ (excited\ dye)} \qquad (94)$$

$$\mathrm{AQ^{*} + Cell{=}C(OH^{d})(H) \longrightarrow AQH^{\bullet b} + Cell{=}\dot{C}(OH)} \qquad (95)$$

$$\mathrm{(2AQH \longrightarrow AQH_2^{c} + AQ)} \qquad (95a)$$

$$\mathrm{Cell{=}\dot{C}(OH) + MH^{e} \longrightarrow Cell{=}C(OH)(MH^{\bullet}) \longrightarrow Graft} \qquad (96)$$

$$\mathrm{Cell{=}\dot{C}(OH) + MH\,(or\ SH^{f}) \longrightarrow Cell{=}C(OH)(H) + M^{\bullet}\,(or\ S^{\bullet})} \qquad (97)$$

$$\mathrm{MH + M^{\bullet}\ or\ (S^{\bullet}) \longrightarrow homopolymer} \qquad (98)$$

$$\mathrm{Cell{=}C(OH)(MH)_n^{\bullet} + SH \longrightarrow Cell{=}C(OH)(MH)_nH + S^{\bullet}} \qquad (99)$$

$$\mathrm{AQ^{*} + MH \longrightarrow AQH^{\bullet} + M^{\bullet} \longrightarrow homopolymer} \qquad (100)$$

(ii) Presence of atmospheric oxygen:

$$\mathrm{AQH^{\bullet} + O_2 \longrightarrow HOO^{\bullet} + AQ} \qquad (101)$$

$$\mathrm{Cell{=}\dot{C}(OH) + O_2 \longrightarrow Cell{=}C(OH)(OO^{\bullet})} \qquad (102)$$

$$\mathrm{Cell{=}C(OH)(OO^{\bullet}) + {}^{\bullet}OOH \longrightarrow O_2 + Cell{=}C{=}O + H_2O_2} \qquad (103)$$

$$\mathrm{Cell{=}C(OH)(H) + {}^{\bullet}OOH \longrightarrow Cell{=}\dot{C}(OH) + H_2O_2} \qquad (104)$$

$$\mathrm{Cell{=}C(OH)(H) + {}^{\bullet}OH \longrightarrow Cell{=}\dot{C}(OH) + H_2O} \qquad (105)$$

$$\mathrm{MH + {}^{\bullet}OOH \longrightarrow M^{\bullet} + H_2O_2 \longrightarrow homopolymer} \qquad (106)$$

$$\mathrm{MH + {}^{\bullet}OH \longrightarrow M^{\bullet} + H_2O \longrightarrow homopolymer} \qquad (107)$$

$$\mathrm{MH + {}^{\bullet}OH \longrightarrow HOMH^{\bullet} \longrightarrow homopolymer} \qquad (108)$$

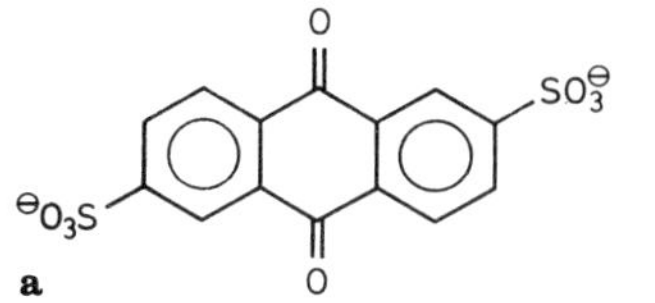

a

d Cellulose

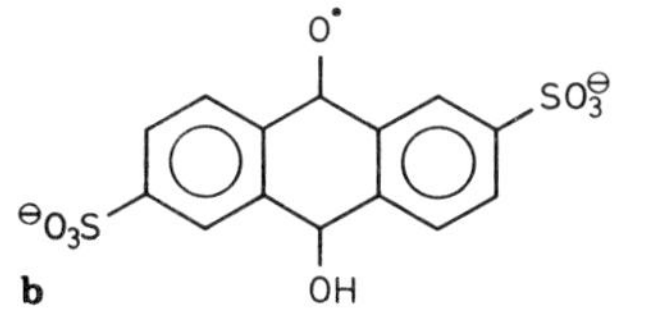

b

e Monomer

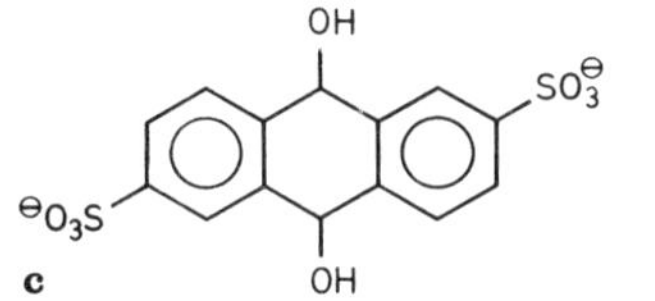

c

f Solvent, possible hydrogen donor

At first sight, it is apparent that the entire process is initiated by step (94). Grafting is initiated by (95) and (96). Homopolymerization can be initiated by (97–100). Under the conditions reported[328] reaction (100) appears to be predominant, since the amount of homopolymer produced is large and not dependent on the presence of cellulose. The dye molecule may be regenerated by the bimolecular combination of two dye semiquinones. However, the mechanism is more complicated in the presence of oxygen, since the number of possible reactions increases. The following differences seem to be important when oxygen is present;

1. The amount of dye, AQ, capable of further hydrogen abstraction may be increased by step (101).
2. The possibilities for initiation of grafting are increased by steps (104) and (105), but decreased by step (101).
3. The number of ways leading to homopolymerization is increased by reactions (106–108).

Figure 30 shows that the amount of grafting in the presence of air is twenty percent higher than that in vacuo after a twelve-hour irradiation has been exceeded. It has also been reported[328] that the extent of monomer conversion was up to three times as great in the presence of air as when carried out in vacuo. When air was bubbled through the reaction mixture, both grafting and homopolymerization were inhibited. In the mechanism proposed, this would indicate that under these conditions, reactions (102) and (103) predominate and inhibit grafting. Homopolymerization is inhibited by a similar mechanism.

With regard to the attendant photo-tendering reactions, two mechanisms have been proposed for the primary events leading to photo-tendering, one based on hydrogen abstraction reactions and the other on direct attack by singlet oxygen. Davies et al.[330–331] have shown that hydrogen abstraction reactions are

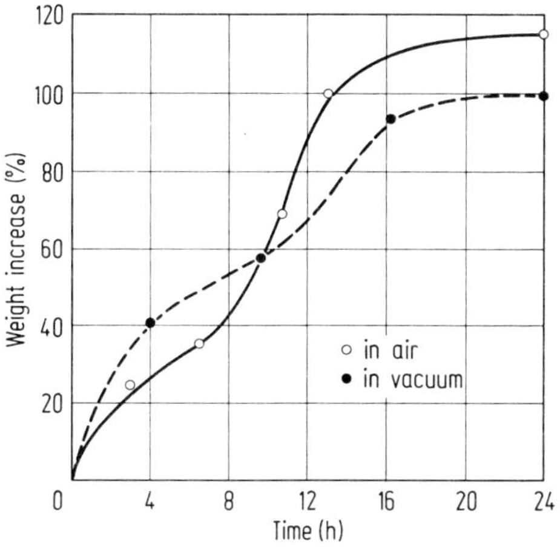

Fig. 30. Effect of irradiation time on amount of grafting. Reprinted from Ref.[328], p. 56, by courtesy of John Wiley and Sons, Inc.

implicated in phototendering. This observation is supported by the fact that grafting can be made to occur during phototendering using active radical sites[330]. The oxygen activation (singlet oxygen) mechanism of phototendering was proposed by Egerton[324,332], though until recently this mechanism was without realistic corroboration. Work by Griffiths and Hawkins[333,334], has now provided substantiation for the singlet oxygen mechanisms role as at least one factor in the photo-tendering of cellulose. In reality, it is difficult to show unambiguously that singlet oxygen will react with cellulose substrates, since almost all efficient sensitizers of singlet oxygen are good hydrogen abstractors as well (quinonoid vat dyes, heterocyclic cationic dyes, and porphins). It follows that photodegradation studies with such compounds will produce results which are ambiguous. Non-photochemical sources of singlet oxygen must also be suspect, as these substances are usually contaminated with other oxidizing species which could play a definite though uncertain role in the degradation of cellulose.

Griffiths and Hawkins[334], in their search for new, stable, singlet-oxygen quenchers that have a minimal tendency to undergo photochemical reduction, obtained the compound

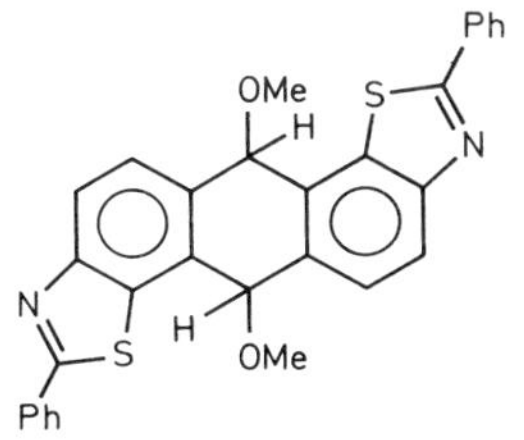

by reductive methylation of Vat Yellow 2 (CI 67300) using zinc and methyl-p-toluene sulphonate. This product of the reductive methylation is particularly resistant to photo-oxidation and is an efficient singlet oxygen generator, though the singlet oxygen generation can be suppressed using diazabicyclo [$2_1 2_1 2$]-octane[335]. Photodegradation of cellulose films was followed viscometrically by measurement of the intrinsic viscosity of solutions in Cadoxen at 30 °C using an Ubbelhode, suspended-level viscometer. Samples were irradiated under standardized conditions for various known times, after treatment with the sensitizer. Unsensitized controls were used. The levels of degradation with sensitized and unsensitized regenerated cellulose film were 30% and 13% respectively after 530 hours of irradiation with an 8 W fluorescent lamp. Thus, it can be stated that singlet oxygen plays a direct part in the photodegradation of cellulose. Griffiths and Hawkins conclude, from this and further experimental observations, that it is evident that singlet oxygen can cause degradation of cellulose at a rate which is a least equal to that of the hydrogen abstraction mechanism.

Perhaps one should state that the main point from this controversy is the danger which may lie in interpreting experimental results, considering any particular mechanism to the exclusion of others. It appears quite feasible that photo-tendering reactions involving cellulose operate through both hydrogen abstraction and singlet oxygen mechanisms. Indeed, these two ought not to be considered sacrosanct since other aspects of photo-tendering may be of importance.

3.3.2. Photoinitiated Grafting Reactions

In photoinitiated grafting reactions, a distinct advantage lies in the fact that a much lesser extent of degradation occurs than with high energy radiation for given levels of grafting. On the whole, graft yields obtained in this way compare very favorably with yields obtained using other radical dependent routes. For instance, irradiation of cellophane, together with a 20% aqueous solution of acrylamide at 0.01 M concentration of dye for 24 hrs in presence of air, gives a weight increase of cellophane on the order of 150%, based on its original weight. Under similar conditions, a 45% solution of acrylonitrile, 45% of dimethylformamide, and 10% water leads to a 120% weight increase. Replacing about two-thirds of the acrylonitrile by styrene leads to a 365% weight increase. In addition, films of cellulose derivatives, which are soluble in organic solutions, can be easily grafted in this system. For example, acrylamide has been grafted onto cellulose acetate and ethyl cellulose[328].

Reine and Arthur have successfully grafted methacrylamide onto cotton cellulose using light of 350 nm. This photoinitiated grafting was unsensitized[336]. Oxygen acted as an efficient inhibitor. The extent of copolymerization was shown to be dependent on the irradiation time and monomer concentration. These findings agree well with those of Needles and Walsey, though these researchers used a riboflavin sensitized graft polymerization procedure[337] and achieved significant grafting levels with a range of monomers. Comprehensive studies of photoinitiated graft polymerization have been carried out by Kubota et al.[181, 338–340], who reported on the decay behavior of cellulose radicals generated by photoirradiation at room temperature. The characteristics of subsequent grafting with methylmethacrylate were also examined. E.s.r. spectra, taken of untreated swollen, oximated, and Fe^{III} ion-sensitized samples, irradiated at room temperature, provided mainly a singlet absorption line with a line width of 20 to 22 gauss and a g value of 2.003. They assert that this spectrum is evidence of radical location on the C_1 or C_4 atoms arising from scission of glucosidic bonds. This assertion is supported by observations made in e.s.r. measurements of gamma-irradiated cellulose.

The decay of these radicals was seen to be accelerated by contact of the samples with various solvents[340], the activity decreasing in the order, water ≈ methanol > acetone > dioxane (Fig. 31, upper). On the other hand, the decay of radicals by vinyl monomers was shown to decrease in the order, methacrylic acid > methyl methacrylate = styrene (Fig. 31, lower). From this work one may speculate that since only radicals at the C_1 and C_4 positions are indicated, as seen in the e.s.r. spectra, arising from scission events, the possibility of block copolymer formation becomes strong. This could be a significant feature of such irradiation procedures and provides the basis of a major difference between high and low energy radiation-induced processes. The ability of U.V., pre-irradiated cellulose samples to initiate graft polymerization with methyl methacrylate is indicated in Table 32.

Such general observations may be extended to various forms of cellulose when grafted with methyl methacrylate. Kubota et al. have studied this point in detail[181,339], using various types of cellulose (untreated samples — adsorbent cotton, sulfite pulp, dissolving pulp, and semichemical pulp; extracted samples — 1% sodium hydroxide, methanol and an ethanol-benzene mixture as extraction solvent; oxidized samples — periodic acid, sodium hypochlorite, hydrogen

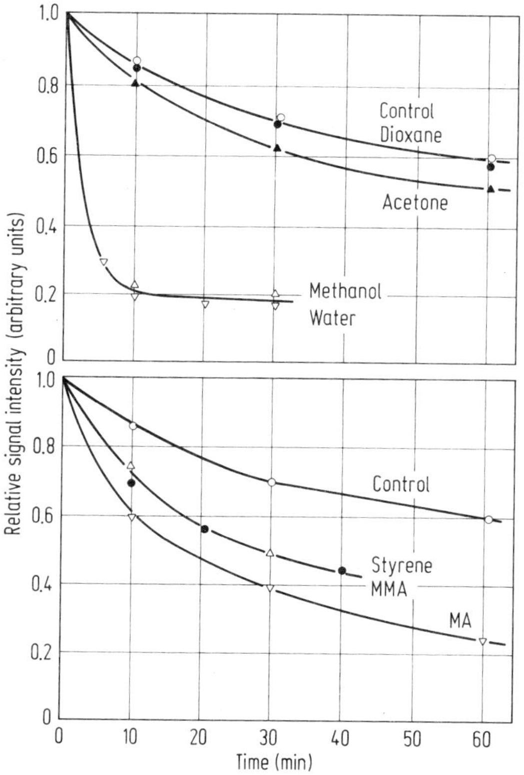

Fig. 31. Effect of various solvents on decay of radicals produced by photo-irradiation. The Fe^{III} ion-sensitized sample was irradiated at room temperature for 30 min in a quartz glass system. Reprinted from Ref.[340], p. 2817, by courtesy of John Wiley and Sons, Inc.

Table 32. Ability of pre-irradiated cellulose to initiate graft copolymerization with methyl methacrylate after photo-irradiation. Reprinted from Ref.[340], p. 2818 by courtesy of John Wiley and Sons, Inc.

Solvent	Grafting (%) at various pre-irradiation times		
	0	30 min	60 min
Water	0	20.0	6.9
Methanol	0	25.1	4.3
Acetone	0	0	0
Dioxane	0	0	0

Graft copolymerization of methyl methacrylate onto an Fe^{III}-ion-sensitized sample carried out at 40 °C for 60 min in the system with a 20% solvent concentration.

peroxide, and nitrogen dioxide; a reduced sample; an oximated sample; and an Fe^{III} ion-adsorbing sample) in experiments. It was clearly shown that impurities within the sample, as well as the oxidized states, noticeably affected the initiation reaction. It was also noted that these factors were especially significant in irradiation systems employing a wavelength >300 nm. However, with sensitizers or light of shorter wavelengths, initiation occurred irrespective of the cellulose characteristics. Work carried out by Kubota et al. on aldehyde celluloses is covered fully in Chapter 6.

Davis et al. have carried out a comparison of photosensitized and gamma ray-induced graft copolymerization of various monomers onto cellulose[341]. Low molecular mass alcohols are found to be particularly effective as grafting solvents in both systems, especially with styrene (Table 33) and methyl metha-

Table 33. Photosensitized grafting of styrene to cellulose by using simple alcohols as solvents. Reprinted from Ref.[341], p. 293 by courtesy of John Wiley and Sons, Inc.

Monomer Concentration % V/V	20	40	60	80	90
Alcohol	% Graft				
Methanol	13	28	34	53	64
Ethanol	9	17	22	50	70
n-Propanol	5	12	18	30	20
Isopropanol	< 5	7	5	14	8
n-Butanol	< 5	< 5	< 5	< 5	< 5
Isobutanol	< 5	< 5	< 5	< 5	< 5
t-Butanol	< 5	< 5	< 5	< 5	< 5
n-Octanol	< 5	< 5	< 5	< 5	< 5

The solutions contained 1% (w/v) of uranyl nitrate and were irradiated for 24 hr, at 24 cm from a 90-W high pressure lamp.

crylate. Unfortunately, polyfunctional acrylates, acrylonitrile and acrylic acid undergo homopolymerization rather than grafting, whereas vinyl acetate and 4-vinyl pyridine graft with little competing homopolymerization, although vinyl acetate requires an appreciable irradiation time to produce a reasonable grafting yield (Table 34). Various sensitizers were used in this work (see Table 35) and a comparison was made between quartz and pyrex vessels (see Table 36). The sensitizers included uranyl nitrate, anthraquinone-2,6-disulphonic acid sodium salt, biacetal, manganese carbonyl, and benzoin ethyl ether. In the absence of a sensitizer, no grafting was obtained with the styrene/methanol/cellulose system. In order of sensitizing efficiency for a given geometrical arrangement and light intensity, we have biacetal > uranyl nirate > benzoin ethyl ether > anthraquinone-2,6-disulphonic acid disodium salt > manganese carbonyl. The standard irradiation conditions involved solutions of styrene (30% v/v) in methanol contained in pyrex tubes with wetted-out cellulose. These were irradiated with a high pressure UV lamp at a nominal sensitizer concentration of 1.0% (W/V). Grafting was carried out on Whatman No 41 filter paper at 21 °C. Replacement of pyrex containers by silica vessels produced very little difference in graft levels, though the slight differences which were observed were toward increased grafting. With styrene, a dramatic effect of alcohol chain length, (alcohol-styrene-cellulose system), on copolymerization, was observed. Thus, there is a progressive decrease in the grafting yield with increasing alcohol chain length, copolymerization ceasing with n-butanol. Such results compare favorably with those obtained via gamma irradiation. However, the addition of small concentrations of higher chain alcohols to the methanol-styrene-cellulose was found to sensitize the system toward photoinitiated grafting. An increase in the sensitizer concentration produced an increase in the yield up to a maximum value.

Table 34. Ultraviolet grafting of acrylic acid, Acrylonitrile, 4-vinyl pyridine, and vinyl acetate in methanol to cellulose. Reprinted from Ref.[341], p. 292, by courtesy of John Wiley and Sons, Inc.

Run No	Monomer	Sensitizer	Time of Irradiation (hr)	Graft (%) 4
1	acrylic acid	uranyl nitrate	3	15.3
2	acrylic acid	benzoin ethyl ether	3	10.6
3	acrylonitrile	uranyl nitrate	3	15.8
4	acrylonitrile	benzoin ethyl ether	3	10.3
5	acrylonitrile	benzophenone	24	7.5
6	4-vinyl pyridine	uranyl nitrate	3	10.1
7	4-vinyl pyridine	benzoin ethyl ether	3	11.1
8	vinyl acetate	uranyl nitrate	3	2.0 (B)

The conditions used a solution of monomer (30% (v/v)) in methanol containing sensitizer and cellulose, irradiated 24 cm from a 90-W high pressure ultraviolet lamp except run number 5, where the monomer concentration was 90% (v/v).

(B) Vinyl acetate was studied at many concentrations in the range 1—100% (v/v) but graft was always less than the 2.0% level observed with the 30% (v/v) monomer concentration.

Table 35. Effect of various sensitizers in the ultraviolet grafting of styrene in methanol to cellulose. Reprinted from Ref.[341], p. 289, courtesy of J. Wiley and Sons, Inc.

Run	Sensitizer	Distance from u-v lamp (cm)	Graft (%)
1	none	24	0.0
2	none	24	0.0 (a)
3	none	12	0.0 (a)
4	uranyl nitrate	24	3.4
5	uranyl nitrate	12	10.9
6	ADS	24	1.8
7	ADS	12	3.7
8	biacetal	12	0.0 (a)
9	biacetal	24	3.8
10	biacetal	12	11.0
11	benzoin ethyl ether	24	1.4
12	benzoin ethyl ether	12	5.6
13	manganese carbonyl	24	0.9 (a)
14	manganese carbonyl	24	1.5

Conditions include solutions of styrene (30% (v/v)) in methanol contained in pyrex tubes with cellulose, irradiated for 3 hours with a 90-W high-pressure ultraviolet lamp, except runs 13 and 14, where styrene (25% (v/v)) solutions were used.
Sensitizer concentration 1% (w/v)
ADS — anthraquinone-2,6-disulfonic acid, disodium salt.
(a) — denotes that the ultraviolet light was excluded.

Table 36. Comparison of quartz with pyrex vessels for the ultraviolet grafting of styrene onto cellulose. Reprinted from Ref.[341], p. 290, by courtesy of J. Wiley and Sons, Inc.

Run No.	Sensitizer	Irradiation Vessel Type	Graft (%)
1	none	quartz	0.0 (a)
2	none	quartz	0.3
3	uranyl nitrate	quartz	1.6 (a)
4	uranyl nitrate	quartz	15.2
5	uranyl nitrate	pyrex	3.4
6	benzoin ethyl ether	quartz	0.0 (a)
7	benzoin ethyl ether	quartz	6.7
8	benzoin ethyl ether	pyrex	1.4

Conditions include solutions of styrene (30% (v/v)) in methanol irradiated for 3 hr at a distance of 24 cm from a 90-W high pressure ultraviolet lamp.
(a) — denotes that the ultraviolet light was excluded.

Thus, for the methyl methacrylate-methanol-cellulose system with uranyl nitrate as the sensitizer, a maximum value of 300% graft was achieved at a sensitizer concentration of 3.0% (w/v). The grafting level at a sensitizer concentration of 10% (w/v) was only 36%.

Davis et al. used the uranyl nitrate system in their proposed mechanism for u-v initiated grafting[341]. The photochemistry of the uranyl ion has been fully reviewed and copolymerizations in two reaction sequences dominate: intermolecular hydrogen atom abstraction and energy transfer[342, 343]. In the styrene-methanol-cellulose system, inter-molecular hydrogen abstraction can be used in several ways to promote copolymerization. Thus, radicals may be formed in the methanol:

$$UO_2^{2+} + h\nu \rightarrow (UO_2^{2+})^* , \tag{109}$$

$$(UO_2^{2+})^* + CH_3OH \rightarrow CH_3O^\cdot + H^+ + UO_2^+ . \tag{110}$$

With pure methanol, the methoxy radical ($CH_3O^\cdot$) is the principal species formed, while with aqueous methanol, $\dot{C}H_2OH$ predominates[344]. These solvent radicals can then abstract hydrogen atoms from the cellulose trunk polymer to provide sites for subsequent grafting. However, the sensitizer may also diffuse into the swollen substrate and either directly abstract hydrogen atoms from the substrate or bring about main chain cleavage[345].

The role of energy transfer reactions in copolymerization reactions of this type are more difficult to evaluate. However, their participation in certain polymerization reactions involving uranyl cation sensitization has been proposed[346, 347]. No doubt some energy transfer must occur in the grafting process, though the level and significance of such a contribution awaits evaluation.

In a comparison between photochemical grafting and high-energy-radiation-induced grafting, one major difference concerns the occurrence of Trommsdorff effects. Using the styrene-methanol-cellulose system as an example[341], with gamma irradiation the peak occurs at approximately 30% (v/v) concentration of monomer in methanol, whereas with the analogous ultraviolet radiation grafting sequence, the peak occurs with the 80–90% (v/v) monomer concentration. This difference may be associated with the location of active sites within the substrate. With ionizing radiation, sites are randomly located volumetrically throughout the substrate. A large proportion of these are accessible to the monomer solution. A charge-transfer mechanism would apply and grafting would occur in confined environments. Although the rates of propagation and termination would both be adversely affected by such a situation, that of termination would be the more seriously influenced. This is because chain termination would become hindered by the viscosity of the localized medium and consequently there would be reduced chain mobility.

In ultraviolet light-sensitized grafting, the active sites will be located at points accessible to the sensitizer or at surfaces. Hence, mobility is enhanced relative to the high energy counterpart, which would mean that termination is not as seriously affected.

However, the actual mechanism of the grafting process produced by ultraviolet light sensitization is probably quite similar to the mechanism proposed for the gamma radiation-induced process.

The current popularity of radiation-induced grafting processes as a means of modifying cellulosics is shown in papers presented at recent A.C.S. conferences. Garnett summarized the work carried out by his group on grafting with u-v and high energy radiation[348]; Phillips et al. presented a detailed account of the interaction of radiation with cellulose in the solid state[349]; and Nakamura and Shimada studied the effect of various solvents on the grafting of styrene with γ-irradiated cellulose[350]. However, despite the interest shown in radiation-induced grafting onto celluloses and other natural and synthetic polymers, there are still several problems requiring solution. These have been summarized by Stannett and Memetea[351].

References

1. Liepins, R. Stannett, V., Walsh, W. K.: Progr. Polym. Sci. *6*, 1 (1978)
2. Brobeck Associates, Particle accelerator safety manual, U.S. Dept. of Health, Education and Welfare publication, PH-86-67-193 Oct. 1973
3. Suchecki, S. M.: Text Inds. (1975) P. 91
4. International Commission on Radiological Units and Measurements, Copenhagen (1953). Nucleonics (1954), 12, 11 and Brit. J. Radiology 27, 243 (1954)
5. International Commission on Radiological Units and Measurements, Geneva (1956). Natl. Bureau of Standard (U.S.) Handbook (1957), 62
6. Boag, J. W. Pilling, F. D., Wilson, T. Brit. J.: Biology *24*, 341 (1951)
7. Laughlin, J. S., Beattie, J. W., Henderson, J., Harvey, R. A.: Amer. J. Roent. *70*, 2 (1953)
8. McEhinney, J., Zendle, B., Domen, S. J.: Res. Natl. Bureau Standards *56*, 9 (1956)
9. Hochanadel, G. R., Weiss, J.: Proc. Roy. Soc., *A 240*, 189 (1957)
10. Lazo, R. M., Dewhurst, H. A., Burton, M. J.: J. Chem. Phys. *22*, 1370 (1954)
11. Hochanadel, C. J.: O.R.N.L. 87 (1950)
12. Weiss, J.: Nucleonics *10*, 28 (1952)
13. Hardwick, J. T.: Canad. J. Chem. *30*, 17 (1952)
14. Shekhtman, Y. L.: Dokl. Akad. Nauk. S.S.S.R. *74*, 767 (1950)
15. Goldblith, S. A., Procter, B. E., Hammerle, O. A.: Industr. Eng. Chem. (Indust.) *44*, 310 (1952)
16. Hartek, P., Dondes, S.: Nucleonics, *14*, 66 (1956)
17. Schulman, J. H., Klick, C. C., Rabin, H.: Nucleonics *13*, 30 (1955)
18. Kreidl, N. J., Blair, G. E.: Nucleonics *14*, 56 (1956)
19. Balestio, F., LeClerc, P., Bonnaud, M. J.: J. Appl. Rad. Isotopes *2*, 252 (1957)
20. Fowler, J. F., Day, M. J.: Nucleonics, *13*, 52 (1955)
21. Day, M. J., Stein, G.: Nucleonics, *8*, 34 (1950)
22. Henley, E. J.: Nucleonics *12*, 62 (1954)
23. Schuler, R. H., Allen, A. O.: J. Chem. Phys. *24*, 56 (1956)
24. Miller, N.: In Actions Chimiques et Biologiques des Radiations (A. Haissinsky, ed) Paris: Masson 1956
25. Day, M. J., Brit. J. Radiol.: *29*, 358 (1956)
26. Haybittle, J. L., Saunders, R. D., Swallow, A. J.: J. Chem. Phys. *25*, 1213 (1956)
27. Burton, M.: Disc. Faraday Soc. *12*, 317 (1952)
28. Chapiro, A.: Radiation Chemistry of Polymeric Systems New York: Interscience 1962
29. Weiss, J.: J. Polymer Sci., *29*, 425 (1958)
30. Guthrie, J. T., Huglin, M. B., Phillips, G. O.: J. Polymer Sci. C., *37*, 205 (1972)

30. a Henley, D. Arkiv Chemie, *18*, 328 (1961)
30. b Alexander, P., Black, P. M., Charlesby, A.: Proc. Roy. Soc. Ser. A. (London) *232*, 31 (1955)
30. c Alexander, P., Charlesby, A., Ross, M.: Proc. Roy. Soc. Ser. A. (London) *223*, 392 (1954)

31. Saeman, J. F., Millett, M. P., Lawton, E. J.: Ind. Eng. Chem. *44*, 2848 (1952)
32. Burton, M.: Discussions Faraday Soc. *36*, 7 (1963)
33. Phillips, G. O., Baugh, P., Nature: *198*, 282 (1963)
34. Phillips, G. O.: Discussions Faraday Soc. *36*, 281 (1963)
35. Phillips, G. O., Blouin, F. A., Arthur, J. C. Jr.: Nature *202*, 1328 (1964)
36. Phillips, G. O., Blouin, F. A., Arthur, J. C. Jr.: Radiation Res. *23*, 527 (1964)
37. Moore, P. W.: Rev. Pure and Appl. Chem. *20*, 139 (1970)
38. Kobayashi, Y.: J. Polymer Sci. *51*, 359 (1961)
39. Huang, R. Y-M., Immergut, B., Immergut, E. H., Rapson, W. H.: J. Polymer Sci. A. *1*, 1257 (1963)
40. Arthur, J. C., Diagle, D. J.: Textile Res. J. *34*, 653 (1964)
41. Teszler, O., Rutherford, H. A.: Nucl. Eng. Sci. Conf. Chicago, Preprint 48, V. (Amer. Inst. Chem. Eng.) 1958
42. Lawton, J. E., Bellamy, W. D., Hungate, R. E., Bryant, M. P., Hall, E.: Science, *113*, 380 (1951)
43. Gilfillan, E. S., Linden, L.: Textile Res. J. *27*, 87 (1957)
44. Charlesby, A. J.: J. Polymer Sci. *15*, 79 (1955), 15, 263 (1955)
45. Horio, M., Imamura, R., Mizukami, H.: Bull. Inst. Chem. Res. Kyoto Univ. *41*, 1, 17 (1963); through C.A., *60*, 9468
46. Arthur, J. C. Jr.: In Energy Transfer in Radiation Processes (Phillips, G. O., ed.) Amsterdam: Elsevier 1966, P. 29
47. Arthur, J. C. Jr., Mares, T.: J. Appl. Polym. Sci. *9*, 2581 (1965)
48. Arthur, J. C. Jr., Mares, T., George, M.: Text. Res. J. *35*, 1116 (1965)
49. Sarkar, I. M., Arthur, J. C. Jr., George, M.: Text. Res. J. *38*, 1145 (1968)
50. Chidambareswaran, P. K., Sundaram, V., Prakash, J., Verma, N. C., Singh, B. B.: C.T.R.L. Publ. No. 25 1971 also J. Polymer Sci. (A-12) *9*, 2651 (1971)
51. Guthrie, J. T., Huglin, M. B., Phillips, G. O.: European Polymer. J., *8*, 747 (1972)
52. Singh, S., Hinojosa, O., Arthur, J. C. Jr.: J. Appl. Polym. Sci., *14*, 1591 (1970)
53. Campbell, D., Williams, J. L., Stannett, V. T.: J. Polymer Sci. (A-1) *7*, 429 (1969)
54. Chapiro, A.: Radiat. Phys. Chems., *9*, 55 (1977)
55. Ang. C. H., Davis, N. P., Garnett, J. L., Naguyen, T. Y.: Int. J. Radiat. Phys. Chem. *4–6*, 831 (1977)
56. Tabata, Y.: Radiat. Phys. Chem., *9*, 31 (1977)
57. Winogradoff, N. N.: Nature *165*, 72 (1950)
58. Arthur, J. C. Jr.: J. Macromol. Sci. Chem. A. *4*, 1057 (1970)
59. Blouin, F. A., Arthur, J. C. Jr.: Textile Res. J. *28*, 198 (1958)
60. Blouin, F. A., Arthur, J. C. Jr.: Paper presented at the 136th National Meeting of the Amer. Chem. Soc., Atlantic City, N.Y. Sept. 13th–18th 1959
61. Arthur, J. C. Jr., Blouin, F. A., Demint, R. J.: Am. Dystuff Reptr. *49*, 383 (1960)
62. Guthrie, J. T., Huglin, M. B., Phillips, G. O.: J. Appl. Polym. Sci. *16*, 1017 (1972)
63. Blouin, F. A., Arthur, J. C. Jr., Orr, R. S., Ott, V. J.: Textile Res. J. *31*, 597 (1961)
64. Gilfillan, E. S., Linden, L.: Textile Res. J. *27*, 87 (1957)
65. Pan, H. P., Proctor, &. E., Goldblith, S. A., Morgan, H. M., Naar, R. Z.: TextileRes. J. *29*, 415 (1959)
66. Leavitt, F. C.: J. Polymer Sci., *45*, 536 (1960)
67. Mares, T., Arthur, J. C. Jr.: J. Polymer Sci. (B) *7*, 419 (1969)
68. Johanson, F., Back, E. L.: Svensk Pappstid: *69*, 199 (1966)
69. Arthur, J. C. Jr., Demint, R. J.: Textile Res. J. *32*, 108 (1962)
70. Porter, B. P., Tripp, V. W., de Gruy, I. V., Rollins, M. L.: Textile, Res. J. *30*, 510 (1960)
71. Guthrie, J. T., Percival, J.: Polymer *18*, 531 (1977)
72. Arthur, J. C. Jr., Hinojosa, O.: J. Polymer Sci., C. *36*, 53 (1971)
73. Panchenkov, G. M., Kaushanski, D. A., Gel'tser, G. A.: Dokl. Akad. Nauk S.S.S.R. *16*, 42 (1972)
74. Baugh, P. J., Hinojosa, O., Arthur, J. C. Jr.: J. Appl. Polym. Sci. *11*, 1139 (1967)
75. Arthur, J. C. Jr., Hinojosa, O., Tripp, V. W.: J. Appl. Polym. Sci. *13*, 1497 (1969)
76. Lawton, E. J., Bellamy, W. D., Hungate, R. E., Bryant, M. P., Hall, E.: Science *113*, 380 (1951)

77. Demint, R. J., Arthur J. C. Jr.: Textile Res. J. *29*, 276 (1959)
78. Arthur, J. C. Jr.: Advances in Chem. Series No. 91 (1969) P. 574
79. Ueno, T., Yamauchi, T., Imamura, R.: Kami Pa Gi-yoshi *25*, 242 (1971)
80. Imamura, R., Ueno, T., Murakami, K.: Bull. Inst. Chem. Res. Kyoto Univ. *50*, 51 (1972)
81. Imamura, R., Ueno, T.: Kami Pa Gi-yoshi *25*, 121 (1971)
82. Ueno, T. Murakami, M., Imamura, R.: Kami Pa. Gi-yoshi *26*, 164 (1972)
83. Irklei, V. M., Alexsandrovich, I. F., Inozemtseva, S. I., Goldinets, N. L., Nosov, V. E.: Khim. Volokna *2*, 40 (1978)
84. Kunz, N. D., Grainer, J. L., Kelly, J. L.: Nucl. Technol. *16*, 556 (1972)
85. Florin, R. E., Wall, L. A.: J. Polymer Sci. *1*, 1163 (1963)
86. Abraham, R. J., Whiffen, D. H.: Trans. Faraday Soc. *54*, 1291 (1958)
87. Park, G. S., Ward, J. C.: Nature *202*, 389 (1964)
88. Guthrie, J. T.: Polymer *16*, 134 (1975)
89. Guthrie, J. T.: Ph. D. Thesis, Univ. of Salford U.K. 1971
90. Glegg, R. F., Kertesz, Z. I.: J. Polymer Sci. *26*, 289 (1957)
91. Arthur, J. C. Jr., Blouin, F. A.: Textile Res. J. *34*, 733 (1964)
92. Sakurada, I., Okada, T., Kaji, K.: J. Polymer Sci., C. *37*, 1 (1972)
93. Urquhart, A. R.: In Recent Advances in the Chemistry of Cellulose and Starch (J. Honeyman, ed.) New York: Interscience 1959, P. 240
94. Mann, J., Marrinan, H. J.: Trans. Faraday Soc. *52*, 481 (1956)
95. King, T., Quellet, C.: Can. J. Res. *14 B*, 444 (1936).
96. Valentine, L.: J. Polymer Sci. *27*, 313 (1958)
97. Dilli, S., Garnett, J. L., Martin, E. C., Phuoc, D. H.: J. Polymer Sci. C. *37*, 57 (1972)
98. Dilli, S., Ernst, I. T., Garnett, J. L.: Australian J. Chem. *20*, 911 (1967)
99. Rosenberg, B.: J. Chem. Phys. 36, 816 (1962)
100. Calkins, C. R.: Tappi. *33*, 278 (1950)
101. Dakin, T. W.: Natl. Res. Council, Div. of Eng. Ind. Res., Ann. Rep. 92 (1950)
102. Arthur, J. C., Jr.: in Cellulose and Cellulose Derivatives (Bikales, N. M. and Segal, L. eds.), Vol. 5. New York: Wiley-Interscience 1971, P. 937
103. Hinojosa, O., Makamura, Y., Arthur, J. C. Jr.: J. Polymer Sci. C *37*, 27 (1972)
104. Howsmon, J. A., Sission, S. W.: In Cellulose and Cellulose Derivatives (Ott, E., Spurlin, H. M., Grafflin, M. W. eds.) 2nd Edn., Part 1, New York: Interscience 1954, P. 231
105. Nakamura, Y., Hinojosa, O., Arthur, J. C. Jr.: J. Polymer Sci., Ç. *37*, 47 (1972)
106. Collins, M. A.: Nature, *193*, 1061 (1962)
107. Bailey, A. J., Barker, S. A., Brimacombe, J. S., Pooley, D., Spence, D. H.: Nature, *190*, 259 (1961)
108. Ueda, H., Kuri, Z., Shida, S.: J. Chem. Phys., *35*, 2145 (1961)
109. Williams, D., Geusic, J. D., Wolfrom, M. L., McCabe, L. J.: Proc. Natl. Acad. Sci., U.S. *44*, 1128 (1958)
110. Williams, D., Schmidt, B., Wolfrom, M. L., Michelakis, A., McCabe, L. J.: Proc. Nat'l. Acad. Sci., U.S. *45*, 1744 (1959)
111. Arthur, J. C. Jr., Mares, T., Hinojosa, O.: Textive Res. J. *36*, 630 (1966)
112. Neubacher, H.: Biophysic, *6*, 161 (1969)
113. Kesting, R. E., Stannett, V.: Makromol Chem. *55*, 1 (1962)
114. Okamura, S., Iwasaki, T., Kobayashi, Y., Hayashi, K.: Paper presented at the International Atomic Energy Conference on the Application of Large Radiation Sources in Industry Specially to Chemical Processes. Warsaw, Poland Sept. 1959
115. Sakurada, I., Okada, T., Akiyama, Y.: Doitai to Hoshasen *3*, 35 (1960)
116. Sakurada, I., Okada, T., Kugo, E.: Doitai to Hoshasen *21*, 297 (1959).
117. Sakurada, I., Okada, T., Kugo, E.: Doitai to Hoshasen *2*, 306 (1959)
118. Sakurada, I., Okada, T., Kugo E.: Doitai to Hoshasen *2*, 316 (1959)
119. Sakurada, I., Okada, T., Uchida, M., Kugo, E.: Doitai to Hoshasen *2*, 581 (1959)
120. Chapiro, A., Stannett, V.: Intern. J. Appl. Radiation Isotopes *S*, 164 (1960)
121. Shinohara, Y.: Kobunshi Kagaku *17*, 197 (1960)
122. Kobayashi, Y.: J. Polymer Sci. *51*, 359 (1961)
123. Arthur, J. C. Jr., Demint, R. J., McSherry, W. F., Jugens, J. F.: Text. Research J. *29*, 759 (1959)

124. Usmanov, Kh. U., Aikhodzhaev, B. I., Azizov, V.O.: Vysokomolekulyarnye Soedineniya *1*, 1570 (1959)
125. Hendricks, R. W.: Canad. Pat *644*, 643
126. Arthur, J. C. Jr., Blouin, F. A.: Am. Dyestuff Reptr. *51*, 1024 (1962)
127. Arthur, J. C. Jr., Blouin, F. A.: U.S. At. Energy Document, TID-7643, 319 (1962)
128. Arthur, J. C. Jr., Demint, R. J.: Text. Research J., *30*, 505 (1960)
129. Arthur, J. C. Jr., Demint, R. J.: Text. Research J. *31*, 988 (1961)
130. Arthur, J. C. Jr., Demint, R. J.: U.S. Pat. *3*, 109, 798 (1963)
131. Arthur, J. C. Jr., Demint, R. J.: U.S. Pat. *3*, 157, 460 (1964)
132. Arthur, J. C. Jr., Grant, J. N.: Text Research J. *36*, 934 (1966)
133. Arthur, J. C. Jr., Markezich, A. R., McSherry, W. F.: Text. Research J. *33*, 896 (1963)
134. Blouin, F. A., Arthur, J. C. Jr.: Text. Research J. *33*, 727 (1963)
135. Blouin, F. A., Morriss, N. J., Arthur, J. C. Jr.: Text. Research J. *36*, 309 (1966)
136. Arthur, J. C. Jr., Blouin, F. A.: J. Appl. Polymer Sci. *8*, 2813 (1964)
137. Demint, R. J., Arthur, J. C. Jr., McSherry, W. F.: Text Research J. *31*, 821 (1961)
138. Ramalingham, K. V., Werezak, G. N., Hodgkin, J. W.: J. Polymer Sci. (C) *2*, 153 (1963)
139. Wellons, J. D., Stannett, V.: J. Polymer Sci. (A) *3*, 847 (1965)
140. Putman, E. S.: J. Polymer Sci. (C) *2*, 377 (1963)
141. Majumdar, S. K., Rapson, W. H.: Text. Res. J. *34*, 1007 (1964)
142. Rapson, W. H.: Amer. Chem. Soc. 138th Meeting. New York, N.Y. Sept. 14th 1960
143. Huang, R. Y-M.: J. Polymer Sci. *51*, 368 (1961)
144. Huang, R. Y-M., Rapson, W. H.: J. Polymer Sci., C *2*, 169 (1963)
145. Kenaga, D. L., Fennessey, J. P., Stannett, V. T.: Forest Prod. J. *12*, 161 (1962)
146. Demint, R. J., Arthur, J. C. Jr., Markezich, A. R., McSherry, W. F.: Textile Res. J. *32*, 918 (1962)
147. Odian, G., Kruse, R. L., Kho, J. H.: J. Polymer Sci. (A-1) *9*, 71 (1971)
148. Detrick, C. A., Kelly, J. L.: Nucl. Appl. Technol. *7*, 472 (1969)
149. Guthrie, J. T., Haq, Z.: Polymer *15*, 133 (1974)
150. Dilli, S., Garnett, J. L.: Austral. J. Chem. *23*, 1767 (1970)
151. Sakurada, I.: Cellulose Chem. Technol., *6*, 35 (1972)
152. Vlagiv, I., Petrus, I., Niculescu, E.: Rev. Roum. Chim. *15*, 1077 (1970)
153. Sakurada, I., Ikada, Y., Nishizaki, Y.: J. Polymer Sci., C *37*, 265 (1972)
154. Dillı, S. et al.: J. Macromol. Sci. (A) *6*, 719 (1972)
155. Dilli, S., Garnett, J. L.: Austral. J. Chem. *21*, 1827 (1968)
156. Garnett, J. L., Martin, E. C.: J. Macromol. Sci., A. *4*, 1193 (1970)
157. Dilli, S., Garnett, J. L.: Austral. J. Chem. *23*, 1163 (1970)
158. Dilli, S., Garnett, J. L.: Austral. J. Chem. *21*, 397 (1968)
159. Dilli, S., Garnett, J. L.: J. Appl. Polym. Sci. *11*, 839 (1967)
160. Dilli, S., Garnett, J. L.: J. Appl. Polym. Sci. *11*, 859 (1967)
161. Byrne, G. A., Arthur, J. C. Jr.: J. Appl. Polym. Sci. *14*, 3093 (1970)
162. Mares, T., Arthur, J. C. Jr.: J. Polymer Sci. C. *37*, 349 (1972)
163. Das Gupta, S.: J. Polymer Sci. C. *37*, 333 (1972)
164. Guthrie, J. T., Huglin, M. B., Phillips, G. O.: Int. Symp. on Macromolecules, Preprints Vol. 1, Paper 11, 26 Helsinki 1972
165. Guthrie, J. T., Huglin, M. B., Phillips, G. O.: Polymer *18*, 521 (1977)
166. Campbell, D., Williams, J. L., Stannett, V. T.: J. Polymer. Sci., A. *17*, 429 (1969)
167. Dilli, S., Garnett, J. L., Phuoc, D. H.: J. Polymer Sci., B. *11*, 711 (1973)
168. Garnett, J. L., Phuoc, D. H., Airey, P. L., Sangster, D. F.: Austral. J. Chem. *29*, 1459 (1976)
169. Garnett, J. L., Phuoc, D. H.: J. Macromol. Sci. Chem., A. *10*, 709 (1976)
170. Dilli, S., Garnett, J. L.: Austral. J. Chem. *24*, 981 (1971)
171. Sakurada, I.: Chem. Techol. *2*, 376 (1972)
172. Matsuzaki, K., Nakamura, S., Shindo, S.: J. Appl. Polym. Sci. *61*, 1339 (1972)
173. Yasukawa, T., Sasaki, Y., Murakami, K.: J. Appl. Polym. Sci. (A-1) *11*, 2547 (1973)
174. Hopfenberg, H. B., Stannett, V. T., Kimura, F., Rigney, P. T.: J. Appl. Polym. Sci. Polym. Symp. *13*, 139 (1970)

175. Borisevich, V. A., Rodzyalovskaya, L. A., Pinchuk, V. M.: Westi-Akad Navuk B. S.S.R. Ser. Khim. Navuk *13*, 138 (1976); through C.A. *85*, 79478 (1976)
176. Arthur, J. C. Jr.: J. Macromol. Sci. Chem. A. *10*, 653 (1976)
177. Stannett, V. T., Williams, J. L.: J. Macromol. Sci. Chem. A. *10*, 637 (1976)
178. Davis, N. P., Garnett, J. L., Urquhart, R. G.: J. Polymer Sci. *B 14*, 537 (1976)
179. Davis, N. P., Garnett, J. L., Urquhart, R. G.: J. Polymer. Sci. *C 55*, 287 (1976)
180. Murakami, K., Sasaki, Y., Yasukawa, T.: J. Polymer Sci. *13*, 17 (1975)
181. Kubota H., Ogiwara, Y., Yamoto, H.: J. Appl. Polym. Sci. *18*, 887 (1974)
182. Bentvelzen, J. M., Kimura-Teh, F., Hopfenberg, H. B., Stannett, V. T.: J. Appl. Polym. Sci. *17*, 809 (1973)
183. Kubota, H., Murata, Y., Ogiwara, Y.: J. Polymer Sci. *11*, 485 (1973)
184. Vandana, I. Ya., Zoldners, J., Surna, J.: Khim. Drev. *4*, 39 (1976)
185. Zoldners, Z., Paelglitis, D., Surna, J.: Khim Drev *5*, 115 (1976)
186. Zoldners, J., Pauls, R., Surna, J.: Khim Drev. *4*, 43 (1976)
187. Yoneho, T.: Kasen Geppo *29*, 60 (1976)
188. Ley, F. J.: J. Soc. Cosmet. Chem. *27*, 483 (1976)
189. Yasakawa, T., Sasaki, Y., Murakami, K.: J. Polymer Sci. (Chem.) *13*, 7 (1975)
190. Gamova, I. A., Natkina, L. N., Shiryaeva, G. V., Prishcepa, N. D.: Isv. Uchebn. Zaed. Lesn. Zh. *19*, 112 (1976)
191. Charlesby, A.: Int. J. Radiat. Phys. Chem. *9*, 17 (1977)
192. Kalontarov, I. Ya., Makhkamov, K. M., In, O. A., Smirnov, R. F.: Dokl. Akad. Nauk. Tadzh. S.S.R. *20*, 34 (1977); through CA *87*, 69992 (1977)
193. Ermolenki, I. N.: Vysokomol. Soed. B. *17*, 538 (1975)
194. Cunanan, S. A., Bunoan, L. S., Verceluz, F. P., Azucena, E. A.: Phillippine A.E.C. Report 1976 PAEC (D)-76032. Available through INIS Atomindex 8. Abst. No. 299354 (1977)
195. Usmanov, Kh., Yunusov, M., Azizov, U., Sadykov, M.: Radiats. Graftpolimerizatsiya 26–32 (1976)
196. Azizov, U., Mirkamilov, I., Usmanov, Kh.: Radiats. Graftpolimerizatstya 42–47 (1976)
197. Hinojosa, O., A.C.S. Symp. Ser. *49*, 46 (1977)
198. Gorskii, G. M., Zhonina, T. A.: Tsellyul., Bum Karton *11*, 7 (1977)
199. Meleshevich, A. P., Vishev, Yu. V., Gordienko, L. L., Richko, N. T., Yurganova, L. G., Voitsekhovskii, R. V.: Khim. Technol. (Kiev) *4*, 59 (1977)
200. Needles, H. L.: J. Consum. Prod. Flammability. *4*, 156 (1977)
201. Stannett, V. T., Walsh, W. K., Bittencourt, E., Liepens, R., Surles, J. R.: Appl. Polym. Symp. *31*, 201 (1977)
202. Lebedev, V. T., Suminov, S. I., Voronina, T. A.: Plast. Massy. *7*, 66 (1977)
203. Dziedziela, W., Wiesielowska, R.: Radiats. Graftpolymerizatsiya, 69 (1976)
204. Onetiu, M., Bacalu, M., Belcetianu, N.: Radiats. Graftpolimerizatsiya 301 (1976)
205. Mirkamilova, G., Sadykov, M., Azizov, U. A., Usmanov, Kh. U.: Radiats. Graftpolimerizatsiya 58 (1976)
206. Pesek, M., Jarkovsky, J., Pultar, F.: Radiats. Graftpolimerizatsiya 208 (1976)
207. Al-Siddique, F. R., Zaman. M. A.: Nucl. Sci. Appl. Ser. B *9*, 80 (1976)
208. Onetiu, M., Baicon, R., Andreikovich, G.: Radiats. Graftpolimerizatsiya 293 (1976)
209. Das Gupta, S.: Can. Spect. *10*, 9 (1965)
210. Sakurada, I., Okada, N., Hayatama, N.: AEC-tr-6316 *2*, 84 (1960)
211. Das Gupta, S.: Amer. Dyestuff Reptr. *55*, 211 (1966)
212. Sakurada, I., Okada, N., Kimura, F., Minegawa, N.: AEC-tr-6372 *3*, 103 (1961)
213. Usmanov, Kh. U., Aikhodzhaev, B. I., Azizov, U.: AEC-tr-6398 *1*, 386 (1961)
214. Sakurada, I., Sakaguchi, Y.: Sen-i-Gakkaishi *18*, 834 (1962)
215. Das Gupta, S.: Can. Textile J. *18*, 57 (1964)
216. Sakura, I., Okada, K., Akiyama, Y.: AEC-tr-6231 *1*, 102 (1958–1959)
217. Huang, R. Y-M., Immergut, B., Immergut, H. E., Rapson, W. H.: J. Polymer Sci. A *1*, 1257 (1963)
218. Ekstrom, A., Garnett, J. L.: J. Amer. Chem. Soc. *86*, 5028 (1964)
219. Ekstrom, A., Garnett, J. L.: J. Phys. Chem. *70*, 324 (1966)
220. Sakurada, I., Okada, N., Kimura, F., Minegawa, N.: AEC-tr-6372 *3*, 109 (1961)

221. Sakurada, I., Okada, N., Kimura, F.: Sen-i-Gakkaishi *18*, 692 (1962)
222. Sakurada, I., Okada, T., Hatakeyama, S., Kimura, F.: J. Polymer Sci. *C 4*, 1233 (1964)
223. Moore, P. W.: AAEC/TM540 (1970)
224. Staudinger, H., Eicher, T.: Makromol. Chem. *10*, 254 (1953)
225. Krässig, H., Schrott, E.: Makromol. Chem. *13*, 179 (1954)
226. Miyama, H.: J. Chem. Soc. Japan (Pure Chem. Sect.) *76*, 658 (1955)
227. Miyama, H.: Bull. Chem. Soc. Japan *2*, 711 (1956)
228. Chen, C. S. H., Stamm, R. F.: J. Polymer Sci. *58*, 369 (1962)
229. Chapiro, A.: J. Polymer Sci. *29*, 321 (1958)
230. Sakurada, I., Okada, N., Hatakeyama, N.: AEC-tr-6372 *3*, 83 (1961)
231. Amamiya, A., Ogawa, S.: AEC-tr-6372 *3*, 593 (1961)
232. Sakurada, I., Okada, N., Minegawa, N.: AEC-tr-6372 *3*, 117 (1961)
233. Immergut, E. H.: In Enclopaedia of Polymer Science and Technology, (Mark, H. E., Gaylord, N. G., Bikales, N. M., eds.) Vol. 3, Interscience, New York, N.Y. 91965, P. 242
234. Selected Constants-Radiolytic Yields, Tables des Constantes Vol. 13, I.U.P.A.C., New York: 1963 Pergamon Press
235. Chapiro, A.: Radiation Chemistry of Polymeric Systems, New York: Interscience Wiley 1962, P. 566
236. Centre Nationale de la Recherche Scientifique, British Pat. 809, 838 (1959)
237. Dupont de Nemours and Company, British Pat. 845, 890 (1960)
238. Smith, R. R.: (to Box Plastics, Ltd.) British Pat. 862, 610 (1961)
239. Smith, R. R.: (to Box Plastics, Ltd.) British Pat. 863, 211 (1961)
240. Chapiro, A.: J. Polymer Sci. *23*, 377 (1957)
241. Chapiro, A.: J. Polymer Sci. *34*, 439 (1959)
242. Ballantine, S., Glines, A., Adler, G., Metz, D. J.: J. Polymer Sci. *34*, 419 (1959)
243. Shinohara, Y., Tomioka, K.: J. Polymer Sci. *44*, 195 (1960)
244. Sakurada, I., Okada, T., Kimura, F., Minakawa, N.: Ann. Rept. Japanese Assoc. Radiat. Chem. *2*, 62 (1960)
245. Atlas, S. M., Mark, H.: Cellulose Chem. Technol. *3*, 325 (1969)
246. Kesting, R. E., Stannett, V. T.: Makromol Chem. *55*, 1 (1962)
247. Battaerd, H. A. J., Tregear, G. W.: Graft Copolymers, New York: Interscience 1967
248. Bryant, G. M., Walker, A. T.: Textile Res. J. *29*, 211 (1959)
249. Okada, T., Kohdera, K., Sakurada, I.: Ann. Rept. Osaka Lab. Radiat. Chem. *2*, 69 (1969)
250. Majumdar, S. K., Rapson, W. H.: Textile Res. J. *34*, 1015 (1964)
251. Kesting, R. E., Stannett, V. T.: Makromol. Chem. *65*, 248 (1963)
252. Imrisova, D., Maryska, S.: J. Appl. Polym. Sci., *12*, 2007 (1968)
253. Imrisova, D., Maryska, S.: J. Appl. Polym. Sci. *11*, 901 (1967)
254. Nakamura, Y., Hinojosa, O., Arthur, J. C. Jr.: J. Appl. Polym. Sci. *13*, 2633 (1969)
255. Nakamura, Y., Hinojosa, O., Arthur, J. C. Jr.: J. Appl. Polym. Sci. *14*, 789 (1970)
256. Nakamura, Y., Hinojosa, O., Arthur, J. C. Jr.: J. Appl. Polym. Sci. *15*, 391 (1971)
257. Hinojosa, O., Arthur, J. C. Jr.: J. Polymer Sci. *B 10*, 161 (1972)
258. Abraham, R. J., Melville, H. W., Overnall, D. W., Whiffen, D. H.: Trans. Faraday Soc. *54*, 1133 (1958)
259. Harris, J. A., Arthur, J. C. Jr.: J. Polymer Sci. *B 14*, 3113 (1970)
260. Okada, T., Kohdera, K., Sakurada, I.: Jap. Atom. Energy Res. Inst. Osaka Rep. JAERI-5022, P. 59
261. Reine, A. H., Arthur, J. C. Jr., George, M.: J. Appl. Polym. Sci. *16*, 1893 (1972)
262. Ishanov, M. M., Nikonovich, G. V., Usmanov, Kh. U.: Vysokomol Soed. *B 11*, 763 (1969)
263. Yunusov, M. Yu., Azizov, U. A.: Uzb. Khim. Zh. *5*, 64 (1971)
264. Harris, J. A., Carra, J. H., DeGruy, I. V., Arthur, J. C. Jr.: Textile Res. J. *42*, 14 (1972)
265. Yul'chibaeva, S. G., Yusupalaev, R. M., Usmanov, Kh. U.: Uzb. Khim. Zh. *7*, 60 (1973)
266. Yusupalaev, R. M.: Nauch. Trudy Tashkent, Univ., *403*, 75 (1971)
267. Goynes, W. R., Harris, J. A.: J. Polymer Sci., C *37*, 277 (1972)
268. Sakurada, I.: Chem. Technol. *2*, 376 (1972)
269. Matsuzaki, K., Kanai, T., Morita, N.: J. Appl. Polym. Sci. *16*, 15 (1972)

270. Williams, J. L., Woods, D. K., Stannett, V. T., Roldan, L. G., Sello, S. B., Stevens, C. V.: Textile Res. J. *43*, 205 (1973)
271. Matsuzaki, K., Nakamura, S., Shindo, S.: J. Appl. Polym. Sci. *16*, 1339 (1972)
272. Yasukawa, T., Sasaki, Y., Murakami, K.: J. Polymer Sci. (A-1) *11*, 2547 (1973)
273. Hayakawa, K., Yamakita, H., Kawase, K.: J. Polymer Sci. (A-1), *8*, 1227 (1970)
274. Chapiro, A., Foëx, M., Jendrychowska-Bonamour, A. M.: Europ. Polymer J. *13*, 637 (1977)
275. Kuri, Z., Fujiwara, Y., Ueda, H., Shida, S.: J. Chem. Phys. *33*, 1884 (1960)
276. Dilli, S., Garnett, J. L.: J. Polymer Sci. (A-1) *4*, 2323 (1966)
277. McClellan, W. R., Hoehn, H. H., Cripps, H. M., Muetterties, E. L., Howk, B. W.: J. Amer. Chem. Soc. *83*, 1601 (1961)
278. Garnett, J. L., Sollich-Baumgartner, W. A.: Adv. in Catalysis *16*, 95 (1966)
279. Garnett, J. L., Sollich-Baumgartner, W. A.: J. Phys. Chem. *69*, 1850 (1965)
280. Chen, C. S. H., Stamm, R. F.: J. Polymer Sci. *58*, 369 (1962)
281. Arifov, V. A., Klein, G. A., Filipov, A. N., Amirova, N. Yu., Adikhodzhaeva, G. A., Okun, J. S., Osipova, L. K.: Izv. Akad. Nauk. Uzbek. SSR. Ser. F12. Mat. Nauk. No. *4*, 59 (1960); Through C.A. *55*, 7848 (1961)
282. Okamura, S., Iwasaki, T., Kobayashi, Y., Hayashi, K.: I.A.E.A. Conf. Appl. Large Radiation Sources Ind. Spec. Chem. Proc. Warsaw (1959)
283. Arthur, J. C. Jr., Mares, T.: U.S. Pat., *3*, 443, 879 (1969); Abdel-Hay, F. I., Guthrie, J. T., Ryder, M.: Polymer Bulletin *1*, 509 (1979)
284. Choudhuri, B. K.: Labdev, Part A (Kanpur, India) *6*, 26 (1968)
285. Zhbankov, R. G., Zhdanovskii, V. Z., Kiselevskii, L. I., Snopko, V. N., Firsov, S. P.: Zh. Prikl. Spektrosk. *20*, 317 (1974)
286. Harris, J. A., Arthur, J. C. Jr., Carra, J. H.: J. Appl. Polym. Sci. *22*, 905 (1978)
287. Ogiwara, Y., Kubota, H.: J. Polymer Sci., Chem. *11*, 3243 (1973)
288. Baugh, P. J., Phillips, G. O., Worthington, N. J.: J. Soc. Dyers Colourists *85*, 241 (1969)
289. Baugh, P. J., Phillips, G. O., Worthington, N. J.: J. Soc. Dyers Colourists *86*, 19 (1970)
290. Virnik, R. B., Ershov, Y. A.: Khim Tekhnol. Proizvod. Tsellyul., 333 (1971)
291. Kudrna, S. K.: Vysokomol. Soedin. *7*, 557 (1965)
292. Kudrna, S. K.: Vysokomol. Soedin. *8*, 1823 (1966)
293. Kudrna, S. K.: Vysokomol. Soedin. *8*, 1828 (1966)
294. Daruwalla, E. H., Moonim, S. M., Arthur, J. C. Jr.: Textile Res. J. *45*, 592 (1972)
295. Akhmedova, Kh. R., Tulyaganov, M. M., Gafurov, T. G., Usmanov, Kh. U.: Dokl. Akad. Nauk. Uzb. SSR. *28*, 24 (1971)
296. Arthur, J. C. Jr., Hinojosa, O.: Appl. Polym. Symp. *26*, 147 (1975)
297. Bertoniere, N., Frankling, W. E., Rowland, S. P.: J. Appl. Polym. Sci. *15*, 1743 (1971)
298. Bertoniere, N., Franklin, W. E., Rowland, S. P.: Textile Res. J. *1*, 41 (1971)
299. Semchuk, R. D., Gordinskii, B. Y.: Nov. Poligr. Tekh. *73* (1971)
300. Portnoy, N. A., Nelson, M. C., Margavio, M. F., Arthur, J. C. Jr.: Textile Res. J. *44*, 449 (1974)
301. Arthur, J. C. Jr.: J. Macromol. Sci. Chem. A. *10*, 653 (1976)
302. Cremonesi, P., Focher, B., D'Angiuro, L.: Chim. Ind. (Milan) *54*, 871 (1972)
303. Focher, B., D'Angiuro, L., Cremonesi, P.: Cell. Chem. Technol. *6*, 277 (1972)
304. Arthur, J. C. Jr.: Sen'i To Kogyo *4*, 212 (1971)
305. Ichimura, S.: Japan. Pat. 72, 12560 (1968)
306. Lagache, M., Chateau, M., Pomey, J.: Compte. Rendu *251*, 2353 (1960)
307. Morimoto, S., Inoshita, K., Toda, T., Masai, N.: Japan. Pat. 7876 (1963)
308. Morimoto, S., Okada, T., Toda, T., Nakagawa, H.: Japan. Pat. *13*, 881 (1962)
309. Naito, M., Sakiyama, S., Kubota, M.: Japan. Pat. 7900 (1962)
310. Oster, G.: Brit. Pat. *856*, 884 (1960)
311. Reine, A. H., Arthur, J. C. Jr., Portnoy, N. A.: U.S. Pat. *3*, 926, 555 (1975)
312. Reine, A. H., Hinojosa, O., Arthur, J. C. Jr.: J. Appl. Polym. Sci. *17*, 3337 (1973)
313. Reine, A. H., Portnoy, N. A., Arthur, J. C. Jr.: Textile Res. J. *43*, 638 (1973)
314. Simionescu, C., Feldman, D., Oprea, S.: Rev. Roum. Chim. *9*, 65 (1964)
315. Tonami, H., Watamoto, H.: Nippon Kagaku Kaischi *4*, 789 (1974)
316. Symposium of Photochemistry in Relation to Textiles, J. Soc. Dyers Colourists *65*, 585 (1949)

317. Bridge, N. K.: J. Soc. Dyers Colourists *76*, 484 (1960)
318. Bamford, C. H., Dewar, M. J. S.: J. Soc. Dyers Colourists *65*, 674 (1949)
319. Rabek, J. F.: Photochem. and Photobiol. *7*, 5 (1968); also G. Scott in 'Singlet Oxygen' (Randby, B. and Rabek, J. F. eds.) New York: J. Wiley and Sons Inc. 1978, P. 230
320. Egerton, G. S.: Nature *204*, 1153 (1964)
321. Egerton, G. S.: J. Soc. Dyers Colourists *64*, 659 (1948)
322. Egerton, G. S.: J. Textile Inst. *39*, T305 (1948)
323. Egerton, G. S.: J. Soc. Dyers Colourists *63*, 161 (1947)
324. Egerton, G. S.: Amer. Dyestuff Reptr. *36*, 561 (1947)
325. Phillips, G. O., Arthur, J. C. Jr.: Textile Res. J., *34*, 47 (1964)
326. Egerton, G. S., Assaad, H. E. N., Uffindell, N. D.: J. Soc. Dyers Colourists *84*, 220 (1968)
327. Geacintov, N., Stannett, V. T., Abrahamson, E. W.: Makromol. Chem. *36*, 52 (1959)
328. Geacintov, N., Stannett, V. T., Abrahamson, E. W., Hermans, J. J.: J. Appl. Polymer Sci. *3*, 54 (1960); Allen, N. S., McKellar, J. F.: In Developments in Polymer Photochemistry (Allen, N. S., ed.) Appl. Sci. Publ. Ltd., London 1980, P. 191
329. Stannett, V. T., Hermans, J. J., Geacintov, N.: 138th Meeting, Amer. Chem. Soc., Cell. Division New York (1960)
330. Davies, A. K., Gee, G. A., McKellar, J. F., Phillips, G. O.: Chemistry and Industry 431 (1973)
331. Davies, A. K., Ford, R., Gee, G. A., McKellar, J. F., Phillips, G. O.: Chem. Commun. 373 (1972)
332. Egerton, G. S.: J. Soc. Dyers Colourists *65*, 679, 764 (1949)
333. Griffiths, J., Hawkins, C.: J. Soc. Dyers Colourists *89*, 173 (1973)
334. Griffiths, J., Hawkins, C.: Polymer *17*, 1113 (1976); Griffiths, J.: In Developments in Polymer Photochemistry (Allen, N.S., ed.) Appl. Sci. Publ. Ltd., London 1980, P. 145
335. Carlsson, D. J., Mendenhall, G. D., Suprunchuk, T., Wiles, D. M.: J. Amer. Chem. Soc. *94*, 8960 (1972)
336. Reine, A. H., Arthur, J. C. Jr.: Textile Res. J. *42*, 155 (1972)
337. Needles, H. L., Walsey, W. L.: Textile Res. J. *39*, 97 (1969)
338. Ogiwara, Y., Hayase, K., Kubota, H.: J. Appl. Polym. Sci. *23*, 1 (1979)
339. Kubota, H., Murata, Y., Ogiwara, Y.: J. Polymer Sci., Chem. *11*, 485 (1973); Hon, D. N-S., J. Polymer Sci., Chem. *17*, 441 (1979)
340. Kubota, H., Ogiwara, Y., Matsuzaki, K.: J. Polymer Sci., Chem. *12*, 2809 (1974); Arthur, J. C. Jr.: In Developments in Polymer Photochemistry (Allen, N. S., ed.) Appl. Sci. Publ. Ltd., London 1980, P. 69
341. Davis, N. P., Garnett, J. L., Urquhart, R. G.: J. Polymer Sci. C *55*, 287 (1976)
342. Burrows, H. D., Kemp, T. J.: Chem. Rev. *3*, 139 (1974)
343. Naite, N., Sakiyama, S., Kubota, M.: Jap. Pat. 7900 (1962)
344. Ledwith, A., Russell, R. J., Sutcliffe, L. H.: Proc. Roy. Soc. (London) *A 332*, 151 (1973)
345. Greatorex, D., Hill, R. J., Kemp, T. J., Stone, T. J.: J. Chem. Soc. Faraday 1 *68*, 2059 (1972)
346. Venkatarao, K., Santappa, M.: J. Polymer Sci. (A-1) *8*, 1785 (1970)
347. Venkatarao, K., Santappa, M.: J. Polymer Sci. (A-1) *8*, 3429 (1970)
348. Garnett, J. L.: In Cellulose Chemistry and Technology (Arthur, J. C. Jr., ed.) A.C.S. Symp. Series No. 48 A.C.S. Washington, D.C., P. 334 (1977)
349. Phillips, G. O., Baugh, P. J., McKellar, J. F., Von Sonntag, C.: In Cellulose Chemistry and Technology (Arthur, J. C. Jr., ed.) A.C.S. Symp. Series. No. 48, A.C.S. Washington, D.C., P. 313 (1977)
350. Nakamura, Y., Shimada, M.: In Cellulose Chemistry and Technology (Arthur, J. C. Jr., ed.) A.C.S. Symp. Series, No. 48, A.C.S., Washington, D.C., P. 298 (1977)
351. Stannett, V. T., Memetea, T.: J. Polymer Sci., C., Polymer Symposia *64*, 57 (1978)

4. Grafting by Chemical Activation of Cellulose

4.1. Introduction

Activation of cellulose, that is, introduction of active centers into the cellulose molecule, has been successfully achieved by a number of chemical means. By and large, the mechanisms proposed for graft initiation and chain growth have been either free radical or ionically based, the radical methods being the more common. Most of the recent work in this field has featured reactions in special systems where oxidation of cellulose precedes grafting via a free radical intermediate[1]. In the homolytic oxidation of cellulose, a hydrogen atom is abstracted, not by a growing chain, but by an independent chemical reaction. The formation of active centers in cellulose can also be achieved by the influence of certain metallic halides such as BF_3, i.e., cationic initiation, or in the presence of a variety of bases such as sodium alkoxides, i.e., anionic initiation. In this section acrylic and vinyl graft copolymerization by chemical activation of cellulose, by introduction of free radicals, and by ionic mechanisms are independently discussed.

4.2. Free-Radical Initiation

Free radicals capable of initiating graft polymerization may be introduced into cellulose by various chemical methods. In addition to well-established initiators such as peroxides and persulfates, the Ce^{IV} salt system has been widely exploited. Other chemical methods of initiating grafting onto cellulose require chemical modification of the cellulosic substrates so that they can participate in initiation reactions, for example, through the introduction of diazonium groups or unsaturated side chains. Such modifications may be permanent or temporary.

4.2.1. Persulfate Initiation

Kolthoff and Miller[2] have shown that when heated, persulfate ions in aqueous solution decompose to produce sulfate ion-radicals:

$$S_2O_8^{2-} \rightarrow 2\,SO_4^{\overline{\cdot}}, \quad \text{rate constant } k_0 \tag{111}$$

This is a first order reaction. The sulfate radicals may react with water to produce hydroxyl radicals and finally oxygen:

$$SO_4^{\overline{\cdot}} + H_2O \rightarrow HSO_4^- + HO^{\cdot} \tag{112}$$

$$2\,HO^{\cdot} \rightarrow H_2O_2 \rightarrow H_2O + {}^1/_2\,O_2 \tag{113}$$

In addition, a decomposition takes place that does not produce radicals, but the rate of this second process is very small compared with that of the first when the pH = 4. Vinyl polymerization may thus be initiated by either the $SO_4^{\cdot -}$ radicals or the $HO^{\cdot}$ radicals. That $SO_4^{\cdot -}$ radicals take part in the initiation is shown by the fact that sulfur is present in the polymer[3].

It has been reported that methanol accelerates the decomposition of persulfate salts[45]. Bartlett et al.[5] attributed this to a reaction between persulfate ions and alcohol radicals (Reaction 115).

$$SO_4^{\cdot -} + Alc^{\cdot} \rightarrow HSO_4^- + Alc^{\cdot}; \quad \text{rate constant } k_1 \tag{114}$$

$$Alc^{\cdot} + S_2O_8^{2-} \rightarrow \text{Aldehyde} + HSO_4^- + SO_4^{\cdot -} \tag{115}$$

$$2\, SO_4^{\cdot -} \rightarrow \text{End products; rate constant, } \lambda\,. \tag{116}$$

In the absence of reducing agents, the $SO_4^{\cdot -}$ radicals disappear (according to reaction 116), whereas in the presence of an alcohol, the initiation of free radicals appears to be a continuous process.

In the presence of a vinyl monomer, the rate of persulfate decomposition is reduced to the much lower value observed in the absence of methanol. This suggests that a reaction occurs between persulfate ions and alcohol radicals[6]. In this case, the $SO_4^{\cdot -}$ radicals, generated by reaction (111) react with the monomer:

$$SO_4^{\cdot -} + M \rightarrow P^{\cdot}; \quad \text{rate constant, } k_i \tag{117}$$

where M represents a vinyl monomer and $P^{\cdot}$ is the polymer radical.

Obviously, then, reaction (114) and, therefore, also reaction (115) are suppressed. It is also possible, however, that only reaction (115) is suppressed and that the alcohol radicals react with the monomer. Thus, vinyl polymerization initiated by persulfate in the presence of cellulose would be expected to lead to grafting onto cellulose.

Some experiments were done with cotton in acrylonitrile vapor by Haydel et al.[7], but these researchers did not investigate whether real grafting occurred and referred to the process as deposition in (and coating of) the cotton. Hayes[8] reported that grafting took place when the (emulsion) polymerization was initiated by persulfate in the presence of highly crosslinked polybutadiene and attributed this result to chain transfer. Although chain transfer from growing polymer chains to cellulose molecules cannot be ruled out, reaction (114) between $SO_4^{\cdot -}$ radicals and methanol suggests that a similar reaction occurs between the $SO_4^{\cdot -}$ radicals and cellulose to form cellulosic radicals directly. The direct formation of cellulose radicals may also be substantiated by the fact that cellulose is a very poor transfer agent[9], as discussed earlier (see Chapter 2). At any rate, however, the cellulose radical formed could then initiate polymerization and form a graft.

Based on the reactions considered by both Bartlett and Kolthoff[5,6], the efficiency of grafting of a vinyl monomer to cellulose using persulfate ions is determined by the relative rates of grafting and homopolymer formation. To achieve maximum grafting, it is necessary that the reaction between the $SO_4^{\cdot -}$ radicals and cellulose take place to an appreciable extent even when monomer is present and that the cellulose radicals formed take part in the initiation of the polymerization. Consequently, the ultimate graft formation will depend on the conditions of the experiment.

In trying to achieve favorable experimental conditions for grafting, Sankalia et al.[10] soaked gel cellophane in an aqueous solution of persulfate and subsequently brought the material into contact with pure monomer. Since the monomer is a poor solvent for the persulfate, only small amounts of catalyst will diffuse from the gel into the monomer phase, so that little, if any, homopolymer is formed outside the cellophane. Since the concentration of cellulose in the gel is high, the rate of cellulose radical formation is large, so that the monomer which diffuses into the gel finds a high concentration of cellulose radicals to react with.

These authors performed all experiments in a nitrogen atmosphere. After completion of the reaction, the film was thoroughly extracted with a solvent for the homopolymer, then dried in a vacuum oven at 60 °C for 48 hrs and weighed. It is possible that part of the weight increase observed is due to homopolymer that could not be removed by solvent extraction. However, when the product containing 80 g poly(vinyl acetate) per 100 g dry cellulose was hydrolyzed with 0.25 M alcoholic sodium hydroxide solution at about 50 °C for 20 hr, under nitrogen to minimize cellulose degradation, the poly(vinyl acetate) chain was converted to poly(vinyl alcohol), and the product was completely soluble in cupriethylenediamine. Acidification of this solution with 4% HCl resulted in a precipitate, which was washed with Rochelle salt solution and then water. The infrared spectrum of this product (KBr pellet) showed absorption in the region 12–13 μ, due to the presence of poly(vinyl alcohol) in the precipitate. This absorption is absent when the same procedure is followed with a physical mixture of 1 g dry cellophane and 0.8 g poly(vinyl alcohol), indicating that at least part of the weight increase was due to grafting.

In order to shed more insight on the mechanism of persulfate-initiated grafting onto cellulose, studies were made on model compounds, such as serine methyl ester, pinacol, and D-glucose[10]. Table 37 shows the rate of decomposition of persulfate in the presence and absence of pinacol and acrylonitrile. Apparently,

Table 37. Rate of decomposition (% per hour) of 0.01 M persulfate in various mixtures at 50 °C, pH = 7. Adapted from Ref.[10], p. 2252, by courtesy of the National Research Council of Canada

Pinacol (mol/l)	Acrylonitrille (mol/l)	Rate (% per hr)
0	0	0.79
0.05	0	1.17
0.10	0	1.30
0	0.3	0.72
0.05	0.3	0.71
0.10	0.3	0.67
~0.30	0.3	0.65

the presence of pinacol increases the rate of decomposition when no monomer is present. On the other hand, the rate of persulfate decomposition is reduced to the original value or even slightly less when the monomer is added with the pinacol.

Fig. 32 shows the monomer conversion as a function of time under the conditions listed in Table 37. The inset in Figure 32 gives the amount of polymer formed after 4 hours reaction time as a function of pinacol content, showing that the rate of polymerization first increases but subsequently decreases with increasing pinacol concentration.

That the rate of persulfate decomposition is increased by the presence of pinacol is substantiated by reaction (115). This effect is not very significant and is suppressed by acrylonitrile. Reaction (114) is not suppressed, since addition of alcohol causes an enhancement in the rate of polymerization. Free radicals are produced from the alcohol and these alcohol radicals progressively participate in the initiation step,

$$\mathrm{Alc}^{\cdot} + \mathrm{M} \rightarrow \mathrm{P}^{\cdot}\,; \quad \text{rate constant } k_i' \,. \tag{118}$$

However, at higher radical concentrations, destruction of these alcohol radicals would be expected to occur,

$$2\,\mathrm{Alc}^{\cdot} \rightarrow \text{end products, rate constant } \mu \,. \tag{119}$$

As a result, the rate of polymerization decreases again. This would account for the lower rate of polymerization at high pinacol content. This argument can be given in more quantitative terms as follows

$$\mathrm{P}^{\cdot} + \mathrm{P}^{\cdot} \rightarrow \text{polymer}; \quad \text{rate constant } k_t \,. \tag{120}$$

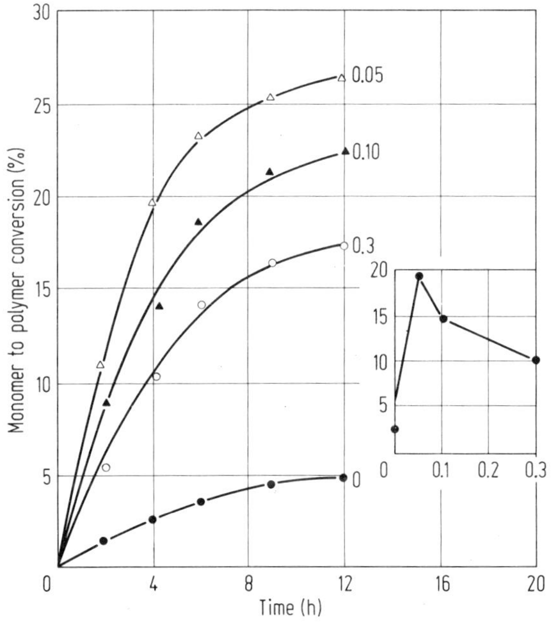

Fig. 32. Percent conversion of monomer as a function of time for pinacol concentrations 0→0.05→0.2 and about 0.3 M. Inset: % conversion after 4 hours as a function of pinacol concentration. Reprinted from Ref.[10], p. 2252, by courtesy of the National Council of Canada.

If we add the termination step, the kinetics are described by the reactions (111 to 120). Of these reactions (112) and (113) will not be considered. Also, reaction (115) will be omitted due to its suppression in the presence of monomer. Then, if C is the concentration of persulfate, A that of the alcohol, M that of monomer, and if x, y, and r represent the concentrations of $SO_4^{\cdot -}$ radicals, alcohol radicals, and polymer radicals respectively, we have

$$2k_0C - 2\lambda x^2 - k_1 Ax - k_i M_x = 0 , \tag{121}$$

$$-2\mu y^2 + k_1 Ax - k_i' My = 0 , \tag{122}$$

$$K_i Mx + k_i' My - 2k_t r^2 = 0 , \tag{123}$$

$$-dM/dt = M(k_i x + k_i' y + k_p r) , \tag{124}$$

where k_p is the propagation rate constant. The Eqs. (121–123) give the expressions for dx/dt, dy/dt, and dr/dt respectively. They have been equated to zero on the (usual) assumption that 'quasi steady state' conditions prevail. Eliminating r from Eqs. (123) and (124), it is seen that the rate of polymerization for a given value of M is determined by the quantity

$$R = k_i x + k_i' y , \tag{125}$$

and thus

$$-dM/Mdt = R + CR^{1/2} ,$$

where

$$C = k_p(M/2k_t)^{1/2} .$$

It will be shown that R increases on addition of alcohol but may decrease again on the addition of more alcohol. To this end, Eqs. (121) and (122) are differentiated with respect to A at constant C and M. From the equations obtained, one can solve for dx/dA and dy/dA, with the result:

$$dx/dA = -(k_1 x/N)(4\mu y + k_i' M) , \tag{126}$$

$$dy/dA = (k_1 x/N)(4\lambda x + k_i M) , \tag{127}$$

$$N = (4\mu y + k_i' M)(4\lambda x + k_i M + k_1 A) .$$

This shows that dx/dA is negative and dy/dA positive for all values of A. Using the definition of R given in Eq. (125), it follows from Eqs. (126) and (127) that:

$$dR/dA = (4k_1 x/N)(\lambda k_i' x - \mu k_i y) . \tag{128}$$

At low concentrations of alcohol, y is small, so that (dR/dA) is positive. However, on increasing A, we know that x decreases and y increases, so that dR/dA will become smaller and may eventually become negative.

Kilkarni et al.[11] grafted polyacrylonitrile onto cotton fabrics using the persulfate-thiosulfate redox system as initiator for grafting under a variety of conditions. They reported that in the absence of any reducing agents, the grafting reaction is accompanied by a relatively long induction period, with homopolymer formation visible to a considerable extent. The induction period as well as homopolymerization were greatly reduced when a reducing agent, such as potassium thiosulfate, was added to the grafting reaction. The rate of grafting was also higher in the presence of reducing agent than in its absence. A further increase in the rate of grafting could be achieved by including copper sulfate traces in the reaction mixture. This may be seen in Figure 33. These authors observed also that when the fabrics were first padded with the reducing solution, polymerization was initiated mainly in the fabrics, with comparatively lesser quantities of homopolymer visible. The electrografting of acrylonitrile onto cellophane pellicles in the presence of Fe^{III} sulfate and potassium peroxy-disulfate has been reported[12]. Fe^{II} ions formed at the cathode participate in the mechanism of electrografting. It was also shown that much homopolymer was formed.

Gaylord and Tomono[13] have examined the preparation of alternating copolymers in aqueous media by the polymerization of a mixture of a donor and an acceptor monomer, for example, styrene and acrylonitrile either in the presence of a complexing agent, such as zinc chloride or magnesium chloride[14] or in the presence of cellulose[15]. It has been shown that alternating copolymers and graft copolymers are produced in the presence of cellulose, in the absence[15,17] as well as in the presence of a metal halide[16,18,19]. The formation of alternating copolymers in the presence of cellulose and the absence of a metal halide is assumed by Gaylord and Tomono to involve the complexation of the acceptor monomer by cellulose, followed by the generation of an alignment of comonomer charge-transfer complexes on the cellulose matrix[20,21]. Grafted as well as ungrafted alternating copolymers are formed in the presence of a water soluble catalyst, provided the water and cellulose contents of the system are maintained within critical ranges[17]. The sodium metabisulfite — potassium persulfate system has proven to be effective, as have ammonium persulfate and cerium^{IV} ammonium nitrate[18]. The copolymerization process was shown to operate at temperatures as high as 70 °C and to require efficient stirring.

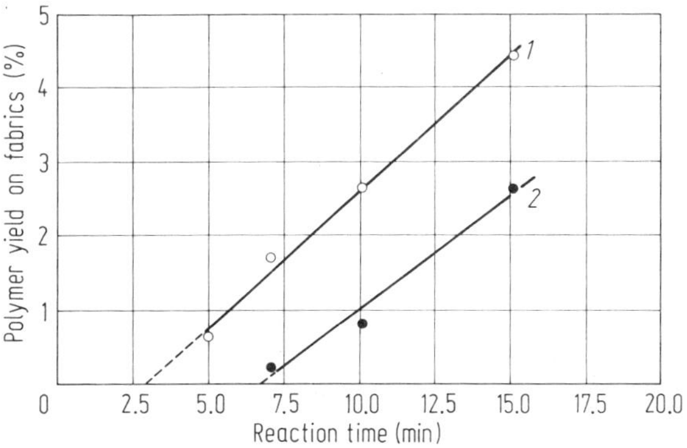

Fig. 33. Effect of copper sulfate traces on AN polymerization. Material/liquor ratio 1:50; $K_2S_2O_8$, 0.2%; AN on $K_2S_2O_8$, 5% (v/v); sodium thiosulfate, 2%; temperature 65 °C; *1* with copper sulfate; *2* without copper sulfate. Reprinted from Ref.[11], p. 1585, by courtesy of John Wiley & Sons, Inc.

Rowland et al.[22] have used ammonium persulfate in their study of the cross-linking of N-methylol acrylamide in cotton fabric. It has been shown that vinyl polymerization and etherification of the monomer occur simultaneously in dry pad-cure operations at elevated temperatures. This provides a facile route to the modification of properties in that the tensile strength, abrasion resistance, and durability of the treated cotton were comparable to those obtained with conventional agents.

Bardhan et al.[23] have examined the grafting of acrylamide onto methyl cellulose in aqueous solution. Grafting achieved using conditions described by Mukhopadhyay et al.[24] was found to be critically dependent on the initiator concentration, with maximum grafting observed at an initiator (potassium persulfate — added to the reaction medium at the desired temperature) concentration of 3.7×10^{-4} mol/l. Grafting was found to increase steadily with an increase in the monomer/polymer ratio and with temperatures up to 40 °C. Grafting was carried out under apparently homogeneous conditions in a nitrogen atmosphere. Purification of the copolymer, often a problem with cellulosic copolymers, was facilitated by the insolubility of the copolymer above 70 °C. Since the homopolymer remained soluble at this temperature, its removal was easily accomplished. The influence of urea on the precipitation temperature was measured and shown to progressively increase the precipitation temperature through its ability to form hydrogen bonds with methyl cellulose in aqueous solution, thus retarding the formation of intra-molecular hydrogen bonds.

4.2.2. Hydroxyl Radical Initiation

When Fe^{II} ions are mixed with H_2O_2, $HO^{\cdot}$, radicals are generated[25–27]. Faber and Weiss[28, 29] have developed the following reaction scheme for the decomposition of hydrogen peroxide in the presence of Fe^{II} ions:

$$H_2O_2 + Fe^{II} \rightarrow HO^- + HO^{\cdot} + Fe^{III}, \quad (129)$$

$$HO^{\cdot} + Fe^{III} \rightarrow HO^- + Fe^{II}. \quad (130)$$

The $HO^{\cdot}$ radicals further decompose H_2O_2

$$HO^{\cdot} + H_2O_2 \rightarrow H_2O + HO_2^{\cdot}, \quad (131)$$

$$HO_2^{\cdot} + H_2O_2 \rightarrow HO^{\cdot} + H_2O + O_2. \quad (132)$$

However, decomposition of H_2O_2 by the $HO^{\cdot}$ radicals takes place to an appreciable extent only if H_2O_2 is present in large excess over Fe^{II} ions[30].

In order to detect the free radical formation in the cellulose upon Fe^{II}-H_2O_2 treatment, fibrous cellulose was immersed in 0.1 M $FeSO_4$ for about 15 minutes, in order to wet the fiber thoroughly, and then dried in a flowing stream of nitrogen. This dried cellulose was placed in a quartz tube; then, 0.3 M H_2O_2 was drawn onto the cellulose. Immediately, the tube and its contents were immersed in liquid nitrogen. The e.s.r. spectrum generated was recorded at −110 °C.

The e.s.r. signal was very weak and short-lived. The spectrometer was operated at high amplification, and the signal was accumulated by a time-averaging computer to obtain the spectrum shown. The three intense lines appeared to be a triplet and could be attributed to a radical formed on abstraction of a hydrogen atom from the hydroxyl group on carbon C_6. The two equivalent hydrogen atoms on Carbon-6 would split the e.s.r. spectrum to form a triplet. The e.s.r. spectrum was very sensitive to temperature; even at temperatures as low as —60 °C, the spectrum became very weak and undefined within a few minutes[26].

The graft copolymerization reaction produced the e.s.r. spectrum shown in Figure 34. The intensity of this spectrum was several times greater than that generated in cellulose by the Fe^{II}-H_2O_2 system alone. The triplet spectrum disappeared during the copolymerization reaction, and an e.s.r. spectrum, generated by the free radical on the growing polymer chain and probably by a free radical formed on oxidative depolymerization of the cellulose molecule, was recorded. The e.s.r. spectrum appeared to be a combination of two singlet spectra[26].

Merz and Waters were among the first to suggest the use of hydroxyl radicals, formed in the Fe^{II}/H_2O_2 reaction, to abstract a hydrogen atom from cellulose in the presence of vinyl monomers[31]. Later, Landells and Whewell reported the successful grafting of several vinyl monomers onto regenerated cellulosic fibers by using Fe^{II}/H_2O_2 in solutions of either acetic or sulfuric acid[32,33].

Kim et al. have used a Fe^{II}-redox system in a study of the grafting of viscose-type cord fibers with various vinyl and diene monomers[34]. Here, rayon was grafted with acrylonitrile using diazonium salts, while isoprene was grafted using an Fe^{II}-cumene hydroperoxide-initiating stytem. The radical mechanism of this process was monitored by e.s.r. It was found that grafting and crosslinking reactions occurred simultaneously and that the thermal stability of the fibers increased with increased grafting.

Focher et al. have begun work on a series of functional cellulose copolymers[35]. Part of this study concerns the synthesis and evaluation of cellulose-poly(p-nitrophenyl acrylate) copolymers and their reactions with amino acids and enzy-

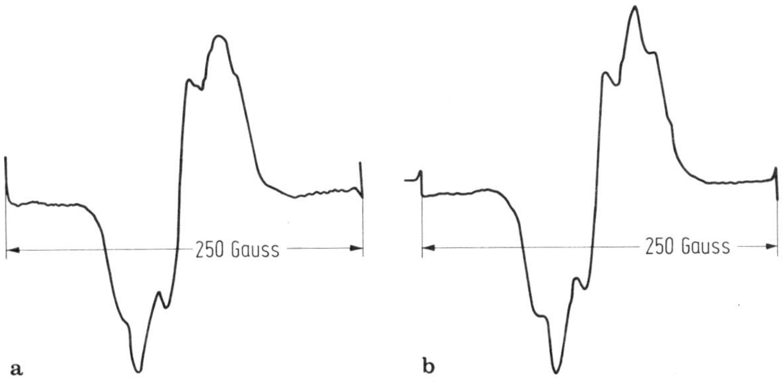

Fig. 34 a, b. E.s.r. spectra of cellulose after copolymerization with acrylonitrile initiated by ·OH radicals formed in Fe^{2+} —H_2O_2 system recorded at —40 °C. **a** Attenuation 10 db, **b** Attenuation 13 db. Reprinted from Ref.[26], p. 1418, by courtesy of John Wiley and Sons, Inc.

mes. This work forms part of a novel area in immobilized enzyme systems. The copolymers are prepared via the decomposition of peroxidized cellulose with Fe^{II} in a dimethylsulfoxide water medium at room temperature. A maximum conversion of 11% was observed after 24 hours' immersion. The couplings of the copolymers with amino acids show apparent first order kinetics, through the rate increase with increasing ionic strength of the reaction medium. These rates were also shown to depend on the particular salt used, KCl or $(CH_3)_4NCl$ respectively. A catalytic effect was observed which was dependent on the pH (7.65 to 8.95) but not on the concentration of amino acid used. A decreased reaction coupling rate was found with albumin compared with glycine. The enzymic activity of the copolymer-enzyme systems, when treated to α-chymotrypsin, was found to be relatively low.

Hamada and Maeda[36] have used the Fe^{II}/H_2O_2 initiating medium in their grafting of methyl methacrylate onto wood pulp. Successful grafting was followed by molding operations at elevated temperature (220°), over an extended period (5 min), and under load (20–30 kg/cm²). The physical properties of the modified boards were markedly influenced by the poly(methyl methacrylate) content.

Mikolajczyk has examined the effect of initiation methods on the composition of cellulose-methyl methacrylate copolymers[37]. He has shown that the molecular mass of the pendant chains of the grafted cellulose can be reduced, while the number of chains is increased, by increasing the concentration of reactive sites in the cellulose, prior to grafting. A similar effect was observed by increasing the time of heating of cellulose, previously impregnated with a Fe^{II}/H_2O_2 solution, before grafting. This markedly increases the concentration of peroxy groups within the substrate, which are available for eventual grafting. Lawniczak has issued a cautionary note regarding thermally induced grafting via hydroperoxide groups. Although this approach leads to high grafting levels, degradation of the cellulosic may be excessive. This leads to mixed block/graft copolymer formation and a deterioration in the mechanical properties[38]. Lawniczak has taken great interest in timber modification, an interest which is shared by other groups, using the Fe^{II}/H_2O_2 system. Young has examined the grafting of acrylonitrile, styrene, and acrylamide onto xanthated pulp in the presence of Fe^{II} [39]. Chemical pulps were found to produce greater grafting levels than mechanical pulps, an indication of the greater accessibility of chemical pulps. Styrene/acrylonitrile mixtures were found to be less efficient than was the grafting of acrylonotrile alone.

Baxendale et al.[4] showed that HO˙radicals can initiate vinyl polymerization; at sufficiently high monomer concentrations, the termination of HO˙ radicals by Fe^{II} ions can largely be suppressed so that most of the HO˙ radicals are captured by monomer molecules leading to initiation of polymerization. Cellulose and a vinyl monomer react in the presence of H_2O_2 and Fe^{II} ions to produce a graft copolymer[41,42]. The reaction which is believed to be responsible for this is the abstraction of a hydrogen atom from cellulose.

In the Fe^{II}-H_2O_2 technique, the grafting reaction depends to a large extent on the conditions of the experiment. The best technique appears to be first to subject the cellulose to a solution of Fe^{II} salt. The Fe^{II} ions are absorbed by the cellulose, and the subsequent reaction with H_2O_2 in the presence of monomer will therefore take place in close proximity to the cellulose molecules[43–45] as is clearly shown by Ogiwara et al.[44,45]. These reachers examined the decomposition of hydrogen peroxide by Fe^{III} ions absorbed on cellulose, the lowering of the degree of polymerization of the bleached softwood sulfite pulp used for grafting, and the resultant grafting in systems containing methyl methacrylate. It was found that an increase in the concentration of Fe^{III} ions adsorbed on the cellulose, or in the concentration of hydrogen peroxide, resulted in an increase in both the extent of decomposition of hydrogen peroxide and the number of scissions of cellulose chains with higher rates at the initial stages of reaction. A relationship was found between the concentration of Fe^{III} ions adsorbed on the cellulose and the initiation efficiency of hydrogen peroxide in graft polymerization[44].

Synthesis of cellulose graft copolymers from the vinyl monomer mixtures — N-vinylpyrrolidone and butyl methacrylate, N-vinylcaprolactam and butyl methacrylate — has been made possible using the Fe^{II}-H_2O_2 redox system. An interesting feature is that the solubility and composition of the vinylpyrrolidone-butyl methacrylate and vinyl caprolactam-butyl methacrylate grafted chains are similar to those of the bulk copolymers which were obtained under identical conditions[46].

It should be pointed out that with the Fe^{II}-H_2O_2 and persulfate systems, the redox reactions generate radicals unattached to cellulose. Grafting thus depends on the abstraction of hydrogen atoms from the cellulose by $HO^{\cdot}$ or $SO_4^{\cdot-}$ radicals. Since these radicals are also capable of attacking the monomer, homopolymer formation is more extensive in these systems than with the Ce^{IV} ion, where the free radicals are formed exclusively on the cellulose backbone (see below).

4.2.3. Ce^{IV} Ion Initiation

Due to its ease of application as well as its grafting efficiency, the Ce^{IV} ion method has gained considerable importance in the grafting reaction. This technique is based on the fact that when cellulose is oxidized by a Ce^{IV} salt such as Ce^{IV} ammonium nitrate, free radicals capable of initiating vinyl polymerization are formed on the cellulose by a single electron transfer[47–96]. Since the radicals produced are almost exclusively on the cellulose, homopolymer formation is minimized.

Early research[47] showed that Ce^{IV} ammonium nitrate exists as a complex salt and is ionized in solution, yielding free ammonium ions and nitratocerate ions. The nitratocerate ions subsequently decompose to yield Ce^{IV} cations and nitrate anions.

$$(NH_4)_2Ce(NO_3)_6 \rightleftharpoons Ce(NO_3)_6^{2-} + 2\,NH_4^+ \,,$$
$$Ce(NO_3)_6^{2-} \rightleftharpoons Ce^{4-} + 6\,NO_3^- \,.$$

In the presence of glycol groups, tetravalent cerium is capable of forming mono-, di- and triglycolated intermediates.

$$Ce^{IV}(X)_6 + G \rightleftharpoons Ce^{IV}G(X)_4 + 2\,X \,,$$
$$Ce^{IV}G(X)_4 + G \rightleftharpoons Ce^{IV}G_2(X)_2 + 2\,X \,,$$
$$Ce^{IV}G_2(X)_2 + G \rightleftharpoons Ce^{IV}G_3 + 2\,X \,,$$

where G represents the glycol, and X, the groups coordinated with the tetravalent cerium, e.g., HO^-, NO_3^-, or H_2O.

There is considerable evidence[48-55] to indicate that disproportionation of these intermediate complexes results in the formation of free radicals capable of initiating vinyl polymerization.

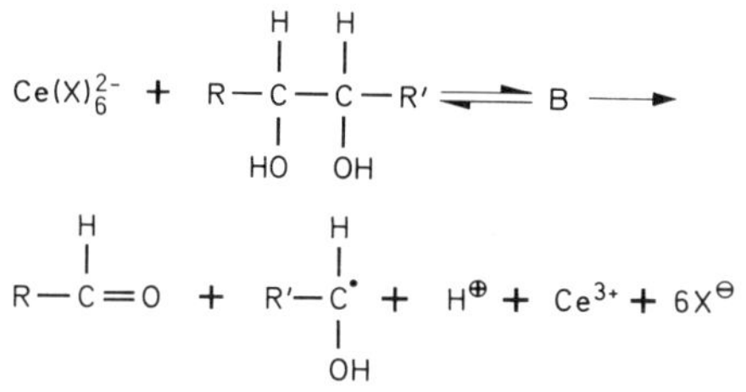

The rate-determining step is presumably the disproportionation of the cerato-glycol complex, B.

Of great interest in the understanding of grafting is the study of oxidation of pinacols by Ce^{IV} sulfate-published by Mino et al.[50]. The Ce^{IV} salt concentration varied from about 2×10^{-3} to 10×10^{-3} M, and the pinacol concentration from 0.03 to 0.16 M. The rate at which Ce^{IV} was reduced was proportional to both the Ce^{IV} and the pinacol content, and 2 mols of acetone were produced for every 2 mols of Ce^{IV} reduced. The reactions describing this are:

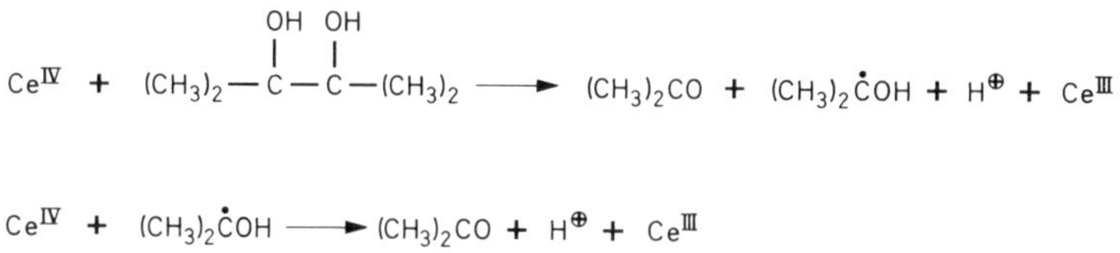

where the first of the two is rate-determining.

In the presence of (excess) acrylamide, only one mol of acetone was produced per mol of Ce^{IV} reduced. Since it is known from kinetic studies of a polymerization that the termination step involves a reaction with Ce^{IV}, the following reactions are believed to take place.

$$(CH_3)_2\dot{C}OH + M \rightarrow RM^{\cdot} \quad \text{(initiation)},$$

$$RM^{\cdot} + nM \rightarrow RM^{\cdot}_{(n+1)} \quad \text{(propagation)},$$

$$RM^{\cdot}_{(n+1)} + Ce^{IV} \rightarrow \text{Polymer} + H^+ + Ce^{III} \quad \text{(termination)}.$$

It is clear that if the initiation step involves the addition of the alcohol radical, and if the cellulose reacts similarly, one may expect grafting[48].

The mechanism and kinetics of polymerization involving Ce^{IV} and alcohol systems have been studied in detail[48,50,51,54]. The proposed mechanism and kinetic scheme have been based on an important simplification of the systems studied, viz., that the rate of initiation is equal to one-half the rate of disappearance of Ce^{IV}, as described earlier in the case of pinacol. There is evidence, however, that other glycol-reducing agents do not allow a similar simplification in their kinetic treatment. Studies on ethylene glycol[54], for example, showed that the reaction scheme,

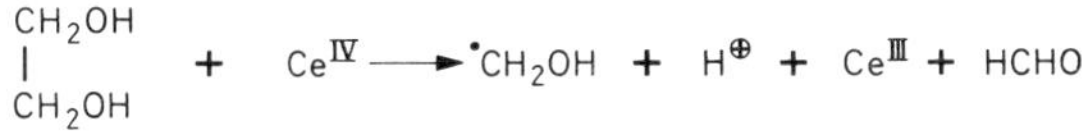

where the ratio of Ce^{III} to formaldehyde is 1:1, was not obeyed.

The observed value was nearly 2:1, suggesting that the formaldehyde had been further oxidized to formic acid[52,55].

The polymerization of acrylonitrile in the presence of an aqueous Ce^{IV} sulfate-ethylene glycol solution was investigated as a model system for cellulose graft copolymerization[54]. Results of this work reveal that the rate of polymerization, R_p, is directly proportional to the square of the monomer concentration; that it is linear with alcohol concentration; and that it is inversely related to the concentration of Ce^{IV} ion. The rate of Ce^{IV} ion reduction was proportional to the initial Ce^{IV} ion concentration and linear with the initial concentration of monomer and alcohol. To account for this, the following reaction scheme was proposed:

$$Ce^{IV} + A \overset{k}{\rightleftharpoons} B \overset{k_d}{\rightarrow} A^{\cdot} + HCHO + Ce^{III} + H^+, \qquad (133)$$

$$A^{\cdot} + M \overset{k}{\rightarrow} M^{\cdot}, \qquad (134)$$

$$Ce^{IV} + M \overset{k_i'}{\rightarrow} M^{\cdot} + Ce^{III} + H^+, \qquad (135)$$

$$M^{\cdot} + M_n \overset{k_p}{\rightarrow} M^{\cdot}_{(n+1)}, \qquad (136)$$

$$M^{\cdot}_n + Ce^{IV} \overset{k_t}{\rightarrow} M_n + Ce^{III} + H^+, \qquad (137)$$

$$A^{\cdot} + Ce^{IV} \overset{k_O}{\rightarrow} \text{Oxidation products} + Ce^{III} + H^+. \qquad (138)$$

Ce^{IV} denotes a cerium ion of high oxidation state, or any inorganic complex thereof; M is the monomer (acrylonitrile); A is an alcohol (ethylene glycol); k (with subscripts) denotes the rate constants for various reactions; k is an equilibrium constant; and B denotes a Ce^{IV} ion-alcohol complex.

Data in Table 38[54] indicate that the amount of polyacrylonitrile formed in the absence of alcohol is by no means negligible, and that the ratio of the

Table 38. Rate of polymerization of acrylonitrile in the presence and absence of ethylene glycol at $[Ce^{IV}] = 0.025$ mol/l at 35 °C. Reprinted from Ref.[54], p. 405, by courtesy of John Wiley & Sons, Inc.

[M]	[A] = 0.5 mol/l		[A] = 0		R_p/R_p'
mol/l	R_p (g/l hr)	$R_p[M]^2$	R_p (g/l hr)	$R_p'/[M]^2$	
0.10	0.12	12	0.060	6.0	2.0
0.25	1.04	17	0.364	5.8	2.9
0.50	6.40	26	1.38	7.5	3.4

rates of propagation (rate of polymerization, R_p) of comparable systems in the presence and absence of alcohol is about 3. Previous results[50)] showed that the termination reaction, i.e., Eq. (137), involves the Ce^{IV} ion. Thus, the kinetics of Ce^{IV}-initiated polymerization of acrylonitrile can best be represented by reactions (133–135) as initiation steps, reaction (136) as propagation, and reaction (137) as termination. This leads to the expressions:

$$d[M^{\cdot}]_{Ce}/dt = k_i'[M]\,[Ce^{IV}]\,, \qquad (139)$$

$$d[M^{\cdot}]_{A}/dt = k_i[M]\,[A^{\cdot}]\,, \qquad (140)$$

$$-d[M]/dt = R_p = k_p[M]\,[M^{\cdot}]\,, \qquad (141)$$

$$-d[M^{\cdot}]/dt = k_t[M^{\cdot}]\,[Ce^{IV}]\,. \qquad (142)$$

Applying the steady state assumption to both ($M^{\cdot}$) and ($A^{\cdot}$) separately, and assuming that the equilibrium between Ce^{IV} and alcohol to form the complex is established rapidly, as compared to the rate at which the complex is decomposed, one obtains the following expression for the rate of polymerization:

$$R_p = \frac{k_p}{k_t}[M]^2\left\{k_i' + \frac{k\,dk(A)}{[M] + (k_0/k_i)\,[Ce^{IV}]}\right\}. \qquad (143)$$

Eq. (143) shows a square dependence on the monomer concentration in the absence of alcohol. However, when alcohol is present the extent of deviation from this square dependence is obtained by the magnitude of [M] versus (k_0/k_i) $[Ce^{IV}]$. The initial rate of polymerization is directly proportional to the alcohol concentration; therefore, an initial R_p versus [A] plot at a particular monomer concentration should yield a straight line. It is noteworthy that the rate of polymerization is inversely related to the Ce^{IV} ion concentration. This leads to the interesting situation where the rate of polymerization at low Ce^{IV} ion concentrations may actually be greater than at high Ce^{IV} ion concentrations.

Kinetic equations involving the disappearance of the Ce^{IV} ion due to reaction with alcohol, alcohol radical, monomer, and polymer radical lead to the following expression:

$$(-1/[Ce^{IV}])\,d[Ce^{IV}]/dt = 2[k\,dk[A] + k_i'[M]]\,. \qquad (144)$$

Eq. (144) represents the relationship that governs the Ce^{IV} ion disappearance in a relatively easily verifiable form. In addition, a more comprehensive relationship was derived taking into consideration the constants governing the various equilibria involving Ce^{IV} ions and sulfate ions. This equation can be written as follows:

$$-d[Ce^{IV}]/dt = 2[A]\,[Ce^{IV}]_0\,f[SO_4^{2-}]\,,$$

where $[Ce^{IV}]_0$ refers to the concentration of uncomplexed Ce^{IV} ion, and $f[SO_4^{2-}]$ is a cubic function of the sulfate concentration.

Evaluation of the ratio k_p/k_t (where k_p is rate constant for propagation and k_t is the termination constant) showed that these two constants were of similar magnitude, contrary to the usual results for free radical systems, wherein k_t is several orders of magnitude greater than k_p. It was then concluded that propagation and termination are similar processes[54)].

According to Mino et al.[50)], the termination of free radicals during oxidation and that of growing polymer chains during grafting occur through conversion of Ce^{IV} ions to Ce^{III} (reactions 133 and 137) respectively, where A represents cellulose. Taking this into consideration, Kulkarni and Mehta[56)] assumed that the total consumption of Ce^{IV} ions should be more or less equal for oxidation (reactions 133 and 138) and grafting (reactions 133 and 137), if all the monomer is effectively grafted onto cellulose. On the other hand, the Ce^{IV} ions consumed during grafting will be more than that during oxidation of cellulose if some homopolymer is formed [reactions (135) and (139)].

To test the validity of this assumption, Hebeish and Mehta[57)] have determined the Ce^{IV} consumption during grafting and oxidation of various cellulosic substrates, such as native cotton, cotton treated with sodium hydroxide (NaOH-Cotton), viscose, and ramie. Some of the results obtained are given in Figures 35 and 36. It is apparent that under similar conditions of reaction, the Ce^{IV} ion consumption during grafting is higher than that consumed during oxidation. This is observed in all the substrates studied, in accordance with the assumption of Kulkarni and Mehta[56)]. The latter was obtained by measuring the cummulative difference in area under the two appropriate curves up to the particular reaction time. There is a direct relation (linear in almost cases) between the extent of homopolymerization and the excess Ce^{IV} ions consumed in grafting. It has also been shown that each substrate shows a uniqueness in this relation, and for a given value of excess Ce^{IV} ions consumed in grafting, the amount of homopolymer formed varies widely with different substrates. This is probably a result of the different efficiencies of these

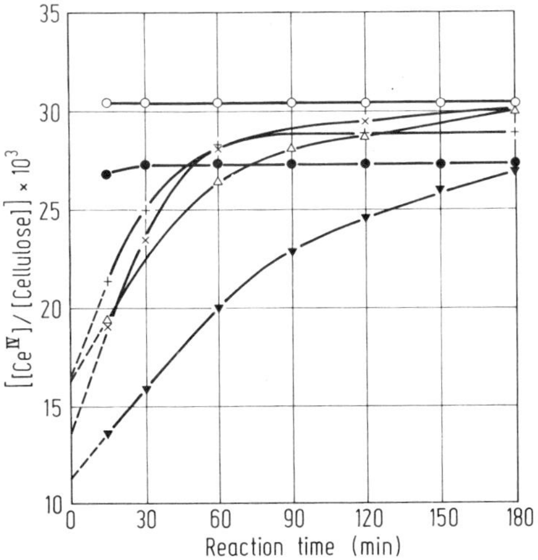

Fig. 35. Consumption of cerium during grafting and oxidation of native cotton at different temperatures. Grafting at (°C): (○) 20; (×) 40; (+) 60; oxidation at (°C), (▼) 20; (△) 40; (●) 60; reaction conditions as in Figure 3. Reprinted from Ref.[57)], p. 1631, by courtesy of John Wiley & Sons, Inc.

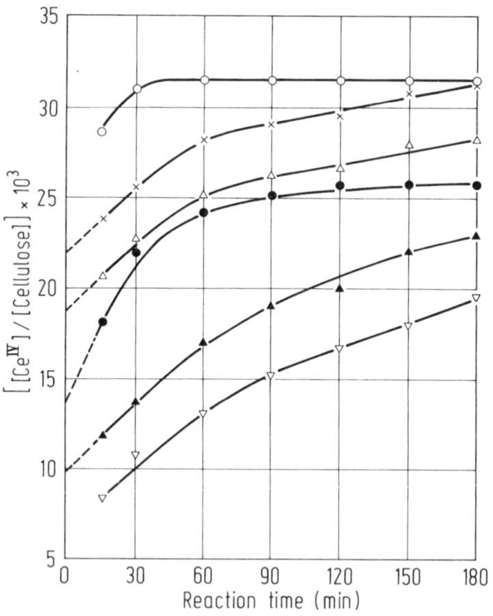

Fig. 36. Consumption of cerium during grafting and oxidation of viscose at different temperatures. Grafting at (°C); (○) 20; (×) 40; (●) 60; oxidation at (°C): (▲) 20; (△) 40; (▽) 60; reaction conditions as in figure 3. Reprinted from Ref.[57], p. 1633, by courtesy of John Wiley & Sons, Inc.

substrates with respect to grafting that leads to different amounts of monomer available in solution for homopolymerization.

Various monomers, such as acrylamide, acrylonitrile, and certain esters of acrylic and methacrylic acid, have been successfully grafted onto cellulosic materials using the Ce^{IV} ion type catalyst[56–83]. Kaizerman et al.[61] have polymerized acrylonitrile, acrylamide, and methyl acrylate within the fibers of cotton and viscose rayon. They reported that the maximum amount of polymer which could be deposited in cotton or rayon varied according to the nature of the monomer. Details of the treatment and the resulting increase due to polymer deposition are presented. The researchers have also mentioned that similar weight increases were obtained under comparable treatment conditions with viscose rayon.

Cumberbirch and Holker[62] grafted poly(acrylic acid) onto cotton, mercerized cotton, viscose, and oxycellulose. Mercerized cotton, in spite of its higher reactivity, showed lower graft yields than native cotton. Brief hydrolysis lowered the cellulose reactivity toward grafting. They also reported that the introduction into rayon fabric of a large number of aldehyde groups, which are more readily oxidized by Ce^{IV} salts than alcoholic hydroxyls, did not increase its susceptibility to grafting with acrylic acid.

Hebeish and Mehta[80,81] reported a detailed study on grafting of various vinyl monomers and their binary mixtures on different cellulosic materials using the Ce^{IV}-cellulose redox system. Table 39 contains the results obtained on grafting of acrylonitrile (AN), methyl methacrylate (MMA), and methyl acrylate (MA) to native cotton, hydrocellulose, viscose, ramie, and cottons swollen with sodium hydroxide (NaOH-cotton), ethylene-diamine (EDA-cotton), and zinc chloride ($ZnCl_2$-cotton).

It is clear that the graft yields achieved with swollen cottons, on grafting with MA and MMA, are marginally higher than those with native cotton. This

Table 39. Grafting various vinyl monomers onto different cellulosic substrates. Reprinted from Ref.[80], p. 177

Substrate	Graft Yield %					
	AN		MA		MMA	
	40 °C	60 °C	40 °C	60 °C	40 °C	60 °C
Native cotton	43.00	27.53	62.54	55.41	43.31	45.72
NaOH-Cotton	52.84	32.05	8.87	56.19	42.59	62.79
EDA-Cotton	48.38	26.60	65.91	55.16	45.00	51.29
$ZnCl_2$-Cotton	47.37	27.68	62.02	54.85	47.83	53.19
Hydrocellulose	37.48	22.47	81.04	68.95	61.45	62.39
Viscose	43.43	19.98	85.05	67.36	65.93	63.07
Ramie	40.84	13.98	70.98	55.34	51.80	52.50

$[Ce^{IV}]$ = 0.006 M; Cellulose, 1 g; Monomer, 1 cm^3; Material: Liquor ratio 1:30; Reaction time, 3 hours.

implies that the effect of cellulose accessibility is not marked under the conditions studied. However, it would be necessary to carry out initiation and grafting separately, in a manner similar to that used with AN, to establish this beyond doubt. Contrary to what would be expected from its higher crystallinity and, therefore, lower reactivity, hydrocellulose showed higher yields of grafting than swollen cottons. This would support the assumption that the process of grafting on different substrates is not necessarily comparable. It might be expected, and with reason, that grafting on substrates such as hydrocellulose, which has very high crystallinity, would be more of a surface phenomenon. This undoubtedly would be due to the fact that the monomer, in the aqueous solution used in this study, cannot penetrate beyond the surface of the fiber.

Samples grafted with AN-MA and AN-MMA mixtures at 40 °C for different times were analyzed for nitrogen[80,81]. The results obtained are given in Table 40. The proportion of AN in the graft has been calculated from the nitrogen content and is shown in the last column. It is evident that the amount of AN grafted is not affected throughout the reaction period by the presence of MA. On the other hand, graft yields of AN from AN-MMA mixtures were significantly lower in the initial stages of the reaction than those obtained with AN alone. Addition of MMA appears to slow down the rate of grafting of AN to cellulose.

Schwab et al.[59] studied the graft polymerization of acrylamide, acrylonitrile, methyl methacrylate, ethyl acrylate, and styrene to paper using the Ce^{IV} ion process. The reactivity of the monomer for graft polymerization was found to be as follows:

ethyl acrylate > methyl methacrylate > acrylonitrile > acrylamide > styrene.

These investigators also reported an optimum temperature (25–40 °C) for grafting initiated by Ce^{IV} ammonium nitrate, beyond which the amount of graft

Table 40. Nitrogen content of AN-MA and AN-MMA grafted cellulose[a]. Adapted from Ref.[80], p. 173

Reaction time (min)	Monomer	Graft[b] %	Nitrogen content %	Graft[c] due to AN %
15	AN	10.4	2.64	10.1
	AN-MA	32.6	2.42	9.1
	AN-MMA	11.5	1.02	3.8
30	AN	15.8	3.97	15.0
	AN-MA	45.2	3.87	14.6
	AN-MMA	24.0	1.73	6.5
60	AN	17.1	4.50	17.0
	AN-MA	51.3	4.21	15.9
	AN-MMA	36.4	3.40	12.8
80	AN	18.8	4.77	18.0
	AN-MA	52.5	4.72	17.8
	AN-MMA	41.6	3.80	14.3
180	AN	19.3	5.04	19.0
	AN-MA	53.1	4.96	18.7
	AN-MMA	49.5	4.48	16.9

[a] [CAN] = 0.006 M in 1% HNO_3; cellulose = 1 g; monomer mixture 1:1 (0.5 cm^3 each); temp. 40 °C; liquor ratio 1:30).
[b] Based on weight of cellulose.
[c] Calculated from the nitrogen content.

formed rapidly decreased. (Results for acrylonitrile and methyl acrylate are in agreement with this). The researchers attributed this partly to the instability of Ce^{IV} ammonium nitrate at elevated temperatures. Results for methyl methacrylate do not parallel this observation, perhaps due to its higher solubility at elevated temperatures. Increasing the solubility of the monomer would facilitate its diffusion in the substrate, thus favoring grafting.

Negishi et al.[76] have been able to effect graft copolymerization of acrylate monomers, such as methyl, ethyl, n-propyl, n- or isobutyl, onto cotton fabrics using an aqueous emulsion containing Ce^{IV} ions. They reported that the rate of graft copolymerization decreases with increasing length of the ester group of the monomer.

Figure 37 shows the percent graft yields on cellulose fibers and the corresponding polymerization yields in solution in the absence of cellulose[56]. Obviously, the grafting reaction proceeds much faster than polymerization without cellulose, showing the high grafting efficiency of the Ce^{IV}-cellulose system. The increased polymerization rates could be due either to the large area offered by cellulose or to the reducing action of the cellulose. The fact that the presence of a large surface area, in silica spheres with the same surface area of cellulose, has not increased the polymerization yields indicates that the high polymerization rates in the presence of cellulose must be due to the reducing action of cellulose. Here, too, the effect of adding traces of Ce^{II} sulfate to the grafting and polymerization mixture was studied. Unlike the persulfate-thiosulfate[11] system, in which addition of Cu^{II} ions significantly accelerated the grafting reaction by

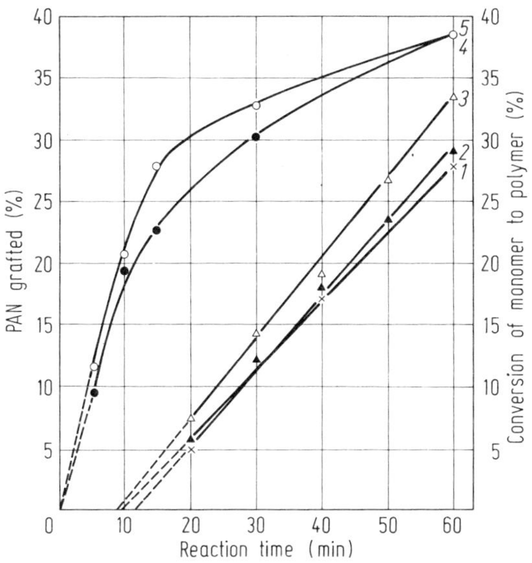

Fig. 37. Effect of surface area and addition of traces of copper sulfate, on acrylonitrile polymerization in presence and absence of cellulose; *1* polymerization in presence of silica spheres; *2* polymerization in absence of cellulose; *3* polymerization with traces of copper sulfate (without cellulose); *4* grafting in presence of copper sulfate traces; *5* grafting without copper sulfate. [Ce^{IV}], 0.006 M; liquor ratio, 30:1; acrylonitrile, 1 cm^3; cellulose, 1 g; temperature of reaction, 40 °C. Reprinted from Ref.[56], p. 1324, by courtesy of John Wiley & Sons, Inc.

reducing the induction period from 7 minutes to 3 minutes (see Fig. 33), the rate of grafting remains unaltered in the Ce^{IV}-cellulose system.

Kubota and Ogiwara[82] have studied the effect of lignin, contained in cellulosic materials, on the vinyl graft copolymerization of lignin containing substrates, using the Ce^{IV} ion technique. Samples containing the highest lignin content showed the lowest graft yield. This is so probably because the Ce^{IV} ions reacted at a faster rate with lignin than with cellulose in wood pulp; but the active sites formed on lignin appeared to participate in the termination reactions, thus increasing the number of grafts but lowering their average molecular weight.

The nature of the cerium^{IV} ion initiation of cellulose grafting has been much studied by Iwakura et al., who have shown that primary alcohols and glycols are more effective than secondary and tertiary alcohols in initiating the polymerization of styrene[60]. Mino et al.[50] have also shown that in the case of glycols, e.g., pinacols, the 1:2 glycol group is cleaved during the oxidation reaction with Ce^{IV} ions. Thus, during grafting on cellulose, too, cleavage of the C_2-C_3 glycol bond is expected in this system of graft initiation. Terasaki and Matsuki[84], considered the hemiacetal group at the end of the cellulose chain to be of primary importance in Ce^{IV} attacks on cellulose, but postulated glycol cleavage as well. Cumberbirch and Holker[62] and Arthur et al.[85] have also postulated glycol bond cleavage in cellulose during grafting initiated by Ce^{IV}. Cumberbirch and Holker[62] have supported this postulate on the finding that adipaldehyde is obtained as a product of oxidation of cyclohexane-1,2-diol by Ce^{IV}, while Arthur et al.[85] have supported it through e.s.r. studies on Ce^{IV} ion-oxidized cellulose.

According to Arthur et al.[85], the Ce^{IV} ion in acidic solution forms a chelate with the cellulose molecule, probably by reaction through the hydroxyl groups on carbon C_2 and C_3 (Eq. 146). There was a transfer of one electron from the cellulose molecule to the Ce^{IV} ions. The Ce^{IV} was reduced to Ce^{III} ions, followed by release of Ce^{III} ions and anhydroglucose ring cleavage at the C_2–C_3 bond (Eqs. 147 and 148). A short-lived free radical was then formed on either carbon C_2 or C_3.

On initiation of graft polymerization at this site (Eq. 149), additional Ce^{IV} ions may chelate with the cellulose, forming additional radical sites. The decay of free radicals, formed on the cellulose molecule by Ce^{IV} ion interaction, is shown in Figure 38. Termination could result from the reaction of Ce^{IV} ions with the free radical to yield Ce^{III} ions and an oxidized product of cellulose (Eq. 150). Reaction of solvent, usually water, with the free radical could also terminate the radical[84].

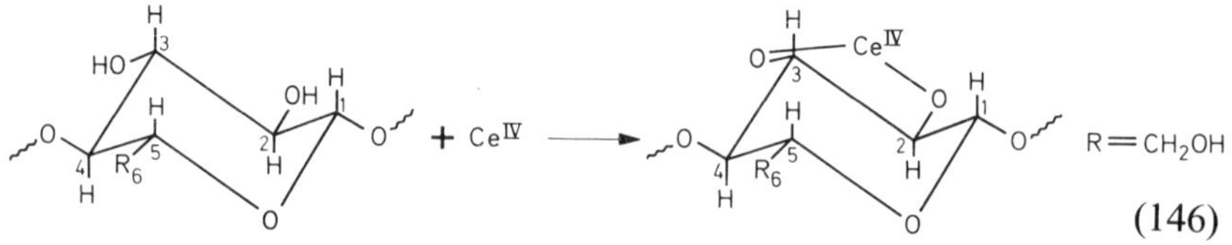

(146)

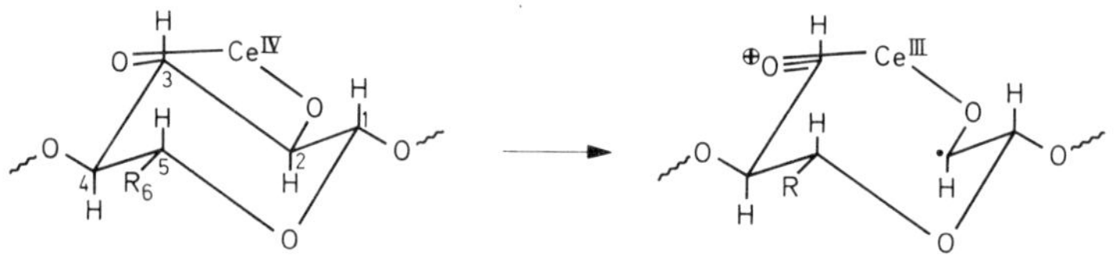

(147)

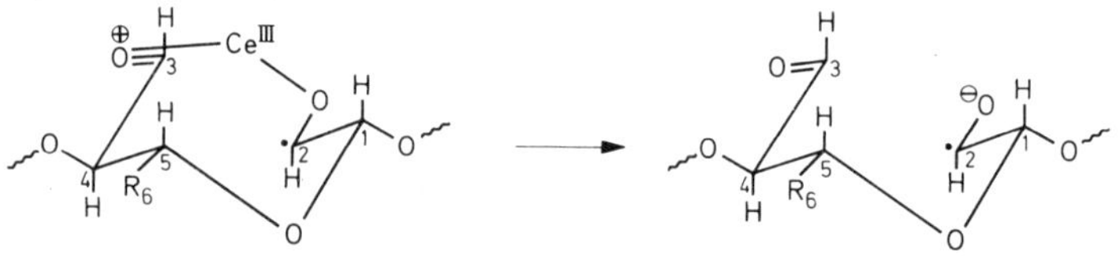

(148)

(a) Radical formation

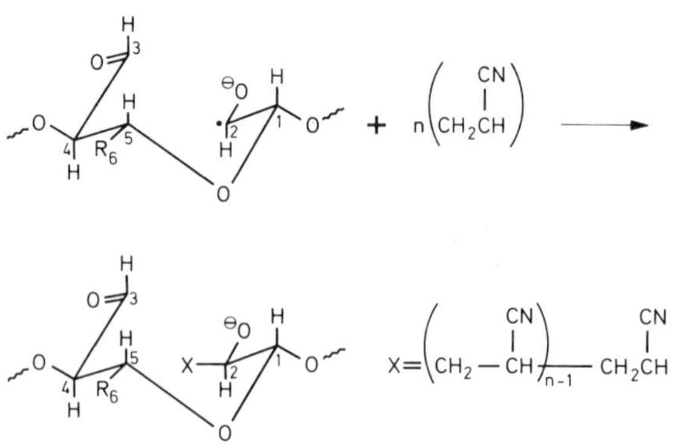

(149)

(b) Graft polymerization

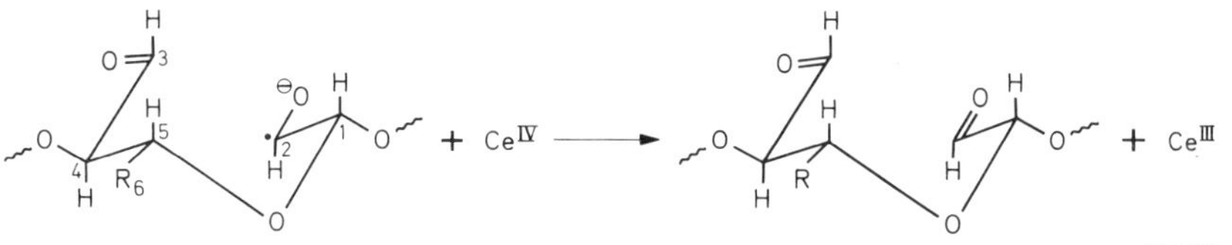

(c) Radical termination (150)

Kulkarni and Mehta[56] examined the question of C_2–C_3 bond cleavage during oxidation of cellulose with Ce^{IV} ions by following a reaction sequence similar to the oxidation of cellulose with sodium metaperiodate and reduction with sodium borohydride. They have shown that in the initial stage of oxidation of cellulose with Ce^{IV} ions, the reaction proceeds very rapidly due to formation of a Ce^{IV}-cellulose complex and oxidation of the hemiacetal group of cellulose. Thereafter, oxidation of the secondary hydroxyl group at carbon C_2 occurs without C_2–C_3 bond cleavage. The glycol bond cleavage takes place at a high Ce^{IV} ion concentration and on prolonged oxidation of cellulose. There are, then, two postulated grafting sites in Ce^{IV}-initiated graft copolymerization onto cellulose[56,62,65,73,75,84,86]: the C_2–C_3 glycol group of the anhydroglucose unit and the hemiacetal group of the end units of cellulose molecule. A number of authors have strongly supported these postulates on the basis of kinetic studies with modified celluloses[66], as well as with model compounds[88,89].

More research[90] indicates that the Ce^{IV} ion oxidizes glucose at the C_1–C_2 bond, forming a free radical on C_2, and that this is presumably the same reaction that occurs at the reducing end group of cellulose. A more startling proposal has been advanced by Gaylord[91], who postulates that the growing polymer chain is initiated on the monomer and terminated by the cellulose, rather than being initiated by a cellulose macroradical (see Chapter 2).

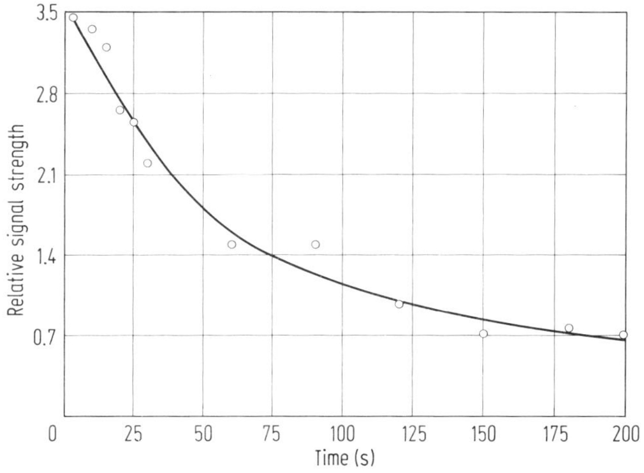

Fig. 38. Decay of free radicals formed on cellulose by Ce^{IV} ion, time measured from saturation of cellulose with Ce^{IV} ions at 25 °C until freezing at −100 °C before recording spectrum. Reprinted from Ref.[85], p. 1599, by courtesy of John Wiley & Sons, Inc.

If a cellulose sample, in which free radicals have been initiated by oxidation with Ce^{IV} ions, is stored, one of two things may happen: the free radical may be terminated, owing to exposure to atmosphere, or there may be a charge transfer, resulting in termination of the radical with simultaneous creation of another active site (in other words, the free radical activity remains unchanged). Thus, the decay of free radical activity of the Ce^{IV}-oxidized cellulose during storage would be a result of these two processes. Hebeish and Mehta[57] have adopted the following experimental procedure for studying this phenomenon. The cellulosic material was oxidized with Ce^{IV} ammonium nitrate at 20 °C, 40 °C, and 60 °C until maximum Ce^{IV} consumption was obtained. The reaction time needed for this was available from data on Ce^{IV} consumption during oxidation at various times of reaction (see Fig. 35). The excess Ce^{IV} solution was removed by suction filtration, and the sample (containing approximately 100% its weight of Ce^{IV} solution) was stored for various intervals of time,ranging from 2 to 300 minutes, before acrylonitrile monomer was added. The percent graft yields shown in Figure 39 were obtained for a 3-hour reaction period after the addition of monomer.

These data (Fig. 39) indicate that the rate of decay of free radical activity increases rapidly with increasing temperatures. The presence of air in the medium also accelerates this decay, since the percent graft yields are much lower in air than in nitrogen. Moreover, it is noticed that on increasing the accessibility of the substrate (e.g., viscose and ethylenediamine treated cotton), the free radical decay is greater. The difference in the percent graft yield in nitrogen and in air atmosphere is also greater in these cases than in that of native cotton.

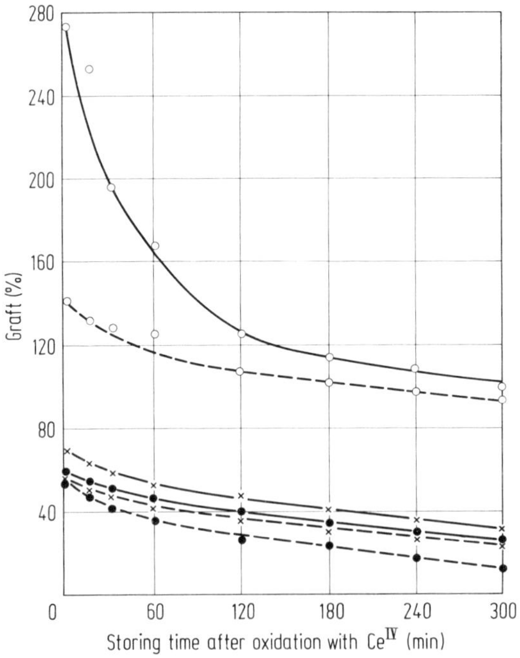

Fig. 39. Rate of decay of free radical activity of Ce^{IV}-oxidized cellulose. Percent graft: (———) in nitrogen; (– – – –) in air. Grafting at (°C); (●) 20; (×) 40; (○) 60. Storing time after oxidation with Ce^{IV}, 2–300 min.; $[Ce^{IV}] = 0.006$ M in 1% HNO_3; cellulose, 1 g; acrylonitrile, 15 cm^3. Reprinted from Ref.[57], p. 1645, by courtesy of John Wiley & Sons, Inc.

Huang and Chandramouli have examined the structure and properties of cellulose — styrene[92)] and cellulose — methyl methacrylate[93)] graft copolymers synthesized, using the Ce^{IV} ion method. In the cellulose-styrene system, the grafting reaction was found to reach a maximum value in a narrow range of concentrations of initiator, 1.0×10^{-3} to 1.8×10^{-3} mol/l at 58 ± 1 °C. The structure of the copolymers were assessed after hydrolyzing away the cellulose backbone to isolate the grafted polystyrene branches. Measurement of the molecular masses and molecular mass distributions of the grafted branches was achieved through gel permeation chromatography. The number average molecular mass, in the above initiator concentration range, varied from 2.3×10^4 to 4.5×10^5, while the polydispersity ratio ($\bar{M}_w/\bar{M}_n$) varied from 2.5 to 8.0. Grafting frequencies of the order of 0.05 to 0.4 polystyrene branches per cellulose chain were obtained. Huang and Chandramouli conclude that the success of initiation depends primarily on the diffusion rate of the monomer onto the cellulose. The initiation step is thought to be favored only when the diffusion rate is at least comparable to the rate of oxidative termination.

For the cellulose-methyl methacrylate system[93)], the number average molecular mass of isolated branches ranged from 3.6×10^4 g/mol to 1.6×10^5 g/mol, and the polydispersity ratio ($\bar{M}_w/\bar{M}_n$) varied from 4.0 to 7.0. The grafting frequency ranged from 0.38 to 3.2. Thus, with methyl methacrylate grafting, greater internal control over the copolymerization reaction is obtained relative to the corresponding cellulose-styrene system. As with the cellulose-styrene system, a critical concentration range for the initiator was observed in methyl methacrylate grafting. This range was from 3.0 to $6.0 + 10^{-3}$ mol/l. A decrease in the percent grafting beyond this range is attributed to the increasing oxidative termination rates. It has been shown that the number average molecular mass of the grafted branches generally decreases with increasing cellulose content in systems of this type.

More recently, general studies of Ce^{IV} initiation include the grafting of acrylonitrile onto sulfated pine and poplar wood pulps. It has been shown that the extent of grafting increases as the lignin content increases. The efficiency of the grafting process using pine was greater than that observed with poplar pulp. However, an induction period is observed with both types of pulp, increasing in length as the concentration of lignin in the pulp increases. This inhibitory character has been confirmed via a kinetic analysis of the process[94)].

Watamoto and Tonami[95)] have examined the graft copolymerization of p-benzoyloxystyrene onto cotton fibers by the pad-cure method using Ce^{IV} nitrate-dimethylformamide. Samples treated in this way showed greater retention of tensile properties when exposed to ultraviolet radiation than did samples of copolymer obtained by non-pad-cure methods. It was found that in the grafting of fibers by the initiation/pad/cure route, the grafting ratio reached a maximum value with increasing temperatures in the range of 80° to 100 °C. However, this limiting value increased as the concentration of p.benzoyloxystyrene in the system increased.

Messiha et al.[96)] have followed the graft copolymerization of α-bromoacrylic acid onto cellulose. Initiation was provided by Ce^{IV} ammonium sulfate and Ce^{IV} ammonium nitrate in aqueous media. HNO_3 was found to retard the grafting process, whereas H_2SO_4, $HClO_4$, and HCl provided an increase in grafting relative to the corresponding acid-free system. These effects were magnified with increasing temperatures in the range of 40 to 80 °C. In each acid system the grafting degree increased rapidly in the initial phase of reaction, passed through a maximum, and then showed only a slow increase in grafting. Infrared spectroscopic examination of the graft copolymers provides evidence of intramolecular lactone formation during the grafting reaction.

The ability of Ce^{IV} to induce graft copolymerization onto chemically modified celluloses has been investigated. Details of these studies will be discussed later (see Chapter 6).

4.2.4. Initiation by Pentavalent Vanadium

Waters[97] has shown that oxidation of alcohols by transition metals, such as Co^{III} ions and pentavalent vanadium ions, takes place through complex formation. The formation of alcohol complexes with Co^{III} and vanadium^{V} is easily noticed due to the immediate color change of the oxidizing solution. These complexes dissociate to yield free radical intermediates. That is, oxidation of the alcohol by such metals proceeds via a free radical mechanism. Thus, if cellulose reacts similarly, initiating free radicals on the cellulose backbone in the presence of a vinyl monomer, one may expect grafting.

Indeed, synthesis of cellulose graft copolymers could be achieved using pentavalent vanadium as an initiator[93]. The reaction has been carried out as follows. Cellulose pulp (3 g) was swollen in distilled water (250 cm^3). Methyl methacrylate (25 cm^3) was added followed by addition of vanadium^{V} nitrate (10 cm^3). Vanadium^{V} nitrate was prepared by adding 3.48 g of ammonium vanadate in 12.5 cm^3 of 69% (w/v) nitric acid and the whole diluted to 100 cm^3. The reaction mixture was stirred for 24 hr. at 30 °C in nitrogen. The reaction product was filtered and washed until free from acid. After drying, the cellulose was extracted with benzene (300 cm^3) for hours. The graft yield obtained amounted to 37.33%.

4.2.5. Initiation by Manganese^{III} Ions

Duke[99] has studied the oxidation of the oxalate ion by manganese ions (Mn^{III}). The oxidation proceeds by an electron transfer reaction via a free radical intermediate. Oxidation of cellulose with Mn^{III} might follow the same mechanism. Free radicals would be formed on the cellulose backbone upon treatment with solutions of manganese salts in high oxidation states. It is believed that the mechanism of the initiation reaction is similar to that of the Ce^{IV} ion/glycol system and can be written as follows[74]:

$$Mn^{III} + RCH_2OH \rightleftharpoons Mn^{II} + H^+ + R\dot{C}HOH \text{ or } RCH_2\dot{O} . \qquad (151)$$

The free radicals produced can initiate graft copolymerization if a vinyl monomer is present.

Using manganese^{III} sulfate in the presence of sulfuric acid as initiator, Singh et al.[100] have been able to graft a number of vinyl monomers onto different substrates containing hydroxyl groups. The grafting levels show a significant dependence on the substrate form.

Obviously, a considerable amount of graft formation was obtained when methyl methacrylate, acrylonitrile, and methyl acrylate were used as monomers. The failure of this system to induce grafting of styrene and vinyl acetate has been interpreted in terms of poor solubility of styrene in water and low reactivity in the case of vinyl acetate. With respect to the substrates, carboxymethyl cellulose showed low graft yields due to the reduction in the number of available hydroxyl groups on the cellulose chain, which act as sites for graft initiation.

4.2.6. Initiation by Manganese^{IV}

Graft copolymerization of vinyl monomers onto textile fibers has been made possible by the use of potassium permanganate as initiator (101–104). Radical formation in this system seems likely to occur through reduction of Mn^{IV} to either Mn^{III} or Mn^{II}. Thus, during vinyl copolymerization on cellulose (cell—OH) under the catalytic influence of potassium permanganate in the presence of an acid (HR), creation of free radicals in the system may be presented as follows[104]:

$$Mn^{IV} + HR \rightarrow Mn^{III} + H^+ + R^{\cdot}, \quad (152)$$

$$Mn^{IV} + H_2O \rightarrow Mn^{III} + H^+ + HO^{\cdot}, \quad (153)$$

$$R^{\cdot} + H_2O \rightarrow RH + HO^{\cdot}. \quad (154)$$

The radical, $R^{\cdot}$, or the hydroxyl radical, $HO^{\cdot}$, attacks cellulose, forming a cellulose macro-radical. The latter may also be formed according to the following reactions:

$$\text{Cell—OH} + Mn^{IV} \rightarrow Mn^{III} + H^+ + \text{Cell—O}^{\cdot}, \quad (155)$$

$$\text{Cell—H} + Mn^{III} \rightarrow Mn^{II} + H^+ + \text{Cell—O}^{\cdot}. \quad (156)$$

A German patent by Teichman and Uhlig[101] outlined the grafting of vinyl monomers onto textile fibers with potassium permanganate as follows: Nylon (lg) was added to 200 cm^3 of potassium permanganate (316 mg $KMnO_4$/200 cm^3 H_2O) at 60 °C. To this was added 15.9 cm^3 acrylonitrile, which resulted in the formation of active MnO_2. The addition of 38 cm^3 of concentrated HNO_3 started the grafting process after one hour. After washing and extracting with N,N'-dimethylformamide at 100 °C, followed by a second washing with water and drying, the polymer showed a 64.5% weight increase due to grafting. In a similar manner, grafting of viscose with ethyl acrylate, and cotton with acrylonitrile, could be achieved. The vapor phase technique was also used in effecting graft copolymerization by this initiator, i.e., potassium permanganate.

Hebeish et al.[104] have extensively studied the copolymerization reactions initiated by manganese^{IV} on a number of cellulosic substrates. Conclusions arrived at from this study may be summarized as follows:

(1) The grafting reaction is characterized by an initial fast rate followed by a slower rate which levels off after a certain time. Leveling off of grafting with time was attributed to a decrease in concentration of both initiator and monomer, as well as to a reduction in the number of sites on the cellulose backbone accessible for grafting as the reaction proceeds.

(2) The effect of raising the temperature of the polymerization system from 30° to 50 °C is to bring about a significant enhancement in the rate of grafting. This is perhaps due to increases in the solubility of the monomer (acrylonitrile and methyl methacrylate), in diffusion of both monomer and the initiator, or to changes in the rates of permanganate decomposition, initiation, and propagation of the graft.

(3) Increasing the permanganate concentration from 0–20 milliequivalents per liter causes a significant increase to the graft yield. Further increases in the initiator concentration cause a decrease in the amount of graft formation, particularly when the polymerization is allowed to proceed for a prolonged time (4–6 hours). The reason for this is that at higher permanganate concentrations, extensive homopolymerization occurs at the expense of grafting.

(4) Increasing the monomer (methyl methacrylate) concentration is accompanied by a significant increase in the graft yield. Moreover, the proportion of monomer converted to grafted polymer is substantially greater at higher monomer concentrations. This accelerated grafting is ascribed to the gel effect.

(5) The graft yield increases with increasing concentration of the acid, attains a maximum, and then falls at a higher concentration. The fall at a higher concentration was attributed to a variety of conditions. First, coagulation of the colloidal homopolymer, in solution and within the fibers during the grafting reaction, increases at lower pH. This retards diffusion of both monomer and initiator into the fiber. Secondly, side reactions, such as $MnO_2 + 2H^+ \rightarrow Mn^{II} + H_2O + 0$, increase at higher acid concentrations. Thus, besides consuming some MnO_2 (required for production of initiating species), this leads to constant production of inhibiting oxygen. Thirdly, the initiating species may be destroyed by side reactions such as, for instance, reaction of the acid radicals with Mn^{III}.

(6) Within the range studied, the effectiveness of the acids follows the order: nitric acid > sulfuric acid > perchloric acid > hydrochloric acid. This has been interpreted in terms of the differences in the side reactions, in the affinity to cellulose of the acid species, and in the half-life of the free radicals formed when these acids were used.

(7) Introduction of carboxymethyl groups into the cellulose molecule prior to grafting enhances the graft rate, as compared with sodium hydroxide-treated cotton. This was anticipated in view of the greater swellability of such modified cotton.

(8) The extent of grafting increases upon increasing the degree of substitution of cyanoethylated cotton. Besides opening up the cellulose structure, the

cyanoethyl groups afford additional sites for grafting, thereby enhancing graft formation, in agreement with the results obtained when the Ce^{IV} ion method was used.

(9) The effect of introducing acetyl groups into the cellulose molecule is to cause a significant decrease in the susceptibility of the cellulose toward grafting. This was ascribed to lack of swellability and to blocking of some of the cellulose hydroxyls.

4.2.7. Initiation by Ozone-Oxygen Mixtures

Treatment of cellulose with an ozone-oxygen mixture[105] introduces hydroperoxide groups. Thermal decomposition of these groups yields cellulose macroradicals capable of initiating grafting. Some crosslinking will result if the termination occurs mainly by coupling. On the other hand, there are many vinyl monomers, especially the acrylates and methacrylates, for which disproportionation is the main termination step and simple grafting will be favored over crosslinking.

The e.s.r. method has been used to study the radical character of cellulose peroxide and hydroperoxide thermal decomposition[106]. It was concluded that the free radical formed arises from a cleavage of the form:

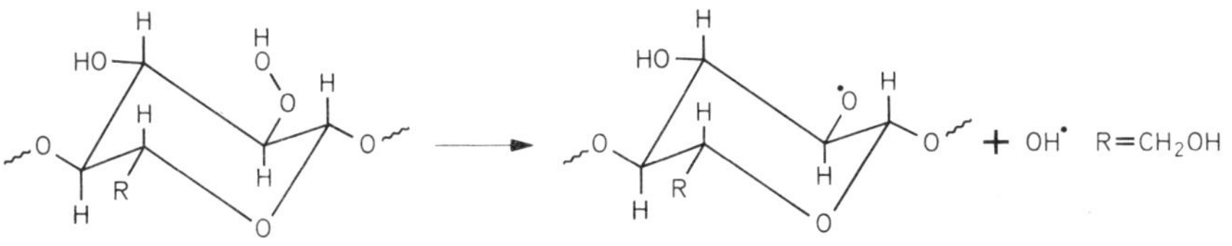

(157)

Ozonizing was carried out experimentally, by placing the cellulose samples in glass cells through which ozone was passed. This hydroperoxidizing process depends on the moisture content of both the ozone and the cellulosic materials, the duration of the ozonizing, the storage time, and the working temperature. Grafting was achieved in hermetically closed isothermal glass tubes, in which monomers and peroxidized cellulose (rayon fiber) were introduced at temperatures between 80–110 °C. The water-soluble monomers (acrylamide and methacrylic acid) were used in aqueous solution, while the water insoluble monomers (the esters of the acrylic and methacrylic acids) were applied as emulsions containing 0.5% sodium oleate in the former case and 1% Dresinat 204 in the latter[106].

The quantity of synthetic polymers grafted onto the cellulose was found to depend on a number of factors. Among these are the time allocated to ozonization and grafting, the monomer concentration, and the working temperature. The degree of grafting is favorably influenced by increasing the magnitude of these factors. The efficiency of grafting of the acrylic monomers follows the order acrylamide > methyl acrylate > ethyl acrylate > butyl acrylate. The methacrylate monomers, despite their longer processing time, have efficiencies in the order methyl methacrylate > methacrylic acid > ethyl methacrylate > butyl methacrylate[106].

4.2.8. Initiation by Periodate Ions

It is well known that sodium periodate oxidizes cellulose specifically at carbon C_2 and carbon C_3 with cleavage of the glycol bond[107,108]. It has been postulated that this reaction proceeds via a free radical mechanism[109,110]. Hence, sodium periodate would be expected to make possible graft copolymerization to cellulose by utilization of such cellulose radicals.

According to a Japanese patent[111], grafting to cellulosic paper was carried out as follows. A 20 × 5 cm filter paper was treated with 16 cm^3 aqueous $NaIO_4$ (0.01 mol/l) saturated with acrylonitrile at 25 °C for 5 hours. The paper increased in weight by 140%. The treated sample was not soluble in an aqueous solution of the standard cupriethylenediamine complex after standing for 24 hours at room temperature. For grafting to viscose, the following procedure was adopted. Viscose rayon (3 parts) was treated with 100 parts aqueous deoxygenated HIO_4 (0.01 mol/l) and 10 parts acrylonitrile at 50 °C, for 60 minutes. After decomposing HIO_4 with KI, a grafted rayon was obtained in an amount of 4.4 parts.

A German patent[112] describes the grafting of vinyl monomers onto polymers containing hydroxyl groups by employing HIO_4 as the initiator. For example, 7 g scoured cotton cellulose was placed in a flask to which 10 cm^3 aqueous HIO_4 (0.2 M) and 10 cm^3 of acrylonitrile were added. The solution was made up to 250 cm^3 and the reaction was conducted under nitrogen for 3 hours at 23 °C. The increase in weight of cotton due to grafting was 5.6%. Satisfactory grafting onto cotton was also obtained with acrylic acid, methyl methacrylate, and tertiary butyroacrylamide.

In an extensive study, Toda[113] has shown that acrylic monomers, such as acrylonitrile and methyl methacrylate, were efficiently grafted to cellulose by sodium periodate. The grafting reaction was carried out in a glass ampoule and sealed under vacuum at 50 °C, with the use of a 7% acrylonitrile solution and a 1:1 mixture of methyl methacrylate and water respectively, at a material to liquor ratio of 1:50. The concentration of sodium periodate varied from 0.5 mmol/l to 10 mmol/l. It was found that with both monomers, a maximum rate of grafting was attained at a sodium periodate concentration of approximately 2.5 mmol. A graft yield of ca. 150% and ca. 170% could be obtained with acrylonitrile and methyl methacrylate respectively, after a reaction time of 7 hours using 2.5 mmol/l sodium periodate. This is compared to yields of ca. 95% and ca. 120% with acrylonitrile and methyl methacrylate when using a concentration of 10 mmol/l sodium periodate under the conditions described above. At higher concentrations, the cellulose radicals formed by the attack of sodium periodate on cellulose are converted to aldehyde groups without contributing to the initiation of graft copolymerization. The grafted chains were also found to be much shorter in cases of higher periodate concentration than in those of lower concentrations, despite the fact that the apparent number of growing graft chains calculated from the percent grafting and the degree of polymerization of poly(methyl methacrylate) were observed to be almost constant. Further consideration of extensively periodate-treated cellulose (aldehyde cellulose) is given in Chapter 6.

4.2.9. Initiation by Aryl Diazonium Groups

Homolytic decomposition of aryl diazonium groups has been used to effect vinyl graft copolymerization onto cellulose. The method consists of introducing aromatic amine groups onto the cellulose backbone[114–118]. This can be achieved by reacting an halogenated amine, such as p-aminophenacyl chloride, with either sodium cellulosate or with the sodium salt of carboxymethyl cellulose; thus,

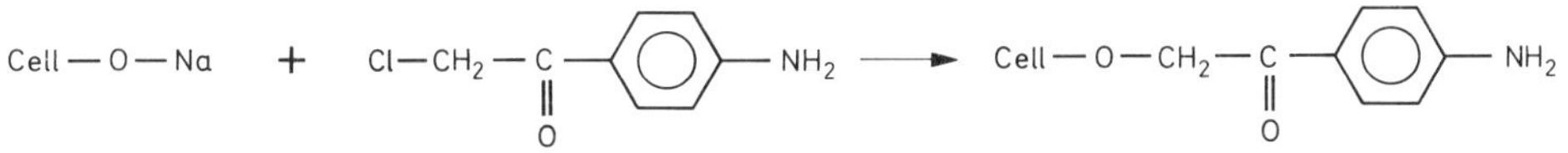

This compound is then reacted with nitrous acid to form a diazonium salt:

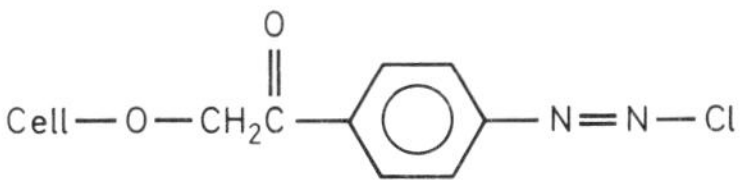

When heated, the diazonium salt splits up to produce free radicals thus:

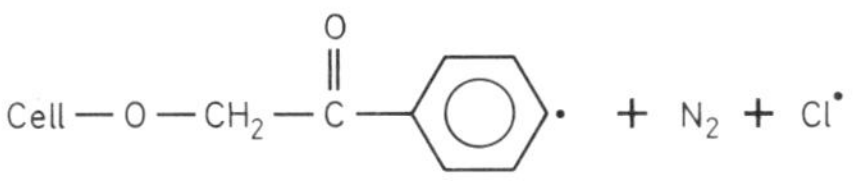

However, in the presence of a vinyl monomer, both graft polymerization through the cellulose derivative radical, as well as homopolymerization through Cl˙ radicals, occur. In order to suppress homopolymerization, a reducing agent such as a Fe^{II} salt is added which quickly destroys the Cl˙ free radicals.

Introduction of aromatic amino groups into the cellulose may be accomplished with reactive dyes according to the following mechanism[117]:

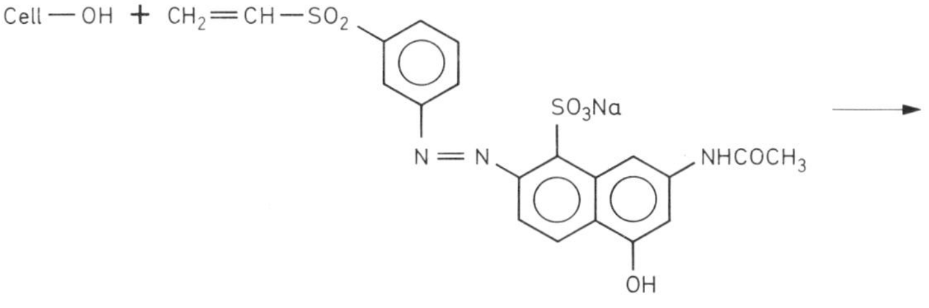

Remazol brilliant orange RR

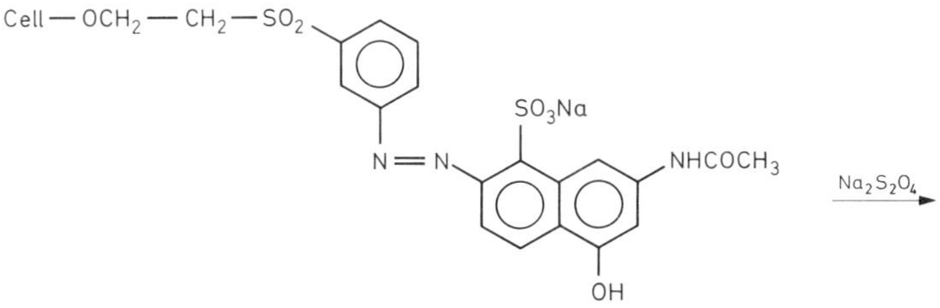

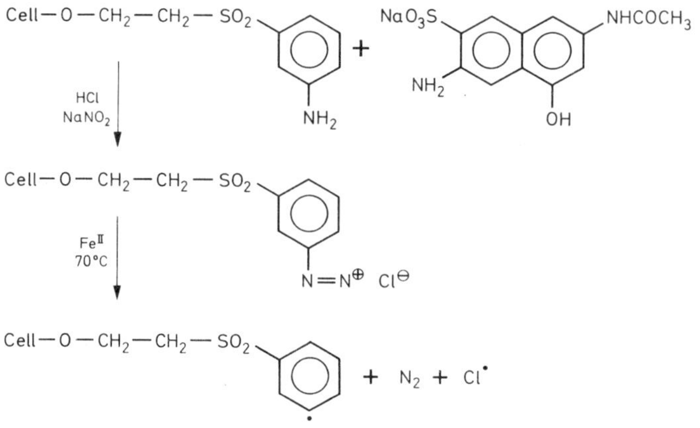

Two classes of dyes have been used, chlorotriazine dyes (Procion, Cibacron) and vinyl sulfonyl dyes (Remazol).

In grafting aromatic amino ethers or esters of cellulose with methacrylic, acrylic, and N-vinylpyrlidone monomers, the degree of grafting depends upon the nature of monomer solution, the pH, the reaction time and temperature, the degree of substitution of the aromatic amino derivative, and the concentration and type of the radical initiating grafting. Increasing the acidity and temperature of the reaction medium has been reported to cause a decrease in grafting. This could be traced back to the behavior of the diazonium salts during their thermal decomposition in media with different pH. In an acidic solution, water is the first to react while decomposition by the radical mechanism is much reduced[117)].

It has also been reported that methacrylic monomers and N-vinyl pyrrolidone show the highest activity in grafting using this method. With each series (methacrylic and acrylic monomers), the grafting degree decreases with the increase in the length of the ester chain[117)].

The nature of the initiating macroradical greatly influences the grafting degree. In a radical where a substituent in the para position has only σ links; its stability is relatively lower than one where the electronic cloud of the valency electron is not localized on a single carbon atom but on a higher number of atoms, so that the radical density is scattered.

Increasing the degree of substitution of aromatic amino-cellulose ethers results in a higher percentage of grafted polymer being obtained. Aromatic amino esters behave differently, but they have a higher degree of substitution and reduced hygroscopicty, which may account for the differences.

In addition, aromatic amino cellulose derivatives, obtained by the agency of the chlorotriazine type of reactive dyes, showed a lower degree of grafting than the aromatic amino cellulose derivatives prepared by the agency of the vinyl sulfone type of reactive dyes. This occurs as a consequence of macroradical stabilization by existing substituents from the aromatic nucleus.

The main advantages of grafting by means of cellulose diazonium salts may be outlined as follows:

(1) The number of free radicals produced on cellulose can be estimated by the nitrogen content of the sample. Furthermore, in the case of sodium carboxy-

methyl cellulose, the grafted polymer chains can be easily removed by hydrolysis, since grafting takes place through an ester linkage. Thus, the estimate of molecular masses of the grafted polymer segments is quite accurate in this case, since the conditions of hydrolysis are mild and no degradation of the polymer chains is expected to occur.

(2) The cellulose macroradical is highly active and provides the opportunity of grafting different polymers, some of which cannot be joined to cellulose using other methods of initiation.

(3) In the presence of a metal with variable valency, homopolymer formation can be eliminated and pure cellulose graft copolymers can be obtained.

This method displays a number of serious disadvantages, including the following: (1) the necessity of first introducing into the cellulose molecule an aromatic amino group; (2) the formation of colored compounds as a result of the interaction of the diazo group with those free aromatic amino groups which do not participate in the diazotization reaction. Commercial exploitation of this route to grafting seems very unlikely in view of the complex nature of the radical producing stages[118)].

4.2.10. Initiation by Radical Attack on Unsaturated Groups

Free radical transfer reactions with unsaturated polymers are most likely to occur through the abstraction of allylic hydrogen atoms. The formation of such resonance-stabilized macroradicals was studied by Schulz et al.[119,120)]. They used poly(methyl methacrylate) terminal double bonds which resulted from termination by disproportionation. The chain transfer constant for AIBN-initiated polystyrene radicals with the allylic hydrogens at the chain ends is $C_{term.} = 11 \times 10^{-2}$; and with hydrogen atoms along the backbone, $C_{interm.} \leq 0.3 \times 10^{-4}$.[119)] Thus, if styrene is polymerized in the presence of such a polymer, the preferred reactivity at the chain ends leads to extensive block copolymer formation.

Grafting of various vinyl monomers, such as methyl methacrylate, styrene, vinyl acetate, acrylonitrile, etc., onto natural and synthetic rubbers has been extensively studied[121)]. Grafting depends on the nature of the initiator. Allen et al.[121,122)] found that good yields of methyl methacrylate-rubber graft copolymers are obtained by the use of benzoyl peroxide as initiator, while AIBN produces only a mixture of homopolymers.

The mechanism of grafting in the methyl methacrylate-polyisoprene system was studied with the aid of C^{14}-labelled initiators[122)]. From the distribution of initiator fragments in the copolymer before and after ozonolysis of the polyisoprene backbone, it was concluded that free radical sites on the rubber molecule are produced by the addition and transfer of the primary radical:

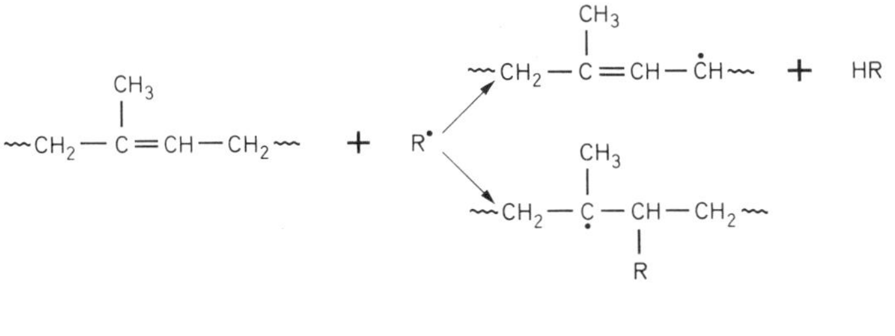

Graft formation has also been reported through addition copolymerization for several polymeric systems containing residual double bonds. By copolymerizing small amounts of ethylidene dimethacrylate with methyl methacrylate, a soluble polymer is obtained that has reactive double bonds[132]:

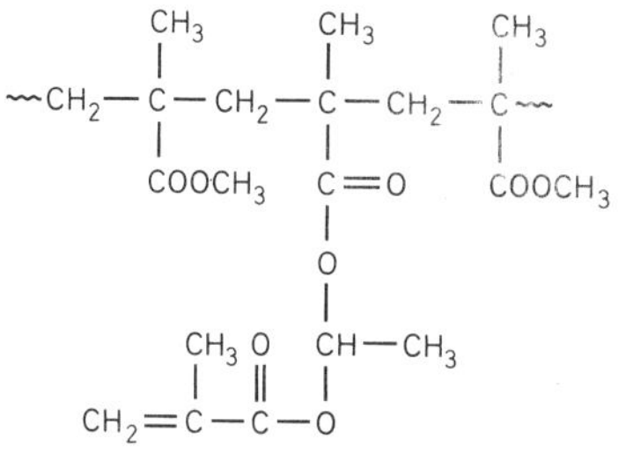

Polymerization of styrene in the presence of this unsaturated polymer produces graft copolymer by addition copolymerization.

Cellulose can be made amenable to grafting if it contains unsaturated groups along its molecule[133-135]. Cellulose containing unsaturated groups may be prepared by reacting it with compounds having two functional groups, one of which is a double bond, such as N-methylol acrylamide and methacryloyl chloride. For example, the partial esterification of cellulose with methacryloyl anhydride allows copolymerization, that is, grafting of cellulose with methyl methacrylate[133].

4.2.11. Initiation by Dimethylaniline

El-Rafie et al.[136] have examined the dimethylaniline (DMA) — Cu^{II} ion system in the graft polymerization of methyl methacrylate onto viscose fiber. DMA has, in fact, been employed as a catalyst with other co-initiators, such as Cu^{II} nitrate[137], benzoyl peroxides[138] or benzyl chloride[139]. The DMA-benzyl chloride system has been shown to be efficient in the grafting of methyl mathacrylate onto woollen substrates[140]. In El-Rafie's study various factors were examined, including the influence of the concentration of DMA, Cu^{II} ion, methyl methacrylate, the reaction time, and the temperature[136]. At 80 °C, the grafting process generally displayed an initial rapid rate followed by a slower one. At 70 °C, however, the graft yield increased directly with the reaction time. It was shown that increasing the monomer concentration did not significantly effect the graft yield during the first forty-five minutes of reaction. Beyond this time, the effect of monomer concentration increases was marked. Figure 40 shows the graft yield as a function of the copper sulfate concentration and indicates that grafting occurs even in the absence of catalyst, but to a considerably reduced extent. The critical nature of the copper sulfate concentration is seen in the maximum grafting which can be achieved at a specific copper sulfate concentration. The occurrence of grafting in the absence of copper may be explained by autoxidation of DMA via a free radical intermediate, which attacks the viscose rayon, producing macro-radicals in the system. Additional radicals, such as the methyl anilinomethyl radical[141] and the $HO_2^{\cdot}$ radical[142], may be created in such a system.

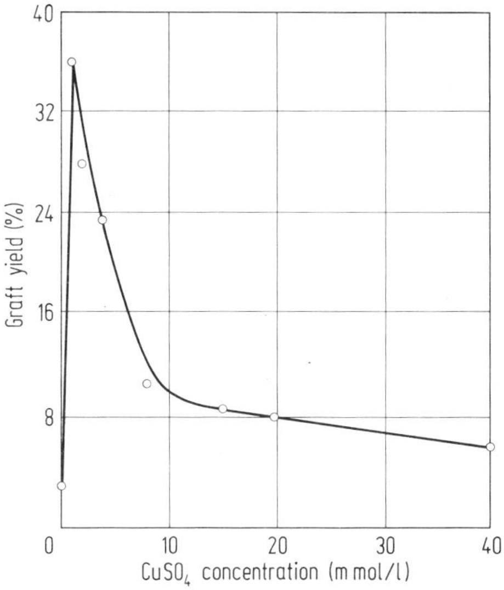

Fig. 40. Effect of the copper sulfate concentration on the graft yield. Temperature 70 °C; methyl methacrylate concentration 12%; DMA concentration 80 mmol l^{-1}. Polymerization time 4 hours; monomer/liquor ratio 1:50. Reprinted from Ref.[136], p. 2904, courtesy of John Wiley and Sons, Inc.

Figure 41 clearly shows the significance of the DMA concentration in the grafting process, with maximal grafting observed at a value of 80 mmol/l. Further increases in the DMA concentration only decrease the yield of grafting. It is supposed that the methyl anilinomethyl radical may take part in termination reactions involving the graft and also in coupling of the type

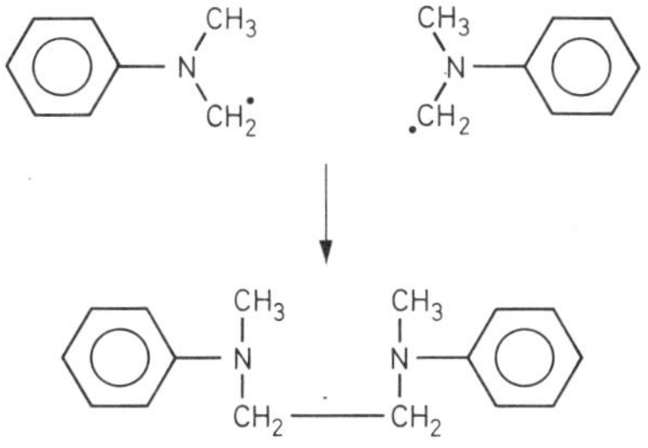

Studies of the monomer dependence of grafting indicate that gel effects operate in this type of system as seen in Figure 42. The gel effect seems to have two functions: it hinders termination of the growing polymer chain radicals by coupling and also by enhancing the viscose rayon swellability; it thus assists diffusion of monomer to growing grafted chains and active sites on the viscose backbone, thereby accelerating grafting. The effect of temperature change on the grafting reaction is clearly shown in Fig. 43. An increase in temperature from 70 to 80 °C causes a significant enhancement in the rate of grafting. With a fourhour reaction time, the grafting reaction shows at 80 °C an initial rapid rate followed by a slower one. At 70 °C, on the other hand, the graft yield increased in proportion to the increase in reaction time within the same duration. El-Rafie concludes that raising the reaction temperature seems to accentuate the

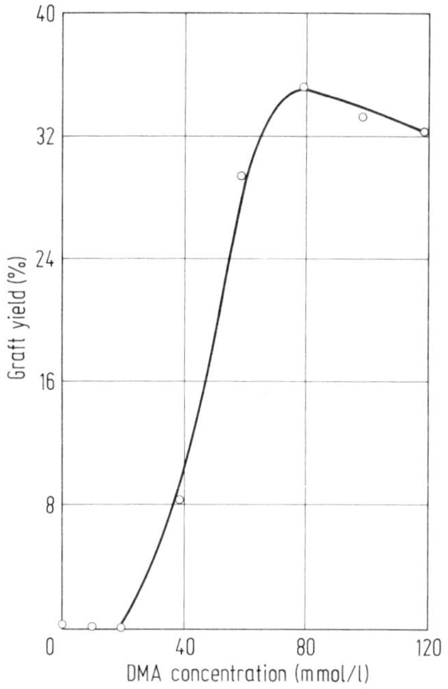

Fig. 41. Effect of the dimethylaniline concentration on percent graft yield. Temperature 70 °C, $CuSO_4$ concentration 1 mmol l^{-1} methyl methacrylate concentration, 12%; time 4 hours; monomer/liquor ratio, 1:50. Reprinted from Ref.[136], p. 2906, courtesy of John Wiley and Sons, Inc.

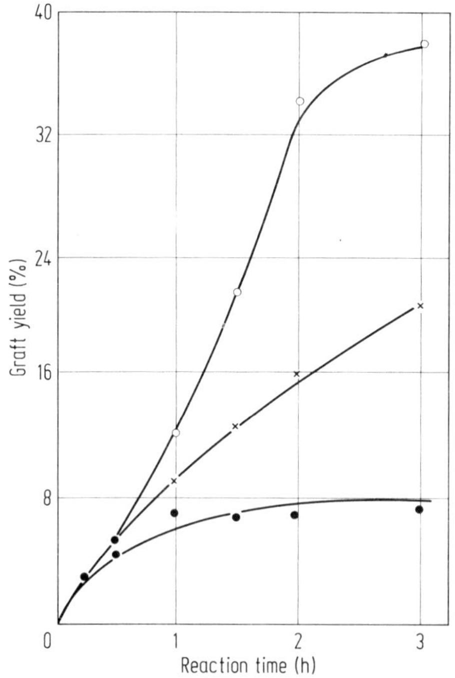

Fig. 42. Influence of the methyl methacrylate concentration on the rate of grafting. (●), 4% MMA, (×), 8% MMA; (○) 12 % MMA; Temperature 80 °C; $CuSO_4$ concentration 1 mmol l^{-1} DMA concentration 80 mmol l^{-1}; Monomer/liquor ratio 1:50. Reprinted from Ref.[136], p. 2907, courtesy of John Wiley and Sons, Inc.

creation of free radicals in the polymerization medium, assist the swellability of the cellulosic substrate, increase the solubility of the monomer in the reaction medium, and enhance the diffusion of the monomer into the fibers. Additional features include an apparent increase in efficiency of initiation and propagation of the grafting reaction when the temperature is raised to the 70 to 80 °C range. The overall effect is one of increased grafting[136].

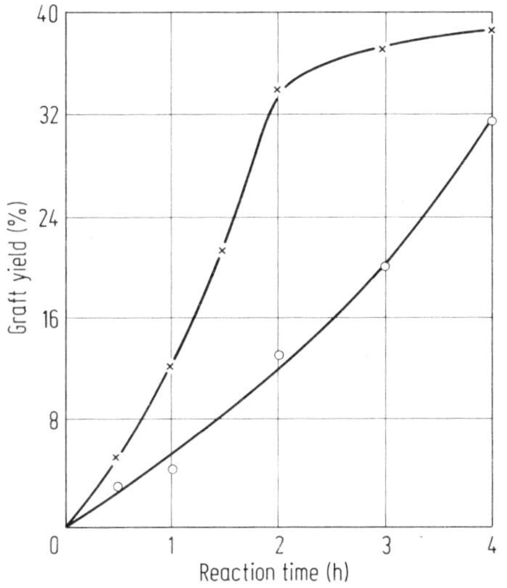

Fig. 43. Effect of temperature on grafting. (○) 70 °C; (×), 80 °C. MMA concentration, 12%; $[CuSO_4]$, 1 mmol [D.M. A], 80 mmol l^{-1}; monomer/liquor ratio 1:50. Reprinted from Ref.[136], p. 2907, courtesy of John Wiley and Sons, Inc.

4.3. Ionic Initiation

Although most of the grafting work on cellulose has been based on radical polymerization systems, ionic polymerization has also been employed[143–146]. Ionic grafting may be divided into two types:

A) Cationic grafting and B) Anionic grafting

4.3.1. Cationic Grafting

Metallic halides of the Friedel-Crafts type, e.g., BF_3, $TiCl_4$, and $AlCl_3$, are well known to effect polymerization of a number of vinyl monomers. The polymerization reaction takes place rapidly at very low temperatures and proceeds through a mechanism involving ions. In almost all cases, the reaction requires the presence of traces of substances, such as water or hydrochloric acid, which act as co-catalysts. The monomers polymerized by these substances contain electron-releasing substituents, so that the doubly bonded carbon atoms tend to share a pair of electrons with these electron acceptors. This would imply that their mode of action involves carbonium ions[147]. One possible mechanism for their mode of action is:

$$BF_3 + CH_2{=}CXY \rightarrow {}^{\ominus}F_3BCH_2C^{\oplus}XY\,, \qquad (158)$$

$${}^{\ominus}F_3BCH_2C^{\oplus}XY + CH_2{=}CXY \rightarrow {}^{\ominus}F_3BCH_2CXYCH_2C^{\oplus}XY\,, \qquad (159)$$

$${}^{\ominus}F_3BCH_2CXYCH_2C^{\oplus}XY + nCH_2{=}CXY \rightarrow \text{Polymer}\,, \qquad (160)$$

or, since a cocatalyst is required, another is possible involving proton transfer[148,144]. BF_3 will initiate polymerization of isobutylene in the presence of water. The first reaction is to form a hydrate $BF_3{-}OH_2$ which will transfer a proton

$$BF_3OH_2 + CH_2{=}C(CH_3)_2 \rightarrow (CH_3)_3C^{\oplus} + BF_3OH^{\ominus}\,, \qquad (161)$$

$$(CH_3)_3C^{\oplus} + CH_2{=}C(CH_3)_2 \rightarrow (CH_3)_3C{-}CH_2{-}C^{\oplus}(CH_3)_2\,, \qquad (162)$$

$$(CH_3)_3C{-}CH_2{-}C^{\oplus}(CH_3)_2 + nCH_2{=}C(CH_3)_2 \rightarrow \text{Polymer}\,. \qquad (163)$$

Based on the principles discussed above, cationic grafting may be represented as follows:

$$\text{Cell}{=}CHOH + BF_3 \longrightarrow \text{Cell}{=}\overset{\displaystyle H}{\overset{|}{C}}{}^{\oplus}\ \overset{\ominus}{H}OBF_3 \qquad (164)$$

$$\text{Cell}{=}\overset{\displaystyle H}{\overset{|}{C}}{}^{\oplus}\ \overset{\ominus}{H}OBF_3 + CH_2{=}C(CH_3)_2 \longrightarrow \text{Cell}{=}CH{-}CH_2{-}\underset{\underset{\displaystyle CH_3}{|}}{\overset{\overset{\displaystyle CH_3}{|}}{C}}{}^{\oplus}{-}\overset{\ominus}{H}OBF_3 \qquad (165)$$

$$\text{Cell}{=}CH{-}CH_2{-}\underset{\underset{\displaystyle CH_3}{|}}{\overset{\overset{\displaystyle CH_3}{|}}{C}}{}^{\oplus}\ \overset{\ominus}{H}OBF_3 + n(CH_2{=}C(CH_3)_2) \longrightarrow \begin{matrix}\text{Graft}\\ \text{Copolymer}\end{matrix} \qquad (166)$$

4.3.2. Anionic Grafting

Many vinyl monomers undergo addition reactions with compound containing active hydrogen atoms. Additions of this general type, known as Michael reactions, require basic catalysts; one of the most powerful catalysts for such reaction is a sodium alkoxide[150]. Thus sodium cellulosate, as well as alkali cellulose, should readily react with certain vinyl compounds. The cyanoethylation of cellulose with acrylonitrile in the presence of dilute sodium hydroxide is a Michael reaction[151].

Thomas[152] has reviewed the polymerization of acrylonitrile and discussed the mechanism of anionic polymerization in forming the homopolymer. The acrylonitrile double bond is activated by the presence of the CN group, which withdraws electrons; in the case of vinyl monomers, $CH_2{=}CHX$, the ease of anionic polymerization can be expected to be a function of the electron-withdrawing power of X. Comparatively recently, the methods of anionic polymerization of vinyl monomers have been adopted for graft polymerization to cellulose. Alkali metal cellulosates were used as initiators for the anionic graft polymerization of acrylonitrile, methacrylonitrile, methyl methacrylate, and other vinyl monomers[144,145]. Grafting was carried out under conditions where rapid termination was not favored following the interaction of a single monomer molecule with the cellulose.

Based on what is known of the literature on anionic polymerization, the anionic graft polymerization of a vinyl monomer, for example, acrylonitrile on sodium cellulosate in aprotic solvents, may be given as follows[144,145,153]:

(i) Initiation

$$\begin{aligned}&\text{Cell—O}^{\ominus}\,\text{Na}^{\oplus} + \text{CH}_2 = \text{CHCN}\\ &\rightarrow [\text{Cell—O—CH}_2\text{—CHC—N}]\,\text{Na}^{\oplus}\,,\end{aligned} \tag{167}$$

(ii) Propagation

$$\begin{aligned}&\text{Na}^{\oplus}[\text{Cell—O—CH}_2\text{CH}^{\ominus}\text{CN}] + \text{n}(\text{CH}_2{=}\text{CHCN}) \rightarrow\\ &\rightarrow \left[\text{Cell—O}\!\left(\text{CH}_2\text{CHCN}\right)_{n-1}\text{CH}_2\overset{\text{CN}^{\ominus}}{\overset{|}{\text{CH}}}\right]\text{Na}^{\oplus}.\end{aligned} \tag{168}$$

(iii) Termination
(By chain transfer to monomer)

$$\begin{aligned}&\text{Cell—O}\!\left(\text{CH}_2\text{—CHCN}\right)_{n-1}\text{CH}_2\text{CH}^{\ominus}\text{CN Na}^{\oplus} + \text{CH}_2 = \text{CHCN}\\ &\rightarrow \text{Cell—O}\!\left(\text{CH}_2\text{—CHCN}\right)_{n-1}\text{CH}_2\text{—CH}_2\text{—CN} + \text{CH}_2\\ &= \text{C—C}^{\ominus}\text{N Na}^{\oplus}\,.\end{aligned} \tag{169}$$

(iv) Termination
(By chain transfer to the free —OH of cellulose)

$$\begin{aligned}&\text{Cell—O}\!\left(\text{CH}_2\text{—CHCN}\right)_{n-1}\text{CH}_2\text{C}^{\ominus}\text{HCN} + \text{Cell—OH}\,,\\ &\text{Cell—O}\!\left(\text{CH}_2\text{—CHCN}\right)_{n-1}\text{CH}_2\text{—CH}_2\text{—CN} + \text{Cell—O}^{-}\end{aligned} \tag{170}$$

(v) Homopolymer Formation
Initiated by the highly reactive species $\text{CH}_2\text{C}^{\ominus}\text{CN}$ formed in step (iii) above

$$\text{CH}_2{=}\overset{\ominus}{\text{C}}\text{—CN} + \text{n}(\text{CH}_2{=}\text{CHCN}) \longrightarrow \text{CH}_2{=}\overset{\text{CN}}{\overset{|}{\text{C}}}\text{—}(\text{CH}_2\text{—}\overset{\text{CN}}{\overset{|}{\text{CH}}})_{n-1}\,\text{CH}_2\overset{\ominus}{\text{C}}\text{HCN} \tag{171}$$

$$\begin{aligned}&\text{CH}_2{=}\overset{\text{CN}}{\overset{|}{\text{C}}}\text{—}(\text{CH}_2\text{—}\overset{\text{CN}}{\overset{|}{\text{CH}}})_{n-1}\text{CH}_2\text{—}\overset{\ominus}{\text{C}}\text{H—CN} + \text{CH}_2{=}\text{CH—CN} \longrightarrow\\ &\text{CH}_2{=}\overset{\text{CN}}{\overset{|}{\text{C}}}\text{—}(\text{CH}_2\text{—}\overset{\text{CN}}{\overset{|}{\text{CH}}})_{n-1}\text{CH}_2\text{—CH}_2\text{—CN} + \text{CH}_2{=}\overset{\ominus}{\text{C}}\text{—CN}\end{aligned} \tag{172}$$

Schwenker and Pacsu[144] have studied the reactions of acrylonitrile and other vinyl monomers with sodium cellulosate and alkali cellulose substrates. It is apparent that the reaction of selected monomers with either sodium cellulosate or alkali cellulose under controlled conditions leads to the formation of graft copolymers of cellulose and the vinyl polymers. According to these investigators the reaction of acrylonitrile with sodium cellulosate appears to result in a mixed product involving primarily (i) a graft copolymer and (ii) acrylonitrile homopolymer, although the possibility of a small amount of cyanoethylation cannot be completely excluded. However, the residue after extraction with dimethylformamide (a solvent for polyacrylonitrile) is believed to be predominantly in the grafted form. Similar products appear to be formed when alkali cellulose is used as the substrate.

It has been well established that low molecular mass polymers are obtained by anionic polymerization of vinyl monomers initiated with alkali metal alkoxides[154, 155]. It has also been shown that the formation of sodium cellulosate with a relatively high alcoholate substitution can be accomplished by an exchange reaction with sodium methoxide, with no chemical degradation and no measurable alteration in the physical structure of the cellulose fiber[161]. Thus, this approach to grafting is particularly promising for the preparation of graft copolymers with low molecular mass side chains and high degrees of substitution[157–161].

Simionescu and Rusan have provided a related route to anionic grafting through their use of metallated cellulose monoethyl malonate as a macromolecular initiator for the grafting of certain vinyl monomers[146]. These researchers have successfully grafted acrylonitrile onto sodium cellulose monoethyl malonate. The copolymerization was carried out in dry N,N-dimethylformamide at —40 °C for 1 hour in an argon atmosphere. After extraction to remove polyacrylonitrile homopolymer, the graft copolymers were obtained as white to yellow/brown products.

Kido et al. have metallated samples of rayon with sodium methoxide[162] and allowed the activated products to react with acrylamide and subsequently graft-copolymerized them with poly-β alanine. Here, the rayon samples were swollen with boiling water and dehydrated by repeated immersions in methanol. Reaction with sodium methoxide at 30° for 24 hours followed removal of the excess methanol. Extraction with dimethylsulfoxide allowed removal of excess methoxide solution. Grafting was achieved by reacting the metallated rayon with acrylamide solution at 100 °C in a nitrogen atmosphere in a sealed glass tube for the required time. The graft copolymer was purified by repeated extraction in 50% aqueous phenol and washing with hot water. The reaction conditions, including the reaction time, the amount of Na combined (mol. per glucose unit), inhibition of radical polymerization, the acrylamide concentration, and the dielectric constant of the solvent were investigated and certain properties of the grafted fiber were measured.

Marchessault et al. have reported developments in a technique of cellulose (paper) graft composite formation in which the monomer (ethylene) is directly grafted within the cellulosic substrate[163]. Marked improvements are thus obtained in most sheet properties. The technique involves absorbing a

Ziegler-Natta catalyst at the external surface of the cellulose fibers and synthesis thereon of a layer of nascent polyethylene having controllable thickness. The catalyst system ($TiCl_4/AlEt_3$), when deposited on the cellulose surface, provides instantaneous grafting when brought into contact with ethylene.

References

1. Bains, M. S.: J. Polymer Sci. *37*, 125 (1972)
2. Kolthoff, I. M., Miller, I. K.: J. Amer. Chem. Soc. *73*, 3055 (1951)
3. Smith, W. V., Cambell, H. N., J.: Chem. Phys. *15*, 388 (1947)
4. Marie, C., Bunel, L. J.: Bull. Soc. Chim. Paris, Ser. 3, *29*, 930 (1903)
5. Bartlett, P. D., Cotman, J. D., Jr.: J. Amer. Chem. Soc. *71*, 1419 (1949)
6. Kolthoff, I. M., Meeham, E. J., Carr, E. M.: J. Amer. Chem. Soc. *75*, 1439 (1953)
7. Haydel, C. H., Janssen, H. J., Seal, J. F., Vix, H. L. E., Gastrick, E. A.: Textile Res. J. *27*, 975 (1957)
8. Hayes, R. A.: J. Polymer Sci. *13*, 585 (1954)
9. Chaudhuri, D. K. R., Hermans, J. J.: J. Polymer Sci. *51*, 373 (1961)
10. Sankalia, S. H., Chaudhuri, D. K., Hermans, J. J.: Canad. J. Chem. *40*, 2249 (1962)
11. Kulkarni, A. Y., Chitale, A. G., Vaidya, B. K., Mehta, P. C.: J. Appl. Polym. Sci. *7*, 1581 (1963)
12. Simionescu, C., Ungureanu, C.: Cell, Chem. Technol. *1*, 33 (1967)
13. Gaylord, N. G., Tomono, T.: J. Polymer Sci. *B 13*, 697 (1975)
14. Gaylord, N. G., Takahashi, A., Anand, L. C.: J. Polymer Sci. *9*, 97 (1971)
15. Gaylord, N. G., Anand, L. C.: J. Polymer Sci. *B 10*, 305 (1972)
16. Gaylord, N. G., Anand, L. C.: J. Polymer Sci. *9*, 617 (1971)
17. Gaylord, N. G., Tomono, T.: J. Polymer Sci. *B 13*, 689 (1975)
18. Gaylord, N. G.: U.S. Patent *3*, 740, 362 (1973)
19. Gaylord, N. G.: U.S., Patent *3*, 781, 232 (1973)
20. Gaylord, N. G., Tomono, T., Mandal, B.: J. Polymer Sci. *B 13*, 693 (1975)
21. Gaylord, N. G.: J. Polymer Sci. *B 13*, 693 (1975)
22. Rowland, S. P., Blouin, F. A., Mason, J. S.: Textile Res. J. *48*, 73 (1978)
23. Bardhan, K., Mukhopadhyay, S., Chatterjee, S. R.: J. Polymer Sci. Chem. *15*, 141 (1977)
24. Mukhopadhyay, S., Prasad, J., Chatterjee, S. R.: Makromol. Chem. *1*, 176 (1975)
25. Uri, N.: Chem. Rev. *50*, 375 (1952)
26. Arthur, J. C. Jr., Hinojosa, O., Bains, M. S.: J. Appl. Polym. Sci. *12*, 1411 (1968)
27. Bains, M. S., Arthur, J. C. Jr., Hinojosa, O.: J. Phys. Chem., *72*, 2250 (1968)
28. Faber, F., Weiss, J.: Naturwissenschaften *20*, 948 (1932)
29. Faber, F., Weiss, J.: J. Proc. Roy. Soc. (London), *A 147*, 332 (1934)
30. Kolthoff, I. M., Parry, E. P.: J. Amer. Chem. Soc. *73*, 3718 (1951)
31. Merz, J. H., Waters, W. A.: J. Chem. Soc. 515 (1949)
32. Landells, G., Whewell, C. S.: J. Soc. Dyers Colourists *67*, 338 (1951)
33. Landells, G., Whewell, C. S.: J. Soc. Dyers Colourists *71*, 171 (1955)
34. Kim, Bu. J., Dumitriu, S., Simoinescu, C. I.: Cell. Chem. Technol. *11*, 4555 (1977)
35. Focher, B., D'Anguiro, L., Cremonesi, P.: Ann. Chim. (Rome) *66*, 283 (1976)
36. Hamada, R., Maeda, E.: Kinki Daigaku Nogakubu Kigo *10*, 71 (1977)
37. Mikolajczyk, T.: Polimery *22*, 11 (1977)
38. Lawniczak, M.: Pr. Kom. Technol. Drewna, Poznan, Tav. Przyj. Nauk. *6*, 49 (1976); Through C.A. *85*, 145003 (1976)
39. Young, R. A.: J. Agric Food Chem. *25*, 738 (1977)
40. Baxendale, J. H., Evans, M. G., Park, G. S.: Trans. Faraday Soc. *42*, 155 (1946)
41. Bridgeford, D. J.: Ind. Eng. Chem. Prod. Res. Develop. *1*, 45 (1962)
42. Richards, G. N.: J. Appl. Polym. Sci. *5*, 539 (1961)
43. Uno, K., Iwakura, Y., Makita, M., Ninomiya, T.: J. Polymer Sci., A-I *5*, 2311 (1967)
44. Ogiwara, Y., Kubota, H.: J. Appl. Polym. Sci. *12*, 2575 (1968)

45. Ogiwara, Y., Kubota, H.: J. Appl. Polym. Sci. *13*, 1613 (1969)
46. Ibragimov, F., Mukhamadaliev, D., Gafurov, T. G.: Vysokomol. Soed *A 12*, 1475 (1970), Ibragimov, F., Mukhamadaliev, D., Gafurov, T. G., Usmanov, Kh. U.: J. Polymer Sci. *C 37*, 239 (1972)
47. Smith, G. F., Sullivan, V. R., Frank, G.: Ind. Eng. Chem. Anal. Ed. *8*, 449 (1936)
48. Mino, G., Kaizerman, S.: J. Polymer Sci. *31*, 242 (1958)
49. Mino, G., Kaizerman, S., Rasmussen, E.: J. Polymer Sci., *38*, 393 (1959)
50. Mino, G., Kaizerman, S., Rasmussen, E.: J. Amer. Chem. Soc., *81*, 1494 (1959)
51. Mino, G., Kaizerman, S., Rasmussen, E.: J. Polymer Sci., *39*, 523 (1959)
52. Hargreaves, G., Sutcliff, L. H.: Trans. Faraday Soc., *51*, 1105 (1955)
53. Ardon, M.: J. Chem. Soc. 1811 (1957)
54. Katai, A. A., Kulshrestha, V. K., Marchessault: J. Polymer Sci. *C 2*, 403 (1963)
55. Smith, G. F., Duke, F. R.: Ind. Eng. Chem. Anal. Ed. *15*, 120 (1943)
56. Kulkarni, A. Y., P. C. Mehta: J. Appl. Polym. Sci. *12*, 1321 (1968)
57. Hebeish, A., Mehta, P. C.: J. Appl. Polym. Sci. *12*, 1625 (1968)
58. Kantouch, A., Hebeish, A., El-Rafie, M. H.: Textile Res. J. *42*, 10 (1972)
59. Schwab, E., Stannett, V. T., Hermans, J. J.: Tappi *44*, 251 (1961)
60. Iwakura, Y., Kurosaki, T., Uno, K., Imai, Y.: J. Polymer Sci. *C 4*, 673 (1963)
61. Kaizerman, S., Mino, G., Meinhold, L. F.: Textile Res. J. *32*, 136 (1962)
62. Cumberbirch, R. J. E., Holker, J. R.: J. Soc. Dyers Colourists. *82*, 59 (1966)
63. Schwab, E., Stannett, V. T., Rokowitz, D. H., Magrane, J. K.: Tappi *45*, 390 (1962)
64. Livshits, R. M., Rogovin, Z. A.: Vysokomol. Soed., Tsellyuloza i ee Proizvodnye, sb. Statein 12–17 (1963); Through C.A. *60*, 13248 (1964)
65. Iwai, S., Kukurai, T., Noguchi, T.: Seni-i Gakkaishi *19*, 547 (1963); Through C.A. *62*, 13369 (1965)
66. Narita, H., Uchino, N., Machida, S.: Sen-i Gakkaishi *19*, 225 (1963) C.A. *62*, 13293 (1965)
67. Hebeish, A., Kantouch, A., El-Rafie, M. H.: J. Appl. Polym. Sci. *15*, 1921 (1971)
68. Bergen, N., Ellefsen, ø., Kringstad, K., Kroden, P.: Norsk Skogind. *17*, 347 (1963)
69. Movsumzade, A. A., Livshits, R. M., Rogovin, Z. A., Konkin, A. A.: Vysokomol, Soed, Tsellyuloza i ee Proizvodnye, Sb. Statein 186 (1963); Through C.A. *61*, 5893 (1964)
70. Ide, F., Handa, R., Makatsuka, K.: Kobunshi Kagaku *21*, 57 (1964); Through C. A. *61*, 1997 (1964)
71. Livshits, R. M., Alachev, V. P., Prokofeva, M. V., Rogovin, Z. A.: Vysokomol. Soed. *6*, 655 (1964); Through C.A. *61*, 5893 (1964)
72. Sun, T., Chou, H. H., Kuo, M. L., Chang, S. H.: Ko Fen Tzu T'ung Hsun *6*, 130 (1964); Through C.A. *63*, 18441 (1965)
73. Neime, L., Sihtola, H.: Paperi Puu., *47* 369 (1965); *47*, 374 (1965); Through C.A. *63*, 15086 (1965)
74. Simionescu, Cr., Feldman, D., Hrihorov-Ciubotariu, M.: Iasi Hirtie, Bucharest, Romenia, *14*, 358 (1965); Through C.A. *63*, 15089 (1965)
75. Iwakura, Y., Imai, Y., Yagi, K.: J. Polymer Sci. A-I *6*, 801 (1968)
76. Negishi, M., Nakamura, Y., Kakinuma, T., Iizuka, Y.: J. Appl. Polym. Sci. *9*, 2227 (1965)
77. Kamagawa, H., Sekiya, T.: Textile Res. *J 31*, 585 (1961)
78. Cornell, R. H.: Tappi *54*, 145 A (1962)
79. Ogiwara, Y., Kubota, H.: J. Polymer Sci., A-I *5*, 2791 (1967)
80. Hebeish, A., Ph. D. Thesis: Gujarat Univeristy India 1967
81. Hebeish, A., Mehta, P. C.: Cell. Chem. Technol. *3*, 469 (1969)
82. Kubota, H., Ogiwara, Y.: J. Appl. Polym. Sci. *13*, 1569 (1969)
83. Immergut, E. H.: In Encyclopaedia of Polymer Science and Technology Vol. 3 (Mark, H. F., Gaylord, N. G., Bikales, N. M., eds.) New York: Interscience 1965, P. 242
84. Teresaki, I., Matsuki, M.: J. Soc. Text. Chem. Ind. Japan. *18*, 147 (1962)
85. Arthur, J. C., Jr., Baugh, P. J., Hinojosa, O.: J. Appl. Polym. Sci. *10*, 1591 (1966)
86. Iwakura, Y., Kurosaki, T., Imai, Y.: J. Polymer Sci. *A* 3, 1185 (1965)
87. Hebeish, A., Kantouch, A., El-Rafie, M. H.: J. Appl. Polym. Sci. *15*, 11 (1971)
88. Imai, Y., Masuhara, E., Iwakura, Y.: J. Polymer Sci. *B 8*, 75 (1970)
89. Pottenger, C. R., Johnson, D. C.: J. Polymer Sci. A-I *8*, 301 (1970)

90. Bhattacharyya, H., Johnson, D. C.: A.C.S. Abstracts 161st National Heating Los Angeles, March 28, Carbohydrate Division, Paper 17
91. Gaylord, N. G.: ibid, Cellulose Division, Paper 59
92. Huang, R. Y-M., Chandramouli, P.: J. Appl. Polym. Sci. *12*, 2549 (1968)
93. Huang, R. Y-M., Chandramouli, P.: J. Appl. Polymer Sci. A-I *7*, 1393 (1969)
94. Erdelyi, D.: Cell. Chem. Technol. *11*, 569 (1977)
95. Watamoto, H., Tonami, T.: Nippon Kagaku Kaishi *6*, 876 (1977)
96. Messiha, N., Shaaban, A. E., Nossier, M. H.: Faserforsch. Textiltech. *29*, 290 (1978)
97. Waters, W. A.: Mechanisms of Oxidation of Organic Compounds, London: Methuen 1964, P. 403
98. Singh, H., Thampy, R. T., Chipalkatti, V. B.: J. Polymer Sci. A *3*, 1247 (1965)
99. Duke, F. R.: J. Amer. Chem. Soc. *69*, 2885 (1947)
100. Singh, H., Thampy, R. T., Chipalkatti, V. B.: J. Polymer Sci. *A 3*, 4289 (1965)
101. Teichmann, R., Uhlig, E.: East German Patent *61*, 532 (1968)
102. Uhlig, E., Teichmann, R.: Faserforsch. Textiltech. *20*, 451 (1969)
103. Kantouch, A., Abdel-Fattah, S., Hebeish, A.: Polymer Journal *3*, 675 (1972)
104. Hebeish, A., Kantouch, A., Khalil, M. I., El-Rafie, M. H.: J. Appl. Polym. Sci. *17*, 2547 (1973)
105. Landler, Y., Lebel, P.: (to Polyplastic, Inc.) German Pat., *1*, 100286 (1958)
106. Simionescu, C. I., Oprea, S.: J. Polymer Sci. *C 37*, 251 (1972)
107. Jackson, E. L., Hudson, C. S.: J. Amer. Chem. Soc. *59*, 2009 (1937)
108. Davidson, G. F.: J. Text Inst. *29*, 195 (1938)
109. Criege, R.: Angew. Chem. *50*, 153 (1937)
110. Waters, W. A.: Trans. Faraday Soc. *42*, 184 (1946)
111. Morimoto, S., Okada, K., Okada, T., Nakagawa, K.: (to Toyo Spinning Company Ltd.), Japan Pat. 2500 (1962)
112. Chemische Werke Huels A-G (by F. Stuerzenhofecker), Ger. Pat. *1*, 132, 724 (1962)
113. Toda, T.: J. Polymer Sci. *58*, 411 (1962)
114. Rogovin, Z. A., Tun, S., Virnik, A. D., Khvostenko, N. M.: Vysokomol, Soed. *4*, 571 (1962)
115. Richards, G. N.: J. Appl. Polym. Sci. *5*, 553 (1961)
116. Simionescu, C. I., Dumitriu, S.: Cell. Chem. Technol. *3*, 47 (1969)
117. Simionescu, C. I., Dumitriu, S.: J. Polymer Sci. *C 37*, 187 (1972)
118. Benson, R., Guthrie, J. T., Lartey, R. B.: Polymer *16*, 903 (1975)
119. Schulz, G. V., Henrici, G., Olivé, S.: J. Polymer Sci. *17*, 45 (1955); Henrici-Olivé, G., Olivé, S., Schulz, G. V.: Makromol. Chem. *23*, 207 (1957)
120. Schulz, G. V., Henrici, G., Olivé, S.: Z. Electrochem. *60*, 296 (1956)
121. Allen, P. W., Merrett, F. M.: J. Polymer Sci. *22*, 193 (1956)
122. Allen, P. W., Ayrey, G., Moore, C. G., Scanlan, J.: J. Polymer Sci. *36*, 55 (1959)
123. Bloomfield, G. F., Swift, P. Mcl.: J. Appl. Chem. *5*, 609 (1955)
124. Bloomfield, G. F., Merrett, F. M., Popham, F. G., Swift, P. McL.: Proc. Rubber Technol. Conf. 3rd, London 1954, P. 1.
125. Merrett, F. M.: Trans. Faraday Soc. *50*, 759 (1954)
126. Koolhaas, D. R., Van der Bie, G. J., Van Essen, W. J.: Arch. Rubbercult, Ned-Ind. *207*, 197 (1950); Through Ceresa R.J. Block and Graft Copolymers, Washington: Butterworths 1962, P. 16
127. Swift, P. M.: J. Appl. Chem. *8*, 803 (1958)
128. Szwarc, M., Levy, M., Milkovich, R.: J. Amer. Chem. Soc. *78*, 2656 (1956)[149]
129. Berlin, A. A.: Russian Chem. Rev. (English Transl.) *29*, 541 (1960)
130. Blanchette, J. A., Nielson, L. E.: J. Polymer Sci. *20*, 317 (1956)
131. Kobryner, W., Banderset, A.: J. Polymer Sci. *34*, 881 (1959)
132. Smets, C., Schmets, J.: Bull. Soc. Chim. Belges. *62*, 358 (1953)
133. Berlin, A. A., Makarova, T. A.: J. Gen. Chem. USSR. *21*, 1383 (1951); Through C.A. *46*, 8017 (1952)
134. Faraone, G., Parasaceo, G., Cogrossi, C.: J. Appl. Polym. Sci. *5*, 16 (1961)
135. Rausing, G., Sunner, S.: Tappi *45* (1), 203 A (1962)
136. El-Rafie, M. H., Waly, A. I., Hebeish, A.: J. Polym. Sci. *14*, 2903 (1976)

137. Sato, T., Takada, M., Otsu, T.: Makromol. Chem. *148*, 2903 (1971)
138. Imoto, M., Otsu, T., Ota, T.: Makromol. Chem. *16*, 10 (1955)
139. Horner, L., Knapp, K. H.: Makromol. Chem. *93*, 69 (1966)
140. Bendak, A., Khalil, M. I., El-Rafie, M. H., Hebeish, A.: J. Appl. Polym. Sci. *19*, 335 (1975)
141. El-Rafie, M. H., Hebeish, A.: J. Appl. Polym. Sci. *19*, 1815 (1975)
142. Nord, H.: Acta Chem. Scand. *9*, 442 (1955)
143. Haas, H. C., Kamath, P. M., Schuler, N. W.: J. Polymer Sci. *24*, 85 (1957)
144. Schwenker, R. F. Jr., Pacsu, E.: Tappi *46*, 665 (1963)
145. Feit, B-A., Bar-Nun, A., Lahav, M., Zilkha, A.: J. Appl. Polym. Sci. *8*, 1869 (1964)
146. Simionescu, C. I., Rusan, V.: J. Polymer Sci. *C 37*, 173 (1972)
147. Eley, D. D., Richards, A. W.: Trans. Faraday Soc. *45*, 425, 436 (1949)
148. Norrish, R. G. W., Russell, K. E.: Trans. Faraday Soc. *48*, 91 (1952)
149. Dainton, F. S., Sutherland, G. B. B. M.: J. Polymer Sci. *4*, 37 (1949)
150. Royals, E. E.: Advanced Organic Chemistry New York: Prentice-Hall 1954
151. MacGregor, J. H.: J. Soc. Dyers Col. *67*, 66 (1951)
152. Thomas, W. M.: Fortschr. Hochpolymer-Forsch. *2*, 401 (1961)
153. Avny, Y., Yom-Tov, B., Zilkha, A.: J. Appl. Polym. Sci. *9*, 3737 (1965)
154. Zilkha, A., Feit, B. H.: J. Appl. Polym. Sci. *5*, 251 (1961)
155. Feit, B. H., Zilkha, A.: J. Appl. Polym. Sci. *7*, 287 (1963)
156. Avny, Y., Rebenfeld, L.: Textile Res. J. *38*, 599 (1968)
157. Avny, Y., Rebenfeld, L.: Textile Res. J. *38*, 684 (1968)
158. Feit, B. A., Bar-Nun, A., Lahav, M., Zilkha, A.: Himia i Technolog. Polimerov 93 (1966)
159. Rusan, V., Asandei, N., Simoinescu, Cr.: Cell. Chem. Technol. *3*, 151 (1969)
160. Pearson, R. G.: J. Amer. Chem. Soc. *71*, 2212 (1949)
161. Tahan, M., Yom-Tov, B., Zolkha, A.: Europ. Polym. J. *5*, 499 (1969)
162. Kido, I., Susuki, K., Ikeda, I., Yama, S., Shunichi, K.: J. Polymer Sci. *B 13*, 767 (1975)
163. Marchessault, R. H., Fisa, B., Revol, J. F.: In Cellulose Technology Research (Turbak, A. F., ed.) A.C.S. Symp. Series No. 10. Washington: A.C.S. 1975, P. 147

5. Grafting of other Types of Monomers onto Cellulose

5.1. Introduction

The formation of copolymers of cellulose and various synthetic polymers via graft polymerization has been extensively studied[1-7]. The chemical modification of different cellulosic materials by this means has been described in detail in Chapters 2, 3, and 4. It has been shown that nearly all the grafting methods have involved the free radical initiation of vinyl and acrylic copolymerization, using cellulose as a macromolecular initiator, to form grafted side chains on the cellulose backbone. Ionic initiation of vinyl graft copolymerization has also been employed. However, little consideration has been given yet to other possibilities of obtaining graft copolymers, such as using other monomers rather than vinyl and acrylic monomers, or by directly linking preformed polymer chains to cellulose. The purpose of this chapter is to summarize our present knowledge of grafting such types of monomers and/or polymers onto cellulose.

5.2. Step-Growth Polymerization of Cyclic Compounds

In these reactions, a cyclic, highly reactive molecule adds to the hydroxyl groups of cellulose, either without a catalyst or under the catalytic influence of hydrogen or hydroxyl ions. It is well known that the ring opening and polymerization of α-epoxides, $\mathrm{R{-}\underbrace{CH{-}CH_2}_{O}}$, may be accomplished with a variety of substances: ammonia, organic acids, amines, alcohols, water, etc. Hence, the synthesis of graft copolymers involving epoxide groups is generally based on the reactions of macromolecules containing active sites, such as $-NH_2$; $-COOH$, $-OH$, with the epoxide.

A series of non-ionic detergents has been made by initiating the polyaddition of ethylene oxide to a poly(propylene oxide) having terminal hydroxyl groups[8]:

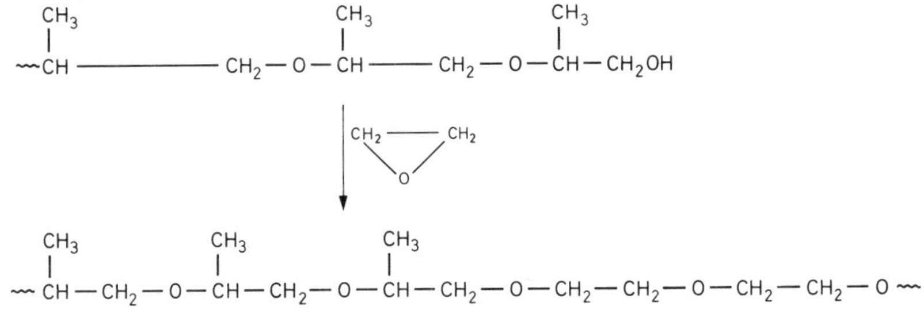

A variation of this technique utilizes a diamine to initiate propylene oxide polymerization, followed by polyaddition of ethylene oxide at the terminal —OH groups of the poly(propylene oxide)[9].

Epoxides are grafted onto polyamides via ring openings initiated by the N-hydrogen[10, 11]

Macromolecules [other than poly(alkylene oxides)] containing pendant or terminal OH groups can also initiate epoxide polymerizations. For example, copolymers of styrene and dimethyl maleate, which have been converted to the polymeric N-ethanolamide, can initiate ethylene oxide polymerization[12]. The average number of ethylene oxide units added to each N-ethanol group varies from 0.8 to 6.8. Similarly, the acetate groups of styrene-allyl acetate copolymers have been saponified to yield polyol sites which can initiate ethylene oxide grafting[12].

Attempts to graft cellulose[13] and poly(vinyl alcohol)[14] with ethylene oxide via the macromolecular —OH groups yield short polyethylene oxide chains due to the high reactivity of the secondary polymeric hydroxyl groups. The reaction of cellulose with ethylene oxide can be written as follows:

Although step-growth grafts of cellulose have not received much attention[15], a commercial product, based on a cellulose-ethylene oxide graft copolymer, has been reported[16].

Grafts of similar cyclic compounds to cellulose have also been prepared[17–19]. Examples of the grafting reactions between cellulose and some of these cyclic compounds are given below:

a. Reaction with β-propiolactone[17]:

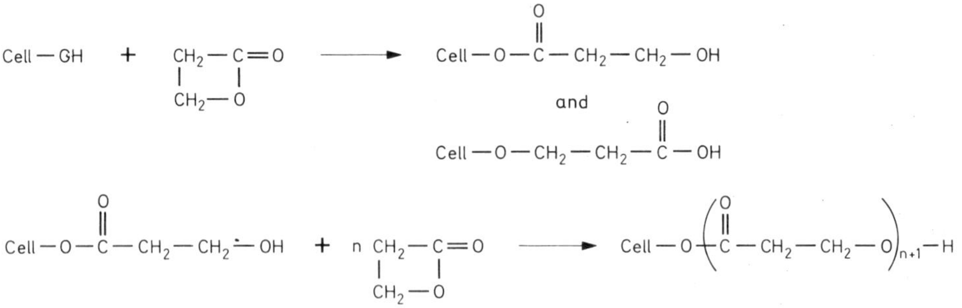

The carboxyethyl compound will also form polymer chains.

b. Treatment of cellulose with ethylenimine[18]:

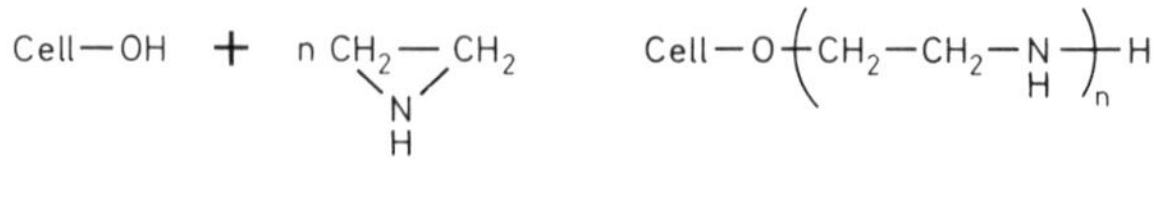

c. Polymerization of caprolactam onto the cellulosic backbone starting with carboxymethyl cellulose[19]:

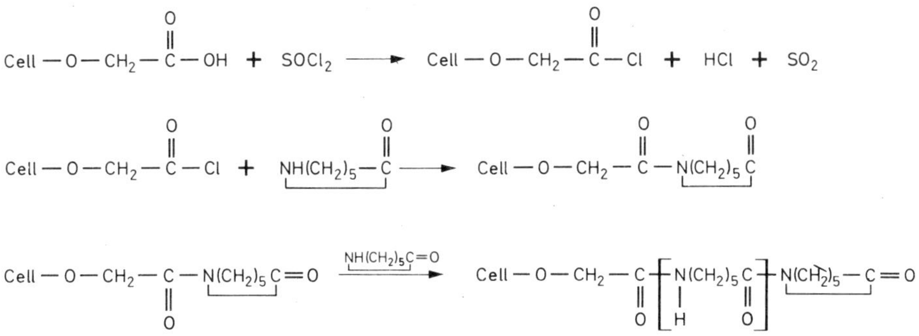

Grafting has also been achieved by making use of the Schotten-Bauman reaction, which involves the step-growth reaction between a hydroxyl group and an acid chloride (—COCl). This reaction can be carried out interfacially, that is, the cellulose and the diamine are dissolved or dispersed in one phase and the acid chloride reagent in the other phase[20]. The following scheme shows the reaction mixture for grafting nylon-6,6 onto cellulose[21]:

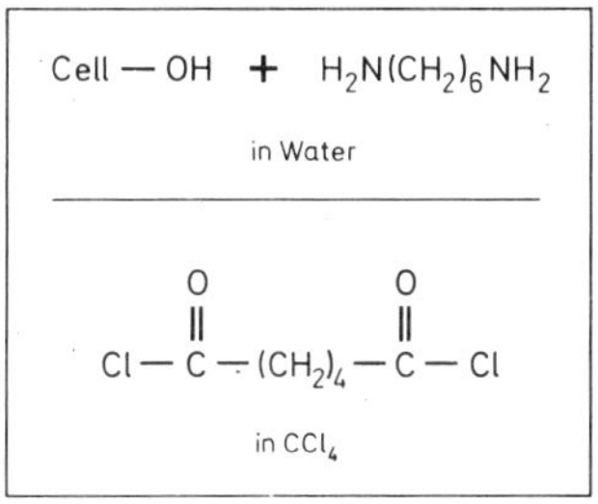

In this reaction, a mixture of graft copolymer and polyamide homopolymer will result because the diacid chloride can react with the diamine without necessarily reacting with a cellulose hydroxyl group. However, it is possible to favor the formation of graft copolymer by adding the diamine after some of the diacid chloride has reacted with the hydroxyl groups of cellulose. It is also possible to carry out this reaction in a single phase by employing an organic base, such as a trialkylamine, as a catalyst

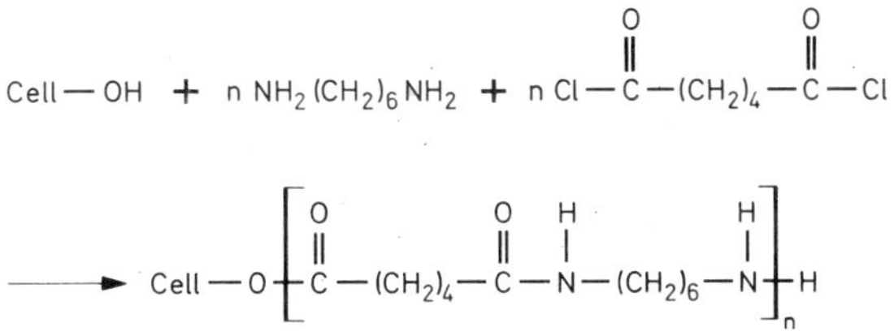

5.3. Preformed Polymers

Grafting of preformed polymers onto cellulose has been made possible. The following reaction[22] was suggested for grafting polyester chains onto a cellulose backbone:

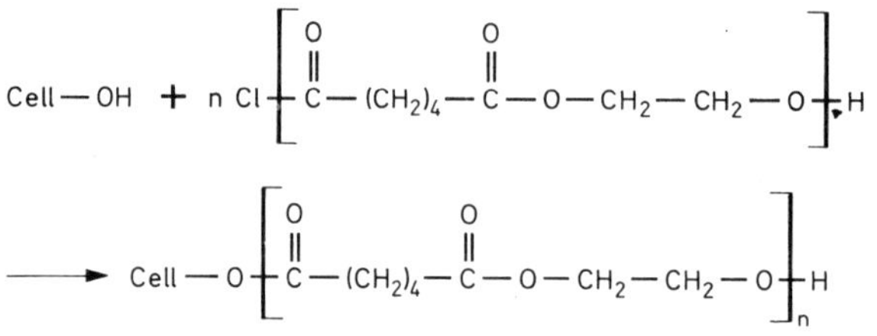

Similarly, it is possible to graft polyamides onto a cellulose backbone[23]:

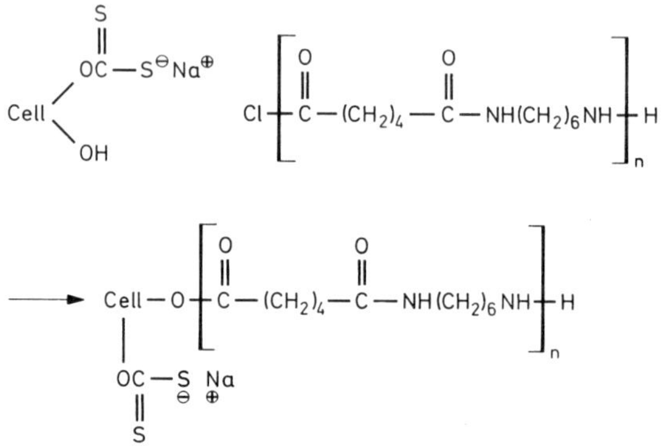

Cellulose graft copolymers may also be based upon the use of epoxy resins, which are composed of macromolocules containing reactive terminal or branch epoxide groups. These resins may be obtained by condensing epichlorohydrin with a polyhydric phenol in the presence of an alkali. The phenol is often "Bisphenol A", [2,2-bis-(p-hydroxyphenyl)propane]. The reaction can be performed in an excess of epichlorohydrin to ensure terminal epoxide groups. The chains are relatively short and may be represented as:

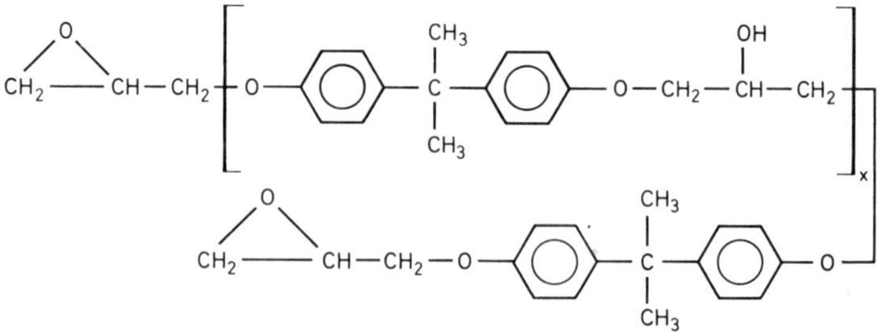

Reaction of cellulose hydroxyls with exposy resins leads to block, graft, and crosslinked copolymers via epoxide group ring opening[24,25].

Obviously, then, for grafting preformed polymers onto cellulose, short-chain synthetic polymers containing reactive terminal groups capable of reaction with the cellulose hydroxyl groups are required. It must also be realized that the heterogeneous reaction of even low molecular weight polymers poses accessibility problems due to steric factors, so that the extent of grafting is somewhat a function of the size of the synthetic polymer molecule.

In an interesting study, Avny and Schwenker[26)] have shown that short-chain vinyl polymers can be synthesized, which contain reactive terminal groups of an acyl chloride type that permit a simple reaction with cellulose to form graft copolymers. Although these researchers grafted only polystyrene onto cellulose, they pointed out that other types of vinyl polymers made by anionic polymerization, such as poly(methyl methacrylate), could also be used: bifunctional carboxyl-terminated polystyrene was obtained in a typical preparation by polymerization conducted in dry tetrahydrofuran (THF) (200 cm^3) using a sodium naphthalene initiator in 1.3 M solution (3 cm^3) and styrene (10 cm^3), introduced in that order. The reaction mixture was maintained at 0 °C for 30 minutes with vigorous stirring, followed by the addition of solid carbon dioxide. The polymer thus produced was isolated by precipitation in water, than filtered, washed, slurried, washed with ethanol, and finally dried in vacuo to yield 8.1 g of product. Monofunctional carboxyl-terminated polystyrene was similarly prepared, except that the initiator was butyl lithium in petroleum ether with the polymerization carried out at room temperature in benzene (200 cm^3) containing THF (0.5 cm^3).

It is clear that two different initiators were used in the preparation of the living polystyrene: butyl lithium in petroleum ether solution and sodium naphthalene in THF solution. Reaction between carbon dioxide and living polystyrene, initiated by butyl lithium, leads to the formation of monofunctional carboxyl-terminated polystyrene, as described below[26)]:

Polymerization

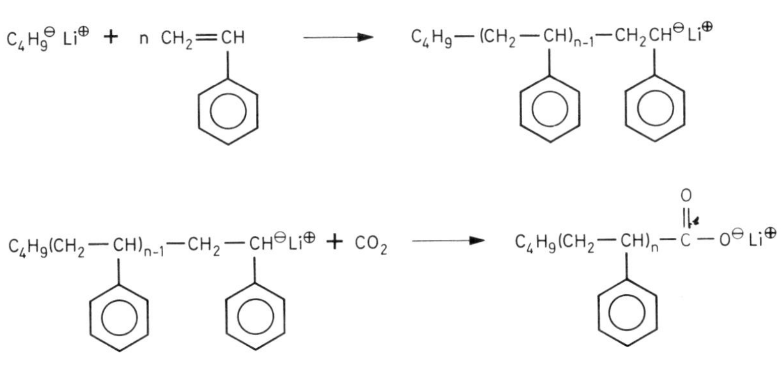

Reaction between carbon dioxide and living polystyrene, initiated by sodium naphthalene, leads to the formation of bifunctional carboxyl-terminated polystyrene in the following manner:

Polymerization

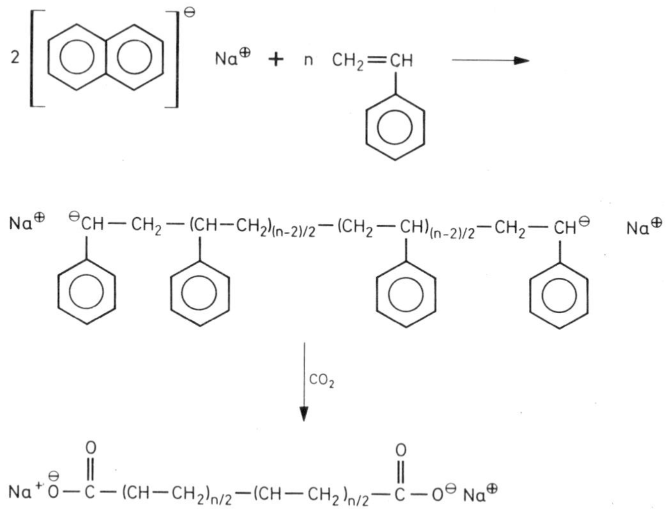

The carboxylate terminal groups of the polystyrene were converted to the acid chloride form by reaction with thionyl chloride. In a typical example, 3.1 g of polystyrene with a carboxyl content of 7.38×10^{-4} mol/g were dissolved in 50 cm^3 of benzene. Redistilled thionyl chloride (50 cm^3) was added and the reaction mixture maintained at 55 °C for 24 hours. Excess thionyl chloride was removed by concentrating the system in vacuo almost to dryness. Three portions (30 cm^3) of dry benzene were then added, distilled, and the acyl chloride terminated polymer dissolved in 40 cm^3 of benzene and used immediately.

A typical preparation involving grafting cotton yarn with polystyrene is as follows. Purified yarns (1.71 g) were swollen in 20% sodium hydroxide solution for 15 minutes at room temperature. The yarns were washed with distilled water, neutralized with 2% acetic acid solution, and washed with water until neutral. Additional swelling of the yarns was accomplished in dimethylformamide (DMF) for 48 hours. The yarns were subsequently solvent-exchanged with dry THF, which had been distilled from sodium benzophenone solution, and then with dry benzene. Fresh benzene (200 cm^3) was distilled into the reaction flask and dry pyridine (10 cm^3) was added. Acyl chloride-terminated polystyrene (3.1 g, carboxyl content 7.38×10^{-4} mol/g) solution in benzene was then injected into the flask. The reaction mixture was refluxed for 24 hours while the yarns were mechanically stirred. After 24 hours, the reaction mixture was removed, the yarns were extracted with dry DMF for 3 days, the DMF having been exchanged three times. The yarns were washed with benzene and dried in vacuo at 50 °C. The modified yarns weighed 1.96 g, which corresponds to a 12.8% polystyrene content.

Table 41 contains the results obtained when cotton fabrics (0.6–1.3 g) were allowed to react with preformed polystyrene (2.0–5.5 g) containing terminal groups in the acid chloride form at 80 °C. Table 42 contains the corresponding

Table 41. Grafting of preformed polystyrene on 80×80 cotton print cloth, Adapted from Ref.[26], p. 820, by courtesy of Textile Research Institute

Swelling agent	Reaction solvent	Reaction time, hr	Yield* %	Polystyrene+ content in %	$\overline{DP}_n$ of preformed polystyrene	Grafting++ D.S.
20% NaOH and DMF	Benzene	24	3.1	15.4	9.3	30.4×10^{-3}
Water	Dioxane	67	8.7	14.0	10.3**	24.6×10^{-3}
86% H_3PO_4	Dioxane	48	5.7	14.8	50.8+++	5.3×10^{-3}
86% H_3PO_4	Dioxane	48	3.9	14.5	50.8+++	5.2×10^{-3}
10% NaOH	Dioxane	48	3.3	10.8	53.6+++	3.5×10^{-3}
Water	DMF	68	10.7	17.3	53.6+++	6.1×10^{-3}
20% NaOH and DMF	Benzene	24	1.2	4.3	107.3+++	0.7×10^{-3}

* Percent by weight of the reacted preformed polymer.
\+ Determined from the weight increase.
++ Calculated from $\overline{DP}_n$ of the grafted preformed polymer and the polystyrene content of the graft.
** Monofunctional polystyrene.
+++ Difunctional polystyrene.

Table 42. Grafting of preformed polystyrene on cotton yarns. Adapted from Ref.[26], p. 821, by courtesy of Textile Research Institute

Swelling agent	Reaction solvent	Reaction time, hr	Yield* %	Polystyrene+ content in graft %	$\overline{DP}_n$ of preformed polystyrene	Grafting++ D.S.
20% NaOH and DMF	benzene	24	8.0	12.8	13.0**	17.5×10^{-3}
86% H_3PO_4	dioxane	48	5.0	5.4	119.3+++	0.7×10^{-3}
86% H_3PO_4	dioxane	120	22.0	22.0	119.3+++	3.2×10^{-3}
10% NaOH	dioxane	120	13.5	19.5	119.3+++	3.7×10^{-3}

* Percent by weight of the reacted preformed polymer.
\+ Determined from the weight increase.
++ Calculated from the $\overline{DP}_n$ of the grafted preformed polymer and the polystyrene content of the graft.
** Monofunctional polystyrene.
+++ Difunctional polystyrene.

data when cotton yarns (1.70 g) were reacted with acid chloride terminated-polystyrene (2.0–3.0 g) under similar conditions[26].

It is obvious that relatively high grafting D.S. values are obtained with low molecular mass polystyrene. The maximum D.S. of 30.4×10^{-3} corresponds to 33 anhydroglucose units per grafted side chain, and the minimum D.S. of 0.7×10^{-3} corresponds to 1430 anhydroglucose units per grafted side chain. The extent of grafting is mainly determined by the $\overline{DP}_n$ of the reacting preformed polymers, increased from a $\overline{DP}_n$ of 9.3 to $\overline{DP}_n$ of 107.3. The same relationship is observed in the modified yarns where a D.S. of 17.5×10^{-3} was

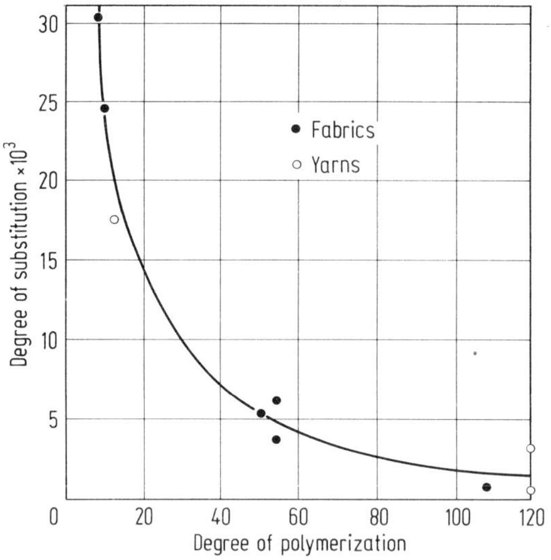

Fig. 44. Relation between grafting degree of substitution and degree of polymerization of preformed polystyrene. Reprinted from Ref.[26], p. 821, by courtesy of Textile Research Institute.

obtained with polystyrene of $\overline{DP}_n = 13.0$ after 24 hours of reaction, while a D.S. of 3.7×10^{-3} was reported for grafting with polystyrene of $\overline{DP}_n = 119.3$ after 120 hours.

Methods of swelling and the reaction solvent do not appear to influence the grafting D.S. The reaction time has some influence on the grafting D.S., with longer times leading to higher grafting D.S., but the changes observed over long periods of time are relatively small (see Tables 41, and 42).

Figure 44 shows the dependence of the grafting D.S. on the $\overline{DP}_n$ of the preformed polymers using the data in Tables 1 and 2. Despite differences in the swelling method, the reaction solvent, and the reaction time, it is apparent that the grafting D.S. is mainly determind by the DP of the reacting preformed polystyrene[26].

It has been claimed that this approach to grafting onto cellulose has many potential advantages over conventional monomer graft polymerization. First, the problem of homopolymer formation is avoided. Second, the grafting D.S. and graft side-chain molecular weight may readily be controlled. Third, though this approach produces relatively low graft yields, significant improvements in the fiber properties would be expected because of the higher grafting D.S. and the short side chains, when compared to those of other grafting methods.

References

1. Hebeish, A.: Koloriszt. Ert. *13* (1–2), 12 (1971)
2. Hermans, J. J.: Pure Appl. Chem. *5*, 147 (1962)
3. Immergut, E. H., Mark, H.: Makromol. Chem. *18*/*19*, 322 (1956); Proc. 1st Cellulose Conf. Syracuse, N.Y. (1958)
4. Krässig, H. A., Stannett, V.: Adv. Polym. Sci. *4*, 111 (1965)
5. Schurz, J.: Papier *16*, 525 (1962)

6. Schwab, E., Stannett, V., Hermans, J. J.: Tappi *44*, 251 (1961)
7. Arthur, J. C. Jr.: Adv. Chem. Series *91*, 547 (1962); also Arthur, J. C. Jr.: Adv. Chem. Series *99*, 321 (1971); Arthur, J. C. Jr.: Adv. Makromol. Chem. *2*, 1 (1970)
8. Vaughn, T. H., Jackson, D. R., Lundsted, L. G.: J. Amer. Oil Chemists Soc. *29*, 240 (1952)
9. Jackson, D. R., Lundsted, L. G.: U.S. Pat. *2*, 677, 700 (1954)
10. Haas, H. C., Cohen, S. G., Oglesby, A. C., Karlin, E. R.: J. Polymer Sci. *15*, 427 (1955)
11. Rafikov, S. R., Chelnokova, G. N., Zhuravleva, I. V., Gribkova, P. N.: J. Polymer Sci. *53*, 75 (1961)
12. Weiss, P., Gerecht, J. F., Krems, I. J.: J. Polymer Sci., *35*, 343 (1959)
13. Cohen, S. G., Haas, H. C.: J. Amer. Chem. Soc. *72*, 3954 (1950)
14. Cohen, S. G., Haas, H. C., Slotnick, H.: J. Polymer Sci. *11*, 193 (1953)
15. Stannett, V. T.: Tappi *47*, 58A (1964)
16. General Article in Chem. and Eng. News *16*, 39 (1961)
17. Daul, G. C., Reinhardt, R. M., Reid, D. J.: Textile Res. J. *24*, 738 (1954)
18. Cooper, W., Smith, R. K.: Makromol. Chem. *40*, 148 (1960)
19. Rogovin, Z. A., Zhui, U. Z.: Vysokomol Soed. *1*, 1630 (1959)
20. Morgan, P. W.: Condensation Polymers by Interfacial and Solution Methods, New York: Interscience Inc. 1965
21. Tun, S., Vey-gan, J., Rogovin, Z. A.: Vysokomol. Soed. *3*, 382 (1961)
22. Rogovin, Z. A.: J. Polymer Sci. *48*, 443 (1960)
23. Rogovin, Z. A.: Unpublished data
24. Bruneau, C.: Compte Rendu *252*, 2413 (1961)
25. Schroeder, C. W., Condo, F. E.: Textile Res. J. *27*, 135 (1957)
26. Avny, Y., Schwenker, R. F. Jr.: Textile Res. J. *37*, 817 (1967)

6. Grafting on Chemically Modified Cellulose

6.1. Introduction

Chemical modification of polymers with the aim of imparting specific, desirable properties is one of the main directions of development in modern macromolecular chemistry. Recently, this approach to tailor-made materials has been successfully applied to synthetic as well as natural polymers in many laboratories in different countries.

In the case of synthetic polymers, this can be achieved either by changing the composition and proportion of the monomeric raw materials used for the synthesis, or by chemical modification of the polymer manufactured in the normal way. With natural polymers and in particular, cellulose, which is one of the most important and widely available, the constitution and chemical composition, as determined by biosynthesis, cannot be modified. One way of solving this problem is by chemical modification. This is indeed a very fascinating field for research with unlimited future possibilities.

The preparation of modified cellulosic materials has been, and still is, made possible by a thorough understanding of the chemistry of cellulose and the development and application of new methods of chemical modification, along with detailed studies of the various properties of the products formed. Currently, much intensive research is being undertaken in different countries with the aim of making cellulosics that have properties superior to those provided by nature.

There are many processes to which cellulose, in particular cotton cellulose, is suggested which modify it chemically. The advantage of treating the native cotton cellulose and maintaining its fibrous form, instead of using a solution of cellulose, is that the cotton fiber is wonderfully built up and has a fine internal structure which is itself fibrillar and helical. Thus, the success of a chemical modification lies essentially in introducing functional groups into the cellulose molecules which confer new properties on the cellulose without destroying its many desirable intrinsic properties.

The cellulose molecule has mainly one type of functional group present, namely, the hydroxyl groups (primary as well as secondary). Hence, one might be tempted to expect that all reactions, normally found to be possible with hydroxyl groups in carbohydrate chemistry, would have potential application in cellulose modification. However, the following two factors have to be considered before such methods can be successfully applied.

1. Chemical reactions with the cellulose hydroxyls take place in heterogeneous media where diffusion of reagents into the fiber structure is involved.

This may lead to considerably slower rates of reaction and to non-uniform reactions in many cases.

2. The 1:4 glucosidic linkages between the neighboring glucose units in the cellulose molecule have very limited stability under acid conditions. Hence, conditions for chemical modifications must be so selected as to prevent or at least to minimize the breakdown of these linkages, which may lead to a decrease in molecular mass and impairment of mechanical properties.

Esterification and etherification reactions of the hydroxyl groups of cellulose have been employed for some time and have led to the development of a number of well-known commercial cellulose derivatives. Among these are cellulose xanthate used for viscose fibers and films; cellulose nitrate for coatings, cinefilms and plastics; cellulose acetate for fiber, cinefilms, fibers and plastics, ethyl cellulose for plastics, paints, inks and coating films; and carboxymethyl cellulose, which has wide application in the textile field as a gum and in drilling muds. The recent advances in transfer printing processes have led to interesting developments in ink technology, some of which utilize cellulose derivatives.

Besides these simple derivatives, methods have been developed with the aim of converting the natural cellulose molecule, which has only one type of functional group, into a modified cellulose with new and different reactive functional groups so that further reactions can be made possible. These investigations have theoretical as well as practical importance and involve solving a number of complex problems, such as the understanding of the theories of the reactions involved, the stability and toxicity of reagents used, prevention of cellulose degradation, and standardizing the optimum conditions for economical development of a new process. A literature survey[1-15)] reveals that the following types of methods have been examined for chemical modifications from the above points of view.

1. Substitution of the hydrogen atom in the hydroxyl groups of the cellulose molecule.
   (a) Esterification
   (b) Etherification
2. Selective and controlled oxidation of the cellulose hydroxyls.
3. Substitution of the hydroxyl groups of the cellulose molecule.
   (a) Intramolecular reactions
   (b) Intermolecular reactions
4. Reactions involving radical additions
5. Reactions involving electrophilic substitution
6. Interaction of cellulose with organometallic compounds
7. Synthesis of block or graft copolymers of cellulose

It is important to emphasize that the same functional groups may be introduced into the cellulose molecule via different reactions. Hence, it must be remembered that introduction of the same functional groups in the cellulose molecule by different methods may result in the production of chemically modified celluloses of diverse reactivities, depending on the position of the functional groups in the cellulose molecule, whether in the anhydroglucose unit of cellulose or in side chains.

Recently, chemically modified celluloses have evoked considerable interest in the grafting area. It is hoped that the grafting of synthetic polymer chains onto modified celluloses will prove of value, both for the achievement of a good understanding of the kinetics and mechanisms of grafting and also for the provision of basic information needed for improvements to be made in the properties of the products.

6.2. Grafting on Cellulose Xanthogenate

The methods given for preparing "standard" cellulose sodium xanthate[16-18] follow the technical process for the production of viscose and yield a product whose average degree of substitution, D.S., is in the range of 0.4 to 0.7. One hundred times the D.S. is sometimes known as the "gamma number". After the cellulose has been steeped in strong sodium hydroxide solution, most of the excess alkali is removed by pressing until the weight of the residual alkali is 2.6 to 3 times the weight of the original air-dry cellulose. Variation in this "press ratio" effects the ease of comminuation (shredding) of the alkali cellulose, the xanthate D.S., and the rate at which the viscose "ripens". The pressed and shredded alkali cellulose is usually slightly depolymerized ("aged") by exposure to atmospheric oxygen before xanthation is brought about by reaction with carbon disulfide.

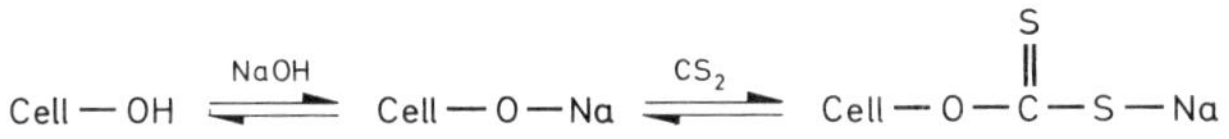

Cellulose sodium xanthate is fairly stable when stored under reduced pressure at room temperature in a pure-dry condition. However, in alkaline solution the equilibria noted in the xanthation reactions are driven slowly to the left as carbon disulfide is consumed in the formation of by-products, such as trithiocarbonate, sulfide, thiosulfate, and carbonate. The decomposition of the cellulose xanthate occurs fairly rapidly near pH 7 and with extreme rapidity in the presence of excess mineral acid.

Xanthate derivatives, such as cellulose xanthate N,N-diethylacetamide[19], cellulose S-methyl xanthate[17,20], and cellulose S-benzyl xanthate[21], are much more stable than the original xanthate salt. For example, the S-methyl esters have been completely benzoylated[21] without a significant decrease in xanthate substitution.

The voluminous literature concerning the preparation, properties, and analysis of cellulose xanthate has been reviewed in detail elsewhere[18,22,23].

Dimov and Pavlov[24] have studied the possibility of obtaining modified cellulosic products by reacting cellulose xanthogenate in solutions containing oxidizers and acrylonitrile (AN). The cellulose xanthogenate was first prepared as follows. Cooked and bleached cotton printed calico was mercerized and then treated with carbon disulfide until the necessary degree of esterification of the cotton cellulose was obtained. The xanthogenate was then treated with a solution containing 300 g/l ammonium chloride and finally washed with a tenfold diluted solution until the yellow color of the wash water disappeared. The xanthogenate retained the structure of the fabric. Next, the ability of the cellulose xanthogenate-hydrogen peroxide redox system to initiate graft copolymerization of acrylonitrile on the cellulose was investigated under a variety of conditions. Variables studied include: pH, temperature of the solution, concentration of hydrogen peroxide, the degree of esterification, the gamma-number, γ, of the cellulose xanthogenate, and duration of the treatment.

The effect of the hydrogen peroxide concentration and the degree of esterification of the cellulose xanthogenate may be observed in Table 43. With increasing degrees of esterification of the initial cellulose xanthogenate, the quantity of AN grafted increases, as the number of centers which can initiate polymerization is increased.

With respect to the H_2O_2 concentration, the quantity of AN grafted increases at first to a concentration of 30 or 50 g/l and then decreases. The initial increase is explained on the basis of an increase in the speed of initiation of the polymerization with increasing concentration of the peroxide. In addition, while the concentration of peroxide is being increased, the number of the xanthogenate groups which have reacted with the oxidant and initiated the polymerization before being hydrolyzed under the action of the strongly acidic medium, also increases[24].

The essential reactions, as postulated by Dimov and Pavlov[24], probably proceed as follows:

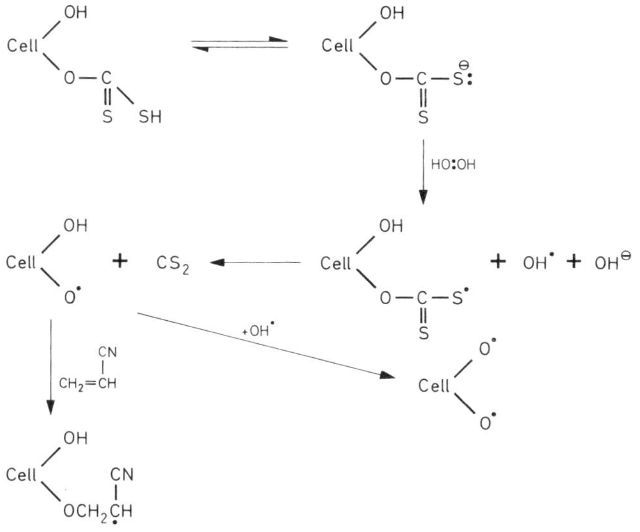

The disintegration of the initially formed radical on the xanthogenate group resembles the disintegration of the radical formed on the carboxyl group under similar circumstances[25].

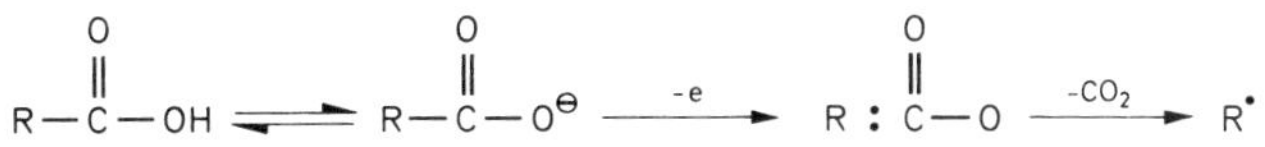

Table 43. Variation of the amount of acrylonitrile (AN) grafted on cellulose xanthogenate with hydrogen peroxide concentration and degree of esterification (γ) of cellulose xanthogenate*. Adapted from Ref.[24], p. 2778, by courtesy of John Wiley & Sons, Inc.

H_2O_2 concentration g/l	AN grafted %					
	$\gamma = 6$	$\gamma = 11$	$\gamma = 21$	$\gamma = 24$	$\gamma = 32$	$\gamma = 36$
1	7.4	11.2	25.6	27.6	28.8	30.0
10	12.3	18.2	27.6	30.9	33.5	35.6
30	20.0	23.7	28.6	32.0	37.6	39.3
50	21.6	24.4	30.0	32.5	34.5	35.4
100	31.7	14.2	14.8	16.3	18.8	22.2

* The experiments were carried out at a temperature of 20 °C, with an aqueous solution of 7% of AN at pH 1 for 1 hr using a material: liquor ratio 1:50.

Obviously, then, cellulose macroradicals are obtained as a result of both the oxidation of the xanthogenate group and as a consequence of radical or chain transfer. The ˙OH radicals can either recombine among themselves or with other radicals; or they can split off hydrogen atoms from the hydroxyl groups of the cellulose, or from the —SH groups of the xanthogenate; they can cause homopolymerization or stop transfer or propagation of growing chains[24]. At higher peroxide concentrations, the ˙OH radicals seem to participate in terminating the growing chain radicals, as well as in initiating homopolymerization, thereby lowering the amount of AN grafted.

In order to study the effect of pH on the amount of AN grafted, experiments in which the pH of the solution varied from 0.3 to 4.0 were carried out. In each experiment the grafting reaction was performed at 20 °C for 0.5 hr, using 7% of AN and 30 g/l of hydrogen peroxide on cellulose xanthogenate of $\gamma = 55$ at a liquor ratio of 50:1. The results of these experiments revealed that the quantity of AN grafted increases parallel to the increase of pH from 0.3 to 1.5. For instance, graft yields of 3.7%, 25.1%, and 30.7% were reached at pH's of 0.3, 0.8, and 1.5, respectively[24]. This is explained by the speedy disintegration of the xanthogenate groups in the strongly acidic medium before they have interacted with the hydrogen peroxide, and by the reduced degree of dissociation of the xanthogenic acid, due to which this reaction is further limited; moreover, the growth of polyacrylonitrile chains is slower in a strongly acidic medium[26].

Under the same conditions, but with a reduction in the acidity of the medium from pH = 1.5 to pH = 4, the quantity of AN grafted decreases to give a graft yield of 15.5%. This is because the polymerization initiated by the redox system, hydrogen peroxide-xanthogenate, begins later at an increased pH. Since the time allowed for the reaction to proceed was too short, i.e., one-half hour, the chains that formed later at the higher pH had less time for polymerization. This was substantiated by the observation that higher graft yields could be obtained at higher pH when the reaction time was 2.5 hours instead of one half hour. For example, a graft yield of 57% was obtained at pH 4, as compared with a yield of 38.1% at pH 1.1. This increment in the graft yield is associated with the slower saponification of the xanthogenate groups and the faster rate of AN polymerization at higher pH.

The effect of increasing the temperature from 10 °C to 50 °C was found to bring about a considerable enhancement in the quantity of AN grafted. This means that the positive effect which the increase of the temperature exerts on the initiation of polymerization and the polymerization itself is greater than the negative effect, i.e., increased rates of saponification of the xanthogenate groups in acidic media result. However, an increase in temperature to over 20 °C is not recommended, because at the end of the process, especially at 40 °C and 50 °C, the formation of homopolymer begins[24].

Dimov and Pavlov[24] have also studied the kinetics of depletion of hydrogen peroxide during grafting of acrylonitrile onto cellulose xanthogenate. As expected, the hydrogen peroxide is, in time, exhausted more slowly and to a smaller degree at the lower degrees of esterification of the xanthogenate. At a lower pH of the medium, the peroxide is exhausted faster, but to a lesser degree. This points out again that, in a less acidic medium, the reaction between the hydrogen peroxide and the xanthogenic groups goes on more slowly, and therefore the initiation of the polymerization is delayed. It can also be seen that in a more acidic medium a smaller quantity of the peroxide is consumed, and consequently, a smaller number of xanthate groups participate in the redox reaction because they are saponified more rapidly.

Grafting onto cellulose xanthogenate systems has recently been reviewed by Morin and Rogovin[27], who emphasied the H_2O_2/thiocarbonation redox system. The main approach involved an examination of factors known to be essential to the efficient production of composite materials in the industrial sense. Kraessing[28] has examined the H_2O_2/thiocarbonate system from the viewpoint of changes in general crystal structure as grafting takes place. Pavlov et al.[29] applied these principles in grafting various water insoluble monomers onto the cellulosic substrate. Thus, each research group was able to carry out grafting reactions onto freshly formed, highly swollen and reactive cellulose fibers. The logical conclusion of this type of work is the production of cellulosic fibers, films, and membranes having properties superior to those available with current regenerated celluloses. Morin et al.[30] have examined this point in detail using the Fe^{II}/Redox system in grafting styrene onto cellulose xanthate. The cellulose xanthate solution medium appears to offer an ideal opportunity for volumetric grafting onto cellulosic substances, though the limitations are appreciable. Grafting at the wood pulp stage, before xanthation, leads to problems in solution and filtration through gel formation[31] and inactivation of the grafting process in the inherently alkaline medium leads to wastage, as does the physical incorporation of monomer into the solution of xanthate, which is ready for grafting upon reaching regeneration baths. Much of the fundamental work done on characterizing this system has now been carried out. What is needed is a careful screening of monomers aimed at providing improvements to physical properties, such as moisture regain, dimensional stability, resilience, and toughness. Brickmann[32] has presented a reasoned discussion of the industrial development of thiocarbonation grafting and suggests that the commercial route most suitable follows seven process steps: preparation, thiocarbonation, post-thiocarbonation wash, metathesis, post-metathesis wash, graft polymerization, and finishing. Using selected monomers, new functional groups can be attached in this way which give new properties to the existing materials, such as dyeability, chemical resistance, fire retardancy, soil and rot resistance, and others. However, since the grafts are not truly volumetric, there is usually little to be gained in properties associated with wear and toughness.

6.3. Grafting on Acetylated Cellulose

Acetylation of cellulose, one of the most important modifying processes, has found wide-spread industrial application. Acetylation involves treatment of the cellulose with a mixture of acetic anhydride, acetic acid, and a catalyst such as sulfuric/perchloric acid. The reaction of cellulose with acetic anhydride may be written as:

$$\text{Cell—OH} + (CH_3CO)_2O \rightarrow \text{Cell—O—CO—}CH_3 + CH_3COOH$$

Acetic acid in the acetylating mixture may be replaced by a nonopolar solvent such as benzene. Using such procedures[33,34], it has been possible to produce a modified cellulose. Acetylated celluloses possess several useful properties[33,35,36], such as heat resistance, mildew resistance, and rot resistance.

Hebeish et al.[37] have investigated the grafting reaction of acrylonitrile (AN) and methyl methacrylate (MMA) with acetylated cellulose in order to elucidate (a) some of the important parameters which affect the grafting reaction, (b) the effect of the introduction of —$COCH_3$ groups into the cellulose molecule on the grafting efficiency, and (c) the mechanism of grafting. For this purpose, they have treated cellulose of different acetyl contents with either Ce^{IV} ammonium sulfate (CAS) or Ce^{IV} ammonium nitrate (CAN) and monomer, i.e., grafting, under various reaction conditions. Some of the important findings reported by them are given below.

Figure 45 shows data for the graft yields obtained with acetylated cellulose having 39.5% combined acetic acid and for those obtained with cotton cellulose treated with glacial acetic acid (control). It is obvious that with acetylated cellulose, grafting proceeds very slowly in the early stages of the reaction, and that the graft formation is considerable only after 30 minutes' reaction time. Thereafter, increasing the reaction time causes no substantial increment in the graft yield within the range studied. The introduction of —$COCH_3$ groups along the cellulosic chains greatly reduces its swellability in aqueous media. Diffusion of the monomer would, therefore, by very difficult, especially in the early stages of the reaction. As a result, only a small amount of graft copolymer is formed. On increasing the exposure time, swelling of the fiber and diffusion of both monomer and initiator all increase giving rise to a substantial amount of graft formation. The control also shows an initial fast rate followed by a slower rate. Besides substrate accessibility, other causes for this could be a fall in the concentration of the monomer and the initiator, as well as a decrease in the available sites for grafting on the cellulose molecule.

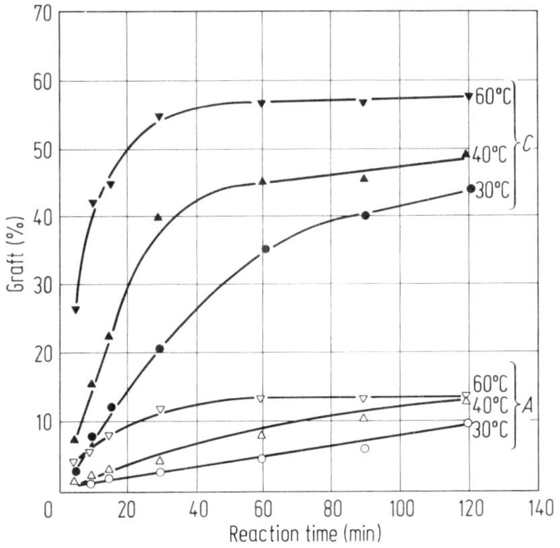

Fig. 45. Effect of temperature on rates of grafting of acetylated cotton having 39.5% combined acetic acid *A* and cotton, treated with acetic acid *C*; 1 g AN; 0.01 M CAS. Reprinted from Ref.[37], p. 16, by courtesy of John Wiley & Sons, Inc.

The data (Fig. 45) also show that with both substrates, the rate of grafting increases upon raising the temperature from 30° to 60 °C. However, the maximum graft yields obtained with acetylated cotton at 40° to 60 °C are comparable, despite the significant difference in the yields during the early stages of the reaction. Grafting to acetylated cotton cellulose appears best at the point of maximum swelling in the Ce^{IV}-AN-water system. Beyond this, the grafted polyacrylonitrile probably forms a diffusion barrier and inhibits the diffusion of acrylonitrile monomer. It is well known that introduction of —$COCH_3$ groups considerably reduces swelling of the cellulose in aqueous media. Thus, the lack of swelling and a rapid slowing down of diffusion of initiator and monomer as grafting proceeds would account for the comparable maximum yield obtained at 40° and 60 °C.

The graft yield obtained with acetylated cotton cellulose is significantly lower than that obtained with the control (cf. Fig. 45). This was also observed when grafting was performed under other reaction conditions, including variations in the concentrations of monomer and initiator, the type of monomer and initiator, the reaction time, and the temperature. The presence of bulky groups, such as —$COCH_3$ in the cellulose, results in two effects: (a) reduction of the swelling capacity of cellulose, thereby decreasing diffusion and adsorption of both monomer and initiator; and (b) blocking of some of the cellulose hydroxyls. Both effects lead to lower grafting[37]. Data presented in Figure 46 corroborate this conclusion; increasing the —$COCH_3$ groups along the cellulosic chains causes a significant decrease in the graft yields.

Like the graft yields, the consumption of Ce^{IV} ions during either grafting or oxidation decreases as the —$COCH_3$ content of acetylated cotton cellulose increases. The explanation given for the graft yield would also apply here. With all the acetylated cottons, the Ce^{IV} ion consumed during grafting is higher than that consumed during oxidation. Hebeish and Mehta[38] noticed the same effect with

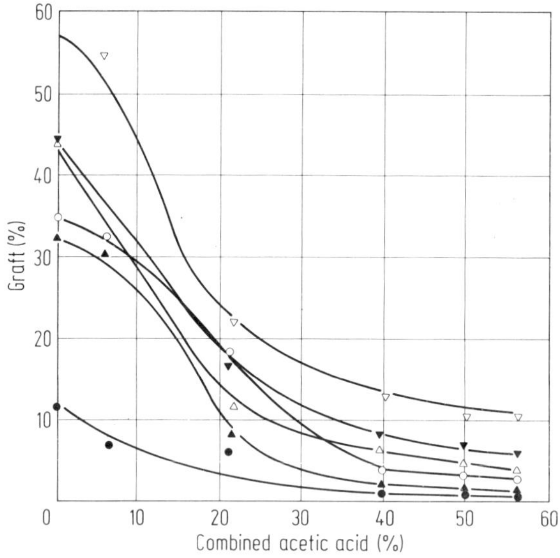

Fig. 46. Acetyl content (expressed as % combined acetic acid) vs. % graft yield using acrylonitrile as a monomer; (●) 30 °C, (▲) 40 °C, (▼) 60 °C at 15 min., (○) 30 °C, (△) 40 °C, (▽) 60 °C at 60 min. 1 g AN; 0.01 M CAS. Reprinted from Ref.[37], p. 16, by courtesy of John Wiley & Sons, Inc.

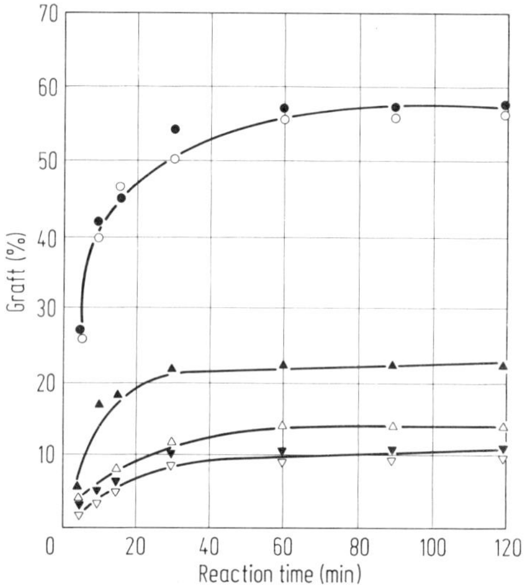

Fig. 47. Rates of grafting of cotton treated with acetic acid and acetylated cottons having different acetyl content (expressed as % combined acetic acid); (●) 0%; (○) 6.48%; (▲) 21.6%; (△) 39.5%; (▼) 49.65%; (▽) 56.2%; 1 g AN; 0.01 M CAS; 60 °C. Reprinted from Ref.[37], p. 21, by courtesy of John Wiley & Sons, Inc.

various cellulosic substrates and attributed this phenomenon to homopolymer formation.

Figure 47 shows the rate of grafting for acetylated cotton celluloses having different acetyl contents; the latter is expressed as percent of combined acetic acid. The rate of grafting of cotton cellulose treated with glacial acetic acid, as a control, is also given in the same figure. Similar trends were also obtained for Ce^{IV} consumption during oxidation.

It is clear that the rate of grafting of acetylated cellulose having the lowest combined acetic acid, i.e., 6.4%, is very rapid at first but subsequently decreases, and a maximum grafting level is reached. An identical rate is found for the control. This indicates that the presence of —$COCH_3$ groups, in small amounts, does not alter the susceptibility of cellulose to Ce^{IV} ion — initiated grafting. Previous studies have shown such grafting to occur at the reducing end of the cellulose molecule and at the C_2 and C_3 glycols[39–46].

Increasing the combined acetic acid to 21.6% considerably lowers the maximum grafting level. However, the grafting shows an initial fast rate followed by a slower rate. This suggests that acetylation might have taken place exclusively at the primary hydroxyl group, i.e., at C_6 of the anhydroglucose unit in the cellulose, without attacking the secondary hydroxyls at C_2 and C_3. It is known, from esterification studies, that the C_6 group is esterified ten times faster than the other groups[47,48], and also that partially acetylated cotton is more resistant to oxidation by nitrogen dioxide, indicating that primary alcohol groups have been preferentially esterified[49]. Thus, the hydroxyls at C_2 and C_3 may be responsible for the initial fast rate, an assumption which is in agreement with previous postulations[39–44]. Lowering of the maximum grafting level is presumably caused by a lack of swelling of the acetylated cotton cellulose.

Allowing the acetylation reaction to proceed, in order to give acetylated celluloses having more than 21% combined acetic acid, would certainly involve the hydroxyls at either C_2, C_3, or both. If these two hydroxyls are the only groups which account for the rapid initial rate, as postulated above, one would expect that their partial or complete blocking would appreciably influence the rate of grafting. Indeed, increasing the combined acetic acid concentration to 39.5% slows down considerably the initial rate as well as the maximum grafting level. Further decreases in the initial rate of grafting and the maximum level are achieved with acetylated cotton having 49.65% combined acetic acid, indicating that partial blocking of the glycol groups affects the rate of grafting. Acetylated cotton with 56.2% combined acetic acid shows a rate of grafting identical to that obtained with acetylated cotton having 49.65% combined acetic acid, despite the fact that nearly all the cellulose hydroxyls are blocked in the former case. This implies that introduction of —$COCH_3$ groups up to a certain limit, i.e., approximately 50% combined acetic acid, in the cellulose adversely affects the rate of grafting. Beyond this value, the influence of —$COCH_3$ groups on the rate of grafting is not marked. It appears that the free hydroxyl groups in highly acetylated cottons are not accessible for grafting, perhaps because of steric hinderance and/or stabilization of the glycol bond with resulting resistance to Ce^{IV} attack. However, a great lack of swellability in these cases, as compared with other acetylated cottons in the range studied, cannot be ruled out.

In any event, an initial fast rate of grafting is observed even when nearly all the cellulose hydroxyls are blocked. This suggests that the hemiacetal groups (corresponding to the C_1 linkage in the end unit of the cellulose molecule) contribute to the initial fast rate of grafting, a conclusion which is in harmony with previous postulations[39,42–46]. Thus, Ce^{IV}-induced polymerization of a vinyl or acrylic monomer in the presence of cellulose yields a mixture of graft copolymer (by Ce^{IV} attack at C_2 or C_3) and block copolymer (by Ce^{IV} attack at the hemiacetal group), as well as homopolymer (monomer attack by Ce^{IV}). That

Ce^{IV}-induced copolymerization produces a mixture of block and graft copolymers has been also reported by Iwakura and co-researchers[46].

Hebeish et al.[37] observed no change in the acetyl content (expressed as the percent combined acetic acid yield) of the Ce^{IV}-oxidized samples. They concluded that the grafting reaction initiated by Ce^{IV} has practically no effect on the acetyl groups of the acetylated cotton cellulose.

6.4. Grafting on Cyanoethylated Cellulose

Modification of cellulose by cyanoethylation has attracted much attention and cyanoethylated cottons are now commercially available. In this procedure, the cellulose is first impregnated with dilute sodium hydroxide and then treated with acrylonitrile at ca. 55 °C, followed by rinsing with dilute acetic acid and washing with water. With care, high values of degree of substitution, D.S., may be achieved[50]. The reaction between the cellulose hydroxyls and acrylonitrile is:

$$\text{Cell—OH} + CH_2{=}CHCN \rightarrow \text{Cell—OCH}_2\text{—CH}_2\text{—CN}$$

This treatment improved resistance to rot, heat, and damage by acids and abrasion,[51-56]. Cyanoethylated cottons are more receptive than cotton to some dyes in all classes of dyestuffs and require more work than untreated cotton for rupture under tension. Furthermore, the dielectric constant of cyanoethylated cellulose is increased proportionally to the extent of reaction. The cyanoethylated cotton can be hydrolyzed to carboxyethyl cellulose.

Hamaya and Yakada have examined the cyanoethylation of cellophane in considerable detail[57]. Reactions were carried out in a heterogeneous system via a one-step process in an NaOH aqueous solution containing known amounts of acrylonitrile. The effects of varying the concentration of acrylonitrile, wetting agent (sodium lauryl sulfate), and alkali were measured, as was the temperature dependence. The aim was to provide uniformity of reaction with retention of film form for the cellophane film.

The radiation-induced interaction of acrylonitrile with cyanoethylated cottons that have different cyanoethyl contents, i.e., different D.S., was studied by Demint et al.[58]. The cotton yarn was cyanoethylated according to a method described by Daul et al.[50] by padding the yarn with 2% sodium hydroxide to a wet pick-up of ca. 85% and then immersing the yarn in acrylonitrile (watersaturated), agitated at 55 °C. The cotton was removed at predetermined times of reaction to give the desired D.S., washed in dilute acetic acid, thoroughly rinsed in distilled water, and dried under ambient conditions.

The grafting reaction was carried out as follows[58]. The yarn was immersed overnight at room temperature in a solution containing 32 parts of acrylonitrile monomer in 68 parts of aqueous 80% zinc chloride and then padded to give about a two-to-fourfold increase in weight of the original yarn. The treated cotton was irradiated in a cobalt-60 source, which had a dose rate of about 0.5×10^6 roentgens/hr. (maximum). The unreacted monomer and homopolymer were removed by a thorough washing in distilled water and extraction with N,N dimethylformamide.

The effect of degree of substitution (D.S.) on the radiation-induced interaction of acrylonitrile with cyanoethylated cotton is considerable. It is clear that increasing the D.S. from about 0 to 1.3 causes about a fourfold increment in the amount of monomer interacted. At 1.3 D.S., the composition of the fibrous product was about 36% cellulose, 15% cyanoethyl group, and 49% interacted monomer[58].

Figure 48 shows the relationship between the interaction of monomer with cyanoethylated cotton and degree of substitution at a dosage of one megaroentgen[58]. Obviously, the amount of interacted monomer per 100 units of cellulose increased linearly over a range from 0.0 to about 1.0 D.S. and then increased more rapidly. The weight ratio of change in interacted monomer to change in cyanoethyl group, as compared with the data for purified cotton containing polymer, both calculated to 100 units of cellulose, was about 1.5. The interacted monomer is composed of the same type of group as that of the cyanoethyl groups. This would indicate that on an average, about 1.5 additional groups for each cyanoethyl group were added by the high-energy radiation-induced interaction.

Here, too, the weight ratio of cyanoethyl groups to interacted monomer increased to about 0.34 at about 0.6 D.S. and then remained relatively constant over the range investigated. This would indicate that on an average about 3 additional groups were added for each cyanoethyl group by the radiation-induced interaction. In an estimate of the value of 1.5 added groups for each cyanoethyl group, it was assumed that the interaction of monomer with the purified cotton occurred in the same manner and to the same extent as with the cyanoethylated cotton. Consequently, the amount of monomer interacted with the cotton was subtracted from the amount interacted with the cyanoethylated cotton before

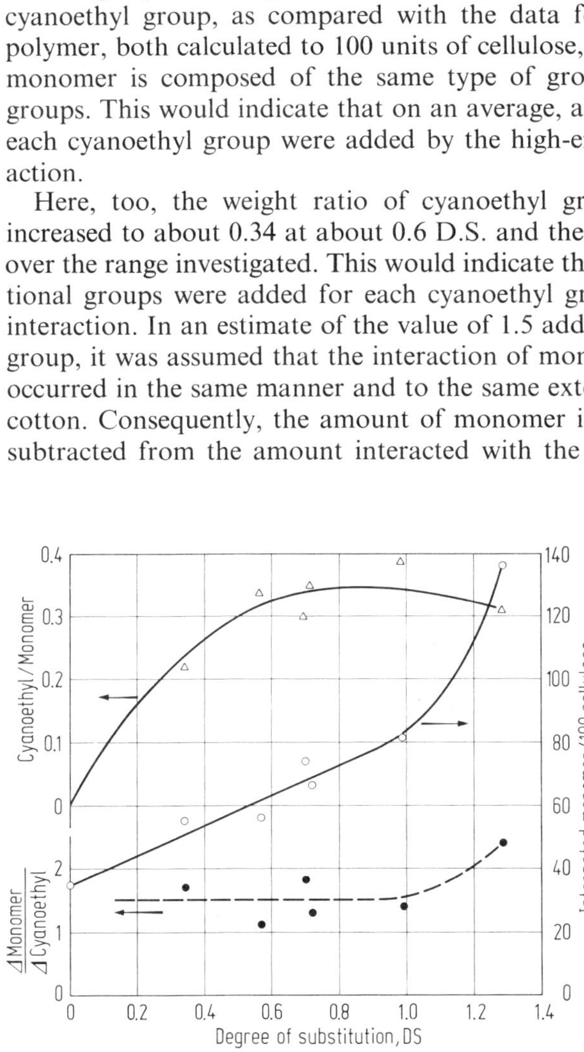

Fig. 48. Relationships between interaction of monomer with cyanoethylated cotton and degree of substitution. Dosage: 1 megaroentgen. Reprinted from Ref.[58], p. 824, by courtesy of Textile Research Institute.

calculating the additional groups. Hence, it was concluded that the interaction of acrylonitrile with cyanoethylated cotton probably involves grafting of side chains of acrylonitrile at the site of cyanoethyl groups[58]. This conclusion was substantiated by infrared absorption studies which reveal that the hydroxyl bond intensities of cyanoethylated cotton were considerably less than the hydroxyl intensity of purified cotton; however, the hydroxyl intensity of cyanoethylated cotton and cyanoethylated-acrylonitrile graft copolymer were almost the same[59].

The interaction of acrylonitrile (AN) and methyl methacrylate (MMA) with cyanoethylated cotton in the presence of Ce^{IV} ions has been studied under a variety of conditions[60]. Variables studied include the D.S. of cyanoethylated cotton; the temperature; the type of initiator, i.e., Ce^{IV} ammonium nitrate (CAN) and Ce^{IV} ammonium sulfate (CAS); the reaction time; and the concentration of both monomer and initiator.

Figure 49 shows the effect of the D.S. of cyanoethylated cotton on the rate of grafting at 40 °C. It is apparent that increasing the D.S. causes an appreciable increment, in the graft yield. However, with all substrates, the grafting reaction is very fast, especially in the earlier stages of the reaction. This holds true for both monomers, i.e., AN and MMA. However, the rate of grafting of AN is much higher than that of MMA. Low solubility in the reaction medium, as well as the effect of the two methyl groups in MMA, making growth of the polymer chain sterically more difficult as compared to AN, would account for this.

Hebeish et al.[61] have examined the grafting of methyl methacrylate on cellulosic materials by chain transfer routes under the influence of azobisisobutyronitrile (AZBN). The study dealt partly with cyanoethylated cotton in a comparision with partly carboxymethylated cotton, cotton treated with 12 M sodium hydroxide, cotton treated with 0.5 M sodium hydroxide, and purified cotton. Proof of grafting was provided by fractional precipitaion. In this study cyanoethylation was achieved by impregnating the cotton with 0.5 M sodium hydroxide for one hour followed by padding. Samples were treated with acrylonitrile at 60 °C for varying time

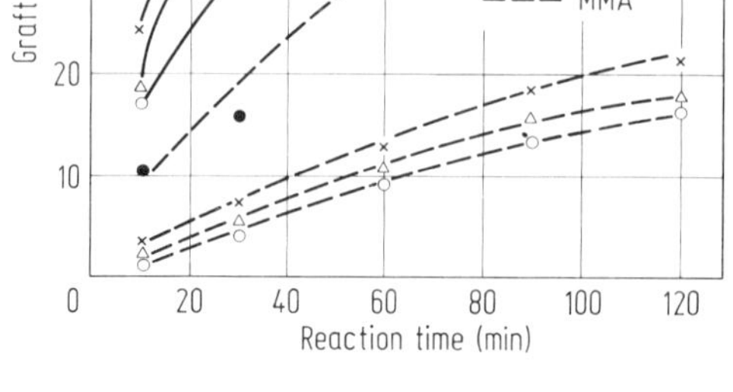

Fig. 49. Effect of D.S. of cyanoethylated cotton (CEC) on rate of grafting: D.S. = (○) 0.0; (△) 0.2; (×) 0.337; (●) 0.515. CAS Concn. = 0.01 M; temp. = 40 °C; M/L ratio = 1:30; AN, MMA, 1 g. Reprinted from Ref.[60], p. 1009, by courtesy of John Wiley & Sons, Inc.

intervals, before washing in dilute acetic acid and water and drying under ambient conditions. Grafting was carried out as follows: The conditioned samples were immersed in 100 cm^3 of a mixture containing water/methanol/dimethylformamide (75:24:1) having a known concentration of the initator (0.01 M) at 60 °C. Reaction proceeded for known times and the products were repeatedly extracted with acetone to constant weight. With all the substrates, an induction period was observed. For cyanoethylated cotton this was 30 minutes. The grafting pattern followed the same trend for all substrates, but for a given reaction time the graft yield follows the order: partly carboxymethylated cotton > cotton treated with 12 M sodium hydroxide > cyanoethylated cotton (D.S. 0.126) > cotton treated with 0.5 M sodium hydroxide > purified cotton. This reflects the effect of the changes in physical and/or chemical structure caused by alkali in the presence and absence of the alkylating agent. Studies of the variation in grafting levels with a degree of cyanoethylation show that there is little to be gained from increased substitution in the D.S. range 0–0.15. This may be partly explained in terms of location of grafted sites, whether surface cyanoethylation or volumetric cyanoethylation.

With the Ce^{IV} initiation system, regardless of the conditions used, it was found that an increase in the D.S. from about 0.0 to 0.3 is accompanied by an increase in the graft yield. This increase becomes more significant when cyanoethylated cotton of higher D.S. i.e., about 0.5, is grafted[60)]. This again suggests that the presence of cyanoethyl groups on the cellulose chain offers sites for increased interaction of monomer with the cellulose upon Ce^{IV} treatment. This will be discussed later in a more detailed manner.

Kantouch et al.[60)] observed that increasing the reaction time, the monomer concentration, and the temperature favorably influences the degree of grafting on cyanoethylated cotton.

An increase in the monomer concentration from 0.5 g to 2.0 g of monomer increased the maximum graft yield obtained with cyanoethylated cotton (D.S. = 0.515) by about three-fold. An increase in the initiator concentration (CAS) up to 0.05 M is accompanied by an increase in the graft yield, while a further increase brings about lower yields. This is probably because of the fast termination of the growing polymer chains owing to the presence of excess Ce^{IV} ions[62)]. On the other hand, raising the temperature brings about a considerable increase in the graft yield.

The rate of grafting of AN and MMA on cyanoethylated cotton (D.S. = 0.515) follows the order 60 °C > 40 °C > 30 °C. Increasing the temperature seems to cause (a) a higher rate of dissociation of initiator, (b) increased solubility of monomer, (c) increased diffusion of the monomer from the aqueous phase to the cellulosic phase, and (d) enhancement of the rate of initiation and propagation of the grafted chains. All these factors would be expected to improve the graft yield.

The rates of the Ce^{IV} ion consumption during grafting and oxidation of cyanoethylated cottons having different D.S. are given in Table 44. With all substrates studied, the Ce^{IV} consumption increases with increasing reaction time. However, the consumption during grafting is much greater than during oxidation. The same phenomenon was reported by Hebeish and Mehta[38)], who attributed

Table 44. Ce^{IV} ion consumption during oxidation and grafting of cyanoethylated cotton having different D.S.[a] Reprinted from Ref.[60], p. 1014, by courtesy of John Wiley & Sons, Inc.

Reaction time, mins.	Oxidation				Grafting (AN)				Grafting (MMA)			
	D.S. 0.0	0.2	0.337	0.515	0.0	0.2	0.337	0.515	0.0	0.2	0.337	0.515
Ce^{IV} Concumption, 30 °C												
10	2.4	0.9	1.34	1.5	3.3	2.7	3.3	3.8	2.7	2.4	2.4	2.4
30	4.2	1.8	2.7	3.0	4.8	5.5	5.7	4.8	4.2	3.3	3.3	3.5
60	5.4	2.7	2.7	3.3	6.2	5.7	6.0	6.0	6.0	4.5	4.2	4.8
90	6.3	3.0	3.0	3.6	7.5	6.3	8.7	7.8	6.6	5.3	5.4	5.4
120	6.3	3.0	3.3	3.8	8.1	7.8	8.7	7.8	7.8	6.3	6.3	6.6
Ce^{IV} Consumption, 60 °C												
10	3.0	1.8	2.4	2.4	4.2	3.6	4.2	3.6	2.7	2.7	3.3	3.0
30	5.7	2.4	2.7	3.0	6.9	5.5	5.4	5.4	6.0	4.2	4.9	4.2
60	6.9	3.0	3.6	3.9	8.4	7.8	6.9	6.9	7.8	6.0	7.5	7.2
90	9.0	3.6	3.6	4.0	11.4	9.9	10.2	9.3	9.6	8.7	8.4	9.3
120	9.0	3.6	3.6	4.8	11.3	10.8	11.0	10.2	11.3	10.2	10.1	10.7
Ce^{IV} Consumption, 60 °C												
10	6.6	3.6	3.6	4.5	8.7	6.9	7.8	7.8	7.2	5.1	5.4	5.4
30	14.7	6.0	6.6	8.4	21.9	14.9	21.3	19.2	24.0	19.8	20.4	19.2
90	18.0	9.0	9.0	9.0	28.8	24.6	25.2	25.8	28.8	25.2	25.2	25.2
120	24.0	20.1	10.5	10.0	30.0	28.8	29.4	29.4	30.0	30.0	30.0	30.0

[a] CAS concentration 0.01 M; monomer, 1 g; M/l ratio, 1:30; Ce^{IV} consumption expressed in mmol/100 g cellulose.

it to initiation and termination of homopolymer by Ce^{IV}. The Ce^{IV} consumption during oxidation of cyanoethylated cotton, regardless of the D.S., is much lower than that of the unmodified cellulose (control). However, the difference in the consumption is marginally decreased at higher D.S. This makes it evident that changing the chemical structure of cellulose by cyanoethylation could certainly affect the susceptibility to graft copolymerization initiated by Ce^{IV} ions[60)].

The presence of cyanoethyl groups on the cellulose seems to bring about an opening up of the cellulose structure, as well as the addition of new functional groups on the cellulose molecule, thereby promoting the Ce^{IV} attack. On the other hand, the cyanoethyl groups block some of the cellulose hydroxyls and determine the eventual location of the Ce^{IV} attack. Work on oxidation of cellulose with Ce^{IV}[37,39,41,42,44)] showed that the total consumption of Ce^{IV} is due to (i) formation of a Ce^{IV}-cellulose complex at the glycol, (ii) termination of free radicals produced as a result of the partial dissociation of this complex, and (iii) initiation and termination of free radicals formed on the hemiacetal groups at the end unit of the cellulose molecule. Thus, the excess consumption of Ce^{IV} in the case of the control could be ascribed to the easier formation of the Ce^{IV}-cellulose complex, as compared to cyanoethylated cotton where this may be impeded due to blocking of some of the cellulose hydroxyls[60)].

Another interesting feature is that, despite the fact that the Ce^{IV} consumption during oxidation of the control is higher than that of cyanoethylated cotton, cyanoethylated cotton showed higher grafting yields (cf. Fig. 63). The graft yields also increase upon increasing the D.S. of the cyanoethylated cotton[60)]. This suggests that the free radicals formed at the cyanoethyl groups are more easily captured by the monomer during the grafting reaction than those brought about via dissociation of the Ce^{IV}-cellulose complex. It seems that in the case of the control, destruction of some of these free radicals by Ce^{IV} ions occurred even in the presence of the monomer. That is, oxidation of the control by Ce^{IV} ions during grafting is much higher than that of cyanoethylated cotton. This is substantiated by the fact that the Ce^{IV} consumptions during grafting of the control and of the cyanoethylated cotton are comparable (see Table 44).

Direct study of the cyanoethylation of cotton indicates a possible preferential reaction at the C_2 position, from which one can assume that most of the cyanoethyl groups are on that hydroxyl group[63)]. The presence of such groups at this position would definitely inhibit the formation of the Ce^{IV}-cellulose complex[60)], which has been reported[39,42–46)] to involve the glycol groups. It is for this reason the rate of oxidation of the control is much greater than that of cyanoethylated cotton. However, the consumption increases with increased D.S. of cyanoethylated cotton. The graft yields obtained with the latter are also higher than those of the control, which are much greater at higher D.S. (Fig. 49). It is possible that the Ce^{IV} ion attacks the cyanoethylated cotton exclusively at the cyanoethyl group. This is reasonable, since the presence of the $-C\equiv N$ grouping with its negative inductive effect would facilitate hydrogen transfer. That is, the presence of —CN groups, due to resonance stabilization, favors the abstraction of hydrogen from the carbon atom next to the —CN group. Consequently, the free radicals are very likely formed at the carbon atom of the cyanoethyl group. The following reaction scheme seems to be plausible for the interaction of a vinyl or acrylic monomer with cyanoethylated cotton in the presence of Ce^{IV} [60]:

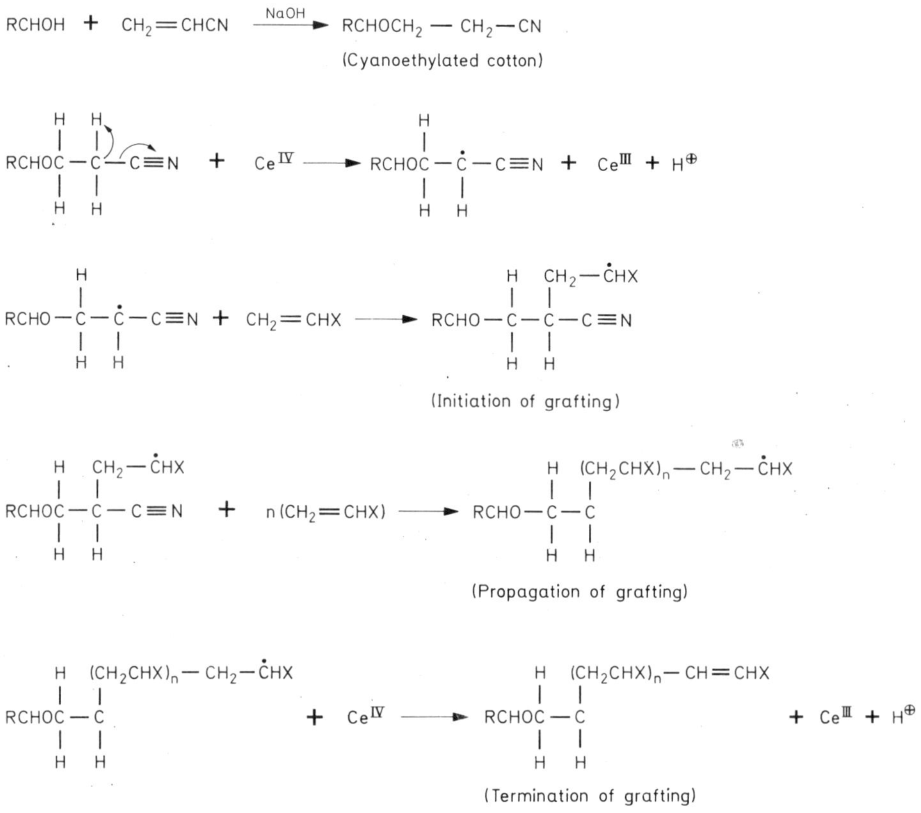

RCHOH is cellulose with the hydroxyl group at C_2 of the anhydroglucose unit and $CH_2{=}CHX$ is a vinyl or acrylic monomer. It is interesting to note that this mechanism is in agreement with the postulation pointed out earlier concerning the interaction of a vinyl monomer with cyanoethylated cotton upon high-energy irradiation.

6.5. Grafting on Partially Carboxymethylated Cellulose

Carboxymethylation is among the more common methods of chemical modification of cellulose. Partially carboxymethylated cotton may be simply prepared by padding the cellulose with monochloroacetic acid or its sodium salt, followed by padding it with sodium hydroxide[61, 63–67]. The overall reaction may be written as follows:

$$\text{Cell—OH} + \text{NaOH} \rightarrow \text{Cell—O—Na} + H_2O$$
$$\text{Cell—ONa} + ClCH_2COONa \rightarrow \text{Cell—}OCH_2COONa + NaCl$$

This is an example of the classical Williamson synthesis for the preparation of ethers. However, mechanism studies on the reactions of α-halocarboxylic acids in aqueous alkaline solutions suggest that a nucleophilic attack may not occur in one step, but that highly reactive α-lactones may be formed as transient intermediates[68].

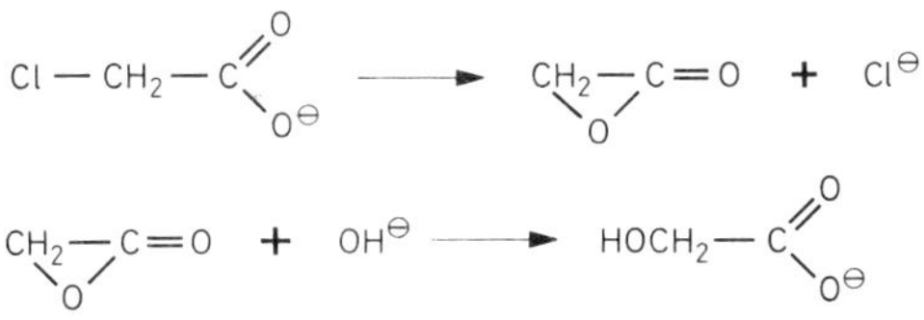

Such transient intermediates may also be involved in the carboxymethylation of cellulose.

Partially carboxymethylated cotton with a D.S. of about 0.05—0.15 retains the original fibrous nature and exhibits a number of potentially valuable properties, such as a crisp hand with a slightly starched feel, increased moisture regain, water absorbancy, water permeability, changed dyeing characteristics, increased resistance to soiling from aqueous dispersions, greater ease of soil removal, cation exchange properties, high water swellability, etc.[69–76]. The presence of two functional groups, hydroxyl and carboxyl, in partially carboxymethylated cellulose affords attractive sites for further chemical modifications, such as hydrolysis[77–78], oxidation[77–80], etherification[79], esterification[78–79], enzyme binding[80], and grafting[81].

Kantouch et al.[82] have reported a detailed study of grafting of various vinyl and acrylic monomers: acrylonitrile (AN), methyl methacrylate (MMA), and acrylamide (AA) onto partially carboxymethylated cotton (PCMC) using tetravalent cerium as initiator. For comparison, they have grafted the same monomers onto native cotton and alkali-treated cotton (NaOH-Cotton). It is clear that the grafting reaction is characterized by an initial fast rate followed by a slower rate which levels off after a certain time, depending on the nature of the substrate and the reactivity of the monomer. The leveling off of grafting with time could be attributed to a decrease in concentration for both initiator and monomer. The data also reveal that the graft yield obtained with PCMC is significantly higher than that obtained with native cotton. The presence of bulky groups, such as-CH_2COOH in the cellulose, may open up the cellulose structure, thereby increasing the diffusion of the initiator and the monomer into the cellulose. In other words, during swelling of PCMC in the reaction medium, there is breakdown of many of the hydrogen bonds in the amorphous regions of the cellulose, and the chain molecules are then able to move apart. Hence, there will be a tendency for the reaction molecules of monomer and initiator to approach the crystalline regions better than when the reaction is performed on unmodified cellulose. The cellulose is held in an open state to give a highly swollen product, in which the distribution of the graft may be quite different from that which occurs in unmodified cotton. The graft yield for NaOH-cotton is substantially higher than that for native cotton. This is expected since NaOH-cotton has greater accessibility than native cotton. However, the graft yield observed with PCMC is higher still than that for NaOH-cotton.

The results also show that with all the substrates studies, the graft yield obtained using AN is much higher than that for MMA. However, AA produces the lowest graft yield. This discrepancy could be attributed to differences in monomer reactivities, solubility, polarity, molecular size, chemical nature, etc. As pointed out earlier, the lower reactivity of MMA, as compared with AN, could be attributed to its lower solubility in the reaction medium, as well as the effect of the two methyl groups in MMA, thereby making growth of the polymer chain sterically more difficult. The lower graft yield for AA could be attributed to inhibition by occluded oxygen; the polymerization of AA is known to be very sensitive to oxygen.

The effect of the monomer concentration, the initiator concentration, the temperature, and the D.S. of PCMC on grafting of AN, MMA, and AA was investigated[82]. It was found that increasing the monomer concentration caused a substantial increase in the graft yield, particularly when AN or MMA was used. The same effect was found upon increasing the initiator up to a certain concentration, but beyond that, there was a reversal. The effect to temperature is shown in Table 45. It is apparent that grafting at 30 °C gives lower yields than at 40° and 60 °C, particularly in the initial stages. At 30 °, no maximum graft yield was achieved, even after 2 hours' reaction time: at 40 °C the reaction proceeds more quickly than that at 30 °C, though the yield is still rising at 2 hours. At 60 °C the grafting proceeds even more quickly. With CAN the yield continues to rise up to 1 hour, and with CAS the yield continues to rise up to 1.5 hours. The graft yields obtained after 2 hours' reaction at the three temperatures are comparable, in contrast with previous results[38] on the grafting of vinyl monomers on cellulosic material using CAN as the initiator. Hebeish and Mehta[38] have observed that the graft yields obtained at 20 °C were higher than those of 60 °C, following the order 40° > 20° > 60 °C. Fast termination was thought to be responsible. In the case of PCMC, ionization of —CH_2COOH groups along the cellulose chain molecules introduces negative charges which attract Ce^{IV} ions to the cellulose molecules. This leads to the formation on the cellulose backbone

Table 45. Graft yield (%) of AN onto PCMC (D.S.: 0.124) using different Ce^{IV} salts (0.01 g Ce^{IV} salt, 1 g monomer/100 g PCMC)*. Reprinted from Ref.[82], p. 1579, by courtesy of Microforms International Marketing Corporation

Reaction time (mins)	30 °C		40 °C		60 °C	
	CAN	CAS	CAN	CAS	CAN	CAS
5	30.28	14.6	45.06	27.60	60.0	38.4
10	44.30	23.4	52.50	40.50	71.9	54.6
15	51.40	37.3	60.50	51.50	76.0	74.4
30	59.60	56.8	81.00	73.00	85.5	82.7
60	73.00	74.7	82.00	84.50	87.7	87.0
90	80.00	79.0	83.00	86.00	87.7	90.3
120	84.80	82.9	86.00	87.40	87.7	90.3

* AN, acrylonitrile; PCMC, partially carboxymethylated cotton, CAN, Ce^{IV} ammonium nitrate; CAS, Ce^{IV} ammonium sulfate.

of more active sites which are easily available to the monomer. Higher temperatures increase both ionization of carboxyl groups and mobility of monomer. The effect is to produce a substantial formation of graft, outweighing the decrease in graft due to fast termination at high temperature.

The effect of D.S. on the graft yield of PCMC is shown in Figure 50. It is obvious that increasing the D.S. up to 0.124 is accompanied by an increase in the maximum graft yield of AN, but that beyond this point, increasing the D.S. does not cause a significant increase in the maximum graft yields; the maximum graft yield obtained with D.S. 0.152 is substantially lower than with D.S. 0.0647 or 0.124. For MMA and AA, the maximum graft yields increase with increasing DS of PCMC over the range studied. It appears that introduction of $-CH_2-COOH$ groups into the cellulose molecules results in two opposing effects: (a) opening up the cellulose structure, thereby enhancing diffusion and adsorption of both the monomer and the initiator, and (b) blocking some of the hydroxyl groups along the cellulose chains. Effect (a) leads to an increase in the graft yields, whereas effect (b) gives rise to a decrease. Which of the two effects predominates would be the main controlling factor of the graft yields. The latter, in turn, would depend on the changes in the substrate composition and the reagent's participation in the process as the grafting reaction proceeds. It is likely that for AN, effect (a) predominates up to a D.S. of 0.124, after which effect (b) becomes much more prenounced. With MMA and AA, on the other hand, effect (a) still outweighs effect (b) up to the highest D.S. studied.

The consumption of Ce^{IV} during grafting and oxidation of PCMC has been studied[82]. These studies reveal that the consumption increases with increasing reaction time. In the case of AN, the Ce^{IV} consumed during grafting is lower than that consumed during PCMC oxidation. MMA and AA give the opposite

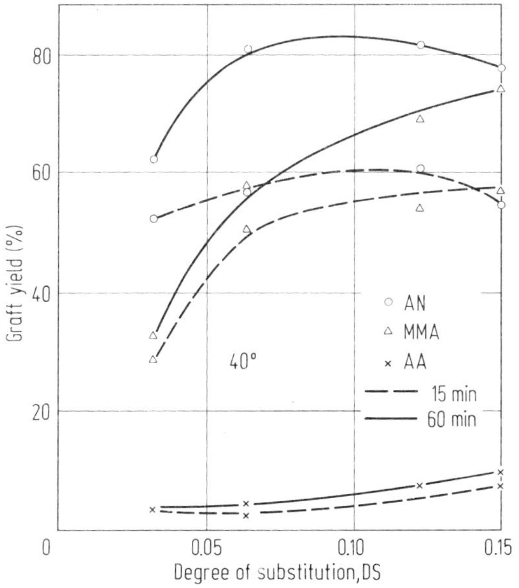

Fig. 50. Percent graft yield vs. degree of substitution (D.S.); cellulose = 1 g; monomer = 1 g; liquor ratio 30. Reprinted from Ref.[82], p. 1580, by courtesy of Microforms International Marketing Corporation.

effect; Ce^{IV} consumption during grafting is higher than that consumed during oxidation.

Taking into consideration the reaction scheme represented by Eqs.[133–138] (as laid out in section 4.2.3.), and assuming that A stands for cellulose, the following conclusions may be drawn:

(i) Total consumption should be more or less equal for oxidation [reactions (134) and (138)] and grafting [reactions (133) and (137)], if all the monomer is effectively grafted onto cellulose.

(ii) Total consumption during grafting will be more than during oxidation of cellulose if some homopolymer is formed [reactions (135) and (137)[38]].

(iii) Total consumption during grafting will be less than during oxidation if the grafted polymer forms a barrier and inhibits diffusion of Ce^{IV} ions.

The second conclusion may be correct in the case of MMA and AA, whereas the third accounts for the excess Ce^{IV} ions consumed during oxidation, as compared with those consumed during grafting of AN. This is substantiated by the fact that Ce^{IV} consumption during grafting of native or alkali-treated cotton is significantly higher than that consumed during oxidation up to 30 minutes.

The effect of D.S. on Ce^{IV} consumption during grafting of various monomers and on that consumed during oxidation of PCMC has been examined. The data indicate that the Ce^{IV} consumed during grafting of the three monomers and that consumed during oxidation increases as the D.S. of PCMC is increased up to 0.124. A further increase in the D.S. lowers the Ce^{IV} consumption during both oxidation and grafting; particularly in the initial stages of the reaction. A similar trend has been observed with the graft yield (cf. Fig. 50). As previously mentioned, introduction of carboxymethyl groups into the cellulose molecule increases the susceptibility of cellulose to Ce^{IV} attack (by attracting Ce^{IV} ions from the reaction medium) and at the same time decreases the number of cellulose hydroxyls. It is likely that increasing the D.S. of PCMC up to 0.124 favors the Ce^{IV} attraction, while further increases operate adversely because of the reduction in the number of free hydroxyl groups in cellulose.

It is not out of place to refer to the study reported by Kantouch et al.[82] on the grafting behaviour of PCMC (D.S. 0.124)when CAS and CAN were used as initiators. Table 45 contains data for percentage graft yields at 30 °C, 40°, and 60 °C with CAN and CAS for PCMC of D.S. 0.124, while Table 46 shows the Ce^{IV} consumption during grafting and oxidation of the same substrate. Table 45 shows that the graft yields obtained with CAN are much higher than those with CAS in the initial stages of the reaction, and at the three temperatures studied. The maximum graft yields obtained with CAN and CAS at 30 °C and 40 °C are comparable, but at 60 °C, the yield obtained with CAS is significantly higher than that with CAN. The fast dissociation of CAN may account for its higher grafting efficiency in the initial stages of the reaction, since less Ce^{IV} would be availabe for initiation. This is substantiated by the data in Table 46, where consumption of Ce^{IV} by PCMC during its oxidation or grafting is much higher for CAN than for CAS, particularly in the initial stages of reaction. With CAN, the Ce^{IV} is completely consumed during grafting after 2 hours at the three temperatures studied. This is the case with CAS at 60 °C only. Schwab et al.[83] have shown that CAN is more efficient at lower temperatures than CAS. It is interesting to note

Table 46. Ce^{IV} consumption (mmol/100 g) during oxidation and grafting of PCMC (D.S.: 0.124) using 1 g AN at different temperatures. Reprinted from Ref.[82], p. 1584, by courtesy of Microforms International Marketing Corporation

Reaction time (mins)	Temperature					
	30 °C		40 °C		60 °C	
	Oxid.	Graft	Oxid.	Graft	Oxid.	Graft
CAN 0.01 M						
5	12.0	17.4	20.4	17.5	21.0	24.0
10	16.0	18.5	23.5	19.0	24.5	24.0
15	19.6	20.5	24.5	19.5	30.0	29.5
30	22.8	21.2	26.7	28.0	30.0	29.6
60	25.5	24.4	29.7	28.0	30.0	30.0
90	29.2	26.6	30.0	28.0	30.0	30.0
120	30.0	27.1	30.0	28.2	30.0	30.0
CAS 0.01 M						
5	0.00	1.62	1.30	2.16	2.92	7.14
10	0.54	2.16	1.95	4.32	3.90	11.50
15	0.54	3.24	2.60	6.49	4.54	18.50
30	1.08	4.86	2.60	11.25	5.84	27.00
60	1.62	8.10	3.90	17.84	7.78	29.50
90	2.70	11.35	4.22	20.00	12.00	30.00
120	2.70	12.86	4.54	26.50	16.22	30.00

AN, acrylonitrile; PCMC, partially carboxymethylated cotton; CAN, Ce^{IV} ammonium nitrate; CAS, Ce^{IV} ammonium sulfate.

that with CAN and CAS, the maximum Ce^{IV} consumption during oxidation of PCMC follows the order 60° > 40° > 30 °C.

Table 46 also shows that the Ce^{IV} consumed during grafting with CAS is greater than that consumed during oxidation at 30°, 40°, and 60 °C, which is in full agreement with previous results[38,61]. In the case of CAN, the Ce^{IV} ion consumption at 60 °C during grafting is slightly higher than that during oxidation, while at 40 °C the Ce^{IV} consumed during oxidation is slightly higher than that during grafting. With CAN, it is apparent that more than 60 percent of the Ce^{IV} in the reaction was used in the first five minutes at 60 °C, most of it for grafting, since homopolymer formation is negligible during this time. On the other hand, CAS is consumed from the reaction medium gradually and at a much slower rate than CAN. This means that appreciable amounts of Ce^{IV} would be available, in the case of CAS, throughout the reaction. Although diffusion might be impeded by the grafted branches on the substrate, it would still be possible due to greater availability, as compared with CAN. Here also, at the same graft yield the Ce^{IV} consumption during grafting differs considerably with CAN and CAS, especially in the case of CAN. This could be attributed to the difference in rates of initiation and termination with the two initiators.

The effect of the grafting reaction, initiated by Ce^{IV}, on the carboxymethyl groups of the modified cotton was also studied[82]. Results of this study indicate that grafting had practically no effect on the carboxymethyl groups (expressed as carboxyl groups) of the PCMC.

As part of a comparative study on grafting by chain transfer mechanisms, Hebeish et al.[61] grafted methyl methacrylate onto partly carboxymethylated cotton cellulose using azobisisobutyronitrile as the initiator. The degree of substitution was carried out in water/solvent mixtures. The solvents used included methanol, ethanol, propanol, butanol, acetone, and carbon tetrachloride. The methyl methacrylate concentration was held at 6%, and grafting was carried out in a monomer/liquor ratio of 1:100 using a temperature of 60 °C and a reaction time of four hours. Table 47 gives the influence of the reaction medium on grafting with this system. It is clear that regardless of the water/solvent system employed, a substantial amount of grafting is obtained. A regular decrease also occurs with increases in the solvent ratio in the water/solvent mixture. However, the magnitude of grafting differs significantly with the type and ratio of solvent used in the mixture. For the alcohols the graft yield follows the order: propanol > ethanol > methanol > butanol when the solvent constitutes 25% of the water/solvent mixture. On the other hand, the graft yield follows the order: methanol > ethanol > propanol > butanol at 50 and 75% solvent in the water/solvent mixture.

Acetone is particularly attractive, possibly because of ketone-enol tautomerism, whereby it is somewhat similar in structure to the alcohols. Acetone shows graft yields which are comparable to those of propanol, particularly when used in a ratio of 25 and 50% in solvent/water mixture, whereas the graft yields obtained with carbon tetrachloride are the lowest compared with those of other solvents in the range studied.

Using a water/solvent mixture containing 25% of either methanol, ethanol, propanol, or acetone seems to constitute a reaction medium where the monomer and initiator are completely miscible and the swelling of cellulose by water is not hindered by the presence of these solvents. This certainly simplifies access and diffusion of monomer and initiator, thereby promoting grafting. The greater yield obtained by using either propanol or acetone not only substantiates this, but also reflects the effect of the molecular size of the solvent on grafting. Propanol and

Table 47. Influence of reaction medium on grafting of methyl methacrylate onto partially carboxymethylated cotton (PCMC). Reprinted from Ref.[61], p. 155, by courtesy of Verlag Chem. GMBH, Berlin

Solvent	Graft Yield (%) Water/Solvent ratio		
	75:25	50:50	25:75
Methanol	135	98	74
Ethanol	150	68	34
Propanol	260	64	31
Butanol	124	34	18
Acetone	264	60	30
Carbon Tetrachloride	73	30	15

acetone, both larger than methanol and ethanol, can disrupt the cellulose microstructure more extensively after having diffused into the lattice, thus rendering the cellulose more accessible to monomer and initiation. With butanol, however, the lower graft yields could be ascribed to its adverse effect on the swelling of cellulose by water. Associated with this factor is the hydrophobic character of the discussed alcohols, which decreases from butanol to methanol. In combination, these effects seriously hinder monomer access to reactive sites on cellulose. This is further supported by the significant fall in grafting when using carbon tetrachloride.

The decrease in the graft yield, which results from an increase in the ratio of solvent in the water/solvent mixture, could be interpreted in terms of the adverse effect of the solvent on the swelling properties of cellulose in water. The data suggest that the relative sorptive and swelling properties follow the order: methanol > ethanol > propanol > acetone > butanol > carbon tetrachloride. However, the difference in capability of these solvents as terminators for the graft polymer radical and the cellulose macroradical via chain transfer cannot be ruled out as a possible reason for this decrease.

Bhattacharjee and Perlin[84)] have examined the potential of crosslinked carboxymethyl cellulose (CLCMC) as a highly absorbent material after modification through grafting with polystyrene. One treatment involves radiation-induced grafting of polystyrene onto CLCMC fibers. It has been shown that, at a critical level of grafting of approximately 25%, the retentivity of the fibers is almost doubled, despite an initial decrease in retentivity on grafting. It is proposed that grafted polystyrene becomes associated preferentially with the residual ordered regions of the CLCMC through the localization of radicals in these regions[85)]. Despite the hydrophobic character of polystyrene-grafted CLCMC, it is possible that this is more than compensated for by an overall disorganization of the fiber structure, which leads to a greater uptake of water by the hydroxyl groups in the matrix. This increase in hydrophilic character has been demonstrated by observed increases in enzymic activity[86)] exhibited by cellulose, which indicates[87)] that the newly accessible regions of the cellulose phase are neither highly substituted, nor extensively crosslinked. This interesting approach to the characterization of cellulosic graft copolymers highlights several difficulties associated with using partially substituted derivatives. The advantages of such partial substitution in terms of product application are obvious. However, mechanistic and kinetic interpretations are hindered by lack of knowledge of the true nature of the substrate, i.e. problems associated with the location of grafts, and difficulties in assigning likely active sites. This topic is covered further in Chapter 7, which deals with cellulosic copolymer characterization. Bhattacharjee and Perlin have shown that an increase in grafting of CLCMC with polystyrene beyond the 30% level leads to predominant water repellency and a marked fall of any water retention value (WRV)[87)]. This supports the findings of Ogiwara et al. in studies on other cellulose grafted systems[88)]. The level of water retention was measured using the method of Lepoutre et al.[89)], which involves soaking in excess water before centrifuging at 0.1 to 0.2 g. The retention values are based on the material which did not pass through the screen of the basket supported one-half inch above the bottom of the metal centrifuge tube.

In an extension of their work on carboxymethylated cotton, Hebeish et al. have examined the dimethylaniline — benzyl chloride (DMA-BC) route to induced grafting using methyl methacrylate[90]. This study is part of an extensive program involving grafting onto modified cottons[91–99]. In studies of the influence of the initiator concentration (DMA-BC) on grafting, it has been shown that a concentration of 0.08 to 0.087 mol/l yields the greatest level of grafting. Beyond this, a decrease in the level of grafting is observed.

This decrease in grafting followed by an increase in the initiator concentration could possibly be associated with a faster rate of termination, probably owing to an abundance of free-radical species in the polymerization system. It is also possible that at higher initiator concentrations DMA reacts with BC to yield a quaternary ammonium salt which is not capable of initiating polymerization. A third possibility is that homopolymerization may prevail over grafting at higher initiator concentrations. There is, however, little reason to believe that any one of the three possibilities is operating to the exclusion of the others.

Upon increasing the monomer concentration in the bulk medium, there is an obvious increase in the grafting levels in the polymerization system. The same situation is also encountered with respect to the total conversion. At high concentrations of MMA, the gel effect brought about by the solubility of poly(methyl methacrylate) in its own monomer seems to be more pronounced. As a result, termination of the growing grafted chain radicals by coupling is hindered, while the swellability of PCMC is enhanced. The ultimate effect of this is increased grafting.

Besides the gel effect, complex formation between PCMC and MMA is more favorable at higher concentrations of MMA. Complexation of PCMC with MMA activates the latter through formation of a donor-acceptor complex in which the uncomplexed MMA, though normally an electron acceptor, behaves as a donor relative to the complexed MMA, which has been converted to a stronger acceptor. Enhanced monomer reactivity is also apparently due to association of complexed monomer into organized arrays, since PCMC acts as a matrix for such alignment. Naturally, the effect of the enhanced monomer reactivity is increased grafting.

Hebeish et al. have also studied the effect of temperature and reaction time on grafting using this system[90]. At 50 °C, the grafting reaction is accompanied by a long induction period (150 min), whereas at a temperature of 60 °C or higher, the reaction proceeds without any induction period. The rate of grafting also increases on increasing the temperature up to 70 °C. Above this temperature, the increment in the rate of grafting is not so striking.

The enhancement in grafting suggests that raising the temperature performs the following functions: (a) increasing the concentration of active species, i.e., free radicals, in the polymerization system; (b) accentuating the swelling properties of PCMC; (c) improving the solubility of MMA; (d) accelerating diffusion of MMA from the polymerization medium to PCMC; and (e) enhancing the rate of initiation and propagation of the graft. There is no doubt that all these effects give rise to higher grafting levels. The effect of increasing the polymerization time is to bring about a significant enhancement in the graft yield. The same holds true for total conversion. However, the magnitude of this enhancement is governed by the polymerization temperature and the monomer concentration, as well as the initiator concentration. Kubota et al.[100] have investigated the decomposition of the peroxide group on fibrous, peroxidized carboxymethyl cellulose[101] and the ability of the decomposition products to initiate grafting. The peroxide groups on the carboxymethyl cellulose liberated hydrogen peroxide when the sample was heated in an aqueous medium. The decomposition of peroxide was found to increase markedly in the presence of an Fe^{II} salt and also on irradiation with light of $\lambda > 300$ nm[102]. Ethyl methacrylate was grafted onto carboxymethyl cellulose peroxide after initiation by heat or irradiation with light. The rate of grafting and the reciprocal average molecular mass of the grafts were found to be proportional to the square root of the peroxide content of the sample, which was itself closely related to the number of grafts initiated. Figure 51 shows the thermal stability of the peroxide groups on carboxymethyl cellulose in an aqueous medium. In Figure 51, the terms a, b, and c represent the amounts of peroxide in the total system, the fibrous material, and the aqueous medium, respectively. The amount of peroxide on the fibrous material (b) shows a tendency to decrease with the passage of time, while an increase in the formation of H_2O_2 was observed in the aqueous medium (c), thereby showing an extended tendency at higher temperatures. The total amount of peroxide, the sum of those on fibrous material and in the aqueous medium, is shown in curve A. The thermal decomposition of peroxide on CMC in a methanolic medium and the formation of H_2O_2 in the medium are slower than in aqueous media. In the

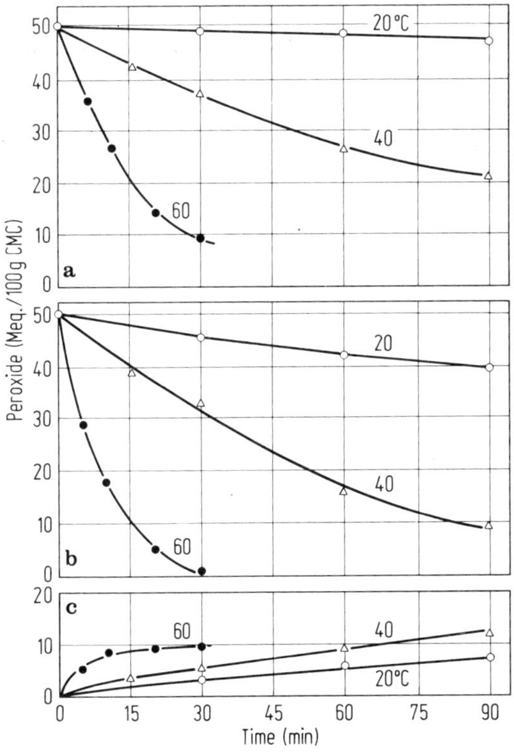

Fig. 51 a–c. The thermal decomposition of peroxide on CMC in an aqueous medium. **a**, **b**, and **c** represent the amounts of peroxide in the total system, in the fibrous material and in the aqueous medium, respectively. Temperatures as indicated. Reprinted from Ref.[100], p. 242, courtesy of John Wiley & Sons, Inc.

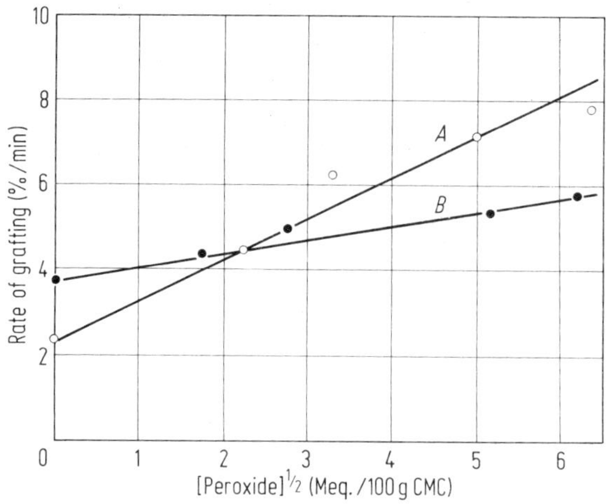

Fig. 52. The relationship between the rate of grafting and the peroxide content in the thermal grafting of MMA at 60 °C *A* and photografting of MMA at 30 °C *B*. Reprinted from Ref.[100], p. 244, courtesy of John Wiley & Sons, Inc.

grafting of methyl methacrylate on carboxymethyl cellulose peroxide, when initiated thermally or by light, various interesting features were seen. Figure 52 shows the relationship between the rate of grafting and the peroxide content of the carboxymethyl cellulose. Carboxymethyl cellulose, though it does not contain a peroxide group, initiates grafting of methyl methacrylate by heat or light. A higher grafting rate in the same initiation systems was observed for the carboxymethyl cellulose peroxide, which shows a nearly linear relationship with the square root of the peroxide content of the sample. The graft efficiencies were of the order 80%–90%, and graftings were characterized by a relatively low formation of homopolymer.

The average molecular mass of grafted poly(MAA) showed a temdency to decrease with increasing amounts of peroxide. A lower molecular mass for grafts

Table 48. The photoinitiated grafting of various monomers on carboxymethyl cellulose peroxide. Reprinted from Ref.[100], p. 246, by courtesy of John Wiley and Sons, Inc.

Monomer	Peroxide meq/100 g CMC	Total conversion %	Percent grafting %	Graft efficiency %
Acrylamide	0	61.9	214.3	84.8
	34.9	100	365.3	89.5
Acrylic acid	0	—	88.0	—
	34.2	—	205.1	—
Methyl methacrylate	0	29.7	102.4	90.3
	34.6	41.1	152.2	96.9
Acrylonitrile	0	15.7	42.9	83.0
	31.5	69.4	191.6	84.0
Vinyl acetate	0	10.5	32.0	80.1
	35.0	54.0	178.8	86.9

Photoirradiation was carried out at 30 °C for 14 min in a Pyrex glass tube.

was obtained for the photoinitiation system, as compared to the thermal system. A straight-line relationship was obtained between the reciprocal of the average molecular mass of the grafts and the square root of the peroxide content of the substrate.

The number of grafts increased in proportion to the peroxide content. However, the number of grafts was higher for the copolymer obtained in the photoinitiation system than in the thermal system. The number of grafts on the CMC substrate increased by introducing the peroxide group. Values of 17 and 6 times were recorded for CMC peroxide with thermal initiation and photoinitiation, respectively.

The level of grafting with various monomers, through photoinitiation on carboxymethyl cellulose peroxide, is given in Table 48. The high grafting levels and grafting efficiencies are immediately apparent.

6.6. Grafting on Acrylamidomethylated Cellulose

Acrylamidomethylated cellulose can be prepared by reacting cellulose with N-methylolacrylamide under the catalytic influence of hydrogen ions[91]:

$$\text{Cell—OH} + CH_2{=}CHCONHCH_2OH \rightarrow \text{Cell—O—}CH_2\text{—}NHCOCH{=}CH_2$$

Acrylamidomethylation may be carried out by impregnating cellulose in a solution containing a suitable concentration of N-methylolacrylamide (e.g., 20%) and an acid-liberating salt, such as ammonium chloride as a catalyst. After being squeezed to a wet pickup of 95–100%, the cellulose is placed for a few minutes in an oven at 150 °C. The cellulose is then treated with an aqueous solution of 2% sodium nitrate at 80 °C, followed by washing in soap solution and finally rinsed with cold water. Acrylamidomethylated cellulose has a pendant double bond which may react, under certain conditions, with various compounds having labile hydrogen atoms, or it may be homopolymerized in the presence of an initiator. Thus, acrylamidomethylated cellulose on subsequent chemical treatments gives a derivative with different reactivity than the unmodified cellulose after subsequent chemical treatment.

Hebeish et al.[91] have studied the effect of the change in the physical and chemical structure of cellulose by acrylamidomethylation on the susceptibility of cellulose toward grafting. They followed the rate of grafting of acrylamidomethylated cellulose and that of unmodified cellulose (control) using either Ce^{IV} ammonium sulfate (CAS) or Ce^{IV} ammonium nitrate (CAN) as initiators. It is clear that with both substrates the grafting reaction is characterized by an initial fast rate followed by a slower rate. However, the initial fast rate observed with the modified cellulose is significantly lower than that of the control. Also, the maximum graft yield obtained with the modified cotton is less than that of the control. The lower graft yields obtained with acrylamidomethylated cellulose, as compared with the control, could be interpreted in terms of the functions of the $—CH_2—O—CONH—CH{=}CH_2$ group; which causes (a) blocking of some of the cellulose hydroxyls; (b) decreasing of the swellability of cellulose due to its hydrophobic character; (c) termination of the growing chain of the graft, hence, decreasing its molecular mass; and (d) crosslinking of cellulose by homopolymerization of the double bonds or addition of a cellulose chain radical to the double bond present on an adjacent cellulose chain. Effects (a), (b), and (c) would bring about a decrease in the graft yield. The same holds true for effect (d), since crosslinked cellulose would be less amenable to grafting due to its lower swellability, as well as the decrease in the number of available cellulose hydroxyl groups (sites for grafting).

A comparison between the efficiency of CAS and CAN in inducing grafting was made. In the case of acrylamidomethylated cellulose, there is no difference in the graft yields obtained with CAN and CAS. On the other hand, with the control, the yields obtained with the two initiators differ considerably; CAN is more efficient than CAS in the range studied. The identical graft yield obtained with both initiators in the case of the modified cellulose could be ascribed to crosslinking through homopolymerization of the double bond during the grafting reaction. Graft yields of a crosslinked cellulose would be expected to be much lower than those of the acrylamidomethylated cellulose and the control, perhaps because of the relatively higher lack of swellability. This is substantiated by the fact that with both initiators, i.e., CAN and CAS, leveling off of grafting occurs after about 30 minutes in the case of the modified cellulose, whereas no maximum graft yield was observed with control even after 120 minutes using CAS[91].

Figure 53 shows the graft yield (percent) as a function of degree of substitution (D.S.) of acrylamidomethylated cellulose. These data make it evident that the graft yield decreases on increasing the D.S. This suggests that the pendant double bonds appear to be homopolymerized by the catalytic action of Ce^{IV}. This conversion increases with the acrylamidomethyl content of cellulose because increasing the acrylamidomethyl content increases the probability that there will be another double bond close enough to any given double bond to permit reaction.

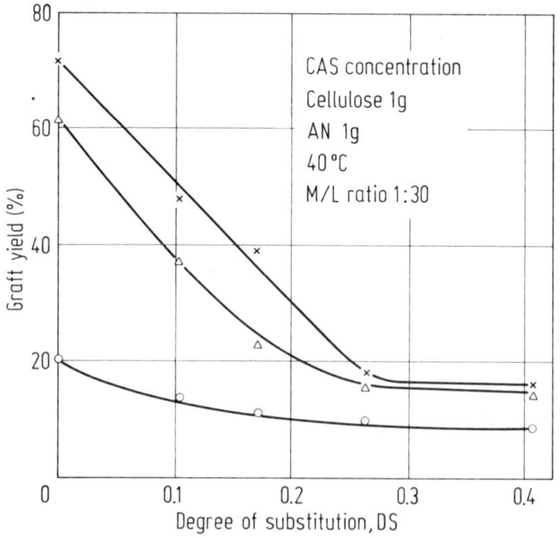

Fig. 53. Graft yield % as a function of the degree of substitution of acrylamidomethylated cellulose. ○, △, × for a reaction time of 10, 60, and 120 minutes, respectively. Reprinted from Ref.[91], p. 1925, by courtesy of John Wiley & Sons, Inc.

Using CAN and CAS, the Ce^{IV} consumption during grafting and during oxidation of acrylamidomethylated cellulose and the control has been followed. Apparently, the consumption increases with increasing reaction time. It is more significant in the initial stages of the reaction, particularly in the case of CAN, than in the later stages, perhaps due to lowering of the Ce^{IV} ion concentration, as well as to the shortage of reactive sites on the substrate backbone as the reaction proceeds. This is observed with both substrates, i.e., acrylamidomethylated cellulose and the control. Nevertheless, the Ce^{IV} consumption during oxidation of acrylamidomethylated cellulose is higher than the corresponding consumption in the case of the control when CAS was used. The opposite holds true for CAN.

It follows from this that introducing acrylamidomethyl groups into the cellulose molecule significantly effects its behavior with regard to Ce^{IV} attack. Besides the direct attack of Ce^{IV} at the cellulose hydroxyl, attack at the pendant double bond seems to occur. As a result, homopolymerization of the double bonds takes place to produce a crosslinked cellulose. Formation of the latter would be expected to depend mainly on the number and distribution of the double bonds, as well as on the availability of Ce^{IV} in the reaction medium. Since the rate of dissociation of CAN is much higher than that of CAS[83], the availability of Ce^{IV} for oxidation of acrylamidomethylated cellulose would be greater in the case of CAN than with CAS. As pointed out above, this would result in a crosslinked cellulose. The lower rate of oxidation of acrylamidomethylated cellulose at the later stages of the reaction in the case of CAN suggests that the crosslinked cellulose is less susceptible to Ce^{IV} attack than the control. On the other hand, formation of crosslinked cellulose seems to occur at a relatively lower rate in the case of CAS due to its slower rate of dissociation. Consequently, the change in the chemical structure of cellulose via crosslinking would not be so striking. This would account for the high Ce^{IV} consumption by acrylamidomethylated cellulose in the case of CAS[91], as compared to the control.

6.7. Grafting on Cellulose Treated with N-Methylolacrylamide in an Alkaline Medium

Like other etherification reactions of cellulose such as cyanoethylation, the reaction of cellulose with N-methylolacrylamide (MAM) in an alkaline medium will proceed via addition of the cellulose hydroxyl to the double bond of MAM as follows[91]:

$$\text{Cell—OH} + CH_2{=}CHCONHCH_2OH \rightarrow$$
$$\text{Cell—O—}CH_2CH_2CONHCH_2OH$$

This reaction may be carried out by impregnation of cellulose in a solution containing about 25% N-methylolacrylamide and 2% potassium hydroxide. After squeezing the sample to a wet pickup of ca. 80%, it is placed in an oven at 150 °C for a few minutes. The cellulose is then neutralized with dilute acetic acid, washed thoroughly with water, and air-dried.

Introducing this type of bulky group, i.e., $—CH_2CH_2CONHCH_2OH$, along the cellulose chains would be expected to improve the reactivity of cellulose to grafting with Ce^{IV}, perhaps by opening up the cellulose structure or by the presence of the methylol groups in the cellulose molecule. Hebeish et al.[91] have shown, however, that the graft yields obtained with cellulose treated with MAM in an alkaline medium are substantially lower than those of the control, thereby indicating that the above expectation is not valid. This would suggest that MAM reacts with cellulose in an alkaline medium as a bifunctional compound and yields a crosslinked cellulose. The reaction may be represented as follows:

$$2\,\text{Cell—OH} + CH_2{=}CHCONHCH_2OH \rightarrow$$

$$\text{Cell—O—}CH_2\text{—}CH_2\text{—}CONHCH_2O\text{—Cell}$$

It is worth pointing out that treating cotton fabric with MAM in an alkaline medium significantly improves its easy care properties, suggesting a cross-linking reaction[103]. Besides its lower swellability, the crosslinked cellulose has less available hydroxyl groups. This adversely affects diffusion of the monomer and initiator and reduces the number of grafting sites on the cellulose backbone, thereby lowering the graft yield.

Data for Ce^{IV} consumption show that the consumption of Ce^{IV} during grafting is generally higher than that consumed during oxidation, regardless of the substrate employed and the temperature used, in accordance with previous results[38]. The lower consumption of Ce^{IV} observed during oxidation of MAM-treated cellulose at 40 °C, as compared with the corresponding consumption for the control, tallies with the trend of the grafting yield. The higher Ce^{IV} consumption observed with the MAM-treated cotton at 60 °C, in comparison with that of the control during oxidation, suggests that at higher temperatures, i.e. 60 °C, and after prolonging the reaction time, breaking of the crosslinks seems to occur.

$$\text{Cell—O—}CH_2\text{—}CH_2CONHCH_2\text{—O—Cell} \xrightarrow[Ce^{IV}]{H_2SO_4}$$

$$\text{Cell—O}CH_2\text{—}CH_2\text{—}CONHCH_2OH + \text{Cell—OH}.$$

It follows from this that, during oxidation with Ce^{IV} at 60 °C, there is a significant change in the physical as well as the chemical structure of cellulose treated with MAM. In addition to the molecular rearrangements of the cellulose, $—CH_2CH_2CONHCH_2OH$ groups are created as side groups along the cellulose molecules. As a result, susceptibility of cellulose toward Ce^{IV} attack increases. Since the graft yields obtained with cellulose treated with MAM are lower than the control even at 60 °C, the excess Ce^{IV} consumed during oxidation of the modified cellulose, is presumably due to the fast initiation and termination rates of free radicals formed on the side group of the modified cellulose[91].

6.8. Grafting on Carbamoylethylated Cellulose

Cellulose reacts with acrylamide in an alkaline medium to give carbamoylethylated cellulose[101, 102]. The reaction may be represented as follows:

$$\text{Cell—OH} + CH_2{=}CHCONH_2 \rightarrow \text{Cell—O—}CH_2CH_2CONH_2$$

In this process, cellulose is treated with a solution containing 12.6% acrylamide and 3.8% sodium hydroxide. After being squeezed to a wet pickup of 85%, the cellulose is heated at 125 °C for 6 minutes, followed by a thorough washing. The carbamoylethylated cellulose possesses good retention of strip breaking strength, a slight degree of rot resistance, and substantial heat resistance[104, 105].

The vinyl graft copolymerization of carbamoylethylated cellulose has been studied[91]. Figure 54 shows the results of the grafting of acrylonitrile (AN) onto carbamoylethylated cellulose, as well as those obtained with cellulose treated with 3.8% NaOH (control) using Ce^{IV} ammonium sulfate (CAS) as initiator. It is obvious that for a given reaction time and at a particular temperature, the graft yield obtained with the modified cellulose is substantially higher than that of the control. There seems to be an attack of Ce^{IV} at the carbamoylethyl groups, as well as at the cellulose hydroxyl groups. The presence of the $—CONH_2$ grouping with its negative inductive effect would facilitate hydrogen transfer. Thus, besides the free radical formation at the hydroxyl groups of cellulose, free radicals at the α-carbon atom of the carbamoylethyl group are very likely to occur. The following reaction scheme was suggested for initiation, propagation, and termination of the vinyl graft polymerization on the modified cellulose that takes place at the carbamoylethyl group[91]:

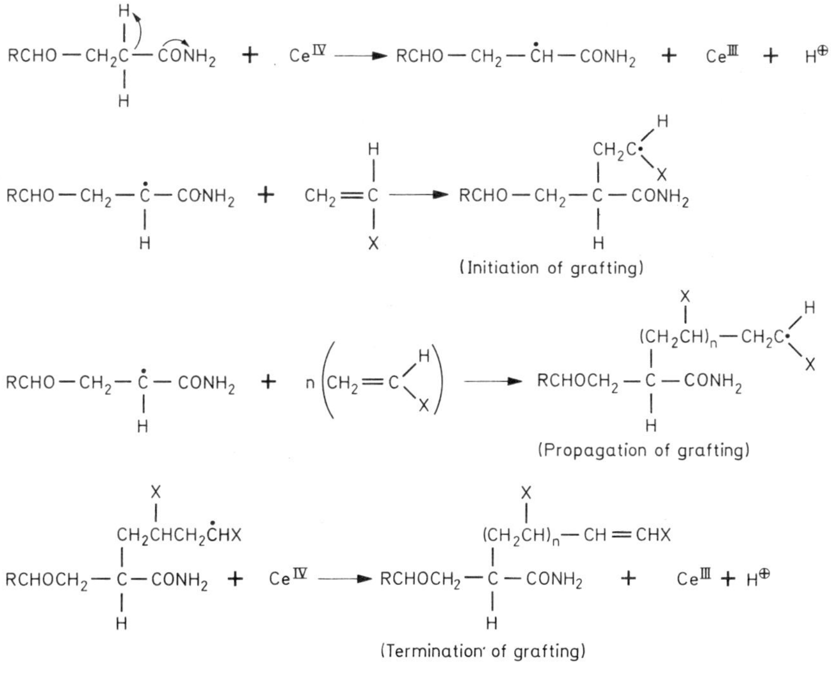

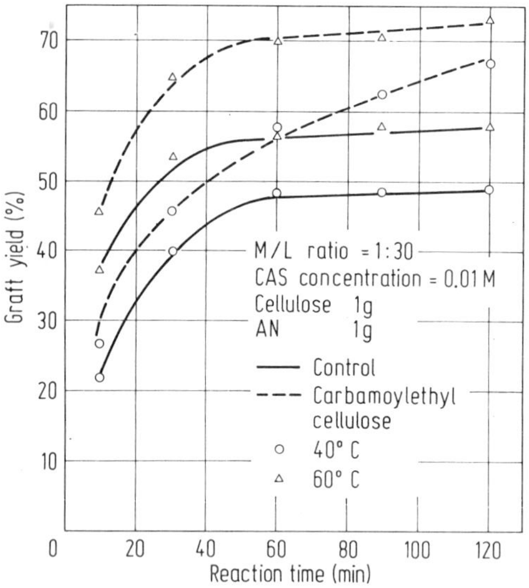

Fig. 54. Graft yield obtained with carbamoylethylated cellulose (D.S. = 0.122) and those obtained with cellulose treated with 3.8% sodium hydroxide (control) when AN was used as a monomer and CAS as initiator. Reprinted from Ref.[91], p. 1935, by courtesy of John Wiley & Sons, Inc.

where $RCHOCH_2CH_2CONH_2$ is carbamoylethylated cellulose and $CH_2{=}CHX$ is a vinyl monomer.

With respect to Ce^{IV} consumption, Hebeish et al.[90] have reported that despite the change in the chemical structure of cellulose by carbamoylethylation, the susceptibility toward Ce^{IV} oxidation remains practically unaltered. Probably the decreased attack due to blocking some of the cellulose hydroxyls in the modified cellulose is counterbalanced by Ce^{IV} attack at the carbamoylethyl groups.

6.9. Grafting on Cellulose Bearing Carboxymethyl and Cyanoethyl Groups

It has been pointed out that changing the chemical structure of cellulose by carboxymethylation up to a D.S. of 0.124 (PCMC) enhances the susceptibility of cellulose to grafting by means of Ce^{IV} ions. The same holds true when the chemical modification is performed via cyanoethylation (CEC), but the graft yield increases on increasing the D.S. In order to study the behavior of cellulose bearing both groups, i.e., carboxymethyl and cyanoethyl (PCMC-CEC) toward grafting, partially carboxymethylated cellulose of D.S. 0.0324 was cyanoethylated up to a D.S. of 1.425 and then grafted with acrylonitrile (AN) and methyl methacrylate (MMA), using CAS as the initiator[91]. Figure 55 shows that for given conditions, the graft yield of PCMC is significantly higher than that of cellulose bearing the same amount of carboxymethyl groups along with cyanoethyl groups. This implies that further modification of partially carboxymethylated cellulose by cyanoethylation greatly reduces its susceptibility to grafting.

The consumption of Ce^{IV} during grafting and oxidation of PCMC-CEC and PCMC (as a control) had been studied in detail. Apparently, the Ce^{IV} consump-

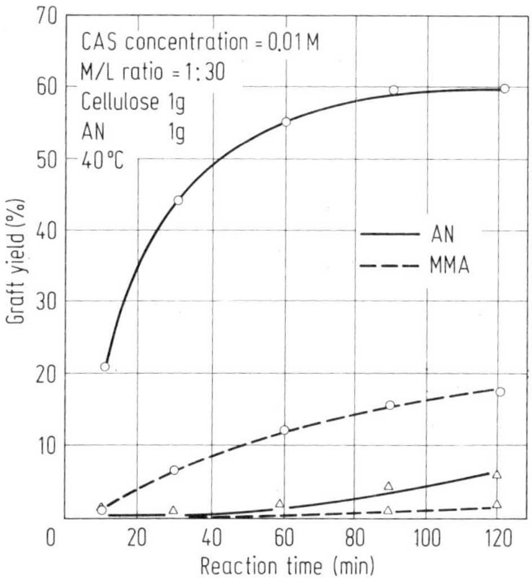

Fig. 55. Grafting of acrylonitrile and methyl methacrylate on cellulose bearing cyanoethyl groups (D.S. = 1.425) along with carboxymethyl groups (D.S.: 0.0324) using CAS as initiator. (△), PCMC-CEC; (○), PCMC. Reprinted from Ref.[91], p. 1929, by courtesy of John Wiley & Sons, Inc.

tion during oxidation of PCMC is very close to that of PCMC-CEC, which indicates that introducing cyanoethyl groups along with carboxymethyl groups in the cellulose molecule has practically no effect on the susceptibility of cellulose to Ce^{IV} attack. This is rather a surprise, since the graft yields obtained with PCMC-CEC are significantly lower than those of PCMC. There seems to be a considerable difference between the mode of Ce^{IV} attack on both substrates, as well as in the rate of initiation and termination of the free radicals on the substrate backbone.

An additional point of interest is that the consumption of Ce^{IV} during grafting of PCMC is considerably higher than that consumed during grafting of PCMC-CEC[75].

This is perhaps due to the difference in the rates of initiation, propagation, and termination of homopolymer formation. The magnitude of the latter was reported to depend on the nature of the substrate[38].

6.10. Grafting on Cellulose Bearing Cyanoethyl and Carboxyethyl Groups

It has been shown that while cellulose bearing either cyanoethyl or carboxymethyl groups shows higher graft yields than unsubstituted cellulose, cellulose bearing both groups shows lower yields. Cellulose bearing cyanoethyl groups along with carboxyethyl groups is discussed here with respect to its susceptibility towards grafting. Creation of carboxyethyl groups along the cellulose molecule is effected by partial hydrolysis of cyanoethylated cellulose with a D.S. of 0.515. In this range, the cyanoethyl groups are perhaps located at C_2 of the anhydroglucose unit in the cellulose molecules[91]. After being hydrolyzed with a 40% sodium hydroxide, the cellulose was shown to contain cyanoethyl groups (D.S.

= 0.06) and carboxyethyl groups (D.S. = 0.038). The sum of both groups expressed as D.S. amounts to 0.098, a value which is much lower than that of the original cyanoethyl cellulose, i.e., D.S. 0.515. This suggests that ether bond cleavage occurs during alkaline hydrolysis[91].

It is established[42] that Ce^{IV} consumption during oxidation of cellulose is due to (a) formation of a Ce^{IV}-cellulose complex, perhaps through chelation between the Ce^{IV} ions and the cellulose hydroxyl groups; (b) formation of free radicals at the hemiacetal group of the end anhydroglucose unit of the cellulose chain; and (c) termination of free radicals resulting from dissociation of acetal groups. The extent of Ce^{IV}-cellulose complex formation would be much greater for PCMC than for PCMC-CEC owing to the greater availability of cellulose hydroxyls in PCMC. Thus, the fact that the consumption of Ce^{IV} during oxidation of PCMC and PCMC-CEC is very similar suggests that excess Ce^{IV} consumption for complex formation during oxidation of PCMC is outweighed by the direct attack of Ce^{IV} on cyanoethyl groups[91] of PCMC-CEC. The following two possibilities may explain the lower graft yield obtained with PCMC-CEC, as compared with PCMC,

(a) dissociation of the Ce^{IV}-cellulose complex to produce free radicals capable of initiating grafting[42] may be easier with PCMC than with PCMC-CEC.
(b) termination of free radicals formed on cyanoethyl groups may be much faster than when brought about by dissociation of the Ce^{IV}-cellulose complex.

Hebeish et al.[90] grafted acrylonitrile (AN) onto cellulose bearing cyanoethyl and carboxyethyl groups using CAS as initiator. Figure 56 shows these results; together with those obtained with cyanoethyl cellulose and cellulose treated with 2% sodium hydroxide for comparison. Obviously, the graft yields obtained with cellulose bearing cyanoethyl along with carboxyethyl groups are much greater than

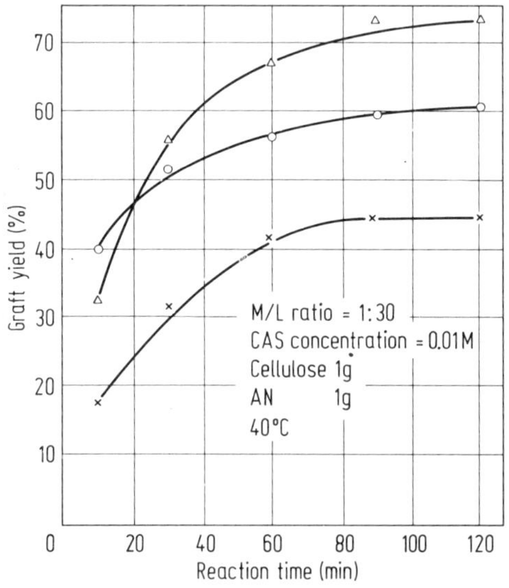

Fig. 56. Graft yield percent vs. reaction time. (△), cellulose bearing cyanoethyl groups (DS: 0.06) along with carboxyethyl groups (DS: 0.0376); (○), cyanoethylated cellulose (D.S. = 0.515); (×), cellulose treated with 2% sodium hydroxide (control for cyanoethylated cellulose). Reprinted from Ref.[91], p. 1931, by courtesy of John Wiley & Sons, Inc.

those obtained with the cyano-ethylated cellulose and the alkali-treated cellulose. These investigators attributed this behaviour to the influence of the following factors taken together: (a) cyanoethyl groups function as additional sites for grafting; (b) carboxyethyl groups increase the swellability of cellulose; thereby facilitating diffusion of monomer and initiator; and (c) the physical and chemical changes in the cellulose structure brought about by the effect of the concentrated alkali used in hydrolyzing the cyanoethyl groups may cause, in particular, a crystal transformation from cellulose I to cellulose II, thus increasing the reactivity of cellulose.

With reqard to the Ce^{IV} consumption, it has been reported[91)] that cellulose bearing cyanoethyl and carboxyethyl groups shows a lower consumption during oxidation when compared with cyanoethylated cellulose and alkali-treated cellulose (control). This has been ascribed to the disappearance of some of the cyanoethyl groups (sites for Ce^{IV} attack) through either hydrolysis to carboxyethyl groups or cleavage of the ether bond. In addition, alkali treatment (used for hydrolysis) produces a substrate with higher grafting efficiency (cf. Figure 56), but lower susceptibility to Ce^{IV} oxidation; this finding is in agreement with previous results[38)], which revealed that cotton treated with concentrated sodium hydroxide showed a higher graft yield and lower Ce^{IV} consumption than did untreated cotton.

6.11. Grafting on Phosphorylated Cellulose

Cellulose phosphate and phosphorylating agents for cellulose have been reviewed[106)]. Reagents include ammonium phosphate, urea-phosphoric acid, phosphorus tri- and oxychloride, monophenyl phosphate, phosphorus pentoxide, and chlorides of partially substituted phosphoric acids[106–108)]. The use of ortho- and condensed phosphate salts for the crosslinking and modification of cellulose has been described[107)]. For example, when a cotton fabric is treated for 10–20 minutes with a phosphoric acid solution at pH 2, which contains 2 to 6 mol of urea per mol of phosphoric acid at a temperature between 140–160 °C, cellulose ammonium hydrogen phosphate is produced.

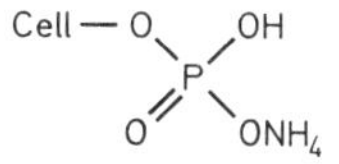

(Cellulose ammonium hydrogen phosphate)

Products with greater than 1.6% phosphorus, obtained by reaction with neutral or slightly acidic alkali metal phosphate or condensed phosphate salts, are insoluble in cupriethylenediamine and are apparently crosslinked[107)].

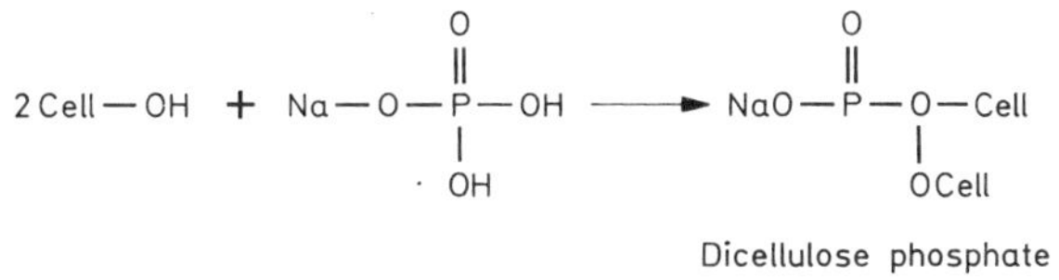

Phosphorylated cellulose esters in the form of ammonium salts resist fire. Phosphorylated cellulose also shows good non-soiling qualities and undergoes rapid cation exchange reactions. The latter property makes phosphorlated cellulose suitable for grafting. This suitability may be due to the number and location of grafting sites in cotton or rayon fibers that can be controlled and directed by making use of the ion-binding capacity of cellulose[109-110]. If at least one component of the redox initiator is cationic, that component can be bound to anionic groups in cellulose, and these become loci at which polymerization takes place. Since homopolymer formation occurs mainly in the treating solution, while grafting can only occur on or within the solid cellulose, the probability of grafting is enhanced when initiator binding sites are present in high concentration in the fibers.

Gallagher[111] has reported that the ion-binding capacity of cotton was readily increased by heating the fabric with alkali metal phosphates to effect phosphorylation. Subsequent contact with a solution of redox polymerization initiators then binds the cationic component of the initiator to the cellulose. Graft polymerization of olefinic monomers in the phosphorylated, ion-exchanged fabric occurs with high speed and efficiency. Table 49 shows the polymerization data for individual vinyl and acrylic monomers. It is clear that with Fe^{II}—H_2O_2 initiation, the polymerizations were considerably accelerated in every instance when phosphorylated fabric was used as a substrate. However, vinyl monomers show greater acceleration than acrylic derivatives. Vinyl acetate produced 900 times more polymer in phosphorylated cotton than in native cotton. After solvent extraction, the phosphorylated cotton retained at least 150 times as much polymer as did native cotton (runs 1, 2). For styrene polymerization, the increase in total polymer deposition caused by phosphorylation was 4–12 times, depending on the level of phosphorus introduced. The increase in polymer remaining after

Table 49. Comparison of phosphorylated and unphosphorylated cotton fabric as grafting hosts. (Redox initiator Fe^{II}-H_2O_2). Adapted from Ref.[111], p. 624, by courtesy of Textile Research Institute

Run No.	P, %	Monomer	Monomer concn.	Time, mins.	Temp., °C	Weight gain, %	
						before extrn.	after extrn.
1	0.00	Vinyl acetate	3.3	60	50	0.3	—
2	1.25	Vinyl acetate	3.3	60	50	274	45.7
3	0.00	Methyl acrylate	4.5	15.5	72	32	15.7
4	0.33	Methyl acrylate	4.5	15.5	72	58.1	35.0
5	0.00	Ethyl acrylate	4.0	15.0	75	51.6	29.0
6	0.33	Ethyl acrylate	4.0	15	75	118	94.0
7	0.00	Styrene	34[a]	60	50	2.2	0.8
8	0.33	Styrene	34[a]	60	50	9.2	7.9
9	1.25	Styrene	34[a]	60	50	26.8	18.7
10	0.00	N-Methylolacrylamide	6.6	18	25	3.3	
11	0.33	N-Methylolacrylamide	6.6	18	25	8.3	
12	0.00	Glycidyl acrylate	5.0[b]	5	70	18.5	
13	0.33	Glycidyl acrylate	5.0[b]	5	70	30.8	

[a] In aqueous emulsion

[b] No Fe^{II} ions present. The aqueous emulsion contained 5% monomer, 0.018% KOH; 0.14% KCl; 0.01% tetraethylenepentamine; 0.5% $TiCl_3$; 0.24% H_2O_2 and 0.5% Triton X-100.

extraction was 10–23 times (runs 7, 8, 9). The rates of reaction with the acrylic monomer showed accelerations in the neighborhood of 2–3 times on phosphorylated fabric.

The effect of fabric phosphorylation on polymer formation for several different initiator systems is shown in Table 50. It is obvious that mixtures of monomers could readily be copolymerized on fabric, thus providing a wide range of fiber modifications. Mixtures of acrylic monomers produced 1.7–7.5 times as much polymer on phosphorylated fabric as on native cotton. A ten-fold increase in polymerization was produced in the case of ethyl acrylate mixed with a minor proportion of bis-2-(vinyloxy)ethyl ether; this is a further reflection of the sensitivity of the vinyl component to increased concentrations of initiator sites in the cellulose (runs 26, 27). For the polymerization of mixed monomers, initiation with hydrogen peroxide in combination with reducing ions, such as

Table 50. Graft polymerization of single and mixed monomers on cotton fabric of varied phosphorus contents (Redox initiator M^+-H_2O_2). Reprinted from Ref.[111], p. 625, by courtesy of the Textile Research Institute

Run No.	P, %	Initiator ion, M^+	Compd.[a]	Concn., %	Mol ratio	Time, min	Temp., °C	Weight gain	
								Before extn.	After extn.
1	0	$N_2H_5^+$	AN	5	—	10	79	5.1	3.6
2	3.1	$N_2H_5^+$	AN	5	—	10	79	15.4	10.0
3	3.1	$N_2H_5^+$	AN	5	—	35	82	53.0	23.4
4	0.3	$N_2H_5^+$	AN	8.3	—	6	71	13.2	9.9
5	0	$N_2H_5^+$	AN/HPMA	2.8/6.4	1.2	9	66	2.6	
6	0.9	$N_2H_5^+$	AN/HPMA	2.8/6.4	1.2	9	66	12.6	
7	0	$N_2H_5^+$	AN/HPMA	2.8/6.4	1.2	31	71	29.7	
8	0.9	$N_2H_5^+$	AN/HPMA	2.8/6.4	1.2	31	71	78.4	57.1
9	0	Fe^{+2}	AN	8	—	20	25	2.8	
10	0.2	Fe^{+2}	AN	8	—	20	25	5.6	
11	0.2	Fe^{+2}	AN	8	—	20	66	40.3	37.5
12	0	Fe^{+2}	AN/MAA	7.1/2.7	5.0	30	25	1.2	
13	0.3	Fe^{+2}	AN/MAA	7.1/2.7	5.0	30	25	9.1	
14	0	Fe^{+2}	AN/MAA	5.9/1.1	10	30	25	3.5	
15	0.3	Fe^{+2}	AN/MAA	5.9/1.1	10	30	25	10.5	
16	0.3	Fe^{+2}	AN/MAA	6.6/2.0	6.4	30	25	4.7	
17	0.3	Ti^{+3}	AN/MAA	6.6/2.0	6.4	30	25	6.0	
18	0.3	Fe^{+2}	AN/MAA	6.6/2.0	6.4	60	25	20.0	
19	0	Fe^{+2}	AN/MAA	6.6/2.0	6.4	30	70	42.0	
20	0.3	Fe^{+2}	AN/MAA	6.6/2.0	6.4	30	70	70.0	
21	0.3	Fe^{+2}	EA	3.7	—	30	25	25.3	
22	0.3	Fe^{+2}	MAA	5.0	—	27	25	9.6	
23	0.5	Fe^{+2}	MAA	6.1	—	18	25	22.4	20.2
24	0	Fe^{+2}	EA/MAA	5.0/0.5	10	4	75	14.6	13.0
25	0.3	Fe^{+2}	EA/MAA	5.0/0.5	10	4	75	30.3	25.9
26	0	Fe^{+2}	EA/DVG	3.4/1.5	3.6	30	25	4.6	
27	0.3	Fe^{+2}	EA/DVG	3.4/1.5	3.6	30	25	45	
28	0.3	Fe^{+2}	MA	4.5	—	2	65	29	
29	0.3	Fe^{+2}	MA	4.5	—	15	65	51.7	46.9
30	0	Fe^{+2}	MA/MAA	4.3/1.0	5.0	1.25	77	2.8	
31	0.3	Fe^{+2}	MA/MAA	4.3/1.0	5.0	1.25	77	5.9	

Table 50. (Continued)

Run No.	P, %	Initiator ion, M^+	Compd.[a]	Concn., %	Mol ratio	Time, min	Temp., °C	Weight gain	
								Before extn.	After extn.
32	0	Fe^{+2}	MA/MAA	4.5/0.53	10	4	73	6.6	
33	0.3	Fe^{+2}	MA/MAA	4.5/0.53	10	4	73	12.9	
34	0.3	Fe^{+2}	S	35	—	60	50	13.1	9.3
35	0.3	Fe^{+2}	S	35	—	90	50	21.3	16.2
36	0.3	Fe^{+2}	S	35	—	22	75	28.5	14.3
37	0.3	Fe^{+2}	S	35	—	32	75	54.3	24.8
38	0.3	Fe^{+2}	VA	3	—	15	62	2.7	
39	1.0	Fe^{+2}	VA	3	—	15	62	15.9	9.4
40	0.3	Fe^{+2}	VA	5	—	10	68	18.3	
41	0.3	Fe^{+2}	VA	5	—	20	69	39.8	
42	0	Ce^{+4}	AN[b]	5	—	15	25	2.8	
43	3.1	Ce^{+1}	AN[b]	5	—	15	25	6.3	
44	0.3	Cr^{+2}	GA[c]	10	—	4	70	7.9	
45	0.3	$N_2H_5^+$	GA[d]	10	—	24	60	10.1	
46	0.3	Sn^{+2}	HPMA/MAA	10/0.6	12	5	75	4.1	

[a] AN = acrylonitrile; DVG = divinyl ether of diethylene glycol; EA = ethyl acrylate; GA = glycidyl acrylate; HPMA = hydroxypropyl methacrylate; MA = methyl acrylate; MAA = N-methylolacrylamide; S = styrene; VA = vinyl acetate.

[b] No peroxide used in this run. Fabric was converted to Ce^{IV}-ion form by treatment with 2.5% $Ce(HSO_4)_4$.

[c] Sodium form of phosphorylated fabric treated with emulsion of 10% GA, 0.02% H_2O_2, 0.018% KOH, 0.14% KCl, 0.02% tetraethylenepentamine, 0.4% $CrCl_3$, 0.3% Triton X-100, and water.

[d] Hydrazine form treated as in (c) except that $CrCl_3$ was omitted.

Fe^{II}, Ti^{III}, or hydrazinium ions could be used. Ce^{IV} ion initiation appeared to be moderately effective[111].

According to Gallaher, the proportion of monomer grafted to cellulose in these experiments can be estimated approximately from the fraction of polymer not removable by known polymer solvents. Consequently, for Fe^{II}—H_2O_2 initiation of such monomers as methyl acrylate, ethyl acrylate, and styrene (Table 50), the durably bound fraction comprised 60–86% of the total polymer formed in the phosphorylated fabric. The relative percentages of bound and unbound polymer varied widely, however, with the monomer and initiator used as well as with polymerization temperature and monomer concentration. Polymerization of acrylonitrile (Table 66, runs 1–4 and 9–11) produced 44–93% yields of bound polymer, showing that a large number of critical variables exist for a given monomer.

It has also been reported that alkaline hydrolysis of phosphorylated cotton failed to remove phosphate groups, making this derivative one of the most alkali-stable cellulose esters available. Bound polyacrylonitrile also resisted removal by alkaline hydrolysis, even though the nitrile groups were apparently converted to carboxyl groups. Hydrolysis in acid, under conditions known to cleave phosphate groups, also failed to remove the polymer[111].

6.12. Grafting on Crosslinked Cellulose

The interaction of cellulose and formaldehyde has received considerable attention over a period of many years. As a consequence, the conditions of reaction and the types of products obtained may be said generally to be understood[112)]. The acid-catalyzed reaction between formaldehyde and cellulose at high temperatures (ca. 150 °C) has been studied extensively, both from theoretical and practical standpoints. In this reaction, the formation of methylene-ether crosslinks takes place predominantly between the secondary hydroxyl groups of cellulose[112–116)]. The same reaction has been used in industry for improving the dimensional stability of viscose fabrics. However, attempts at applying this treatment to cotton fabrics on a commercial scale have not been successful since it causes large losses in strength[117)]. Interest has been revived in the reaction of formaldehyde with cotton in a swollen condition. This is due to the observations that (ii) the drip-dry characteristics of cotton fabric depend to a large extent on the crease recovery of wet fabrics, and that (ii) crosslinking treatments of the swollen fiber produce large increases in wet crease recovery. It has been postulated that the crosslinking of cellulose in a swollen condition occurs largely in the regions having reasonable lateral order, while crosslinks introduced in the earlier treatments at high temperatures (when the fiber is in a collapsed state) are located in the noncrystalline regions[118–120)].

The effect of the change in the cellulose structure, brought about by crosslinking cotton cellulose in the dry and wet states, on grafting of cellulose with acrylonitrile (AN) and methyl methacrylate (MMA), using the Ce^{IV} ion technique was examined[121)]. The graft copolymerization reaction was studied with respect to the graft yield and the Ce^{IV} consumption. Figure 57 shows the rate of grafting of cellulose, crosslinked in the dry state, and that of the control. It is clear that with both monomers, low grafting rates were obtained with crosslinked cellulose, as compared with the control. This is to be expected since the crosslinking reaction takes place in highly disordered regions. Besides blocking some of the cellulose hydroxyls, such a change results in a measurable decrease in swellability of the fiber, probably by restricting the freedom of movement of the cellulose molecules while in the aqueous grafting medium. The effect of low swellability of the crosslinked cellulose is to impede diffusion of both monomer and initiator, thereby adversely affecting the grafting yield. This is substantiated by the fact that lower grafting rates are obtained with cellulose having higher amounts of bound formaldehyde. It can also be seen (Fig. 57) that in the early stages of the reaction, crosslinked celluloses having lower amounts of bound formaldehyde show higher grafting levels than those with higher amounts; but this difference nearly disappears after one hour's reaction time. The prolonged reaction seems to have two effects: (1) enhancing fiber swellability, as well as diffusion of monomer and initiator; and (b) breaking down some of the crosslinks owing to the high acidity of the treating bath. Both effects lead to increased graft formation. Thus, within the range studied, the differences in the graft yields observed at short grafting times are reduced by increasing the swellability of the fiber and by breaking down some of the crosslinks at prolonged grafting times[121)].

Contrary to cellulose crosslinked in the dry state, cellulose crosslinked in the

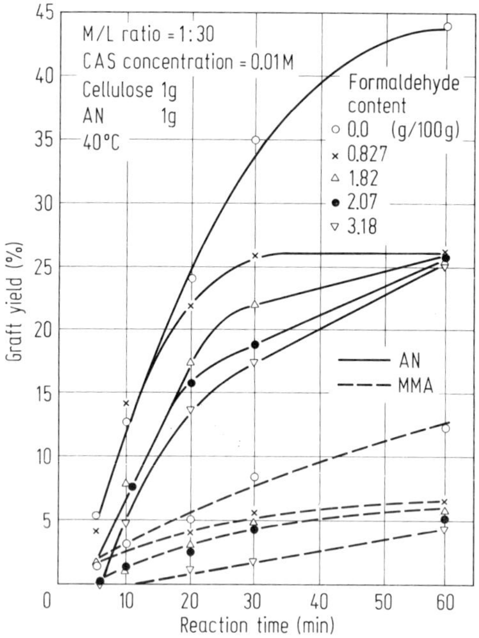

Fig. 57. Graft yield of cellulose containing various amounts of bound formaldehyde vs. reaction time. Reprinted from Ref.[121], p. 11, by courtesy of Textile Research Institute.

wet swollen state shows graft yields which are comparable to those of the control.

In crosslinking cellulose while it is in the uncollapsed state, the crosslinkages appear to be located largely in regions of some lateral order due to the inclusion of the swelling agent (phosphoric acid) in the process. This stabilizes the swollen structure of the fiber, in the wet state, without reducing its swellability. The disappearance of some of the cellulose hydroxyls (possible sites for grafting) as a consequence of crosslinking is compensated for by the possibility that some of the crosslinks introduced may occur in the crystalline, or at least in the near crystalline regions of the cellulose. Hence, there will be a greater tendency for the reacting molecules to approach these regions than would be the case with uncrosslinked cotton. The cellulose is held in a somewhat open state to give a swollen product in which the location of the graft sites may be quite different from those in the control, or even from that of cellulose crosslinked in the dry state[121].

With respect to Ce^{IV} consumption, Kantouch and collaborators[121] have reported changing the chemical structure of cellulose via interaction with formaldehyde in either the dry or wet states has very little effect on the susceptibility of cellulose toward Ce^{IV} oxidation. Although this observation agrees with the graft yields in the case of cellulose crosslinked in the wet state, it does not concur with the graft yields obtained with cellulose crosslinked in the dry state. In the latter instance, the crosslinkages appear largely to be on the surface of the cellulose. These crosslinkages would be more sensitive to Ce^{IV} attack, possibly through their cleavage under the hydrolytic influence of the acidic medium of the reaction, giving rise to Ce^{IV} consumption without necessarily initiating grafting. It is also possible that the discrepancy between the consumption of Ce^{IV} during oxidation of cellulose crosslinked in the dry state and its graft yields could be due to a higher rate of Ce^{IV} diffusion than monomer diffusion. That is, the rate of Ce^{IV} diffusion into the modified and unmodified cellulose may be comparable, whereas the rate of monomer diffusion into both substrates may differ considerably because of its higher molecular size, as compared to Ce^{IV}.

It has been reported that treatment of cellulose, crosslinked in either dry or wet conditions with acidic Ce^{IV} solutions, results in a decrease in the amount of bound formaldehyde. This decrease was very significant with prolonged treatment times, indicating the occurrence of crosslink rupture during grafting initiated by Ce^{IV}.[121)]

6.13. Grafting on Aldehyde Cellulose

As stated in Chapter 4, the chemical reactivity of cellulose is known to be strongly influenced by its oxidized status. The enhanced thermal initiation of graft copolymerization on hydrogen peroxide or sodium hypochlorite-oxidized cellulose[122)] and poly (vinyl alcohol) fiber[123)] without the use of an initiator has been reported. Furthermore, periodic acid-oxidized cellulose[124–125)] or poly (vinyl alcohol) fiber[126)] has sufficient activity to initiate grafting via photo-irradiation.

Ogiwara et al.[127)] have used the periodic acid route to prepare so-called aldehyde cellulose. Substrate preparation involved treating commercial pulps from softwoods, with an aqueous periodic acid solution at 45 °C for 60 minutes (ratio of material to liquid 1:100). After this treatment, the samples were washed, dried under vacuum, and then used to establish both the relationship between the ability of the aldehyde cellulose to inititate grafting under photo-irradiation and the properties of the photoinduced radicals of the aldehyde cellulose[124)]. Aldehyde cellulose was shown to have a high affinity for grafting, though the level of grafting is dependent on the carbonyl content of the aldehyde cellulose. A higher graft efficiency for each monomer is obtained with aldehyde cellulose of higher carbonyl content. The hydrophilic monomers used by Ogiwara et al.[127)] included acrylic acid, N-vinyl-2-pyrrolidone, and acrylamide and were reasonably representative of this class of monomer.

The situation regarding the grafting of hydrophobic monomers was also followed by Ogiwara et al. The monomers used were vinyl acetate, methyl methacrylate, and acrylonitrile. It was found that the percent grafting of vinyl acetate increased with increasing carbonyl content of the cellulose sample, and that acrylonitrile grafting leveled off in the range beyond a critical carbonyl content, while methyl methacrylate grafting gave a maximum value. The hydrophobic monomers gave less grafting than the hydrophilic monomers. The differences are due to water solubility, affinity for the cellulose substrate, and the monomer type. It was shown that the carbonyl group in aldehyde cellulose is very reactive toward light. A consequence of this activity is the production of radicals for grafting on the substrate.

From e.s.r. studies of the photo-irradiated aldehyde cellulose and subsequent decay patterns, Ogiwara et al.[127] concluded that the high activity needed to induce graft copolymerization of various monomers on aldehyde cellulose under photo-irradiation was derived from an acyl radical which originated in the aldehyde group of the sample.

6.14. Grafting onto Methylol Cellulosics

The methylol cellulose (cellulose/DMSO/formaldehyde) system, as a means of cellulose solution, is now reasonably well established[128–130]. This procedure has been used as a preliminary to cellulose regeneration and derivitivization. This type of system offers certain advantages, including relative ease of handling, considerable storage life, and the prospect of greater homogeneity of product modification, resulting from the fact that grafting can be completed in solution.

The term methylol cellulose is used in its loosest sense, since product characterization is one of the areas which require thorough elucidation. Guthrie et al.[131] have reported the sensitized photoinitiated grafting of styrene onto methylol cellulose in DMSO.

Typically, the methylolation procedure was a modification of that given by Johnson et al.[128] as reported by Guthrie and Hardcastle[129]. Cellulose (cotton wool or high-grade wood pulp), DMSO, and paraformaldehyde (1:12:2 by weight) are heated under stirring at 120 °C for five hours, though a clear solution is usually achieved within two hours. Dynamic viscometric measurements of filter-press treated solutions show them to be stable over long periods.

Photosensitized grafting of styrene onto the methylol cellulose solution in DMSO was achieved using 2-chlorothioxanone as a sensitizer and dimethylaminoethanol as a synergist[131]. Irradiation was carried out for various known times and monomer concentrations using a 125 W Primarc medium-pressure ultraviolet lamp. Homopolymer was removed from the bulk polymer (precipitated from chilled methanol) by Soxhlet extraction procedures.

Figure 58 and 59 show the increase in grafting and homopolymerization, respectively, with increasing irradiation time for the various initial monomer concentrations. While the homopolymerization process appears to be reasonably straight forward, the grafting reaction is complicated by the presence of maxima which maximize at an initial monomer concentration of 4.36 mol of styrene/l of bulk solution. These maxima are perhaps best shown by plotting the change in the overall rate of grafting against the original bulk monomer concentration (Fig. 60) for each of the irradiation periods covered. Figure 60 presents some interesting features. The occurrence of maxima in studies of grafting patterns have been observed previously. They are often associated with the Trommsdorff type phenomena or where precipitation of polymer follows as a consequence of the polymerization reaction. In the study by Guthrie et al., the situation is somewhat more complex in that there is little change in the viscosity of the medium as a result of grafting, since the reaction is always carried out to low conversion levels[131]. Also, these deviations from expected behavior are not restricted to extended conversion but apply throughout each kinetic run. One

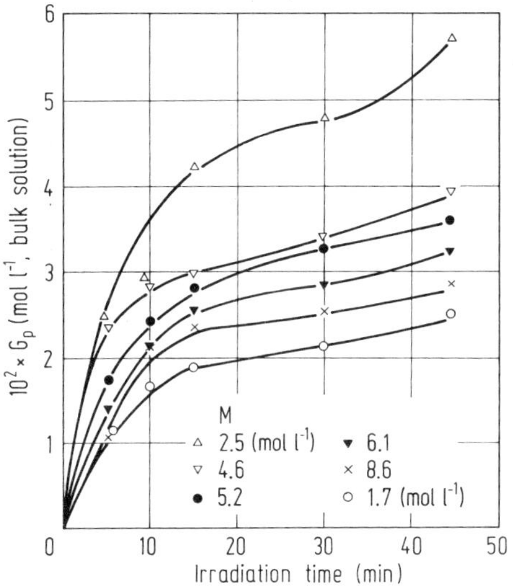

Fig. 58. Variation in extent of grafting with irradiation time for monomer concentrations in the bulk phase of (○), 1.7; (△), 2.5; (▽), 4.6; (●), 5.2; (▼), 6.1; and (×), 8.6 mol/l of bulk phase respectively. Reprinted from Ref.[131], courtesy of Springer-Verlag, Berlin.

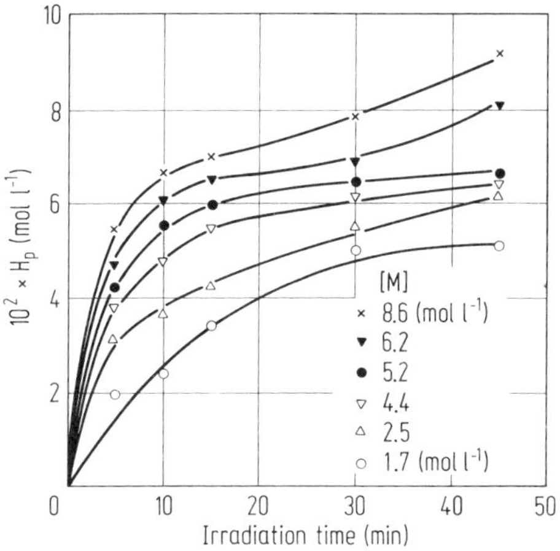

Fig. 59. Variation in the extent of homopolymerization with irradiation time for monomer concentrations of (○), 1.7; (△), 2.5; (▽), 4.6; (●), 5.2; (▼), 6.2; (×), 8.6. Reprinted from Ref.[131], courtesy of Springer-Verlag, Berlin.

possible explanation is that the addition of monomer to the bulk medium brings about changes in the morphological state of the methylol cellulose in the DMSO. Guthrie et al.[131] have proposed that an open gel-like phase is formed which contains trapped monomer as well as the photoinitiator and synergist. Should initiation take place within these transparent 'gels', polymerization would proceed until monomer depletion occurs. An open 'gel phase' would be supported to an upper limit in the monomer concentration (a non-solvent as far as the methylol cellulose/DMSO is concerned), during which the tendency toward reduced grafting, arising from 'gel' compaction, is more than compensated for by the higher monomer concentration within the loosely structured 'gel'. At a critical monomer concentration, the methylol cellulose/DMSO/styrene medium acquires a more rigid form in which methylol cellulose/DMSO domains are produced owing to

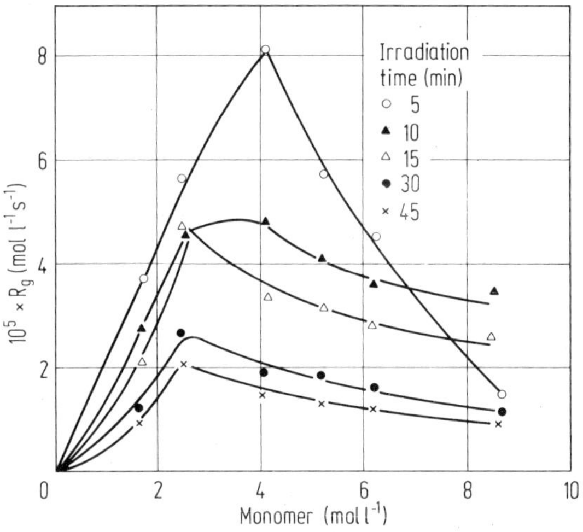

Fig. 60. Variation in the rate of grafting, R_g, with monomer concentration. Irradiation times as stated. Reprinted from Ref.[131], courtesy of Springer-Verlag, Berlin.

preferential adsorption of the DMSO as the styrene concentration is increased. The transport of styrene from the exterior to the interior of these domains then becomes diffusion controlled as a result of the increase in the heterogeneity of the system arising from increases in the bulk monomer (non-solvent) concentration.

Similar reasoning could be put forward for the role of the synergist and photoinitiator. Since the concentration of each of these was held constant throughout this study, Guthrie et al. state that their contribution to the abnormal behavior is relatively minor. However, an alternative cause for this type of behavior could arise from a decrease in the number of grafting sites as the monomer/solvent ratio is changed. There is a possbility that the addition of styrene to the methylol cellulose/DMSO causes partition of the initiator/synergist system away from the vicinity of the methylol cellulose, leading to a decrease in the number of available grafting sites.

In an interesting extension of the immobilized enzyme concept, Beddows et al. have reported the binding of various enzymes onto methylol cellulosic substrates, as well as p-amino carbanilated cellulose derivatives[132]. Here β-galactosidase and other enzymes were bound using the diazo method[132] and through glutaraldehyde[133] to provide a modification of the grafting system. Coupling to methylol cellulose p-amino carbanilate (MCTAC) and p-amino carbanilated cellulose (CTAC) was carried out under heterogeneous conditions. In coupling β galactosidase using the diazo group, the optimum pH was 7 to 7.5, while the optimum concentration of Mg^{II} ion was in the region of 2×10^{-3} to 3×10^{-3} mol/l. Figure 61 shows that most of the activity is observed after three hours. Although these levels of activity in the grafted immobilized enzymes are not excessive, they are sufficient to indicate that this type of grafted cellulose deserves attention. The stability of the E. Coli — C.T.C. system in a 0.1 M, pH = 7 buffer at 0–4 °C is given in Figure 62. The half-life is approximately fourteen days. Beddows et al.[132] have reported the successful binding of glucose oxidase, papain, trypsin, and pepsin. Such

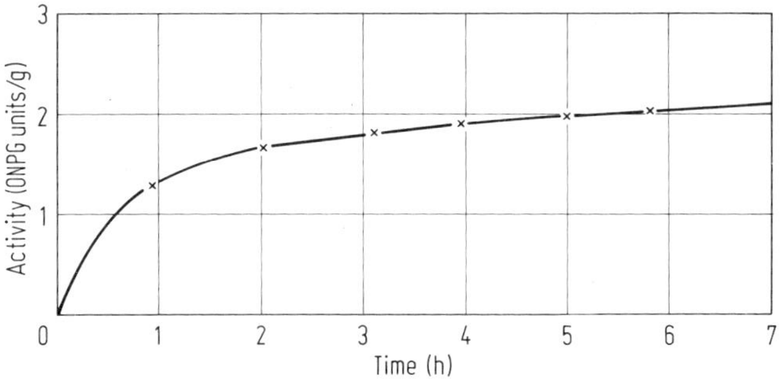

Fig. 61. The effect of time (hours) on the amount of active yeast β-galactosidase coupled to CTAC using the diazo method. (ONPG = o-nitrophenol-β-D-galactopyranoside). Reprinted from Ref.[132], courtesy of John Wiley and Sons, Inc.

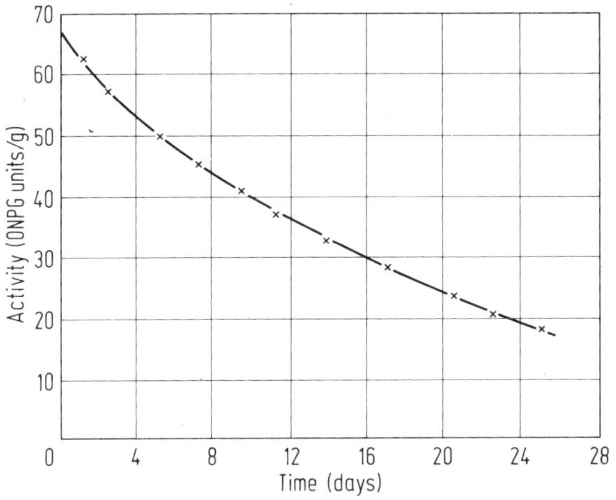

Fig. 62. The effect of time (days) on the activity of immobilized (E-Coli) β-galactosidase coupled to CTAC, in pH = 7.0 buffer at 0–4 °C (ONPG = o-nitrophenol-β-D-galactopyranoside). Reprinted from Ref.[132], courtesy of John Wiley and Sons, Inc.

work is of importance, since it permits recycling of used enzyme, a factor which is of significance when the purer or more expensive enzymes are involved. Cellulosic-bound enzyme systems of the type mentioned above seem to possess the right balance of hydrophobic-hydrophilic character to enable them to work efficiently and yet be easily and inexpensively recovered.

References

1. Achwal, W. B.: Colourage Jan. *29*, 25 (1970)
2. Achwal, W. B.: Colourage Aug. *12*, 28 (1971)
3. Reinhardt, R. M., Reid, J. D., Daul, G. C.: Textile Res. *J 26*, 1 (1956)
4. Achwal, W. B., Deshpande, S. D.: Curr. Sci *38*, 425 (1969)
5. Achwal, W. B.: Colourage Sept. *23*, 32 (1971)
6. Achwal, W. B.: Colourage July *1*, 34 (1971)
7. Achwal, W. B.: Colourage June *3*, 35 (1971)
8. Achwal, W. B.: Colorage, Dec *31*, 22 (1970)

9. Achwal, W. B.: Colourage, April *23*, 37 (1970)
10. Achwal, W. B.: Colourage, Oct. *21*, 34 (1971)
11. Achwal, W. B.: Colourage, March *26*, 31 (1970)
12. Arthur, J. C. Jr.: Adv. Chem. Ser. *91*, 574 (1969)
13. Arthur, J. C. Jr.: Adv. Chem. Ser. *99*, 321 (1971)
14. Arthur, J. C. Jr.: Adv. Macromol. Chem. *2*, 1 (1970)
15. Achwal, W. B.: Colourage, Aug. *13*, 35 (1970)
16. Chen, C.-Y., Montana, R. E., Grove, C. S. Jr.: Tappi *34*, 420 (1951)
17. Sanyal, A. K., Falconer, E. L., Vincent, D. L., Purves, C. B.: Can. J. Chem *35*, 1164 (1957)
18. Phillips, B.: Svensk. Papperstidn. *60*, 88 (1957)
19. Schurz, J., Zimmerl, W.: Papier, *12*, 468 (1958)
20. Adamek, E. G., Purves, C. B.: Can. J. Chem. *35*, 960 (1957)
21. Willard, J. J., Pacsu, E.: J. Amer. Chem. Soc. *82*, 4347, 4350 (1960)
22. Mathes, A.: Faserforsch. Textiltech. *3*, 127 (1952)
23. Schurz, J.: Holzforschung *10*, 1 (1956)
24. Dimov, K., Pavlov, P.: J. Polymer Sci. A-I *7*, 2775 (1969)
25. Smets, G., Poot, A., Mullier, M., Bex, J. P.: J. Polymer Sci. *34*, 287 (1959)
26. Iuguti, S., Watanabe, M.: Kobunshi Kagaku. *16*, 467 (1959)
27. Morin, B. P., Rogovin, Z. A.: Vysokomol. Soedin. *A 18*, 2451 (1976)
28. Kraessig, H. A.: J. Macromol. Sci. Chem. *A 10*, 759 (1976)
29. Pavlov, P., Dimov, K., Simenov, N.: J. Appl. Polym. Sci. *21*, 291 (1977)
30. Morin, B. P., Rogovin, Z. A., Samoilov, V. I.: Vysokomol. Soedin. *B 19*, 1573 (1975)
31. Harnof, V., Kokta, B. V., Valade, J. L.: J. Appl. Polym. Sci. *18*, 772 (1976)
32. Brickman, W. J.: Tappi *56*, 97 (1973); also in Cellulose Chemistry and Technology Research (ed. Turbak, A. F.) A.C.S. Symp. Series No 10. A.C.S. Washington, D.C. 1975 P. 9
33. Anderson, E. V., Cooper, A. S. Jr.: J. Ind Eng. Chem. *51*, 608 (1959)
34. McMillan, O. J. Jr., Decosas, K. M., Cooper, A. S. Jr., Hamalalinen, C., Murphy, A. L., Pollard, E. F.: Amer. Dyestuff Reptr. *49*, 839 (1960)
35. Honold, E., Keating, E. J., Skau, E. L.: Text. Res. J. *27*, 400 (1957)
36. Tripp, V. W., Guiffria, R., de Gruy, I. V.: Text. Res. J. *27*, 14 (1957)
37. Hebeish, A., Kantouch, A., El Rafie, M. H.: J. Appl. Polym. Sci. *15*, 11 (1971)
38. Hebeish, A., Mehta, P. C.: J. Appl. Polym. Sci. *12*, 1625 (1968)
39. Terasaki, I., Matsuki, M.: J. Soc. Text. Cell. Ind. Japan *18*, 147 (1968)
40. Iwai, S., Kakurai, T., Noguchi, T.: Sen-i-Gakkaishi *19*, 547 (1963); Through CA *62*, 13369 (1965)
41. Cumberbirch, R. J. E., Holker, J. R.: J. Soc. Dyers Co. *82*, 59 (1966)
42. Kulkarni, A. Y., Mehta, P. C.: J. Appl. Polym. Sci. *12*, 1321 (1968)
43. Imai, Y., Masuhara, E., Iwakura, Y.: J. Polymer Sci. *B 8*, 75 (1970)
44. Pottenger, C. R., Johnson, D. C.: J. Polymer Sci. A-1 *8*, 301 (1970)
45. Iwakura, Y., Imai, Y., Yagi, K.: J. Polymer Sci. A-1 *6*, 801 (1968)
46. Iwakura, Y., Kurosaki, T., Imai, Y.: J. Polymer Sci. *A 3*, 1185 (1965)
47. Malm, C. J., Tanghe, L. J., Laird, B. C., Smith, G. D.: J. Amer. Chem. Soc. *75*, 80 (1953)
48. Fedorova, A. F., Rogovin, Z. A.: Vysokomol. Soedin. *5*, 519 (1963)
49. Schreiber, W. T., Bullock, A. L., Ward, W. L.: Textile Res. J. *24*, 819 (1954)
50. Daul, G. C., Reinhardt, R. M., Reid, J. D.: Textile Res. J. *25*, 246 (1955)
51. Reinhardt, R. M., Markezich, A. R., Moore, H.-B., Reid, J. D.: Textile Res. J. *26*, 423 (1956)
52. Mazzeno, L. W. Jr., Reinhardt, R. M., Reid, J. D., Dickson, J. D.: Textile Res. J. *26*, 597 (1956)
53. Bikales, N. M., Rapoport, L.: Textile Res. J. *28*, 737 (1958)
54. Conrad, C. M.: Textile Res. J. *29*, 287 (1959)
55. Compton, J., Martin, W. H., Gargarine, D. M.: Textile Res. J. *40*, 813 (1970)
56. Compton, J.: Amer. Dyestuff Reptr. *43*, 103 (1954)
57. Hamaya, T., Yamada, S.: J. Appl. Polym. Sci. *22*, 875 (1978)
58. Demint, R. J., Arthur, J. C. Jr., McSherry, W. F.: Textile Res. J. *31*, 821 (1961)
59. Arthur, J. C. Jr., Demint, R. J.: Textile Res. J. *31*, 988 (1961)
60. Kantouch, A., Hebeish, A., El-Rafie, M. H.: J. Appl. Polym. Sci. *15*, 1007 (1971)

61. Hebeish, A., Khalil, M. I., El-Rafie, M. H.: Angew. Makromol. Chem. *37*, 149 (1974)
62. Mino, G., Kaizerman, S., Rasmusen, R.: J. Amer. Chem. Soc. *81*, 1494 (1959)
63. Carter, M. E.: J. Polymer Sci. *51*, 199 (1961)
64. Kamel, M., Kantouch, A., Hebeish, A.: U.A.R. Pat. 6625 (1962)
65. Kamel, M., Kantouch, A., Hebeish, A.: Textil. Praxis *19*, 1114 (1964)
66. Kamel, M., Kantouch, A., Hebeish, A.: Textil. Praxis *20*, 577 (1965)
67. Reinhardt, R. M., Fenner, T. W.: Ind. Eng. Chem. Prod. Res. Dev. *4*, 82 (1965)
68. Cram, D. J., Hammond, G. S.: Organic Chemistry, 2nd ed., New York: McGraw-Hill 1964, P. 232
69. Conrad, C. M.: J. Textile Inst. *50*, T133 (1959)
70. Daul, D. C., Reid, J. D.: U.S. Pat. *2*, 584, 114 (1952)
71. Daul, D. C., Reinhardt, R. M., Reid, J. D.: Textile Res. J. *22*, 787 (1952)
72. Daul, D. C., Reinhardt, R. M., Reid, J. D.: Textile Res. J. *23*, 719 (1952)
73. Grant, J. N.: Textile Res. J. *26*, 74 (1956)
74. Grant, J. N., McDonald, A. W., Humphreys, G. C.: Textile Res. J. *28*, 60 (1958)
75. Reinhardt, R. M., Fenner, T. W., Reid, J. D.: Textile Res. J. *27*, 873 (1957)
76. Frieser, E.: Z. Ges. Text. Ind. *60*, 977 (1959)
77. Reinhardt, R. M., Fenner, T. W., Reid, J. D.: Amer. Dyestuff Reptr. *50*, 67 (1961)
78. Kamel, M., Kantouch, A., Hebeish, A.: Indian J. Technol. *5*, 58 (1967)
79. Reinhardt, R. M., Reid, J. D., Fenner, T. W., Mayne, R. Y.: Textile Res. J. *29*, 802 (1959)
80. Manecke, G.: Pure Appl. Chem. *4*, 507 (1962)
81. Reinhardt, R. M., Fenner, T. W., Reid, J. D.: Textile Res. J. *32*, 735 (1962)
82. Kantouch, A., Hebeish, A., El-Rafie, M. H.: Europ. Polym. J. *6*, 1575 (1970)
83. Schwab, E., Stannett, V. T., Rakowitz, D. H., Magrane, J. K.: Tappi *45*, 390 (1962)
84. Bhattacharjee, S. S., Perlin, A. S.: J. Polymer Sci. *B 13*, 113 (1975)
85. Ahmed, A. U., Rapson, W. H.: J. Polymer Sci. A-1, *9*, 2129 (1971)
86. Bhattacharjee, S. S., Perlin, A. S.: A.C.S. Meeting Los Angeles: April 1974 Abstracts
87. Bhattacharjee, S. S., Perlin, A. S.: J. Polymer Sci. *C 36*, 509 (1971)
88. Ogiwara, Y., Kubota, H., Murayama, H., Sakamoto, A.: Tappi *53*, 1685 (1970)
89. LePoutre, P., Hui, S. H., Robertson, A. A.: J. Appl. Polym. Sci. *17*, 3143 (1973)
90. Hebeish, A., El-Rafie, M. H., Khalil, M. I., Bendak, A.: J. Appl. Polym. Sci. *21*, 1901 (1977)
91. Kantouch, A., Hebeish, A., El-Rafie, M. H.: J. Appl. Polym. Sci. *15*, 1921 (1971)
92. Ogiwara, Y., Kubota, H.: J. Appl. Polym. Sci. *17*, 2427 (1973)
93. Kantouch, A., Hebeish, A., El-Rafie, M. H.: Textile Res. J. *42*, 10 (1972)
94. Bendak, A., Khalil, M. I., El-Rafie, M. H., Hebeish, A.: J. Appl. Polym. Sci. *19*, 335 (1975)
95. Bendak, A., Hebeish, A.: J. Appl. Polym. Sci. *17*, 1953 (1973)
96. Abdel-Fattah, S. H., Kantouch, A., Hebeish, A.: J. Chem. (Egypt) *17*, 311 (1974)
97. El-Rafie, M. H., Hebeish, A.: J. Appl. Polym. Sci. *19*, 1815 (1975)
98. Hebeish, A., Bendak, A.: J. Appl. Polym. Sci. *18*, 1305 (1974)
99. Hebeish, A., Abdel-Fattah, S. H., Bendak, A.: Angew. Makromol. Chem. *37*, 11 (1974)
100. Kubota, H., Ogiwara, Y.: J. Appl. Polym. Sci. *23*, 241 (1979)
101. Kubota, H., Ogiwara, Y.: J. Appl. Polym. Sci. *22*, 3363 (1978)
102. Klinshpont, E. R., Milinchuk, V. K.: Vysokomol. Soedin. *B 16*, 35 (1974)
103. Mock, U., Mock, W.: Melliand Textilchemie *1*, 82 (1965)
104. Frick, J. W., Reeves, W. A., Guthrie, J. D.: Textile Res. J. *27*, 294 (1957)
105. Frick, J. W., Reeves, W. A., Guthrie, J. D.: Textile Res. J. *27*, 92 (1957)
106. Reid, J. D., Mazzeno, L. W. Jr.: Ind. Eng. Chem. *41*, 2828 (1949)
107. Gallagher, D. M.: Amer. Dyestuff Reptr. *53*, 361 (1964)
108. Schiffner, R., Lange, R.: Faserforsch. Textiltech. *9*, 417 (1958)
109. Bridgeford, D. J.: Ind. Eng. Chem. Prod. Res. Dev. *1*, 45 (1962)
110. Bridgeford, D. J.: (to Tee-Pak Inc.) U.S. *3*, 083, 118 (1963)
111. Gallagher, D. M.: Textile Res. J. *40*, 621 (1970)
112. Roff, W. J.: J. Textile Inst. *49*, T646 (1958)
113. Gruntfest, I. J., Gagliardi, D. D.: Textile Res. J. *18*, 643 (1948)
114. Stelle, R.: Textile Res. J. *25*, 545 (1955)
115. Goldthwait, C. F.: Textile Res. J. *21*, 55 (1951)
116. Wagner, R. E., Pacsu, E.: Textile Res. J. *22*, 12 (1952)

117. Marsh, J. T.: J. Textile Inst. *48*, 374 (1957)
118. Marsh, J. T.: J. Soc. Dyers Col. *75*, 244 (1959)
119. Reeves, W. A., Perkins, R. M., Chance, L. H.: Textile Res. J. *30*, 179 (1960)
120. Hebeish, A.: L'Industrie Textile *415*, 991 (1970)
121. Kantouch, A., Hebeish, A., El-Rafie, M. H.: Textile Res. J. *42*, 10 (1972)
122. Ogiwara, Y., Igeta, K., Kubota, H.: J. Appl. Polym. Sci. *15*, 119 (1971)
123. Ogiwara, Y., Igeta, K., Kubota, H.: J. Appl. Polym. Sci. *10*, 845 (1972)
124. Cremonesi, P., Focher, B., D'Anguiro, L.: Cell. Chem. Technol. *4*, 496 (1970)
125. Kubota, H., Ogiwara, Y., Yamato, H.: J. Appl. Polym. Sci. *18*, 887 (1974)
126. Ogiwara, Y., Yasunaga, T., Kubota, H.: J. Appl. Polym. Sci. *20*, 1413 (1976)
127. Ogiwara, Y., Hayase, K., Kubota, H.: J. Appl. Polym. Sci. *23*, 1 (1979)
128. Johnson, D. C., Nicholson, M. D., Haigh, F. C.: Appl. Polym. Sym. *28*, 931 (1976)
129. Guthrie, J. T., Hardcastle, S. V.: Polymer *18*, 203, (1977)
130. Portnoy, N. A., Anderson, D. P.: In Solvent Spun Rayon, Modified Cellulose Fibres and Derivatives (Turbak, A. F. ed.), A.C.S. Symp. Ser. *58*, 52 (1977)
131. Guthrie, J. T., Ryder, M., Abdel-Hay, F. I.: Polymer Bulletin, *1*, 509 (1979)
132. Beddows, C. G., Guthrie, J. T., Mirauer, R. A.: Biotechnol. Bioeng., *22*, 311 (1980)
133. Regan, D. L., Dunhill, P., Lilly M. D.: Biotech. Bioeng. *16*, 333 (1974)

7. Characterization and Properties of Cellulose Graft Copolymers

7.1. Introduction

During grafting, a certain amount of monomer is not used in the reaction and the determination of "grafting yield"; i.e., percentage of monomer which is grafted to the polymer, is an important factor in the economics of the grafting process. However, in addition to this, the exact analysis of the grafted polymer is important and involves a determination of (i) the homopolymer content, (ii) the total grafted monomer, (iii) the fraction of cellulose or cellulose derivatives not taking part in grafting, (iv) the number of grafts per definite amount of cellulose, (v) the mean length of the grafted chain, and (vi) the polydispersity. Such studies are quite complex, but necessary if the optimum conditions of grafting for each system are to be realized.

When the method of modifying polymers, by grafting side chains of a different type onto the given backbone macromolecules, emerged from the work of several investigators[1)], a great deal of enthusiasm was generated. However, in the early studies dealing with grafting onto cellulose, very few researchers characterized their so-called "grafted products" other than by measuring the "add-on" after some kind of extraction procedure. The products of each grafting reaction are usually extracted exhaustively with a solvent in order to remove the homopolymer. The material which cannot be extracted is usually considered as grafted, although whether the synthetic polymer is held in the system by chemical bonds or by physical entanglement, or by both in particular cases, is still uncertain. As a matter of fact, in the early days of cellulose grafting, there was considerable scepticism as to whether true grafting had taken place. There is no longer any doubt that grafting can occur, but the extent of actual grafting varies considerably with the method and conditions used.

There have been comparatively few published studies of the properties of pure cellulose graft copolymers owing to difficulties associated with freeing the grafted material from ungrafted cellulosic and the attendant homopolymer. Attempts have been made to isolate the grafted branches and examine their properties in a systematic way[2–10)]. Most papers dealing with grafted cellulosics contain details of the attempted removal of homopolymer, where such details are relevant. From a practical or commercial standpoint, the heterogeneous mixture of graft and homopolymers might be deemed to satisfy particular requirements, and thus no further purification stages would be used. Since large-scale production probably often produces heterogeneous composites, a study of their bulk properties is a legitimate goal of applied research. Such studies have been reported on cotton, rayon, paper, board, wood, and various cellulose derivatives.

In search of a deeper understanding of the nature of graft copolymerization processes and the 'true' graft copolymer, several groups require that the graft copolymer be in the purest possible form. Two distinct avenues to graft copolymer characterization have been developed, namely, evaluation through studies of physical properties and elucidation of the cellulosic copolymer molecular structure. An analysis of selected studies[11–15] on the structure of grafted cellulose prepared by free radical initiation methods has revealed that, grafting is generally, limited to a comparatively few high molecular weight branches involving about 0.2 to 2 side chains per cellulose chain, a degree of graft substitution (D.S.) of approximately 0.001. For example, the free radical graft polymerization of methyl methacrylate onto viscose rayon resulted in only 1.4–1.7 PMMA side chains per cellulose chain with molecular masses of 700,000 to 900,000[15], and grafting of polystyrene onto cotton by radiation yielded, on the average, less than 1 side chain of molecular mass of 500,000[11]. The need for better control of the grafting process to yield shorter side chains and higher D.S. levels is apparent. Indeed, approaches to grafting via ionic-initiation or grafting of preformed polymers having relatively low molecular masses show the importance of the D.S. of the graft in producing cellulosic graft copolymers with improved properties[16,17].

The usual dilemma which arises in preparing cellulose graft copolymers involves the choice between having several relatively low molecular mass grafts or a few high molecular mass grafts for a given add-on value. Careful selection of the copolymerization reaction may provide a measure of control, though normally no clear dividing line between the two options can be drawn.

7.2. Proof of Grafting

A number of quantitative tests can be applied to show that the product contains graft copolymer, although they do not give us quantitative information about the ratio of graft to homopolymer. In these tests, the properties of the product believed to be a graft are compared with those of a physical mixture of the two types of macromolecule. This mixture should have the same overall composition as the graft. If possible, the average molecular mass of the polymer grafted onto cellulosic material should be the same as that of the homopolymer in the physical mixture. One way to meet this last requirement is to extract the homopolymer from the substrate on which grafting takes place. This may not always be a sufficient guarantee that the desired goal has been reached, because if the extraction of the homopolymer is incomplete, it is likely that molecules with relatively low molecular mass will be extracted in preference to those of high molecular mass.

Easily measurable properties of the graft copolymer which may be compared with those of a physical mixture are the intrinsic viscosity and the concentration dependence of the viscosity[18]. A difference in these parameters for the 'copolymer' relative to a mixture of homopolymers having the same composition could be indicative of the occurrence of grafting.

Another property is solubility. Separation of the graft copolymer from the homopolymers is based upon the different solubilities of chemically different polymeric species[19-21]. By way of example, mention is made of the following experiment[21]. Acrylamide was grafted onto cellophane. The product was dissolved in cuprammonium hydroxide and the solution acidifed. The amount of precipitate formed was always larger than the amount of cellulose present. When the same experiment was performed with a physical mixture of cellulose and polyacrylamide of the same composition, only the cellulose was precipitated. Further more, infrared spectroscopy showed the absence of polyacrylamide in the precipitate from the physical mixture. Hence, the extra weight of precipitate from the graft must be presumed to originate from the acrylamide grafted.

The fractional precipitation method, as described above, is also readily adapted to graft copolymers of cellulose derivatives, such as cellulose acetate, cellulose nitrate, etc.[22]. The reaction product is dissolved in a suitable solvent, and a precipitant is added slowly to the solution. At various stages the precipitated polymer is removed and the stepwise addition of precipitant to the supernatant liquid is continued. A plot of percent polymer precipitated versus the amount of precipitant added is then prepared in order to demonstrate the presence of graft copolymer. If there is only a physical mixture, the curve shows a sharp difference in the solubility behavior of the two homopolymers. On the other hand, if graft copolymer has been formed, a more or less continuous precipitation curve is obtained. This method has been applied to demonstrate the grafting of polystyrene to cellulose acetate.

Turbidimetric titration, a type of fractional precipitation, is a rapid analytical method for showing the presence of graft copolymer in a polymer mixture. With this technique, one measures the amount of turbidity resulting from the addition of a precipitant to a dilute polymer solution. The turbidity is expressed as the ratio of the amount of light absorption at partial precipitation to the absorption at complete precipitation[23]. The titration curve is obtained by plotting the turbidity against the amount of precipitant added. The curves obtained by recording the increase in solution turbidity upon addition of nonsolvent to a solution of the grafted product and to a physical mixture are analogous to those obtained when the fractional precipitation method is applied[24].

The density-gradient ultracentrifugation method, which is based on the difference in density between untreated cellulose, graft copolymers, and nongrafted homopolymers, has been successfully applied to prove that grafting has occurred. This density difference is sufficient to cause the separation of the reaction product into distinct bands when the material is ultracentrifuged in a mixture of two solvents selected in such a way that a density gradient, covering the density range of the graft and homopolymers, is established. An example of such an examination is shown in Figure 63; the physical mixture shows clearly the two components. The original graft copolymer, which was believed to be pure, shows contamination with polystyrene, and the graft which has been re-extracted with benzene finally reveals only the peak of the graft itself[25].

Another physical method of establishing the formation of a cellulose graft copolymer is differential thermal analysis. By this method, any transition

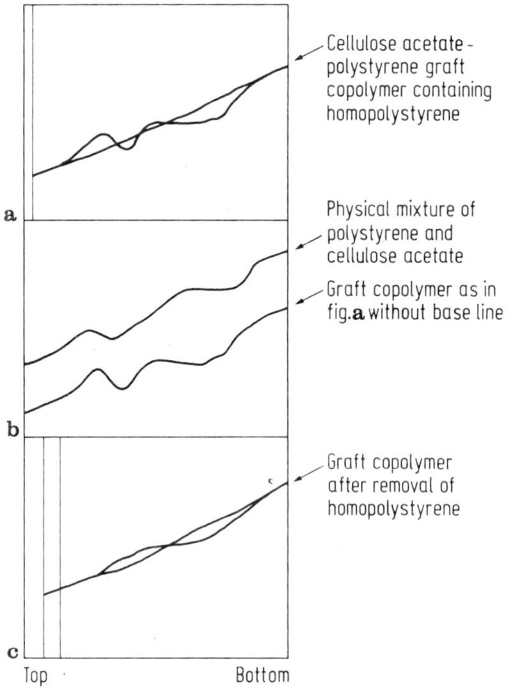

Fig. 63a–c. Density gradient ultracentrifuge for a cellulose acetate-polystyrene graft copolymer; a physical mixture of cellulose acetate and polystyrene (upper) and the graft copolymer containing about 10% occluded polystyrene (lower). The plots are of the first derivative of refractive index with respect to distance (depth) in the cell vs. cell depth. Reprinted from Ref.[25], p. 4049, by courtesy of John Wiley & Sons, Inc.

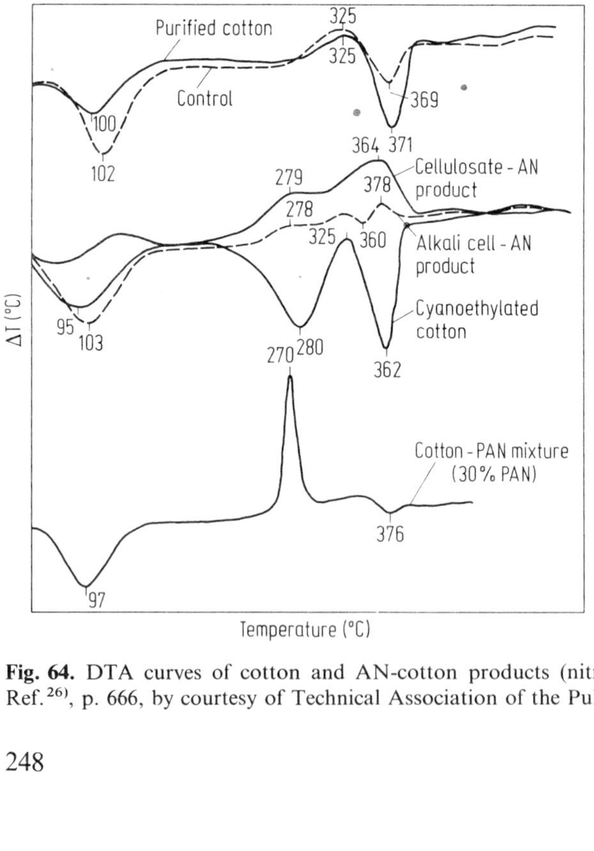

Fig. 64. DTA curves of cotton and AN-cotton products (nitrogen atmosphere). Reprinted from Ref.[26], p. 666, by courtesy of Technical Association of the Pulp and Paper Industry.

or reaction accompanied by evolution or absorption of heat, as the sample and thermally stable reference substance are heated through a cetain temperature interval, are detected in terms of temperature difference between the two substances. Endothermic changes give rise to negative peaks, exothermic to positive peaks. Figure 64 shows that the sharp exotherm at 278 °C, which is characteristic for polyacrylonitrile homopolymer and which is observed with a physical mixture of cellulose and polyacrylonitrile, is practically absent in the grafted materials[26)].

Infrared spectral analysis has been adopted to prove grafting by making use of the variation in the intensity of the hydroxyl infrared absorption bands that may occur because of grafting[27-30)]. For example, a comparison was made of the infrared absorption of cellulose, cyanoethylated cellulose, acrylonitrile-grafted cellulose, cyanoethyl cellulose, and a physical mixture of cellulose and polyacrylonitrile (PAN). It was found that the hydroxyl band intensities of the graft copolymers were considerably less than the hydroxyl intensity of the pure cellulose, but that the hydroxyl intensities of cyanoethylated cotton and the cyanoethylated-acrylonitrile graft copolymer were almost the same; therefore, it was concluded that grafting had probably occurred at the cyanoethyl groups[28,29)]. The infrared spectrum of a PAN-cellulose graft copolymer, measured in the 4000–400 cm^{-1}, region, was compared with the infrared spectrum of pure cotton, as shown in Figure 65, which covers only 2500 to 400 cm^{-1}. It can be seen that the absorption bands in most of the copolymer spectrum remain essentially unchanged; only in the bands corresponding to hydroxyl groups was a certain decrease in intensity observed[30)]. Further information is supplied by the peak at 2249 cm^{-1} corresponding to the —C≡N group of PAN.

Chromatographic analysis has also been used in an extensive investigation of cellulose grafted with vinyl chloride by the Ce^{IV} ion technique[31)]. Homo-

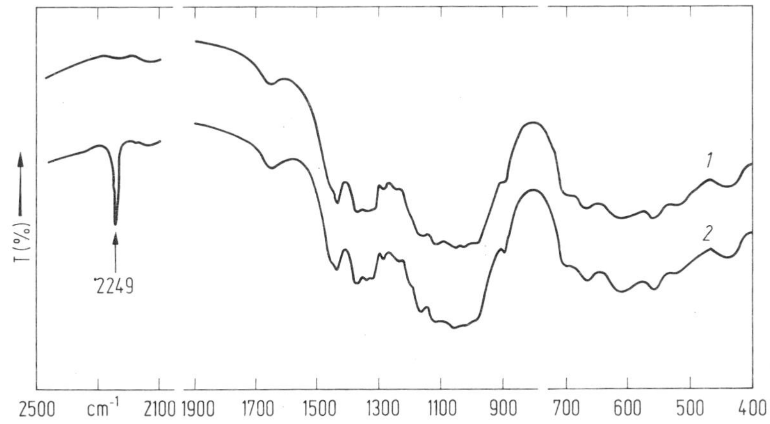

Fig. 65. Infrared spectra of *1* cotton and *2* cotton-PAN copolymer containing 42.9% wt.-% of PAN. Reprinted from Ref.[30)], p. 903 by courtesy of John Wiley & Sons, Inc.

polymer was first removed from the reaction mixture by extraction with tetrahydrofuran, a solvent for poly(vinyl chloride). The residue was then hydrolyzed in 72% H_2SO_4 to remove the cellulose backbone from the poly(vinyl chloride) branches. The product remaining after hydrolysis was dissolved in tetrahydrofuran. Chromatographic analysis of this product showed no sugar content. However, when the material was hydrolyzed again, this time with 98% formic acid, chromatographic examination of the aqueous solution showed the presence of glucose, which indicates that glucose residues must have been chemically attached to the poly(vinyl chloride) branches removed from the cellulose substrate by the first hydrolysis.

Also, radioactive cellulose may be used to show that grafting has taken place. For example, cellulose labeled with C-14 was used as the starting material for grafting, and the resulting graft copolymer was hydrolyzed to remove the cellulose backbone. It was found that each of the remaining graft branches had acquired a radioactive glucose end group, indicating that grafting had occurred and that the grafted branches were attached to cellulose through a C—C bond[31].

Differential refractometry may be applied to soluble cellulose copolymer derivatives, after dialysis in mixed solvent systems, and may provide information regarding the occurrence of preferential adsorption of one of the components of the mixed solvents. This technique has been applied with some success to a series of cellulose tricarbanilate-polystyrene copolymers[32–34].

Finally the grafting process itself sometimes shows features which are incompatible with the assumption that only homopolymer is formed. For example, the rate at which the weight of the substrate increases may become gradually slower with increasing time, while the homopolymerization outside the substrate shows no such tendency. This does not constitute a proof, but could be considered as an indication of grafting. In one case[18] the failure to observe a weight increase in the presence of acetone might be considered as evidence for grafting in the absence of acetone, because the homopolymerization takes place in either case.

7.3. Molecular Structure

The first step in a complete characterization of the cellulose graft copolymer is a proof of grafting. This can be achieved by applying one or more of the methods already described above. Next, the molecular mass of the cellulose backbone and of the grafted branches, the frequency of branching, and the site of branch attachment should also be determined. Recent work has indicated that the molecular mass of the 'true' copolymers can be achieved directly[32].

7.3.1. Molecular Mass of Cellulose Backbone

Under normal conditions it is extremely difficult to follow the changes in the cellulosic backbone during simultaneous graft copolymerization. The graft copolymer can be separated from the homopolymers and its composition and the molecular mass of the side chains determined. However, little knowledge can be gained from the cellulose backbone itself, since it is destroyed during isolation of the synthetic side chains, i.e., grafts, by either acid hydrolysis or acetolysis (see below). It has been reported that formation of free radicals by radiation and redox systems[35)] initiates oxidative degradation of the cellulose molecule in the absence of monomer, thereby decreasing its molecular mass. This degradation is unequivocally due to the formation of cellulose-free radical intermediates which subsequently lead to oxidation products of cellulose. In the presence of monomer, however, one would expect that cellulose degradation by any of the initiating methods would be eliminated or at least minimized, because it is very likely that monomer adds to the active centers before they can undergo the oxidation reaction. It is also possible that the oxidation reaction is inhibited by the presence of the monomer or polymer during grafting. Therefore, the molecular mass of cellulose is determined before grafting, and the values obtained are considered to represent the molecular mass of the cellulose in the graft copolymer on the assumption that no degradation of cellulose occurred during the graft polymerization reaction.

There are two approaches to the determination of the average molecular mass of cellulose: (i) direct molecular mass determination in a solvent for cellulose such as cuprammonium hydroxide, cupriethylene-diamine, Cadoxen, etc.[36–41)], and (ii) conversion of the cellulose to a derivative which is soluble in a standard organic solvent and therefore lends itself to routine molecular mass measurements[33,34,36,42–47)]. For graft copolymers, (ii) offers the only viable approach. Thus, the cellulose component of the graft copolymer would normally be converted into a derivative which has similar solubility characteristics to the grafted backbone. Average molecular mass values are usually obtained from osmotic pressure ($\bar{M}_n$), light scattering ($\bar{M}_w$), or viscosity ($\bar{M}_v$) measurements. For copolymers, the values obtained must be handled with caution. This point will be dealt with later.

7.3.2. Molecular Masses of Grafted Branches

In order to characterize the grafted side chain of the copolymer, the grafted products must be separated. This can be achieved by first isolating the grafted branches from the cellulose backbone. There are three methods of removing the cellulosic component: *hydrolysis*, *acetolysis*, and *enzymic attack*. Stannett et al.[13)] have separated and characterized the cellulose acetate-styrene grafting system as follows: The grafted films were washed first in benzene, then in a 55:45 acetone-water mixture. The infrared spectra of both extracts were checked to determine if any graft copolymer had been removed. The extractions were then repeated and the residue considered as graft copolymer. Occasional samples were double-checked by dissolving them in dimethylformamide and precipitating in benzene, and then repeating the washing with the aqueous acetone mixture; virtually complete extraction was found to have taken place in every case examined. The increase in weight after the first benzene wash was recorded and is equivalent to the percent grafting often reported in grafting studies. The

Table 51. Heterogeneous grafting experiments with styrene. Adapted from Ref.[13], p. 552, by courtesy of John Wiley & Sons, Inc.

Monomer-solvent mixture	Film thickness, mm	Dose rate, r/hr	Total dose Mrad	Increase in weight		Composition of graft		Efficiency of graft			M.W. of PS side chain	M.W. of PS produced in solution
				Actual	Due to true graft %	CA,[a] %	PS,[b] %	CA[a] %	PS[b] %			
90 S-	0.025	7,650	0.45	183.0	93.6	13	87	13.6	51.8	6.65	4,200,000	—
10 DMF	0.025	21,000	0.95	147.0	45.0	27	73	16.7	30.5	5.80	3,500,000	—
	0.025	340,000	1.00	13.9	6.2	—	—	—	—	—	—	70,000
	0.025	340,000	10.00	33.7	—	—	—	—	—	—	—	70,000
	0.0025	340,000	1.00	60.8	51.2	22	78	11.1	84.1	3.81	1,945,000	70,000
	0.0025	340,000	10.00	169.5	157.4	20	80	34.2	92.9	2.90	1.338,000	70,000
80 S-	0.025	340,000	1.00	22.9	17.2	30	70	4.0	75.3	2.42	1,044,000	61,000
20 DMF	0.025	340,000	10.10	64.1	37.8	34	66	19.8	59.1	1.45	518,000	61,000
	0.0025	340,000	1.00	11.1	—	—	—	—	—	—	—	61,000
	0.0025	340,000	10.00	67.2	37.1	41	59	21.5	53.3	0.68	184,000	61,000
95 S- 5 pyridine	0.0025	340,000	10.00	79.9	57.3	12	88	7.9	71.8	0.78	220,000	74,000
90 S- 10 pyridine	0.025	340,000	10.00	53.4	34.0	53	57	19.2	63.7	0.84	245,000	72,000
90 S- 10 pyridine	0.0025	340,000	10.00	104.7	103.2	34	66	13.6	98.5	1.91	755,000	72,000
80 S- 20 pyridine	0.025	340,000	10.00	111.9	92.7	29	71	17.1	90.9	2.28	962,000	66,000
80 S- 20 pyridine	0.0025	340,000	10.00	79.9	73.7	29	71	62.4	92.2	2.11	865,000	66,000
80 S- 20 pyridine	0.0025	340,000	10.00	67.7	65.2	34	66	39.2	96.3	—	—	66,000
70 S- 30 pyridine	0.0025	340,000	10.00	17.5	17.0	61	39	31.0	97.1	0.67	181,000	59,000
70 S- 30 pyridine	0.025	340,000	10.00	23.1	23.1	55	45	28.3	(100)	0.91	272,000	59,000

[a] CA, cellulose acetate.
[b] PS, polystyrene.

difference between this value and the true grafting yield can be seen in Table 51. The grafts prepared in solution were isolated by pouring then into methanol, filtering, washing with methanol, and drying under vacuum. The dry polymer was then extracted with benzene to remove homopolystyrene and the residue further extracted with an acetone-water mixture. The two extractions were then recorded, the product dried under vacuum and weighed. The composition of the graft copolymer produced in solution is given in Table 52.

The hydrolysis of the graft copolymers was achieved by first swelling about 1 g in a mixture consisting of 70% acetone, 25% of dimethylformamide, and 5% water. A 50 cm^3 portion of 72% sulfuric acid was then added and left for 2 hr at room temperature; 500 cm^3 of water were then added and the mixture refluxed for 5 hr after first boiling off the acetone. The remaining polystyrene was then freeze-dried from a benzene solution[13].

An alternative procedure of hydrolysis was developed by Morris et al. to separate and characterize the vinyl side chains from grafted polyacrylonitrile-cotton copolymers[48]. This was modified by Guthrie et al.[2] in their study of the cellulose-polystyrene copolymer system. Here 2 g of copolymer were dispersed overnight (after removal of occluded homopolymer) in 200 cm^3 of 75% (w/w) aqueous zinc chloride with continuous shaking. An equal volume of 6 M hydrochloric acid was added and the mixture stirred for 2 hours, at which point dispersion of the sample was complete. After subsequent heating for 4 hours at 100 °C, the product was poured into an excess of methanol and the precipitated polystyrene filtered. The product was reprecipitated from a suitable solvent and dried under vacuum at 60 °C.

Hebeish[49] described another method for isolation and characterization of cellulose grafted with polyacrylonitrile (PAN). This method is based on removing the cellulose backbone by acetolysis.

The experimental procedure adopted was as follows. An accurately weighed quantity of the PAN grafted sample was transferred to a 100 cm^3 stoppered conical flask and was treated with sufficient solution, consisting of 170 cm^3 glacial acetic acid plus 17 cm^3 acetic anhydride plus 18 cm^3 concentrated sulphuric acid, to give a liquor ratio of 40.1. The acetolysis was carried out at 30 °C for 3 hours, after which the flask containing the reaction mixture was

Table 52. Graft copolymers produced in solution at 25 °C [a]. Reprinted from Ref.[13], p. 560, by courtesy of John Wiley & Sons, Inc.

Styrene — pyridine ratio	Composition of graft		M.W. side chains	M.W. homopolystyrene	Approx. No. of side chains CA[b] molecule
	CA, %	S, %			
30:70			—	15,000	—
40:60	69	31	24,000	19,000	1.0
50:50	66	34	31,000	25,000	0.9
60:40	61	39	39,000	35,000	0.9

[a] Dose rate 340,000 rads/hr., total dose 10 Mrads.
[b] Calculated from the viscosity-average molecular masses.

transferred to a thermostat bath at 90 °C for 10–15 minutes. After cooling, the contents of the flask were filtered through a sintered glass crucible. The residue, after weighing, was dissolved in dimethylformamide, and the insoluble portion was removed by filtration. The filtrate was poured into distilled water, and the precipitated polymer was dissolved again in dimethylformamide and reprecipitated with distilled water. The polymer obtained was dried in vacuo and the molecular mass of the grafted polyacrylonitrile determined. The composition of the graft copolymer (E), as well as the molecular masses of the nongrafted homopolymer (PAN) and the grafted-PAN, are given in Table 53.

This approach has also been used by Huang and Chandramouli[8] with the polyacrylonitrile-cellulose copolymer system. The application of this acetolysis route has been shown to cause difficulties with cellulose-polystyrene graft copolymers[2]. Although the acetolysis process is efficient in cleaving the polystyrene grafts from the cellulose backbone, there is also appreciable involuntary sulfonation of the aromatic groups in the grafted chains arising from the reaction conditions. Viscometric studies of polystyrene grafts, isolated by the acetolysis procedure[8], showed the presence of polyelectrolyte behavior, i.e., a marked increase in η_{sp}/C at low concentrations of grafted branches in solution. This behavior was supressed by the use of 10^{-3} M trichloroacetic acid in chloroform as the solvent in the viscometry. The concentration of sulfuric acid and the conditions recommended for acetolysis are similar to those used in standard methods of the sulfonation of aromatic hydrocarbons. The possibility arises of introducing sulfonic acid groups at the para-position of polystyrene. Elemental analysis gave sulfur concentrations which corresponded to 1 in every 27 styrene segments carrying a sulfonic acid group, which is in good agreement with a value of 1 in 29 obtained from titrimetric analysis. Using lower sulfuric acid concentrations, Schurz et al.[50] have successfully obtained toluene-soluble polystyrene by acetolysis of cellulose tricarbanilate grafted with polystyrene. These products did not exhibit polyelectrolyte behavior.

Arai and Ogiwara[10] have examined a rather novel method of separating branches from poly(vinyl acetate) grafted cellulose involving hydrolysis of

Table 53. Composition of cellulose-acrylonitrile graft copolymer and molecular masses of grafted polyacrylnitrile and Homopolyacrylonitrile. Reprinted from Ref.[49], p. 188

Substrate	Temp. °C	Composition of Graft copolymer (E)		(a)		$M_n \times 10^{-5}$ (b)	
		Cellulose wt. %	PAN wt. %	Graft	Homo-polymer	Graft	Homo-polymer
Native cotton	20	4.77	95.23	2.40	3.14	1.7380	2.5120
NaOH-Cotton	20	3.96	96.04	1.88	2.42	1.2480	1.7620
Viscose	20	5.56	94.44	2.250	2.72	1.5890	2.0650
Ramine	20	4.51	95.49	1.165	2.18	0.6486	1.5210
Ramine	40	4.48	95.52	1.220	3.43	0.6918	2.8310
Ramine	60	8.45	91.55	0.835	1.58	0.4121	0.9840

(a) Determined in DMF at 20 °C.
(b) Calculated from $[\eta] = 3.46 \times 10^{-4}\, \bar{M}_n^{0.733}$.

cellulose. Because of the time-course of deactivation of cellulose and competitive inhibition with the hydrolyzed products of cellulose, the grafted samples were subjected to twelve sequences of cellulose removal and acetone extractions, alternatively. Work of this type is usually carried out in a carefully buffered environment. The degree of polymerization (DP) obtained at each isolation stage showed a leveling-off tendency in the range above 60% of recovered poly(vinyl acetate). The ease with which the poly(vinyl acetate) was recovered was influenced greatly by the extent of grafting in the sample. This method of hydrolysis by cellulase may have much wider application as a separation method of non-cellulosic material from cellulosic-non-cellulosic mixtures of composites, especially when the non-cellulosic is sensitive to extreme conditions.

The average molecular mass of the grafted branches may be determined by the usual methods. Such molecular mass measurements have shown that the grafted branches have a much higher average molecular mass than any homopolymer formed at the same time[11,13,22,50]. Typical examples of this behavior are shown in Table 51. This has been attributed to a reduction in the rate of termination because of low mobility of the grafted chain radicals inside the fibers[51]. Stannett et al.[13] have reported, however, that when grafting was carried out using a solution of styrene in dimethylformamide or pyridine, rather than a swollen but undissolved "included" cellulose sample, the molecular mass of the grafted branches was much lower and almost comparable to that of the homopolymer formed at the same time. This is shown in Table 52. Other investigators, on the other hand, have reported higher molecular masses for the homopolymers than those for the grafted branches[12,52,55], as can be seen in Table 53. Table 53 also shows that the molecular masses of the grafted branches, as well as those of the homopolymer, are greatly influenced by the nature of the substrate and reaction temperature. The relatively low molecular masses at higher temperature are perhaps due to a fast termination rate of the propagating polymer chains.

Morris et al.[48] have shown that the molecular mass relationship between cellulose and grafted copolymer depends on the method of initiation of the free radical sites on the cellulose molecule and the composition of the solution of monomer, particularly as this affects the propagation and termination reactions. The degree of grafting can be as much as 1 or 2 polymer grafts per cellulose molecule when the Ce^{IV} ion and aqueous systems are used, or when high energy radiation combined with grafting from organic solvents is used: molecular masses of the grafted polymer range from 84,000 to 33,000 respectively. When high energy radiation is combined with grafting from aqueous salt solutions, which cause large dimensional changes in the cellulosic fibers, the degree of grafting can be as low as 1 polymer graft per 100 cellulose molecules, with molecular masses of the grafted polymer as high as 1,000,000.

Rogovin[55] has reported that the molecular mass of the grafted synthetic polymer in the heterogeneous medium may be considerably decreased by the addition of a regulator (chain terminator). Indeed, this method is widely used for the regulation of the molecular mass of polymers by radical polymerization. In Table 54 are given some data relating to the molecular masses of the grafted

Table 54. The dependence of the quantity and type of regulator on the molecular mass of the grafted chains. Adapted from Ref.[55], p. 230, by courtesy of John Wiley & Sons, Inc.

Monomer	Regulator	Quantity of regulator (% on the monomer weight)	Degree of polymerization of the grafted chain
Acrylonitrile	CCl_4	0	1350
		11	1180
		42.5	870
		100	560
Acrylonitrile	CCl_4	0	3140
		0.02	660
		0.10	198
		0.20	96
		0.30	56
		0.50	31
		1.0	20
Methylvinylpyridine	Triethylamine	0	2240
		1	1790
		2	1630
		3	1420
		5	980
Methylvinylpyridine	Dodecylmercaptan	0	2240
		0.5	1120
		1.0	690
		5.0	386

Table 55. The influence of the number and molecular mass of grafted chains on the glass transition and temperature of cellulose graft copolymers and their esters. Adapted from Ref.[55], p. 231, by courtesy of John Wiley & Sons, Inc.

Grafted copolymer	Composition of reaction products		Molecular mass of the grafted chain	Glass transition temperature
	Cellulose %	Grafted copolymer %		
1) Cellulose-polystyrene	60	40	158,000	126
	60	40	74,150	102
	74.8	25.2	7,800	104
	72.9	27.1	4,150	96
2) Cellulose tri-acetate-poly (methyl methacrylate)	18.8	81.2	1,350,000	142
	21.7	78.3	770,000	130
	24.9	75.1	420,000	125
	26.4	73.6	168,000	116
3) Secondary cellulose acetate-poly (methacrylic acid)	38	62	800,000	141
	37	63	200,000	132
	37	63	10,000	128

chains and also to the type and amount of regulator used. Also relevant is the fact that a decrease in the average molecular mass of the grafted branches leads to superior mechanical properties of films prepared from graft copolymers of cellulose triacetate and cellulose nitrate. Analogous relationships also exist between the length of the grafted chain and the properties of fibers made up of cellulosic graft copolymers[55)].

The question of the effect of the length of the grafted chain (on the basis of equal quantities of grafted polymer) on the properties of the modified cellulosic material is of considerable scientific and practical interest.

One may well obtain different properties if one grafts 100 long chains or 1000 considerably shorter chains. This important question has not been systematically investigated as yet.

There are available only scattered data which indicate the considerable influence of this factor on the properties of the cellulose graft copolymer. Thus, for example, it has been shown that the length of the grafted chain greatly influences the viscosity of concentrated solutions of grafted cellulose ester copolymers. If the molecular mass of the poly(methyl methacrylate) chain grafted to a molecule of secondary cellulose acetate is increased from 100,000 to 800,000, then the viscosity of solutions with the same concentration increases by 20–25 times[56)], even though there is present in each case the same quantity of grafted polymer. Obviously, the spinning of fibers from such a viscous solution is considerably more difficult.

Upon increasing the molecular mass of the grafted chain, the glass transition temperature of the polymer is increased. This may be realized from the data given in Table 55.

7.3.3. Average Molecular Masses of 'True' Cellulosic Copolymers

Attempts at copolymer characterization using the true copolymer, rather than isolation of backbone and grafts, have been made[32–34, 57–59)]. Some attention has been given to the cellulose-polystyrene graft (Cel-PS) copolymer system[32–34)], produced by radiation-induced grafting. Purification of the copolymer involved efficient extraction of ungrafted cellulose with Cadoxen to constant weight. The resulting copolymer was then rigorously extracted with benzene for 16 hours to remove any occluded homopolymeric polystyrene (P.S.), again to constant weight. The copolymer product was then reacted with phenyl isocyanate in pyridine at 107 °C for 24 hours to convert the cellulose into the carbanilate. This gives a cellulose tricarbanilate-polystyrene (CTC-PS) graft copolymer which is soluble in a variety of organic solvents. Light-scattering techniques were used in the molecular mass measurements. For light scattering, refractometry, and differential refractometry, a wavelength of 436 nm in vacuo was used. The apparent degree of grafting of the cellulose-polystyrene copolymer was 0.108, calculated as $(M - M_0)/M_0$, where M_0 and M are the weights of the ungrafted and grafted films respectively. The true degree of grafting was calculated as $(M - M_0)/(Mg - M + Mo)$; where Mg is the constant residual weight of pure copolymer. After extraction of ungrafted cellulose and polystyrene homopolymer, the true degree of grafting was 1.79. Thus, the fraction of weight, Wg, of cellulose which actually participates in grafting is $[M_g - (M - M_0)]/M_0 = 0.060$.

A consideration of the (CTC-PS) copolymer, obtained by carbanilation of the (Cel-PS) copolymer possessing a true degree of grafting of 1.79, provides an indication of the proportion of (CTC) in the (CTC-PS) copolymer. Using this true degree of grafting in conjunction with the factor 3.20 (= 519/162 = the ratio of the molecular weights of the segments in CTC and Cel), one obtains a value of $3.20\,[M_g - (M - M_0)]/\{3.20\,[M_g - (M - M_0)] + (M - M_0)\} = 0.64$, for the weight fraction, W_{CTC} of the CTC in the CTC-PS copolymer. Alternative figures have been provided by elemental analysis of the CTC-PS copolymer ($W_{CTC} = 0.71$); by thermogravimetric analysis ($W_{CTC} = 0.73$); and by measurement of the specific refractive index increment of the copolymer in given solvents ($W_{CTC} = 0.73$). Although the refractive index increments of CTC and PS in a given solvent are rather similar, the measured increment for the copolymer in the same solvent affords the quoted value of W_{CTC} according to

$$\nu_{copolymer} = W_{CTC}\nu_{CTC} + (1 - W_{CTC})\,\nu_{PS}\,. \tag{173}$$

Measurements were made in methyl ethyl ketone and dioxane to give the quoted value of W_{CTC} of 0.71. There is some disparity in the composition of the copolymer as measured by the different techniques. Table 56 contains values of the specific refractive index increment ν, for the CTC-PS copolymer is a series of single solvents. Values are also given for ν_μ, the specific refractive index increment acquired for solutions of this copolymer in a series of mixed solvents, after attainment of dialysis equilibrium[32]. Such values are used in

Table 56. The effect of solvent refractive index on the specific refractive index increment of a cellulose tricarbanilate-polystyrene (CTC-PS) copolymer. Single and mixed solvents. Reprinted from Ref.[32], pp. 198 and 200

Solvent System	η_0 (30 °C) = 436 nm	ν copolymer at 30 °C ($cm^3\,g^{-1}$)	
		experimental	calculated
Methyl Ethyl Ketone	1.3836	0.202	0.206
Dioxane	1.4269	0.172	0.172
Cyclohexanone	1.4567	—	0.141
Pyridine	1.5294	—	0.090
Nitrobenzene	1.5806	—	0.053
Dioxane/bromonaphthalene (a)		ν_μ values	
96.3: 3.7	1.440	0.140	0.142
90 :10	1.458	0.040	0.023
88.7:11.3	1.461	0	−0.010
83 :17	1.478	−0.150	−0.110

(a) measurements taken after achievement of dialysis equilibrium[32].

Table 57. Values of $10^{-3} \times M$ (kg mol^{-1}) of cellulose tricarbanilate (CTC), polystyrene (PS), and a cellulose tricarbanilate-polystyrene copolymer (CTC-PS), obtained by light scattering studies. Reprinted from Ref.[34], courtesy of IPC, Science and Technology Press, Ltd.

L.S. measurement	C.T.C.	P.S.	CTC-PS
Homopolymer	—	0.46	—
Graft-isolated	—	3.07	—
Copolymer in single solvents	0.43	2.45	1.06
Copolymer in selected mixed solvent	0.63	2.91	1.21
Carbanilated form of initial irradiated cellulose (Cel)	0.93	—	—
Carbanilated form of irradiated, ungrafted extracted cellulose (Cel)	0.61	—	—

subsequent molecular mass determinations involving light-scattering measurements.

The molecular masses obtained by light scattering of the graft copolymer are listed in Table 57[34]; the procedures adopted for their determination are summarized as follows. For PS homopolymer the same value of M (the average molecular mass) was obtained with both toluene and dioxane as solvent. Dioxane was used as a solvent for measurement of M for the isolated PS grafts, irradiated CTC, and the unirradiated CTC. In general, compositional heterogeneity in a copolymer yields an apparent value for its average molecular mass M_C^*, which is related to the true molecular mass of the copolymer (M_c) of its constitutent CTC (M_A) portion and of the constituent PS (M_B) branches[57].

A range of single and mixed solvents was used to determine the apparent average molecular mass of the CTC-PS copolymer, the values of which are shown in Table 57[32], using the two equivalent expressions:

$$M_C^* = (\nu_A \nu_B/\nu_C^2)\, M_c + [\nu_A(\nu_A - \nu_B)/\nu_C]\, W_A M_A - [\nu_B(\nu_A - \nu_B)/\nu_C^2]\,(1 - W_A)\, M_B \tag{174}$$

and

$$M_C^* = M_c + [(\nu_A - \nu_B)/\nu_C]\, 2P + [(\nu_A - \nu_B)/\nu_C]^2\, Q \tag{175}$$

where P and Q are compositional heterogeneity parameters which are defined elsewhere[57]. If $\nu_A = \nu_B$ (i.e., the specific refractive index increment of cellulose tricarbanilate is equal to that of the polystyrene branches), then it can be seen that $M_C^* = M_c$ and that measurements always yield the true molecular mass of the copolymer directly. Figure 66[34] gives the values of ν_A and ν_B as a function of the refractive indes of the pure solvent (η_0) for solution of CTC and PS in several pure and mixed solvents[32]. This illustrates that the values of ν_A and ν_B are quite similar in most solvents. At the intersection of the two lines, $\nu_A = \nu_B$. Since there is no suitable solvent with a refractive index corresponding to the value when $\nu_A = \nu_B$, 1.543, recourse is made to using the more versatile mixed-solvent system. The calculation procedure in obtaining M_c from single solvents when $\nu_A \neq \nu_B$ is somewhat tedious.

With mixed solvents, it has been shown[59] that the relevant expressions governing M_C^* are identical in form to those given above in equations (174) and (175),

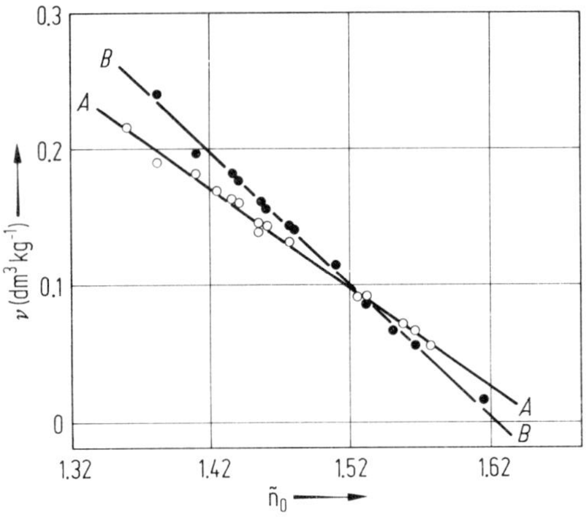

Fig. 66. Specific refractive index increment at constant composition as a function of the refractive index of the solvent for solutions of CTA, *A* and PS, *B* in several pure and mixed solvents CTC — cellulose tricarbanilate; PS — polystyrene. Reprinted from Ref.[34], by courtesy of IPC Science and Technology.

except that the refractive index increments are replaced by the corresponding quantities at a constant chemical potential of solvents, defined as $v_{\mu A}$, $v_{\mu B}$ and $v_{\mu C}$. This is achieved by dialysis procedures[60]. In Figure 61[34], the dependencies of $v_{\mu A}$ and of $v_{\mu B}$ on n_0 are shown for binary solvents comprising various volume fractions Φ_1, of 1-bromonaphthalene in admixture with dioxane. The nature of the selected mixed solvents referred to in Table 57 was dictated by relevant features of Figure 66[32]. Thus, when $v_{\mu B} = 0$, the scattered light results solely from the CTC of a copolymer, and measurements yield the molecular mass of the cellulose tricarbanilate component of the copolymer, CTC-PS. From Figures 66 and 67, we can see that these conditions are given when $v_{\mu A} = -1.60$ dm³/kg, $n_0 = 1.478$, and $\Phi_1 = 0.17$. Similarly, when $v_{\mu A} = 0$, the value of M_B (the grafted polystyrene branches) is produced by the con-

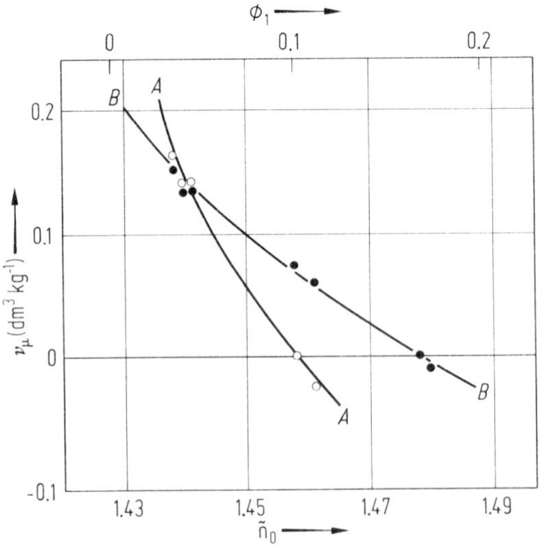

Fig. 67. Specific refractive index increments at constant chemical potential as a function of the refractive index of the solvent for solutions of CTC, *A* and PS, *B* in binary mixtures of 1-bromonaphthalene (volume fraction $\varnothing_1$) and dioxan. Reprinted from Ref.[34], by courtesy of IPC Science and Technology Press, Ltd.

ditions $v_{\mu B} = 0.075\ \mathrm{dm^3/kg}$, $n_0 = 1.458$, and $\Phi_1 = 0.037$. Also, the value of M_c (the true molecular mass of the copolymer) is obtained directly from one measurement in the mixed solvent defined by these conditions. Light-scattering studies of the cellulose tricarbanilate-polystyrene graft copolymer (CTC-PS) in several different single solvents yielded values for the apparent molecular mass of the copolymer, M_C^*, which differ according to the refractive index of the solvent. Thus, in methyl ethyl ketone, dioxane, cyclohexanone, pyridine, and nitrobenzene, M_C^* ranged from 1.28×10^3 to 0.95×10^3 kg/mol. Using previously calculated values of v_{CTC}, v_{PS} and $v_{\mathrm{copolymer}}$, the data were used to supply the true molecular mass via several combinations of simultaneous equations, taking the weight fraction of the cellulose tricarbanilate in the copolymer to be 0.71. The mean values in Table 57 relate to this procedure.

7.3.4. Preferential Adsorption by Cellulosic Copolymers in Mixed Solvents

For a polymer or copolymer in a binary solvent, the greater affinity of one of the solvents for the macromolecule produces a solvent sheath in the immediate vicinity of the polymer which is of a different composition from that of the bulk binary solvent. Differential refractometry provides a means of estimating this effect quantitatively[58], as shown here:

$$\gamma_1 = (v_\mu - v)/\mathrm{d}n_0/\mathrm{d}\Phi_1 \qquad (176)$$

where $\mathrm{d}n_0/\mathrm{d}\Phi_1$ is the dependence of the refractive index of the binary solvent on the volume fraction of solvent 1; v_μ and v are the specific refractive index increments, sometimes denoted $\mathrm{d}n/\mathrm{d}c$, of the copolymer at constant chemical potential and constant concentration respectively; and γ_1 denotes the coefficient of selective adsorption of solvent 1. Measurements of n_0 for sixteen different values of Φ_1 showed that $\mathrm{d}n_0/\mathrm{d}\Phi_1$ is constant and has a value of $+0.270$. Values of γ_1 (where solvent 1 is 1-bromonaphthalene) were calculated via equation (176) for systems encompassing only the rather restricted range of Φ_1 (0.03 to 0.185), which was required to obtain values of v_μ for the light-scattering studies. The relevant data are shown on Figure 67. Within this range, γ_1 ($\mathrm{dm^3/kg}$) varies from 0 to -1.1 for solutions of cellulose tricarbanilate and from -0.1 to -0.6 for solutions of polystyrene. Hence, for both polymers there is preferential desorption of 1-bromonaphthalene, when the content of this solvent in the mixed solvent medium is low. From Figure 67 it can be seen that the cellulose tricarbanilate-polystyrene copolymer behaves in similar manner to the major component of the copolymer, namely, cellulose tricarbanilate ($W_{\mathrm{CTC}} = 0.71$).

Observations obtained from light scattering, osmometry, and differential refractometry[32] indicate that in the cellulose-polystyrene sample being investigated, grafting frequencies of a value less than unity are present. Such values are not without precedent in the literature[61,62], though they preclude a simple backbone/graft model and are somewhat indicative of crosslinking, the presence of ungrafted substrate fragments, and possibly also of block copolymer formation. Much of the evidence favors the last two as being the dominant factors. Approached with care, these techniques can supply valuable information.

From Tables 57 we can see that the molecular mass of the bulk homopolymer is significantly less than that of the grafted branches (0.46×10^6 and 2.45×10^6 respectively in this instance). Here we have further evidence for the presence of Trommsdorff effects in the grafting reactions.

7.3.5. Molecular Mass Distribution of Grafted Branches

Two cellulose graft copolymers with branches having the same average molecular mass may have different physical properties due to differences in the molecular mass distribution of their branches. It may therefore be of interest to fractionate the grafted branches and to plot their molecular mass distribution curve[63)].

Guthrie has shown[32)] that in the radiation-induced grafting of vinyl monomers onto cellulosic substrates, the molecular mass distribution of grafted branches is significantly greater than that of homopolymeric species. This is especially so when the grafting is carried out in a heterogeneous medium[64)].

7.3.6. Frequency of Branching

The same weight of polymer grafted onto a cellulose substrate may correspond to either a few long branches or many short ones. The physical properties of these two different types of grafted copolymer will evidently not be the same. It may therefore be important to calculate the number of grafted branches per cellulose backbone chain[11)].

7.3.7. Site of Branch Attachment

With many of the methods used for the preparation of cellulose graft copolymers, the branches are attached to the oxygen atom of one of the three hydroxyl groups on the anhydroglucose unit. In other cases, such as where grafting occurs through hydrogen abstraction, the branches are attached to one of the three carbon atoms bearing hydroxyl groups. However, when grafting is initiated by irradiation, a number of different branching sites are created, and it may not be possible to determine unequivocally where the branches are attached to the cellulose molecule. Depending on whether a C—H bond, C—C bond, or C—O bond is ionized, different points of attachments will be formed resulting in different products. Breaking a C—H bond or C—C bond will give rise to grafting, whereas breaking an ether, C—O—C sequence will cause formation of block copolymers (see Chapter 3).

7.4. Morphology

Grafting may take place by interactions of monomers at the surface of the cellulosic substrate, with molecules on the surface of the supermolecular structure of the cellulosic, or at random within the cellulose molecules in the substrate. With cellulosic fibers, grafting takes place both in the amorphous regions and on the surface of the oriented regions (crystallites) in the fibers. The distribution of grafted branches on the surface may lead to different effects on the properties from those produced by volumetric grafts located throughout the fibers.

It must be emphasized that the optimum manner of distribution of grafts to get the best results also varies with the property desired. For instance, in order to improve the stability of cellulosic fibers to abrasion or to make them hydrophobic, surface grafting will be quite sufficient. However, if one wishes to impart a wool-like appearance and properties or to alter the dyeing properties of fibers, the grafting must be carried out not superficially, but internally (volumetric grafting).

The morphology of cellulose graft copolymers has been shown to depend on the method of producing the graft, the experimental conditions during copolymerization, the nature of the substrate, and the type of monomer used[65,66]. Various cellulose graft copolymers have been examined by light microscopy in order to determine the location of the grafted polymer. Studies on cotton fibers grafted with β-propiolactone[67] showed that the fibers are swollen and round after a weight increase due to grafting of 35%. With increased grafting, the fibers swell so much that they crack. This may be taken as evidence that grafting takes place inside the fiber rather than on its surface.

Kaizerman et al.[68] dyed cotton and rayon fabrics, grafted with polyacrylonitrile, with a basic dye (Calcozine Green) and examined the cross sections microscopically. Since the dye colors polyacrylonitrile to a fairly deep shade, yet barely stains cellulose, the location of the polymer could be readily observed. According to these researchers the fibers were almost completely penetrated by the polymer deposits, although the polymer concentration appeared to be greater near the surface. Similar observations have been reported by Hebeish[49], who added that the process of grafting onto different cellulosic materials is not necessarily comparable, since grafting onto hydrocellulose was found to be much more in the nature of a surface phenomenon, whereas grafting onto swollen cottons and viscose took place inside the fibers.

Iwakura et al.[69] examined the cross-section of cellulose grafted with glycidyl methacrylate (GMA) after dyeing-on an acid dye in the presence of pyridine. They disclosed that when copolymerization was initiated with Ce^{IV} ions, the grafting of the GMA on cellulose starts from the surface of the fiber and progresses towards the center when the grafting ratio is increased. On the other hand, grafting starts in the core of the fiber with the redox graft polymerization of the Fe^{II}-H_2O_2-cellulose type. This may be seen from Figure 68.

It is also of interest to determine whether or not grafting occurs along the fiber axis, or whether there are centers where grafting has occurred followed by ungrafted sections. A longitudinal view of rayon fibers grafted with butadiene[70] shows that the width of the filament is completely uniform and that grafting is therefore continuous along the entre fiber. Kaizerman et al.[68] have reported that a cellulose-polyacrylonitrile copolymer with a graft of about 30% and a cellulose-polyacrylamide copolymer with a graft of about 20% showed

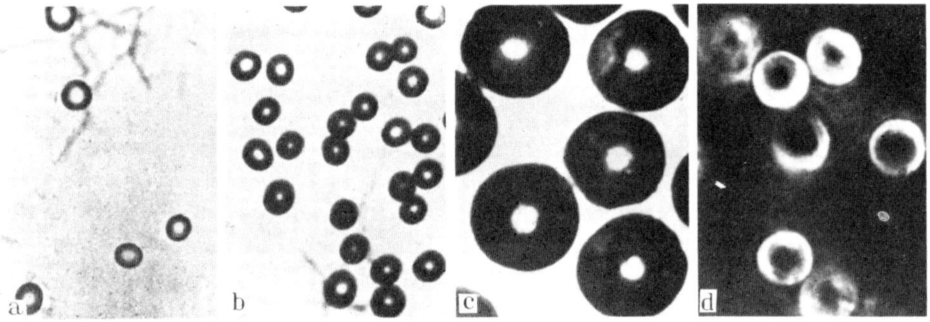

Fig. 68a–d. Cross section of fiber II dyed with Acid Brilliant Scarlet 3 R in the presence of pyridine; **a** Bemberg rayon grafted with GMA in the presence of Ce^{IV} salt, 11% grafting; **b** Bemberg rayon grafted with GMA in the presence of a Ce^{IV} salt, 25% grafting; **c** Bemberg rayon grafted with GMA in the presence of a Ce^{IV} salt, 128% grafting; **d** Bemberg rayon grafted with GMA in the presence of H_2O_2-Mohr salt; ca. 5% grafting. Reprinted from Ref.[69], p. 684, by courtesy of John Wiley & Sons, Inc.

no detectable increase of fiber diameter, whereas grafting at the level of 325% of poly(methyl methacrylate) on viscose fiber had a pronounced swelling effect on the fiber.

In grafting studies[29,71–87], it is possible to determine the morphology and distribution of grafted polymer in the macrostructure of the fiber in a much more detailed manner by means of electron microscopy. For instance, when cotton is pre-irradiated and then treated with acrylonitrile in dimethylformamide, the fibrous cross-section resembles that of a native fiber, and a uniform distribution of the polymer is obtained (Fig. 69B). On the other hand, when the pre-irradiated cotton is treated with acrylonitrile in aqueous zinc chloride (80%), the reaction starts at the outer edge of the fiber and proceeds towards the lumen area[74], as shown in Figure 69A. If more dilute aqueous zinc chloride and a lower concentration of monomer are used, the cross section of the fiber

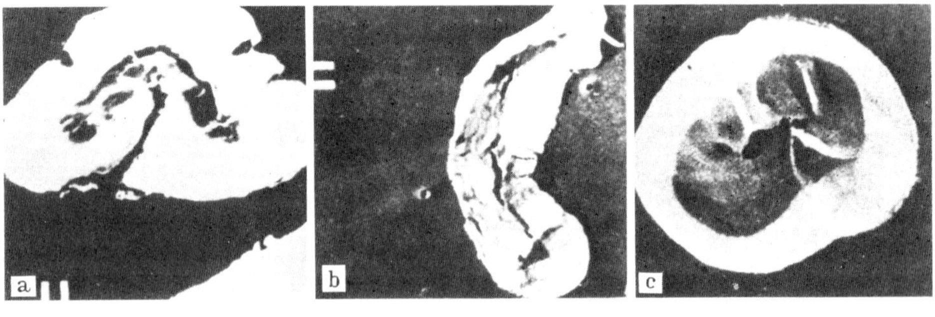

Fig. 69a–c. Poly acrylonitrile — cellulose copolymer using — radiation (consecutive steps procedure). **a** monomer solution: 32% acrylonitrile in 80% aqueous $ZnCl_2$, **b** monomer solution: 32% acrylonitrile in N,N-dimethyl formamide, **c** monomer solution: 15% acrylonitrile in 75% aqueous $ZnCl_2$. Reprinted from Ref.[72], p. 587, by courtesy of the American Chemical Society.

is rounded and the polymer is more concentrated in the outer layers of the fiber[75,83], as shown in Figure 69C.

The possible variations in the location and distribution of the grafted copolymer in the fiber when mutual irradiation grafting is employed may be realized from Figure 70a. It is apparent that the fiber cross-section is rounded and the polymer is concentrated about midway between the inner and outer layers of the fibers[29,83]. When cellulose is cyanoethylated before irradiation and aqueous zinc chloride is the solvent for the acrylonitrile monomer, the fibrous

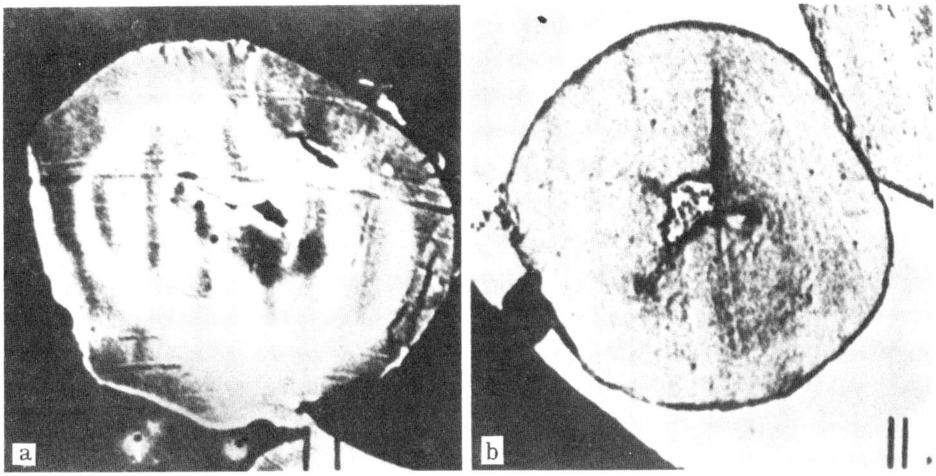

Fig. 70a, b. Poly acrylonitrile-cellulose copolymer using γ-radiation (simultaneous steps procedure). **a** monomer solution: 32% acrylonitrile in 80% aqueous $ZnCl_2$, **b** Cyanoethylated cellulose (D.S. 0.7); monomer solution same as a. Reprinted from Ref.[72], p. 586, by courtesy of American Chemical Society.

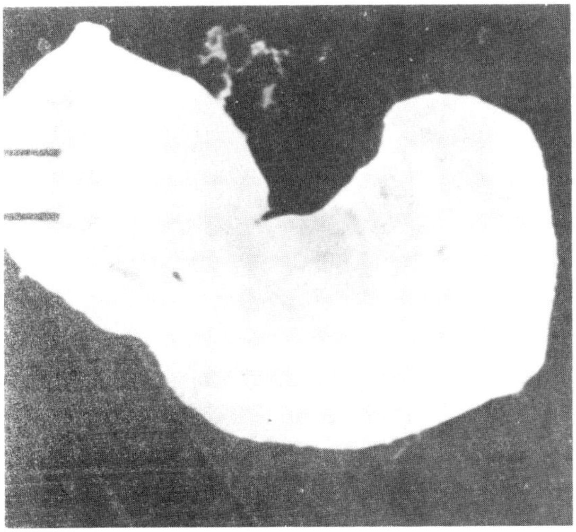

Fig. 71. Poly acrylonitrile-cellulose copolymer using Ce^{IV} ion initiated copolymerization reaction. Reprinted from Ref.[72], p. 588, by courtesy of American Chemical Society.

cross-section is rounded, and a uniform distribution of the polymer is obtained, as shown in Figure 70b. Redox systems give a uniform distribution of graft, as shown in Figure 71.

An additional point of importance is that a layering or opening of the fibrous structure occurs when the ionizing radiation (mutual) procedure is used with styrene in methanol[72,77], as shown in Figure 72a.

By the same procedure, alkyl methacrylate in methanol also causes layering of the fibrous structure. The extent of the layering effect increases as the molecular mass of the alkyl methacrylate increases[72,76]. The effect of poly(hexyl methacrylate) on the fibrous structure is shown in Figure 72b.

Goynes and Harris[88] have investigated the structure of cotton fibers grafted by radiation-induced polymerization with butyl methacrylate and acrylonitrile by light and electron microscopy. They reported that the grafting reaction occurred throughout the fiber and increased in uniformity as the degree of grafting increased. During the reaction, the fibers were swollen into a layered configuration. The graft polymerization prevented the cellulose from dissolving in the cellulose solvent, cupriethylenediamine hydroxide. Scanning electron micrographs indicated that grafting so changed the fibrillar structure of the fibers that, on abrasion, the fibers with high graft content no longer separated into fibrils but into chunks of graft polymer. Surface views of the highly grafted fibers showed that coatings of polymer had built up on the surface. Their fused appearance on abrasion indicated that the samples were thermoplastic.

Guthrie and Percival[89] have presented information relating to the thermal properties of cellulose tricarbanilate-polystyrene (CTC-PS) graft copolymers. The original substrate was regenerated cellulose film. Techniques such as dif-

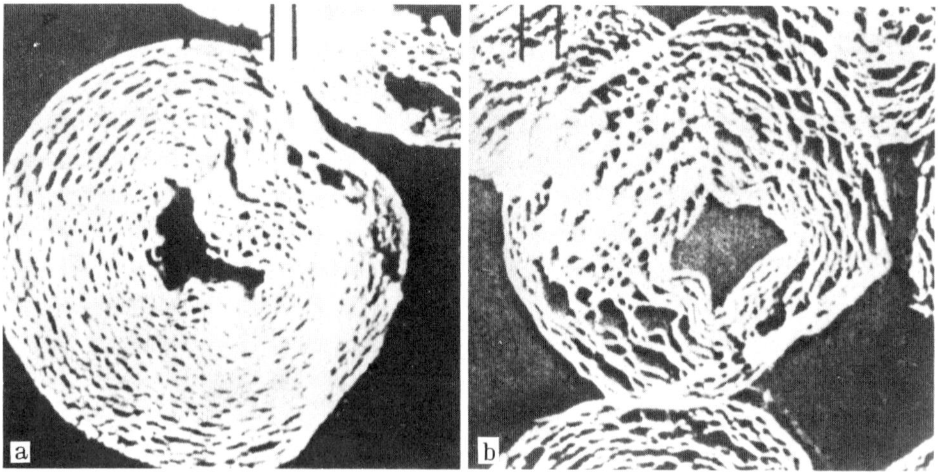

Fig. 72a, b. Layering or opening effects in cellulose copolymers using γ radiation (simultaneous steps procedure). **a** monomer solution: 30% styrene and 70% methanol, **b** monomer solution: 30% hexyl methacrylate and 70% methanol. R°printed from Ref.[72], p. 588, by courtesy of Americal Chemical Society.

ferential thermal analysis (DTA), thermogravimetric analysis (TGA), hot stage optical microscopy (HSOM), and electron microscopy (SEM) were used in the study. A corresponding study of the individual homopolymeric species (CTC and PS) was also undertaken[89]. Cellulose tricarbanilate has been studied extensively for its solution properties[90], its chemical stability[91], its ability to protect cellulosics from radiation induced damage[92], and its value as an intermediate to self-colored polymers[93].

In an earlier related study, Benson et al.[93] examined the thermal properties of cellulose tricarbanilate (A), cellulose p-nitrocarbanilate (homogeneous nitration) (B), and cellulose p-nitrocarbanilate (indirect homogeneous nitration) (C), (Fig. 73). In this study the original source of cellulose was cotton linters. The melting process with cellulose tricarbanilate is seen to commence at 200 °C. This broad endotherm points to the polydisperse nature of the sample. Nitrated products resulting from indirect homogeneous routes are clearly more stable than those produced by direct homogeneous nitration. The thermogram (A) in Figure 74, when compared with (A) in Figure 73, clearly shows the importance of the source and type of cellulose used in carbanilations when consideration is given to the thermal properties. Regenerated cellulose film, converted to the carbanilate and subsequently examined by thermal analysis, gives a much sharper melting profile than does its counterpart derived from cotton linters.

Table 58 contains information related to the thermal decomposition of irradiated CTC, unirradiated CTC, and a CTC-PS copolymer[89]. Three methods of data analysis were used, the relevant merits of each having been fully considered[94–96], and the procedure used given in Table 58. Figure 74 contains differential thermograms from a DTA examination of the three species mentioned above. It should be noted that the symbols A, B, and C, used in Figure 74 have a different significance than when they are used in Figure 73. These thermograms (Fig. 74)

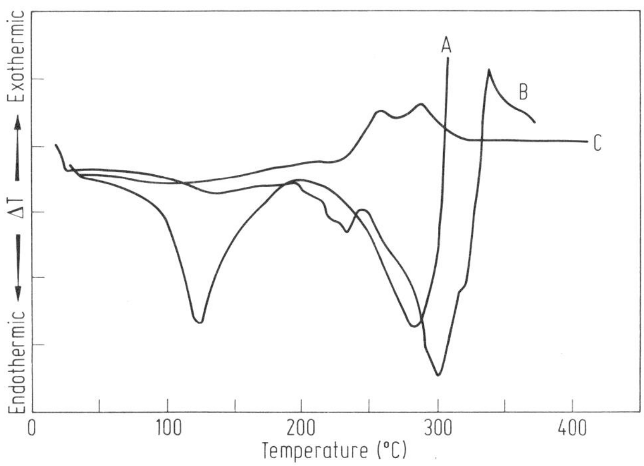

Fig. 73. Differential thermograms from the thermogravimetric analysis of *A* cellulose tricarbanilate, *B* cellulose p-nitrocarbanilate (homogeneous nitration), and *C*, cellulose p-nitrocarbanilate (indirect homogeneous nitration). ΔT = 0.2 °C/in. Reprinted from Ref.[93], p. 904, by courtesy of IPC Science and Technology Press, Ltd.

Table 58. Thermogravimetric examination of the decomposition of a cellulose tricarbanilate-polystyrene copolymer. Reprinted from Ref.[89], p. 533 by courtesy of IPC, Science & Technology Press, Ltd.

Sample and Temperature Range	Method of Analysis	Assumed Reaction Order	Activation Energy of decomposition kJ/mol	Correlation Coefficient
Non-irradiated CTC	Reich and Levi[94]	0	59.2 ± 3.2	—
160–270 °C		1	64.8 ± 4.0	—
	Reich[95,96]	0	57.2	0.96
		1	65.6	0.96
Irradiated CTC	Reich and Levi[94]	0	40.0 ± 1.6	—
160–230 °C		1	44.0 ± 1.2	—
CTC-PS	Reich and Levi[94]	0	48.0	—
Copolymer		1	49.2	—
Section (a) 230–330 °C	Reich and Levy[95,96]	0	26.4	0.86
			44.0	0.95
Section (b) 340–410 °C	Reich and Levy[95,96]	0	81.2 ± 1.2	0.92
		1	112.0 ± 1.6	0.95
Section (c) 410–480 °C	Reich and Levy[95,96]	0	no meaningful result	—
		1	166.0 ± 0.8	1.0

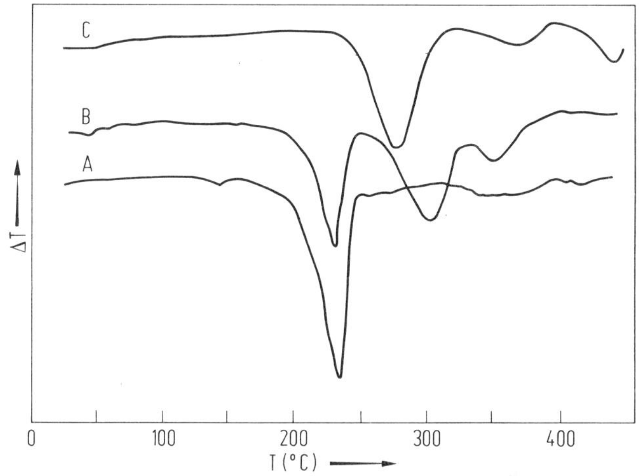

Fig. 74. Differential thermograms from the differential thermal analysis of *A* (unirradiated CTC), *B* (irradiated CTC), and *C* (the CTC-PS copolymer). Reprinted from Ref.[89], p. 533, courtesy of IPC, Science and Technology Press, Ltd.

were subsequently used in the provision of information required for the SEM examination of samples removed from the DT analyzer at temperatures associated with thermally induced changes in the sample. Again, the similarities between the irradiated CTC, (B), and the CTC-PS copolymer (C) are apparent, as is the added thermal stability of the copolymer. The primary endotherms shown in

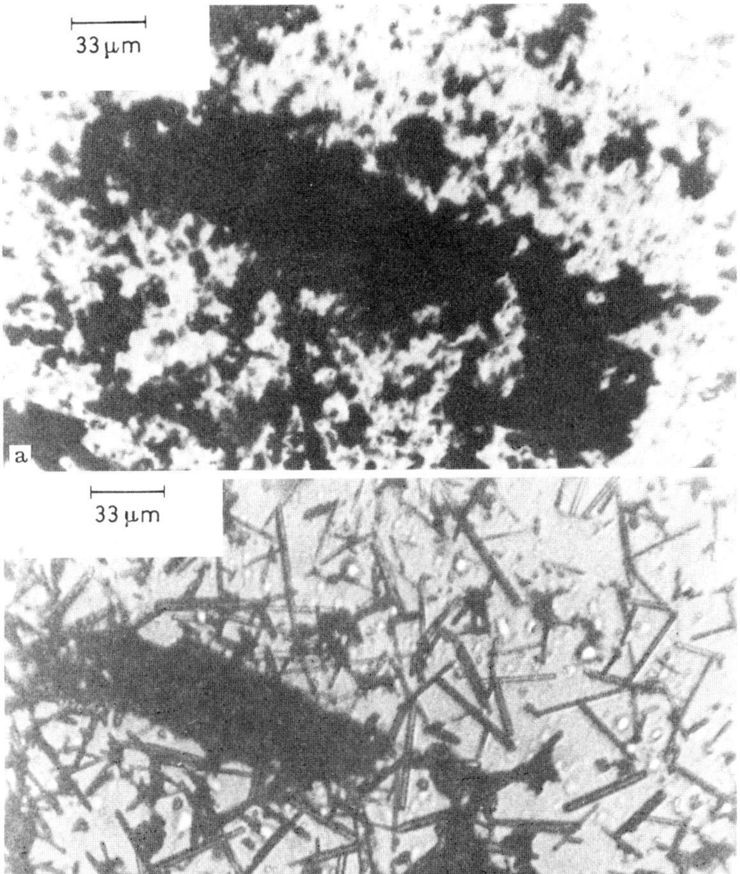

Fig. 75 a–c. Hot-stage optical microscopic examinations of unirradiated CTC: **a** unheated, **b** at 164 °C, and **c** at 187 °C. Prints 3X (X200). Reprinted from Ref.[89], p. 534, courtesy of IPC, Science and Technology Press, Ltd.

Figure 74 correspond to the initial weight losses observed in Figure 73. H.S.O.M. enables a visual appraisal of the complex decomposition processes to be made.

Figure 75 portrays the change in physical characteristics in non-irradiated CTC under the conditions given in the Figure legend. On heating above 160 °C, breakdown occurs to produce a vapor which immediately condenses on the cover-slide of the microscope as very sharply defined needle-like crystals (Fig. 75 (b)). On holding the temperature at 185°–187 °C, these crystals are seen to grow at a relatively rapid rate, supplied by vapor from the decomposing CTC (Fig. 75 (c)). An increase in the temperature beyond 203 °C leads to rapid melting of these crystals. At this point little of the original CTC remains, though what does appears to undergo a decomposition process normally associated with unsubstituted cellulose. There are indications that the continued growth of the large crystals occurs at the expense of the smaller ones.

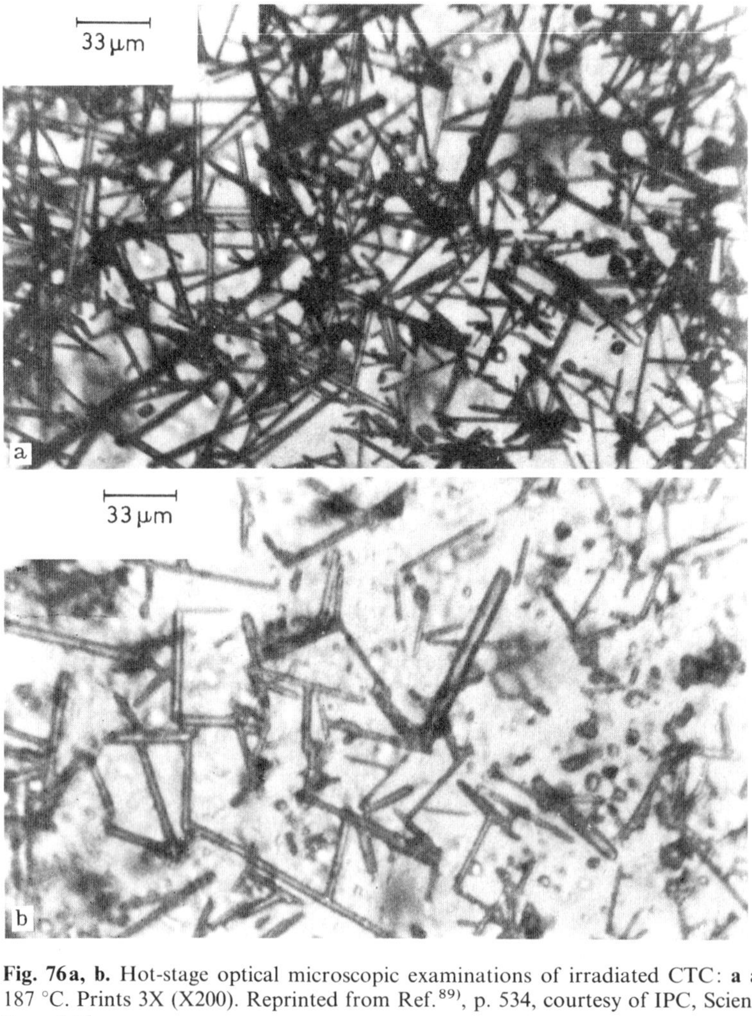

Fig. 76a, b. Hot-stage optical microscopic examinations of irradiated CTC: **a** at 165 °C and **b** at 187 °C. Prints 3X (X200). Reprinted from Ref.[89], p. 534, courtesy of IPC, Science and Technology Press, Ltd.

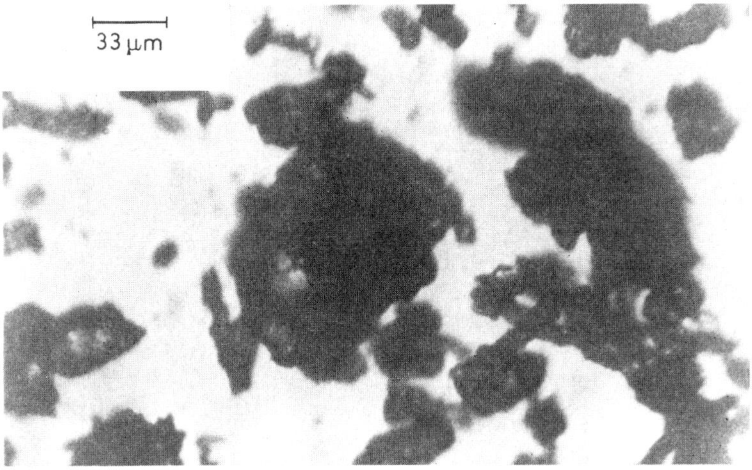

Fig. 77. Hot-stage optical microscopic examination of a CTC-PS copolymer at 420C. Prints 3X (X200). Reprinted from Ref.[89], p. 534, courtesy of IPC, Science and Technology Press, Ltd.

Similar observations were made with the irradiated CTC, though here the extent of crystallization was reduced. The product, as seen after the initial decomposition (165 °C), is shown in Figure 76 (a). Figure 76 (b) shows the sample at 187 °C when crystal growth is taking place. The second and third endotherms shown in Figure 76 (b) must be associated with changes taking place in cellulosic residues. Examination of the CTC-PS copolymers showed no vaporization and crystallization in the temperature range studied (25–425 °C). A partial melt process was seen at 236 °C, which produced a swollen ash that underwent little change until 415 °C when an ill-defined melting process occurred. This provides direct visual evidence (Fig. 77) of the effect of polystyrene grafts on the thermal stability of CTC. The composite is stable through melting and beyond the temperature normally associated with the decomposition of styrene itself. Electron micrographs were taken of each of the three samples, the irradiated and non-irradiated CTC, and the CTC-PS copolymer, after heating to the given temperature. The micrographs relate to residues, unless otherwise specified. The details are given in the accompanying legends where the term exo or endo refers to the changes noted on the D.T.A. trace at the particular temperature.

Figure 78 deals with non-irradiated CTC at various temperatures. Figure 78 (a) shows the sample before heating. Here we see the amorphous nature of the sample, each section having a semi-independent existence. After heating to 132 °C, [Fig. 78 (b)], there is evidence of phase change during which the fibers expand and lose their separate identity. Figures 78 (c) and 78 (d) show the residue after the 250 °C endotherm. In these micrographs we see the smooth nature of the highly expanded sections, which appear as a continuous melt residue.

Figure 79 displays the micrographs from the CTC-PS copolymer. The unheated sample [79 (a)] has a scaled-fiber appearance when seen on the TV monitor and shows evidence of film formation. Similar micrographs for cellulose copolymers have been recorded elsewhere[65]. Controlled heating to 200 °C produced the

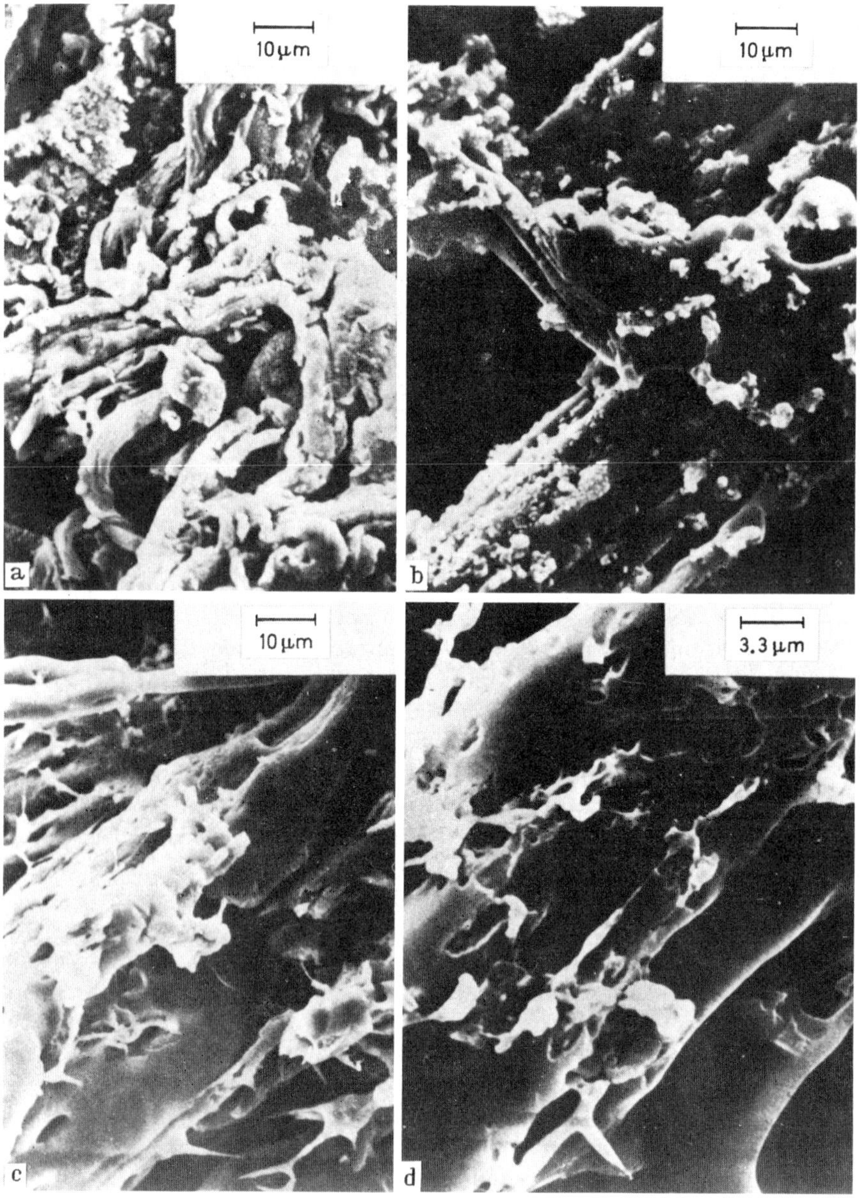

Fig. 78 a–d. Electron micrographs. Non-irradiated CTC **a** $\times 10^3$ unheated; **b** $\times 10^3$, after 132 °C endo; **c** $\times 10^3$ after 250 °C ends, and **d** $\times 3 \times 10^3$ after 250 °C endo. The endotherms referred to are those shown in Figure 74. Reprinted from Ref.[89], p. 535, by courtesy of IPC, Science and Technology Press, Ltd.

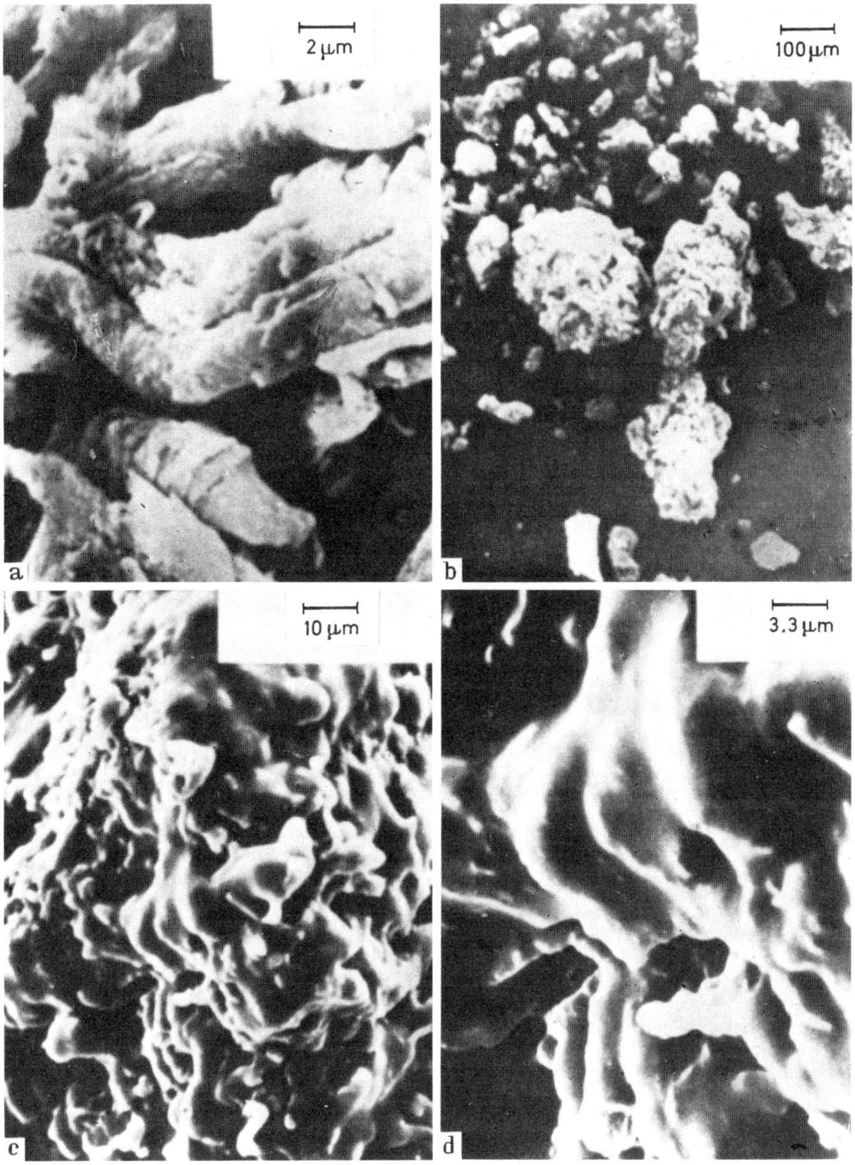

Fig. 79 a–d. Electron micrographs. The CTC-PS copolymer **a** $\times 5 \times 10^3$, unheated **b** $\times 10^2$ at 200 °C; **c** $\times 10^2$ after 325 °C, and **d** $\times 3 \times 10^3$ after 325 °C. Reprinted from Ref.[89], p. 538, by courtesy of IPC, Science and Technology Press, Ltd.

residue whose micrograph is given in Figure 79 (b). Again, there is little evidence of any flow, though the fibrous nature of the granules is still evident. Heating to higher temperatures in excess of 325 °C produces a melt and flow condition which leads to coverage of the inner pores. The initial conclusion is that we have reached the melting temperatures of the CTC-PS chains, and that the PS component serves as a surface coating for the enclosed amorphous network. These flow characteristics are in marked contrast to the irradiated CTC and the unirradiated CTC. Irradiated CTC shows no tendency for flow, and the unirradiated CTC, although flow occurs at relatively low temperatures, maintains a degree of surface order. This order is lost in the CTC-PS copolymer.

On heating to 185 °C, unirradiated CTC undergoes decomposition, though the exact mechanism is as yet uncertain. This decomposition is accompanied by a simultaneous evaporation of the volatile decomposition products, which leads to an overall endotherm in the D.T.A. trace. Analysis of T.G.A. data shows the activation energy of this decomposition to be low (58.0 kJ/mol), so that the heat produced would be small relative to the heat required for the vaporization of volatile products.

Pyrolysis of the irradiated CTC leads to an even lower value of E_a for the 160° to 230 °C region of the thermogram (42.0 kJ/mol), which indicates the lesser stability of the irradiated CTC in this temperature range. Because of difficulties in analysis, values of E_a corresponding to the second stage of the decomposition are not available. However, this change is absent from the unirradiated CTC sample and possesses a higher, though unknown, activation energy. It is possible that this second endotherm arises from decomposition of the cellulose residues or more tightly bonded carbanilate groups.

The CTC-PS possesses significantly greater overall thermal stability than either CTC variety. This is seen in the high temperature endotherm for the primary decomposition. Here we have no evidence for vaporization and subsequent crystallization. The endotherms are associated with melt or swelling processes and weight losses which are difficult to specify. The final melting process (410–480 °C) concerns polystyrene branches in the PS-CTC copolymer. The value of E_a for this process, at 166 kJ/mol, is in agreement with independently published data for polystyrene decompositions[97)], for which a radical chain reaction is assumed to be the mechanism of degradation.

An explanation for the increase in thermal stability of the PS-CTC copolymer which does not involve interference with the mechanism of the subsequent decomposition is not easily found. The possibility of negative catalysis arises, though the pathway could be indirect. Evidence from the literature suggests that the breakdown of cellulose derivatives[98)] and polystyrene[97)] is free radical in character. Hence, the polystyrene must interfere with the pathway which is open to unmodified CTC. Such interference could originate in the radiation-induced graft copolymerization reaction which preceded carbanilation. Evidence is available to indicate that Trommsdorff effects operate during the preparation of graft copolymers which have cellulose as the substrate[99)]. This leads to the production of high molecular mass branches. Further work has revealed the occurrence of intramolecular crosslinking within the graft copolymer. Such crosslinks are maintained during carbanilation and would increase the thermal stability of the carbanilated celluloses relative to unmodified samples. These crosslinks could also explain the reluctance for release of volatile products during decomposition. Upon heating to the temperature at which polystyrene normally breaks down (300 °C), little change is observed in the PS-CTC copolymer. Subsequent heating produces the examples of overall flow shown in the electron micrographs (Fig. 79 (c) and 79 (d)) which occur at a temperature in excess of 450 °C. Again, we have evidence for the presence of crosslinking during the preparation of the cellulose-polystyrene graft copolymer.

The electron micrographs shown in Figures 80 (a) to 80 (d) relate to the residues of irradiated CTC at various temperatures. The amorphous nature of the residues from the irradiated CTC suggests the potential adsorption systems or catalyst supports. However, any work along these lines will require a knowledge of the adsorption characteristics.

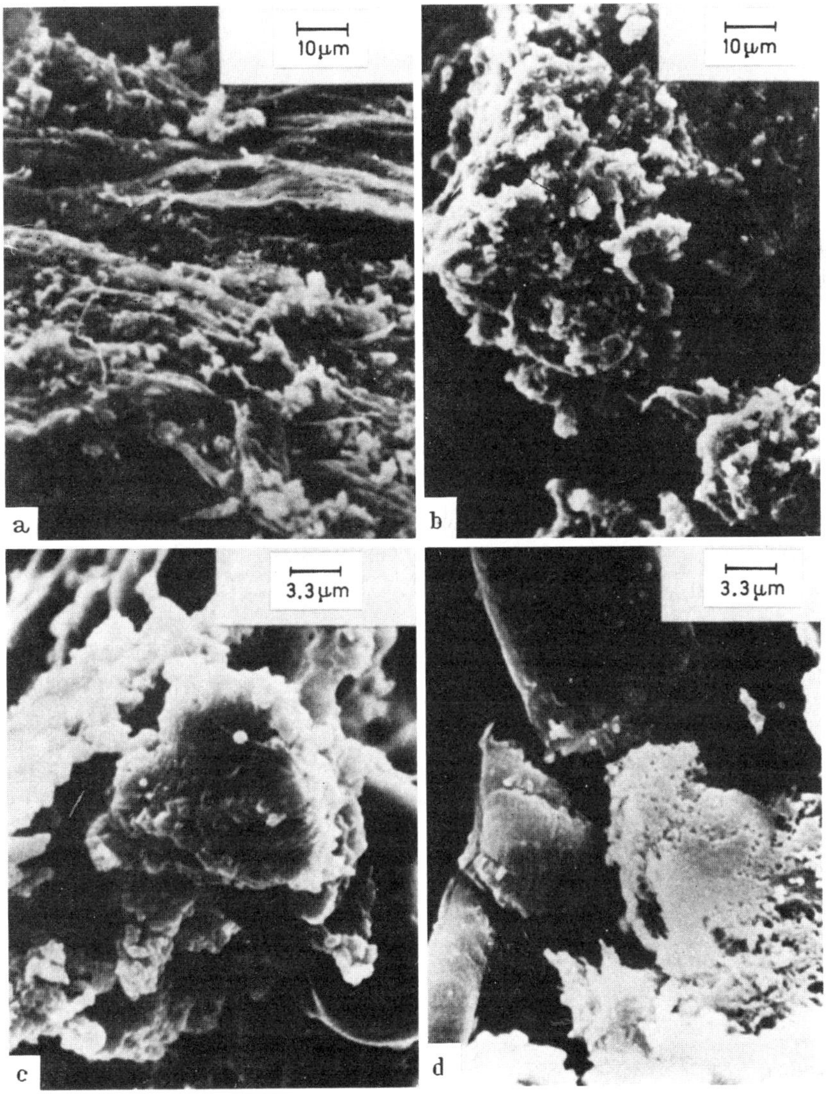

Fig. 80a–d. Electron micrographs. Irradiated CTC **a** $\times 10^3$ unheated; **b** $\times 10^3$ after 200 °C endo; **c** $\times 3 \times 10^3$ 200 °C endo; and **d** $\times 3 \times 10^3$ heated to 325 °C endotherm (Fig. 74). Reprinted from Ref.[89], p. 537, by courtesy of IPC Science and Technology Press, Ltd.

The CTC-PS copolymer micrographs, Figure (79 (c) and 79 (d)), are interesting. They show the surface to be completely covered in molten, frozen polystyrene when seen under high magnification. Two possibilities arise: first that the graft-substrate bond is cleaved with subsequent migration of molten polystyrene to the surface: The second possibility involves melting without decomposition; hence, we may be seeing frozen, melted grafts on the cellulose substrate.

It is clear that organized thermal analytical examination can provide a wealth of information concerning the structural characteristics of cellulosic copolymers. Harris et al.[102] have used electron microscopy to examine the distribution of poly(glycidyl methacrylate) within cotton fibrous structures. The polymerization was photoinduced in an unsensitized system involving aqueous media. When several layers of cotton fabrics were immersed in solutions of glycidyl methacrylate and only the surface layer was exposed to light, polymerization reactions were initiated in this layer and also in inner layers, probably by chain transfer reactions. In this instance, the chain transfer process is seen to be very efficient, in that with a six-layer sample the grafting levels are 29 %, 25 %, 22 %, 20 %, 14 %, and 11 % respectively. Electron microscopic examination[103] shows that the polymer is distributed throughout the cross-section of the fiber, though at the surface of the fibers the polymer trends to be more concentrated than within the cross section of the fibers and to encapsulate them. The photoinitiated reactivities of several vinyl and acrylic monomers in solution with cotton fabrics are compared with those from glycidyl methacrylate as follows: methyl methacrylate > glycidyl methacrylate > diacetone acrylamide > 1,3-butylene dimethacrylate > methacrylic acid > acrylonitrile > divinyl benzene.

This information may be highly significant when considering the properties and application of this type of copolymer system, especially when comparisons are made between genuine copolymers of cellulosics and blended composites[104, 105].

7.5. Fine Structure

Since many grafting reactions involving cellulose are performed in an aqueous medium, and since the intercrystalline swelling of cellulose by water occurs predominantly in the accessible or amorphous regions of cellulose, the diffusion and, therefore, grafting of monomer is probably restricted to these regions. Sakurada et al.[85] have shown that no change in the X-ray diffraction pattern was observed in poly(vinyl alcohol) grafted with large amounts of methyl methacrylate and reported that grafting probably occurred only in the amorphous regions. X-ray analysis of ramie grafted with polyacrylonitrile, as reported by Toda[15], showed no change in degree of crystallinity or orientation; however, the X-ray diffractogram showed a superimposed pattern of polyacrylonitrile.

X-ray photographs of rayon grafted with glycidyl methacrylate, shown in Figure 81, reveals that the crystallinity and orientation of the fiber does not change after graft copolymerization[69]. Obviously then, although grafting onto cellulose causes a significant change in the fibrous structure, it leaves the fine structure of cellulose practically unaltered.

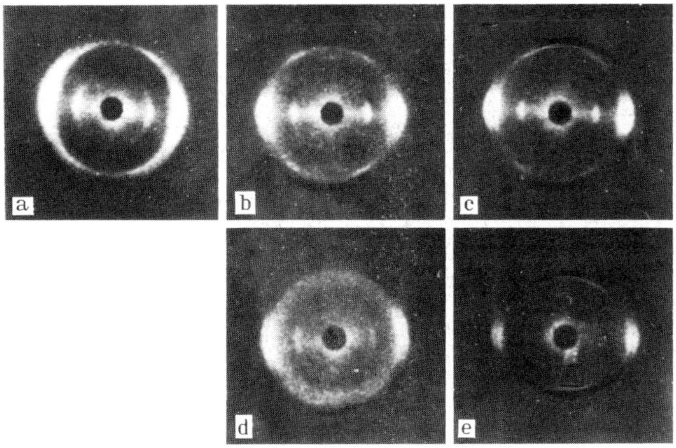

Fig. 81a–e. X-ray diagrams of Bemberg rayon grafted with GMA: **a** ungrafted; **b** 11.8% grafting; **c** 44% grafting; **d** 95% grafting; **e** 155% grafting. Reprinted from Ref.[69], p. 680, by courtesy of John Wiley & Sons, Inc.

7.6. Properties

Although cellulosic materials may be superior to synthetic polymers in certain properties, such as moisture absorption and transmission, ability to withstand high temperatures, etc., they have some outstanding drawbacks. Cellulosic materials are susceptible to rot and mildew. They are not very extensible. They swell in water and are notoriously liable to shrinkage. Cellulosic fibers acquire wrinkles readily and recover with difficulty. By grafting branches of a synthetic polymer onto the cellulosic backbone, some of the drawbacks of the cellulosic materials can be eliminated or at least reduced. It is hoped, in the textile field at least, that fibrous cellulose graft copolymers may offer advantages which will be greater than those obtained by the physical blending of synthetic fibers with cotton or regenerated cellulosic fibers.

Graft copolymerization is one of the more promising methods for the chemical modification of cellulose. Considerable changes in the properties of cellulosic materials (cloth, fibers, and films) can be achieved using this method. However, it must be pointed out that this method is not yet widely used, nor is it the only method available for the production of chemically modified celluloses with novel properties. At present the method of treating cellulosic materials by impregnating them with solutions of different reagents in order to obtain modified celluloses which are non-flammable, oil- and water-resistant, "easy-care", etc., is widely used in a number of countries.

In several procedures, in particular in preparing "easy-care" cellulosic materials, graft copolymerization cannot yet be used for mass production. The methods used at present will retain their importance for some time to come. However, in preparing most of the types of modified cellulosic materials which can be made by graft copolymerization, as well as by impregnation, the former method shows a number of significant engineering and economic advantages. The main advantages of the graft copolymerization over the impregnation method can be summarized as follows:

(1) The achieved effect is stable to repeated water treatment and in particular to washing. The new properties are maintained after 20 or even 50 washings. Using the impregnation method, such stability cannot be achieved as a rule.

(2) The grafting of synthetic polymers does not harm the hygienic properties of the product (in particular its permeability to air), in contrast to adsorbed reagents applied by impregnation as a covering film.

(3) In many cases, the effect achieved by graft copolymerization cannot be attained by other methods of chemical modification. For example, the creation of different types of cellulosic materials which exhibit ion-exchange properties or the conferring of wool-like properties to viscose staple fiber with polyacrylonitrile grafting. This was undertaken in the USSR under the name Mtilon and can be achieved only by graft copolymerization.

(4) The method is simple and more economical when inexpensive monomers are used, when produced on a large scale, and especially when the desired properties can be obtained by grafting comparatively small quantities of monomers. This advantage becomes especially significant in the chemical modification of viscose and cellulose acetate fibers, as there have been developed in the last few years methods of graft copolymerization which can be carried out directly, as one of the technological stages in the production process of these fibers.

The properties of the cellulose graft copolymers would be expected to depend on the type of monomer, the total amount of polymer grafted, the molecular mass and the molecular mass distribution of the grafted polymer, the frequency of branching, and the location and distribution of the grafted polymer in the macrostructure of the fiber. At present, there are no apparent relationships between the sites of attachment on the cellulose, the molecular mass of the grafted polymer, the fraction of cellulose molecules grafted, and the properties of the cellulose graft copolymers, particularly those prepared by free-radical initiation processes. Little or no stress can be applied to the grafted polymer to enhance its properties, because if this is done the fibrous nature of the cellulose may be lost. Consequently, the effects of "amorphous" graft polymer on the properties of cellulose are being investigated. Some of the important changes in cellulose graft copolymers are summarized below.

7.6.1. Textile Properties of Cellulose Graft Copolymers

Mechanical and Physical Properties

The mechanical properties (tensile strength and tear strength, flat and flex abrasion resistance, etc.) of cotton fabrics grafted with polyacrylonitrile, polyacrylamide, and poly(methyl methacrylate) by the Ce^{IV} ion technique were investigated by Kaizerman et al.[68)]. Fabrics grafted with polyacrylonitrile show a significant increase in tensile strength and a very considerable improvement in flat abrasion resistance for a grafting level of 27%. The mechanical behaviour of grafted cotton has also been investigated by Blouin et al.[75)]. The results obtained are shown in Table 59. Cotton yarns and fabrics were pre-irradiated in nitrogen and then treated with monomer solution (vinyl acetate, acrylonitrile, and methyl methacrylate) in 75–80% zinc chloride to ensure considerable swelling of cellulose during grafting. It appears likely that much of the grafted polymer is situated inside the cellulose fibers. An interesting feature was that fabrics grafted with poly(vinyl acetate) and polyacrylonitrile showed an exceptionally high degree of abrasion resistance, while the improvement in tear and tensile strength is not so spectacular. Chitale and Kulkarni[86] however did not observe any improvement in the abrasion resistance of cotton fabrics grafted with polystyrene, poly(methyl methacrylate), and polyacrylonitrile when the grafting reaction was carried out in the absence of a swelling agent. Thus, it appears that swelling of cellulose fabric during grafting may be an important factor contributing to the improvement in the mechanical properties of the grafted cellulose.

Prahl et al.[87] have reported that the primary result of grafting vinyl monomers onto cotton fabric is a substantial increase in elongation and a moderate decrease in tensile strength, with the net result being, in many cases, an increase in toughness. This may be seen in Table 60 and 61. These authors pointed out that for the graft range studied, there appeared to be no dependence of the tensile properties on the graft level.

Table 59. Mechanical properties of grafted cotton fabric*. Adapted from Ref.[75], p. 315, by courtesy of Textile Research Institute

	Untreated control	Irradiated control	15% vinyl acetate in 70% aq. $ZnCl_2$		15% acrylonitrile in 75% aq. $ZnCl_2$		15% methyl methacrylate in 80% aq. $ZnCl_2$	
			Treated Control	Grafted Sample	Treated Control	Grafted Sample	Treated Control	Grafted Sample
Grafted polymer %	—	—	—	23	—	25	—	24
Thread count (F) yarns/in	81	78	83	84	81	84	81	82
Breaking strength (F), lb.	44	41	42.7	46.6	42.5	50.5	42.2	43.2
Elongation at break (F), %	24.5	24.3	31.8	32.5	22.9	25.9	23.9	24.8
Tear resistance (F), g	540	520	540	567	553	507	553	480
Flex abrasion resistance (F) cycles to rupture	228	232	291	951	347	397	274	239
Flat abrasion resistance, cycles	91	86	233	1521	159	551	92	133
Conditioned wrinkle recovery (W + F), degrees	174	181	95	62	188	154	193	204

* Purified cotton print cloth (86 × 80) irradiated dry and in nitrogen to a dosage of 10^6 R, and after irradiation treated with 15% monomer in 70, 75, or 80% aq. $ZnCl_2$ at about 23 °C for sufficient time to yield the grafted polymer indicated.

Table 60. Tensile properties of grafted 80 × 80 cloth (1 replicate). Adapted from Ref.[87], p. 247, by courtesy of Textile Research Institute

Graft	Graft	Warp tensile*	
Type	%	lb.	% Elongation
None	0.0	16.0	4.9
n-butyl acrylate	8.0	9.0	4.5
n-Butyl acrylate	13.5	9.0	5.5
Ethyl acrylate	27.5	17.0	11.9
Methyl acrylate	47.6	10.0	7.6
Methyl methacrylate	23.7	9.0	20.0
Methyl methacrylete/styrene**	7.5	17.0	20.0
Ethyl methacrylate	22.7	10.5	7.8
n-Buthyl methacrylate	17.0	9.5	7.4
Acrylonitrile	11.8	9.5	23.8
Styrene	5.1	10.5	13.2

* $^{1}/_{2} \times 3$ in sample; cross-head rate 0.05 in/min.
** 1:1 in liquid phase; unknown proportion in graft polymer.

Table 61. Tensile properties of ethyl acrylate grafted 80 × 80 cloth. Reprinted from Ref.[87], p. 247, by courtesy of Textile Research Institute

% Graft	Warp tensile		Filling tensile	
	lb.	% Elong.	lb.	% Elong.
0.0	45.0	7.2	37.7	13.4
11.5	48.2	13.5	32.6	22.2
18.3	44.5	14.4	34.3	21.1

Iwakura et al.[69] have examined the strength, elongation, and elastic recovery of cellulose fibers grafted with glycidyl methacrylate. The results obtained are given in Table 62. It is apparent that denier and recovery increased with the increase of the grafting content, but that strength and elongation decreased.

Demint et al.[78] have studied the physico-mechanical properties of cyanoethylated cotton grafted with polyacrylonitrile. Some of their results are shown in Table 63 and 64. It is obvious that with increasing interaction of acrylonitrile in the grafted polymer, the elongation at break and yarn number increases; the average stiffness, the breaking strength, and the toughness also increase slightly. The elastic recovery properties of a product having a graft yield of about 38% are rather interesting (Table 64). At elongations of 25% or less of their elongation at break, decreases in permanent set and increases in delayed recovery of the yarns were observed. There is apparently no effect on the immediate recovery of the yarns. These changes indicate that a stabilizing of molecular bonds occurs (increased delayed recovery and decreased permanent set), but that the bonds involved are apparently low in energy (little or no change in immediate recovery).

Table 62. Mechanical properties of GMA-grafted viscose rayon. Reprinted from Ref.[69], p. 678, by courtesy of John Wiley & Sons, Inc.

No.	Content of grafting (*wt* %)	Denier	Strength g/den.		Elongation %		Knot strength g./den.	Knot elongation %
			dry	wet	dry	wet		
0	0.0	2.00	2.84	1.80	21.0	23.2	1.72	12.8
1	16.2	2.31	2.23	1.48	19.2	22.1	1.16	9.3
2	23.3	2.40	2.15	1.50	19.0	22.6	1.20	10.2
3	44.4	2.74	2.01	1.39	18.8	24.6	1.23	10.3
4	58.9	3.10	1.81	1.25	21.4	25.1	1.09	10.5

Table 63. Effect of radiation dosage on the interaction of acrylonitrile with cyanoethylated cotton and on its properties.* Reprinted from Ref.[78], p. 823, by courtesy of Textile Research Institute

Property of Yarn**	Dosage, megaroentgen							
	0.0***	0.0	0.2	0.4	0.6	0.8	1.0	1.5
Interacted monomer content, %	0.0	0.0	1	4	27	38	38	36
Parts monomer to 100 parts cellulose	0.0	0.0	1	5	45	74	74	69
Breaking strength, lb.	10.1	6.2	7.1	6.7	6.3	7.4	8.2	7.0
Elongation at break, %	13	12	15	17	24	26	28	24
Yarn number, tex	241	298	312	345	513	656	649	622
Breaking toughness, g/tex	0.79	0.40	0.61	0.53	0.48	0.55	0.64	0.49
Average stiffness, g/tex	141	76	67	53	24	20	21	21

* Treating solution: 32 parts of acrylonitrile in 60 parts of aqueous 80% $ZnCl_2$. Degree of substitution of cotton, 0.7.

** Instron properties.

*** Purified, untreated, unirradiated yarn.

Cellulose graft copolymers, in the form of yarns, have been found to be thermoplastic[77,104,105]. This is probably due to lowered molecular cohesion. For instance, polystyrene-cellulose graft copolymers show a pronounced drop in resiliency index and stiffness when the temperature is increased to 100 °C. Polyacrylonitrile-cellulose graft copolymers and polyacrylonitrile-cyanoethylated cellulose graft copolymers show drops in their resiliency index and stiffness at 80°–100 °C. The polystyrene-cellulose graft copolymers also show decreased rates of changes in the resiliency index from 40 °C to 60 °C and from 100° to 150 °C, indicating phase transition. Summations of the relative energies, expended over a temperature range from 21° to 200 °C to elongate the fibrous polymer to 1.5%, are as follows: cyanoethylated cellulose (D.S.: 0.7) > acrylonitrile-cellulose > cellulose > styrene-cellulose > acrylonitrile-cyanoethylated cellulose. Upon relaxation, the recovery of the energy expended is about 40% in all cases, except for acrylonitrile-cyanoethylated cellulose, where about 66% recovery is observed.

Table 64. Effect of radiation-induced interaction of acrylonitrile with cyanoethylated cotton on its elastic recovery properties.* Adapted from Ref.[78], p. 823, by courtesy of Textile Research Institute

Elongation of cotton⁺, % of elongation at break	Recovery⁺, % of actual elongation		
	Purified	Purified cyanoethylated	Cyanoethylated irradiated
Immediate recovery			
5	30	34	34
10	23	29	24
25	9	10	9
50	6	8	5
75	6	7	4
Delayed recovery			
5	49	49	63
10	58	55	71
25	66	69	74
50	46	46	48
75	29	30	29
Permanent set			
5	21	17	3
10	19	16	5
25	25	21	17
50	48	46	47
75	65	63	67

* Degree of substitution, 0.7. Radiation dosage, 1 megaroentgen. See Table 82 for other properties of cottons.
⁺ Instron properties.

Reports[106, 107] on the properties of cotton cloth grafted with polystyrene using the post-irradiation technique reveal the following points. (1) When styrene is grafted by this technique with a dose of $2\text{–}3 \times 10^6$ r, the loss of strength of the cotton fabric is slight. (2) When the percent graft is higher than 40%, the grafted cotton fabric shows excellent thermoplasticity. (3) The tearing strength of cotton fabric and the flex life of cotton fiber are somewhat lowered by grafting, but the flex-abrasion resistance is improved.

Measurements of strength, elongation, and modulus of single cellulosic fibers of various apparent percent grafts have been carried out to elucidate the relation between percent graft and mechanical properties[106, 108, 109]. In Table 65 are shown some of the results obtained when the grafting of styrene onto high-tenacity rayon by mutual irradiation with γ-rays was carried out in the absence of air. The term apparent percent graft means the weight increase of fiber after benzene extraction as a percentage of the initial weight of fiber. With increasing radiation doses, the apparent graft increases up to 189%.

The data in Table 65 make it evident that the conditioned strength of fibers remains almost unaltered despite the large weight increase due to grafting. The wet strength decreases very slightly below a dose of 1 Mrad, and there is a

Table 65. Grafting of styrene onto high tenacity rayon by mutual irradiation with gamma rays. Adapted from Ref.[108], p. 14, by courtesy of John Wiley & Sons, Inc.

Dose 10^6 Rad.	App. graft %	Titer (den.)	Conditioned			Wet		
			Strenght (g)	Tenacity (g/den.)	Elong. %	Strength (g)	Tenacity (g/den.)	Elong. %
0.0	0.0	1.54	5.27	3.42	25.6	3.70	2.40	35.3
0.1	1.6	1.61	5.31	3.30	25.5	3.54	2.20	36.3
0.2	46.1	2.35	5.52	2.35	28.6	3.30	1.43	54.3
0.4	77.4	3.40	5.84	1.72	44.0	3.50	1.03	57.7
0.8	95.0	3.57	5.54	1.55	40.6	3.30	0.93	62.7
1.0	153.0	4.99	5.20	0.96	53.0	3.50	0.72	87.0
1.6	189.0	6.15	5.38	0.88	41.2	2.77	0.45	72.6

Monomer mixture: styrene: methanol: water = 20:72:8; Dose rate: 2.0×10^5 rad/hr; room temperature.

sharp drop at 1.6 Mrad, possibly because of radiation damage. Both the conditioned and wet elongation at break increase with increasing percent graft. Although the strength of a single fiber does not change much with grafting, the tenacity, i.e., the strength per unit thickness, decreases with increasing percent graft. The simplest explanation for this is that the grafted polystyrene is only a dead load as far as the tensile properties are concerned. However, the change of load at the yield point with percent graft is not in agreement with this explanation. It is possible that though cellulosic fibers become hydrophobic through the grafting of styrene, the tendency to absorb water, calculated on the cellulose content basis, increases by increasing the percent graft. This may be the reason for the rather low wet tenacity of the styrene-grafted cellulosic fibers.

Butadiene, butadiene-styrene, or butadiene-acrylonitrile was also grafted onto cotton and rayon[109]. It was found that elasticity, thermoplasticity, and adhesion to rubber markedly improved.

Mares and Arthur have prepared cotton-cellulose graft copolymers and terpolymers by the irradiation techniques. Monomers copolymerized with irradiated cellulose from solutions of methanol-water were: methacrylic acid (MMA), methacrylonitrile (MAN), and hexafluoroisopropyl acrylate (6FIPA). Cotton cellulose, immersed in solutions of N-vinyl carbazole (NVC) in 1,2-dichloroethane, was irradiated to form a copolymer. Terpolymers prepared were: cellulose-poly(MAA)-poly(MAN), cellulose-poly(MAA)-poly(6FIPA), and cellulose-poly(MAN)-poly(6FIPA). The strength and abrasion resistance of the cotton fabric copolymers and terpolymers are shown in Table 66 and 67.

From Tables 66 and 67, it is clear that the breaking strengths of the copolymer fabrics are relatively unchanged when compared with those of untreated fabric. However, for cellulose-poly(MAN) and cellulose-poly(MAA) copolymer fabrics, small increases in breaking strengths were recorded with increased polymer add-on. There were usually increases in the flat abrasion resistance and decreases in the tearing strength of copolymer fabrics with increased polymer add-on. Cellulose-poly(6FIPA) copolymer fabrics exhibited an increase in flex abrasion,

Table 66. Preparation and properties of cotton cellulose graft copolymers as print cloth fabrics. Reprinted from Ref.[111], p. 351, by courtesy of John Wiley & Sons, Inc.

Solution composition			Reaction time (hr)	Polymer add-on (%)	Strength		Abrasion resistance	
Monomer (vol.-%)	Solvent (vol.-%)	Water (vol.-%)			Breaking (lb)	Tearing (g)	Flex	Flat
0.0	0.0	0.0	0.0	0.0	37	770	1.0	1.0
Cotton cellulose-poly(MAN) fabric								
11	71[a]	18	0.58	8	34	640	0.76	0.91
20	56[a]	24	0.75	18	38	630	0.82	1.10
25	60[a]	15	1.00	19	38	580	0.49	1.10
20	55[b]	25	1.50	23	37	550	—	1.3
20	55[b]	25	3.00	31	40	550	—	1.9
20	55[b]	25	4.00	35	43	520	—	2.2
Cotton cellulose-poly(MAA) fabric								
6	75[a]	19	0.50	19	35	540	0.45	1.1
5	76[a]	19	1.00	23	36	560	0.58	1.2
7	74[a]	19	1.00	28	38	540	0.44	1.5
10	72[a]	18	1.00	38	39	520	0.45	1.7
Cotton cellulose-poly(6FIPA) fabric								
7	74[a]	19	8.00	58	36	560	1.7	—

[a] Methanol;
[b] Dimethylsulfoxide.

Table 67. Preparation and properties of cotton cellulose-poly(NVC) print cloth fabrics. Reprinted from Ref.[111], p. 351, by courtesy of John Wiley & Sons, Inc.

Solution composition		Reaction time (hr.)	Polymer add-on	Breaking strength (lb)	Abrasion resistance	
NVC (wt-%)	Solvent[a] (wt-%)				Flex	Flat
0.0	0.0	0.0	0.0	37.0	1.0	1.0
10.0	90.0	1.0	5.0	35.0	0.32	1.5
13.0	87.0	1.74	10.4	37.0	0.35	1.8

[a] 1,2-dichloroethane.

as compared with that of untreated fabrics while all the other copolymer fabrics exhibited decreases in flex abrasion.

The breaking strengths of the grafted block copolymer fabrics (Table 68) slightly increased, the tearing strengths decreased, and the resistance to flat abrasion increased after grafting. At low molar ratios of 6FIPA/MAA in the block copolymer, the flex abrasion resistance of the fabrics was less than that of untreated fabrics. In fabrics containing block copolymers with molar ratios of 6FIPA/MAA of about 1 or greater, the flex abrasion resistance of the fabrics was greater than that of untreated fabric. As the proportion of poly(6FIPA)

Table 68. Preparation and properties of block copolymers of cotton cellulose and poly(6FIPA)(MAA) sateen fabrics[a]. Reprinted from Ref.[111], p. 354, by courtesy of John Wiley & Sons, Inc.

Solution composition (vol.-%)				Reaction time	Terpolymer		Strength		Abrasion Resistance	
FIPA	MAA	MeOH	Water	hr.	Add-on	6FIPA/MAA[b]	Breaking (lb)	Tearing (g)	Flex	Flat
0.0	0.0	0.0	0.0	0.0	0.0	—	99	6300	1.0	1.0
9	8	66	17	1	39	0.02	104	4450	0.58	2.3
10	7	66	17	1	34	0.17	103	4750	0.78	1.8
9	2	71	18	2	31	0.93	94	4700	1.1	1.6
10	2	70	18	2	36	1.00	97	4900	1.2	1.9
10	1	71	18	2	40	3.80	101	4350	1.3	2.3

[a] Irradiated cellulose immersed in solutions containing mixture of monomers to form block copolymers.
[b] Molar ratio estimated from results of separate copolymerization reactions of each monomer with irradiated cellulose under the conditions indicated for the terpolymer.

increased in the cellulose terpolymer, the effective glass transition temperature of the terpolymer probably decreased, resulting in increased flex abrasion resistance of the fabric.

The presence of nitrile and/or methyl groups on a monomer is reported to increase the glass transition temperature of its homopolymer[111] relative to its unsubstituted analogue. Formation of block copolymers from two monomers, one with a high and one with a low glass transition temperature (with a molar ratio of about one or greater), gives a terpolymer with cellulose which has increased flexibility (c.f. Table 68). However, formation of a terpolymer with cellulose by first graft copolymerizing one of the monomer with cellulose and then graft copolymerizing the second monomer gives products with increased

Table 69. Preparation and properties of copolymers of cotton cellulose and poly(6FIPA) plus poly(MAN) sateen fabrics[a]. Reprinted from Ref.[111], p. 355, by courtesy of John Wiley & Sons, Inc.

Solution composition (vol.-%)				Reaction time	Polymer		Strength		Abrasion Resistance	
6FIPA	MAN	MeOH	Water	hr.	Add-on	6FIPA/MAN[b]	Breaking (lb)	Tearing (g)	Flex	Flat
0	(0)	0 (0)	0 (0)	0	0	—	88	6300	1.0	1.0
9		73	18	1.17	12	—	—	—	—	—
	(21)	(63)	(16)	(0.83)	(9)	0.38	89	5470	1.5	1.6
9		73	18	2.17	23	—	—	—	—	—
	(13)	(70)	(17)	(1.25)	(7)	1.00	92	5540	2.0	1.5
9		73	18	3.00	17	—	—	—	—	—
	(13)	(70)	(17)	(1.25)	(4)	1.25	85	5400	2.1	1.6

[a] MAN copolymerized with irradiated cellulose; then 6FIPA copolymerized with irradiated poly(MAN)-cellulose copolymer in first and second terpolymer; order of preparation reversed for third terpolymer.
[b] Molar ratio calculated from polymer add-on.

Table 70. Abrasion resistance[a] of styrene-methyl methacrylate mixed monomer grafted cotton fabric. Reprinted from Ref.[113], p. 345, by courtesy of John Wiley & Sons, Inc.

Percentage weight increase due to total grafting	Tensile strength in lbs./in.			
	Before abrading		After abrading	
	Warp	Filling	Warp	Filling
Untreated control	58.0	30.0	8.0	4.0
5.2	54.6	43.2	25.0	27.0
10.6	55.4	45.6	51.5	42.6
16.3	57.2	49.5	55.0	48.0
25.8	65.0	50.2	62.0	49.5

[a] ASTM D 1175-64T, oscillatory cylinder method, Load: 2 lbs., tension: 2 lbs., 250 continuous cycles.

flexibility at a lower molar ratio of monomer than in the previous case (cf. Table 69). Similarly, the large carbazole group in NVC probably accounts for the high glass transition temperature of its homopolymer and the low flexibility of its cellulose copolymer[111].

The retention of and, in some cases, the increases in, the breaking strength of the cellulose copolymers and terpolymers containing MAA and MAN, with increasing polymer add-on, could, when compared with untreated cellulose, result from increased hydrogen bonding in the products. Methyl and carboxyl groups, if present in the grafted copolymers, would increase the number of hydrogen bonds in the products[110]. This is substantiated by the fact that cellulose-poly (MAA) copolymers exhibit increased aqueous soil-release properties, as compared with those of untreated cotton[112].

DasGupta[113] has reported a substantial increase in tensile strength for cotton fabric grafted with styrene-methyl methacrylate and acrylonitrile-methyl methacrylate mixtures by radiation grafting. Values for resistance to abrasion are also shown in Table 70. From the results it is seen that there is very little change in tensile strength, even after subjecting the grafted samples to a load and tension of 2 pounds for 250 continuous cycles. The superiority over ungrafted controls is very significant.

The same author also reported that radiation grafting of monomers, such as styrene and styrene-methacrylic acid mixtures on to cotton-polyester blends (35:65), makes them superior to the ungrafted controls. An increase of 45% in tensile strengths was observed with 15% grafting[113].

Easy Care Properties

Since grafting is a monofunctional reaction between cellulose and the monomer, no significant improvement in the easy care properties of the grafted fabrics would be expected. However, by using monomers such as methylolated acrylamide and divinyl benzene, it is possible to crosslink the grafted cotton fabrics either in a two-step or a one-step process. Kamogawa and Sekiya[114] have grafted acrylamide to cotton fabrics by the Ce^{IV} ion initiation method. The amido group was then methylolated by reacting the grafted cotton fabrics with formaldehyde at pH 11. This group was then reacted with adjacent cellulose molecules through substances such as ethylene urea, melamine, etc. Kamogawa and Sekiya found that the fabrics grafted and crosslinked by this method showed better retention of tensile and tear strength than did ungrafted fabrics crosslinked with similar crosslinking agents.

This method of grafting and crosslinking was slightly modified by Gardon[115)], who used methylolated acrylamide as the starting monomer. The double bonds, of the vinyl group and the methylol group were reacted with cellulose in two different steps. The methylol was reacted through an acid catalyst step, while the vinyl double bond was reacted under alkaline conditions. This method makes it possible to follow the course of the grafting and crosslinking reactions independently by determining the vinyl double bond content and the methylene group content of the reacted sample. A quantitative measure of the number of crosslinks introduced in the sample is given by the simple derivation:

[crosslinks = methylene content − double bonds.].

Table 71. Two-stage cross-linking of phosphorylated cotton (redox polymerization followed by acid-catalyzed postcuring)[a]. Adapted from Ref.[119)], p. 626, by courtesy of Textile Research Institute.

Run No.	Monomers: Compounds	Monomers: Concentration	Mol Ratio	Monomer applied (°C/min)	Oven precure (°C/min)	Add-on %	Wrinkle recovery after post-cure
1	MA/MAA	4.3/1.0	5.0	75/3	none	11.5	266
2	EA/MAA	9.6/2.0	4.8	70/1	100/3.5	4.4[b]	286
3	EA/MAA	9.6/2.0	4.8	70/6	100/3.5	45.7	
4	EA/MAA	3.4/0.4	8.5	70/1	100/3	7.6[b]	279
5	AN/MAA	7.1/2.7	5.0	25/30	none	9.1	273
6	AN/MAA[c]	7.1/2.7	5.0	25/30	none	1.2	231
7	EA/AN/MAA	3.2/0.1/0.06	52/3.2/1	70/4	none	19	
8	S/MAA[d]	25/9.5	2.6	70/10	none	39	
9	S/MAA[d]	17/1.1	15	70/4	none	6.2	248
10	MA/GA[f]	10/3.0	5.0	70/1	70/4	5.8[b]	225[e]
11	GA[f]	5.0	—	25/1	70/4	10.0	281
12	GA[f]	10.0	—	25/8	none	6.5	239
	(untreated)						179
	(phosphorylated)						195

[a] Glycidyl acrylate (GA) or N-methylolacrylamide (MAA) copolymerized in fabric (0.3% P) with Fe^{II}-H_2O_2 initiation. Fe^{II} exchange was at 25 °C in runs 1, 5, 6, 11 and 12 but at 70 °C in all others. MAA post cure was with 3% catalyst H-7 at 170°/3 min. GA post cure was with 1.33% $Zn(BF_4)_2$ at 145°/5 min. No catalyst was applied for precuring step.
[b] Samples not washed prior to post cure. [c] On unphosphorylated fabric. [d] in water emulsion containing 14% sodium lauryl sulfate. [e] post cured with 0.5% $Zn(BF_4)_2$. [f] monomer solution contained 0.24% H_2O_2, 0.018% KOH, 0.14 KCl, 0.02% tetraethylenepentamine, 0.3% Triton X-100 and water.

Similarly, Walsh et al.[116] have shown that N-methylol acrylamide can be polymerized in cotton fabrics using either persulfate initiation of high energy radiation, and subsequently heat-cured with an acid catalyst to impart increased wrinkle recovery. Similar properties were obtained by Kamel et al. in their work on acrylamidomethylated cottons, as seen through reactions with selected sulfur compounds[117,118].

Obviously, then, grafting of olefinic monomers containing epoxy, N-methylol, or other cellulose-reactive groups onto cotton fabrics is of particular interest, since the grafts are potentially capable of subsequent crosslinking reactions with cellulose. Starting with phosphorylated cotton, Gallagher[119] has shown that redox polymerization of bifunctional monomers, either alone or in admixture with other vinyl or acrylic monomers, is very rapid, especially if monomer application is followed by a precure at 70–100 °C, prior to introduction of the acid catalyst and final curing. As shown in Table 71, polymerization times of 3–10 minutes produce a polymer add-on of 4–45 %. The number of epoxy or N-methylol groups in the grafted polymer and, therefore, the number of potential cross-linking groups are controllable by adjusting the ratio of bifunctional monomer to monofunctional monomer in the copolymerization bath. Wrinkle recovery angles of 265–285° (W + F) are readily obtainable.

Kulkarni et al.[19] used a different method for grafting and crosslinking cotton fabrics. Fabrics grafted with acrylonitrile were treated with a mild NaOH solution (2.5 %) to convert some of the —CN groups of the polymer to the amide groups. These groups were then methylolated by reaction with formaldehyde at a pH of 9–9.5. The methylolated groups were then reacted with cellulose by an acid curing step in the presence of a catalyst, e.g., $MgCl_2 \cdot 6\,H_2O$, at a high temperature (150 °C); thus:

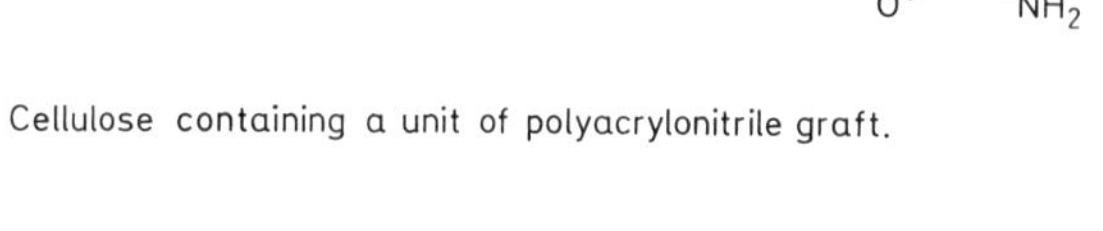

Cellulose containing a unit of polyacrylonitrile graft.

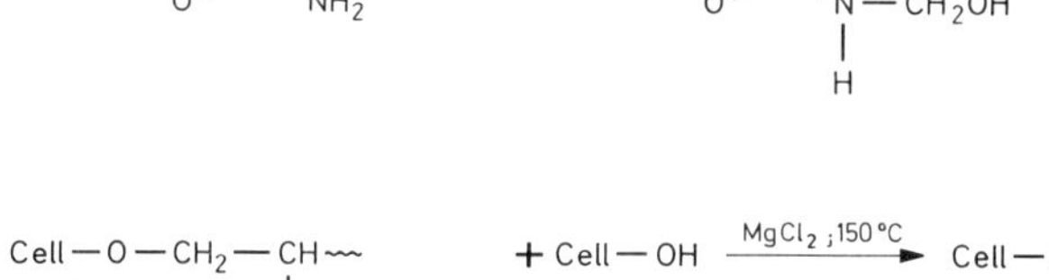

There is also a possibility that the polymer can form crosslinks within itself, if two or more —CO—NH—CH_2OH groups of two modified polyacrylonitrile chains (PAN) interact, as shown below:

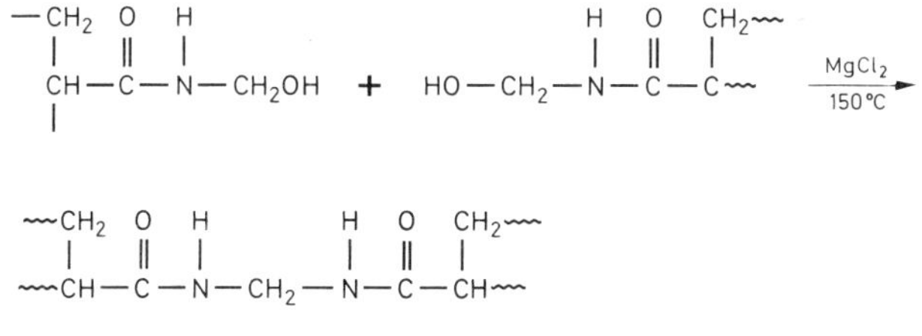

Figure 82 schematically shows some possible formations of cellulose-PAN crosslinks after modification with formaldehyde and acid cure. Fabrics with appreciable crease recovery can be obtained.

Although grafting of monofunctional monomers on cotton fabrics generally produces no significant improvement in wrinkle recovery[120], DasGupta[121] did note an improvement after the radiation grafting of styrene. Negishi et al.[122] have also shown that the dry and wet wrinkle recovery of cotton fabrics can be improved considerably by grafting certain acrylic polymers, such as methyl acrylate, n-propyl acrylate, etc. Prahl et al.[87] observed the same when n-butyl acrylate was grafted onto cotton fabrics by either the vapor phase or liquid phase technique (see Table 72). Furthermore, Avny and Schwenker[17] observed a substantial improvement in the wrinkle recovery of cotton fabrics grafted with preformed polystyrene (see Table 73). The wrinkle recovery of fabrics grafted with bifunctional polystyrene, i.e., a polymer with a reactive terminal group at each end of the chain, is shown to be significantly higher, both wet and dry, than for the control or for cotton grafted with monofunctional polystyrene, i.e., only one reactive terminal group. In addition, the wet wrinkle recovery value

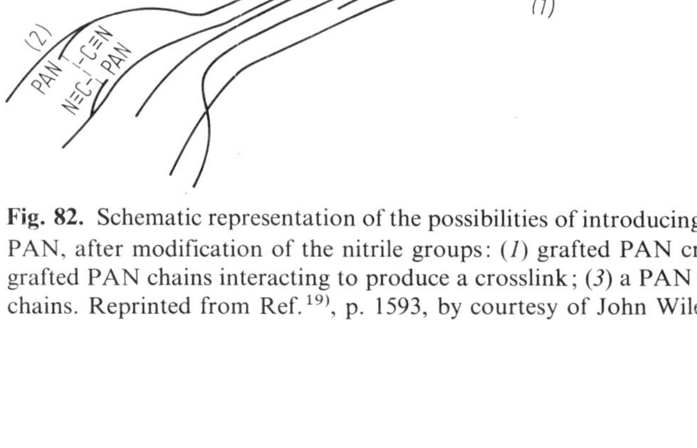

Fig. 82. Schematic representation of the possibilities of introducing crosslinks into fabrics containing PAN, after modification of the nitrile groups: (*1*) grafted PAN crosslinking with cellulose; (*2*) two grafted PAN chains interacting to produce a crosslink; (*3*) a PAN chain bridging adjoining cellulose chains. Reprinted from Ref.[19], p. 1593, by courtesy of John Wiley & Sons, Inc.

Table 72. Effect of vapor-phase n-butyl acrylate grafts on monsanto wrinkle recovery of print cloth. Adapted from Ref.[87], p. 250, by courtesy of Textile Research Institute

n-Butyl Acrylate Graft %	Monsanto Wrinkle Recovery (W + F)
0	173
5–10	192
15–20	202
20–25	212
25	214

Table 73. Wrinkle recovery of polystyrene grafted cotton fabrics adapted from Ref.[17], p. 821, by courtesy of Textile Research Institute

Sample	$\overline{DP}_n$ preformed polystyrene	Grafting D.S.	Polystyrene content %	Wrinkle Recovery (W + F) degrees	
				Dry	Wet
Control	—	—	—	104	131
Monofunctional Graft	9.3	30.4×10^{-3}	15.4	120	150
Bifunctional Graft	107.3	0.7×10^{-3}	4.3	146	236

of 236 (W + F) greatly exceeds the dry wrinkle recovery value, as would be the case for the conventional crosslinking of cotton in the wet, swollen state. It is believed that the increase in wet crease-recovery, in the case of grafting with bifunctional preformed polystyrene, is due to crosslinking in the cotton fabric. It is noteworthy, in this connection, that the improvement in wrinkle recovery of cellulosic fabrics grafted with monofunctional monomers, either alone or in admixture, has been attributed to the restriction of segmental movements of molecules containing both graft and cellulose chains, by entanglement between the graft polymer chains and between these and the cellulose chains.

The relationship between the glass transition temperature of grafted polymers and the wrinkle recovery of cotton cellulose copolymer fabric has recently been examined[111]. Table 74 shows that the wrinkle recovery of the copolymer fabric decreases as the glass transition temperature of the grafted polymer increases.

An additional point of interest is that the grafting of styrene, either alone or together with methacrylic acid, onto cotton-polyester blends (35:65) yields a product with a wrinkle recovery angle (warp + weft) of about 300. This value remains practically unaltered even after the fabric is subjected to 5 washings (each equivalent to 50 ordinary commercial ones)[113].

As indicated earlier, graft copolymerization onto cellulose leads to an increase in the fabric toughness. This increase is reflected[87, 104, 105, 123] in the commonly occurring improved abrasion resistance of the grafted fabric, and also in the wrinkle-resistance of the fabrics derived therefrom by typical crosslinking treat-

Table 74. Effect of glass transition temperature of polymer on the wrinkle recovery angle of cotton cellulose graft copolymers as print cloth fabrics. Adapted from Ref.[111], p. 356, by courtesy of John Wiley & Sons, Inc.

Grafted polymer	Homopolymer glass transition temperature (°K)	Polymer add-on %	Wrinkle recovery angle (conditioned), degree (W + F)
Control	—	0.0	190
Hexafluoroisopropyl acrylate (6FIPA)	288	6.0	220
Methacrylic acid (MAA)	403	7.0	215
N-vinyl carbazole (NVC)	481	5.0	160

Dry fabric was first irradiated at about 25 °C in nitrogen to a dosage of 5.2×10^{19} eV/g, then immersed in solutions of monomer (about 250 cm^3/40 g of fabric) at 25 °C. In the case of NVC, to initiate graft polymerization, fabric was immersed in a 1,2-dichloroethane solution of NVC and irradiated.

Table 75. Abrasion resistance of liquid-phase grafted, crosslinked twill (average of 6 replicates). Adapted from Ref.[87], p. 249, by courtesy of Textile Research Institute

Graft type	% Graft	% resin	MWR⁺	Abrasion cycles
None	0	54	307	37
methyl methacrylate	5.0	4.5	307	42
methyl methacrylate	4.1	4.3	309	44
methyl methacrylate	2.8	4.1	310	56
methyl methacrylate/ethyl acrylate*	1.2	4.9	308	40
methyl methacrylate/ethyl acrylate*	2.2	4.9	311	41
methyl methacrylate/ethyl acrylate*	3.9	5.2	304	58
ethyl acrylate	3.7	4.2	312	44
ethyl acrylate	2.5	4.0	309	50
ethyl acrylate	4.6	4,3	314	60
methyl acrylate/ethyl acrylate*	2.0	4.7	317	53
methyl acrylate/ethyl acrylate*⁺	3.6	4.6	316	54
methyl acrylate/ethyl acrylate*	4.8	4.9	306	65
n-butyl methacrylate	7.5	5.6	308	50
n-butyl methacrylate	1.6	5.8	311	54
n-butyl methacrylate	0.7	5.2	306	66
dimethylaminoethyl methacrylate	1.4	5.1	307	53
dimethylaminoethyl methacrylate	2.4	5.5	299	80
dimethylaminoethyl methacrylate	4.6	6.7	298	82

* 1:1 in liquid phase; unknown proportion in graft polymer.
⁺ MWR, Monsanto Wrinkle Recovery.

ments. The best grafted systems for this purpose, notably those based on diethylaminoethyl acrylate or ethyl acrylate, do not exhibit the high loss in abrasion resistance incurred by conventional crosslinking treatments. Table 75 shows a marked tendency toward improvement in abrasion resistance by grafting one of a series of acrylate monomers, even at low grafting levels. This could mean that there is an optimum graft level for each monomer and that the graft levels of Table 75 tend to fall on either one or the other side of this optimum[87].

Dimensional Stability and Affinity for Water

Kaizerman et al.[68] have shown that the dimensional stability of polyacrylonitrile-grafted viscose rayon is considerably improved when the graft add-on is 20–30 %. These data are shown in Table 76. However, a considerable decrease in tear strength in the dry state accompanies the improvement in dimensional stability. Negishi et al.[122] have reported that the water repellency of cotton fabrics can be considerably improved by grafting acrylate monomers such as n-butyl acrylate, iso-butylacrylate, etc. This may be seen in Table 77. The water repellency was measured by the sinking method with the use of specimens of 1 cm^2 and water containing 1 % wetting agent at 60 °C. The water repellency of the cotton fabrics grafted with n-butyl acrylate seems to be better than that of the styrene-grafted fabric.

Table 76. Properties of viscose rayon challis containing polyacrylonitrile*. Adapted from Ref.[68], p. 140, by courtesy of Textile Research Institute

	Polymer content (based on untreated fabric)			
	13.4%	22.4%	29.2%	Untreated
Tensile strength (W + F) lb. Dry	91	86	82	90
Wet	48	45	41	59
Tear strength (Elmendorf) Dry	6.1	4.6	3.1	10.2
Wet	2.7	2.5	1.6	2.6
Abrasion (Accelerator 2000 rpm.)				
% weight loss after 2 mins	0.4	2.5	4.1	0.4
4 mins	2.6	6.4	8.0	2.0
% Tensile strength retained after soil				
burial 2 weeks	26	66	94	0.0
6 weeks	0	12	26	0.0
% shrinkage on laundering				
(AATCC Test 40–52)	5.0	1.5	0.0	7.5

* Grafted by the vapor saturation technique using cerium^{IV} ions as initiator.

Table 77. Water repellency at about 30% graft-on of various acrylates or styrene. Adapted from Ref.[122], p. 2248, by courtesy of John Wiley & Sons, Inc.

Sample	Extent of grafting, %	Time required to sink, sec.
Untreated cotton	0	1
Cotton grafted with		
n-butyl acrylate	29.4	720
isobutyl acrylate	29.6	480
methyl methacrylate	28.9	120
styrene	30.7	240

Achwal[124)] has reviewed the various methods of introducing halogen, as well as silicon-contaning groups, into cellulose molecules. Emphasis has been placed upon the use of graft copolymerization for introducing such groups. It is claimed that water-repellent fabrics can be obtained by grafting cellulose with vinylidene chloride or fluorine-containing monomers.

The water repellency of cotton cellulose-poly[hexafluoroisopropylacrylate, (6FIPA)] fabrics is shown in Table 78. Copolymerization followed by heat treatment gives copolymer fabrics with useful spray ratings[111)]. The water spray resistance of cotton, grafted with a mixture of styrene and methyl methacrylate, is shown in Table 79. Samples with over 20% grafting showed 100% resistance; ungrafted controls did not show any resistance[113)].

The effects of grafting and wetting agents on water imbibition, calculated on the basis of the cellulose component, is shown in Figure 83 and Tabele 80.

Table 78. Water repellency of cotton cellulose-poly (6FIPA) fabrics. Adapted from Ref.[111)], p. 353, by courtesy of John Wiley & Sons, Inc.

Fabric	Polymer add-on	Spray rating	
	(%)	Not cured	Cured[a]
Print cloth	0	0	—
	27	—	80
	33	—	90
	57	70	—
Sateen	0	0	—
	11	50	—
	22	50	—
	25	—	90
	28	50	—
	30	—	80

[a] Heated at 160 °C for 5 mins.

Table 79. Water spray resistance of cotton grafted with a mixture of styrene and methyl methacrylate. Reprinted from Ref.[113)], p. 340, by courtesy of John Wiley & Sons, Inc.

Percentage weight increase due to total graft	Percentage polystyrene grafted	Percentage poly(methyl-methacrylate) grafted	Resistance to water spray (%)
Ungrafted control	—	—	0
10.6	2.1	8.5	80
16.3	4.6	11.7	80
25.8	7.4	18.4	100
26.9	7.7	18.7	100
29.9	9.7	20.2	100
46.6	11.3	35.3	100

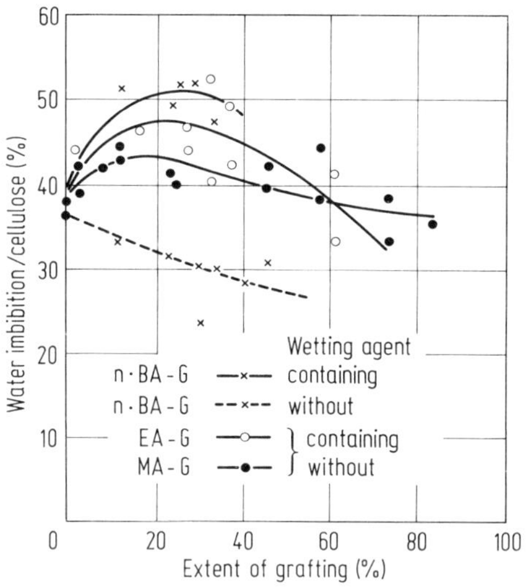

Fig. 83. Effect of graft-on and wetting agent on water imbibition per cellulose component for cotton grafted with methyl, ethyl, and n-butyl acrylates. Reprinted from Ref.[122], p. 2239, by courtesy of John Wiley & Sons, Inc.

Table 80. Water imbibition at about 30% graft-on of various acrylates or styrene. Adapted from Ref.[122], p. 2240, by courtesy of John Wiley & Sons, Inc.

Sample	Extent of grafting %	Water imbibition/cellulose	
		Water	0.5% of nonionic wetting agent
Untreated cotton	0	37.5	37.5
Cotton grafted with			
Methyl acrylate	36.0	40.7	42.8
Ethyl acrylate	34.0	38.8	41.5
n-Propyl acrylate	35.0	30.4	46.3
n-Butyl acrylate	34.0	27.5	54.3
Untreated cotton	0	—	36.6
Cotton grafted with			
n-Butyl acrylate	29.4	—	51.5
Isobutyl acrylate	29.6	—	50.2
Methyl methacrylate	28.9	—	50.6
Styrene	30.7	—	42.5

In the cotton grafted with n-propyl or n-butyl acrylate, the water imbibition is significantly increased by the presence of 0.5% wetting agent, while the cotton grafted with methyl or ethyl acrylate is almost unaffected. A slight maximum of water imbibition appears in most cases, suggesting competition between the opening of the fiber structure and the imparting of a hydrophobic property by grafting (see Fig. 83). Furthermore, the water imbibition seems to increase with the increasing length of the acrylate ester group, indicating that the bulky groups of the ester play a role in opening the fiber structure. The opened structure of the cotton fiber is stabilized in the polymer, especially by bulky ester groups, and consequently the capacity for water imbibition increases. This also may perhaps permit the cellulose molecules to attain greater flexibility in the wet state than untreated cotton. Therefore when the entanglement-crosslinking of softer graft polymers is present to an appropriate extent, swelling elastomeric behavior would be expected, thereby improving the wet wrinkle recovery[122].

Moisture Regain

Vinyl graft copolymerization onto cellulose usually reduces the moisture regain due to the generally hydrophobic nature of the grafted branches. With cotton-polystyrene graft copolymers, prepared by irradiation of cotton yarn in a solution of styrene and methanol[77], the moisture regain ranges from 7.5% for the untreated, unirradiated yarn to 2.7% for a sample with a graft of 165%. However, if the moisture regains are calculated on the basis of the cellulose content only, rather than on the basis of the entire copolymer, the regain values vary from 7.5 to 7.2%. This indicates that the grafted polystyrene branches do not really affect the moisture regain of the cellulosic backbone, probably because the branches are very long but only few in number. Similar cotton-polystyrene graft copolymers were also examined by Kesting and Stannett[125]. They found that the moisture regain based only on the cellulose content actually increased with the extent of grafting. The moisture regain ranged from 10.7 for untreated cotton cellulose to 19.2% for a sample with a graft of 159%. In this case, therefore, the accessibility of the cellulose increases with grafting, presumably owing to the opening of the fiber structure by the polystyrene branches.

Hebeish and Mehta[49, 54] have examined the moisture regains of native cotton, hydrocellulose, viscose, ramie, and cottons swollen with sodium hydroxide (NaOH-Cotton), ethylenediamine (EDA-cotton), and zinc chloride ($ZnCl_2$-cotton) before and after grafting with acrylonitrile by the Ce^{IV} ion method. The results are shown in Table 81.

The data clearly indicate that the regains of the grafted materials progressively decrease as the extent of grafting increases. The theoretical moisture regain values for the treated fibers were also calculated on the assumption that the cellulose and polymer do not interact and that each contributes to the sorption in direct proportion to its concentration in the treated fibers. The theoretical values are shown in parentheses in Table 81. It is clear that the observed regain is less than the theoretical regain. This suggests that fewer sorbing sites are available on the grafted celluloses, due to the production of a hydrophobic environment by graft polymerization which reduces the effectiveness of the cellulose hydroxyls as sorbing sites. It is interesting to note that on calculating the moisture regain on the cellulose content only, that is, after subtracting the regain due to polyacrylonitrile graft from that of the entire graft copolymer, it is found that the moisture regain is less than that of the original untreated celluloses in all cases.

Solubility

The solubility in cupriethylenediamine of the cellulose in certain cotton-derived products, as reported by Demint et al.[78], is shown in Table 82. The solubilities of

Table 81. Moisture regain of cellulose-polyacrylonitrile grafts. Reprinted from Ref.[49], p. 191

Native			NaOH-Cotton			EDA-Cotton			$ZnCl_2$-Cotton		
Graft %	M_0 %	M_c %	Graft %	M_0 %	M_c %	Graft %	M_0 %	M_c %	Graft %	M_0 %	M_c %
Untreated	6.32	6.32	Untreated	9.05	9.05	Untreated	8.02	8.02	Untreated	6.90	6.90
12.63	5.68	6.11	11.92	8.06	8.76	16.40	7.04	7.83	11.80	6.21	6.68
	(5.86)			(8.32)			(7.20)			(6.40)	
14.85	5.59	6.10	19.72	7.57	8.63	21.25	6.75		15.60	6.01	6.60
	(5.79)			(7.92)			(7.00)			(6.27)	
21.14	5.34	6.01	23.54	7.24	8.42	26.95	6.37	7.49	18.23	5.83	6.49
	(5.60)			(7.74)			(6.78)			(6.18)	
34.12	4.78	6.67	33.52	6.73	8.25	29.46	6.23	7.42	21.16	5.58	6.30
	(5.25)			(7.33)			(6.69)			(6.08)	
38.52	4.61	5.54	38.62	6.33	7.93	32.15	6.07	7.32	28.53	5.39	6.30
	(5.17)			(7.14)			(6.60)			(5.86)	
40.00	4.55	5.49	42.85	6.05	7.70	36.85	5.80	7.13	39.69	4.87	5.94
	(5.14)			(6.99)			(6.45)			(5.56)	
43.64	4.46	5.44	52.84	5.40	7.09	48.38	5.41	6.96	47.00	4.57	5.69
	(5.07)			(6.68)			(6.12)			(5.40)	

Hydrocellulose			Viscose			Ramie		
Graft %	M_0 %	M_c %	Graft %	M_0 %	M_c %	Graft %	M_0 %	M_c %
Untreated	5.21	5.21	Untreated	13.12	13.12	Untreated	8.20	8.20
11.07	4.75	5.03	13.6	11.59	12.87	8.00	7.61	8.05
	(4.91)			(11.51)			(7.75)	
14.76	4.64	5.00	17.27	11.22	12.78	13.47	7.28	7.97
	(4.82)			(11.51)			(7.49)	
17.56	4.56	4.96	20.19	10.92	12.68	23.05	6.73	7.78
	(4.76)			(11.29)			(7.08)	
20.11	4.52	4.98	29.78	9.87	12.16	24.06	6.72	7.80
	(4.71)			(11.61)			(7.04)	
28.21	4.24	4.82	33.62	9.53	12.00	30.12	6.36	7.61
	(4.55)			(10.37)			(6.81)	
32.50	4.12	4.74	39.92	8.89	11.56	35.51	6.06	7.42
	(4.47)			(10.00)			(6.63)	
37.61	3.96	4.62	41.86	8.69	11.40	40.80	5.76	7.21
	(4.39)			(9.90)			(6.46)	

Footnote:
(1) M_0: Observed moisture regain based on weight of the entire copolymer.
(2) M_c: Moisture regain calculated on the basis of cellulose content only.
(3) Values in brackets present theoretical values of moisture regain (moisture regain of cellulose + moisture regain of PAN).
(4) Moisture regain of PAN = 2.20%.

Table 82. Solubility of cellulose and some derivatives.* Adapted from Ref.[78], p. 824, by courtesy of Textile Research Institute

Cotton Product⁺	Solubility of cellulose %	Nitrogen %	
		Total	Insoluble fraction
A	99	—	—
B	48	6.6	10.2
C	97	4.8	—
D	0	13.0	11.5

* Solvent: 0.5, M cupriethylenediamine
⁺ A: purified cotton,
B: purified cotton, irradiated 0.8 megaroentgen, containing 25% polymer.
C: cynaoethylated cotton, (DS 0.67).
D: cyanoethylated cotton (DS 0.67), irradiated 0.8 megaroentgen, containing 38% interacted monomer.

purified and cyanoethylated cottons, products A and C, were almost 100%. The solubility of the cellulose in product B, which contained radiation-polymerized acrylonitrile, was about 50%. From the magnitude of the increase in the nitrogen content of the insoluble fraction of product B over that of total product B, it could be concluded that only the cellulose dissolved.

The solubility of cellulose in product C, which contained cyanoethylated cotton, radiation-grafted with acrylonitrile, was zero. The relatively constant nitrogen content indicates little or no change in the product.

Resistance to Rot and Biological Degradation

Under warm and humid conditions, cotton fabrics rot, become unsightly, and lose strength and, hence, their usefulness at a rapid rate. Remarkable resistance to micro-organisms can be achieved by various ways, such as: (1) impregnation of cellulosic materials by different compounds containing salts of different metals and copper in particular; (2) partial esterification or etherification of cellulose (acetylation as well as cyanoethylation has been employed for this purpose); and (3) synthesis of graft copolymers of cellulose using various copolymerants.

Graft copolymerization of vinyl monomers onto cellulose may be regarded as one of the best methods of obtaining material whose resistance to rot and biological degradation is maintained even after prolonged exposure to a moist atmosphere, rain, or the usual washing treatments. In Table 83 are shown the tensile strengths of grafted samples and controls after being subjected to the action of microorganisms for periods up to 2 weeks. It is clear that the

Table 83. Loss in breaking strengths of grafted samples after exposure to fungus growths. Reprinted from Ref.[113)], p. 343, by courtesy of John Wiley ≦ Sons, Inc.

Percentage weight increase due to total grafting	Percentage loss in tensile strength after exposure to micro-organisms
Ungrafted original cotton fabric	100% (fabric rotted)
Fabric coated with polystyrene-poly(methyl methacrylate) mixture	100% (fabric rotted)
19.2% styrene + methyl methacrylate (PS:PMM as 3.5:1)	1%
20.2% acrylonitrile + methyl methacrylate (PAN:PMM as 2:1)	4%
19.5% Acrylic acid + styrene (PS:PAA as 2:1)	3%

Table 84. Effect of grafted polyacrylonitrile on rot resistance of cotton*. Adapted from Ref.[68)], p. 139, by courtesy of Textile Research Institute

Polymer concentration %	Percent Tensile Strength retained after**	
	2 weeks' burial	6 weeks' burial
9.6	86	45
14.2	100	72
18.6	99	100
21.7	100	100

* Grafted by the vapor saturation technique using the Ce^{IV} ion system in initiation.

** Based on untreated fabric.

grafted samples remain practically unaltered, while untreated cotton becomes useless[113)].

Various studies have reported on the improvement of rot resistance in cotton fabrics grafted with polyacrylonitrile[68, 70, 74, 75, 82, 126)]. The effect of grafting polyacrylonitrile onto cotton fabrics to test their susceptibility to rot resistance is shown in Table 84. Apparently, the grafting of polyacrylonitrile onto cotton imparts significant resistance to bacterial attack[68)]. It has been claimed that good rotproofing can be obtained with as little as 3% to 10% polyacrylonitrile reacted[126)].

Resistance to light

Cellulosic materials with a high resistance to light are essential for application in regions where intense exposure to ultraviolet rays is unavoidable, particularly for covering of packages, outdoor wear, curtains, etc. Considerable improvement in resistance to light can be achieved by many techniques. One of the methods of chemical modification involves introduction of nitrile groups into the cellulose molecule by cyanoethylation or graft copolymerization. This method appears at present to be of considerable importance as the introduction of nitrile groups, as already pointed out, simultaneously improves the resistance of the cellulosic materials to the action of microorganisms. For example, on weathering cotton fabric in air for four months, the strength is reduced by 43% and the elongation at break by 42%. Similar fabrics having 16% of polyacrylonitrile graft under the same conditions lose only 15% in strength and their elongation at break decreases by only 30%[127].

The grafting of cellulose with methyl vinyl ketone ($CH_2{=}CHCOCH_3$) introduces keto groups into the cellulose molecule. The presence of such groups increases the resistance of cellulose to ultraviolet radiation but simultaneously decreases the thermal stability[128].

Resistance to fire

The demand for non-inflammable or heat-stable fibrous materials is increasing as new fields of use in coverings and furnishings of aeroplanes, submarines, ships, workers overalls, etc., are developed. The methods generally used to impart a non-flammable character involve impregnation with various chemicals. Most of these methods, however, have certain drawbacks, the most outstanding of which are: (1) considerable decrease in strength, (2) change in appearance of the fabric and in some cases decrease in permeability, and (3) low stability of finish to wet treatments or washing.

Modified cellulosic materials, prepared by grafting, do not usually exhibit such drawbacks. A non-flammable character can be imparted to cellulosic materials by using halogen- and phosphorus-containing monomers[129, 130]. Grafting polyvinylidene chloride to cellulose gives a non-burning product when the amount of chlorine introduced is 30–32%, which corresponds to about 40–45% polymer on the weight of the cellulose. Such a high polymer add-on, however, simultaneously decreases the tearing strength and increases the stiffness of the fabric.

The amount of phosphorus required to impart fire resistance to cellulose is about 2–4%. However, this amount depends on the nature of the phosphorus-containing agent used for finishing. Thus, in the case of agents which form a C-P bond, 2–5% phosphorus is quite adequate. In contrast, double this value is needed for compounds forming the C-O-P bond. It is also important to note that the amount of phosphorus necessary to impart fire resistance also increases with the size of the molecule or organoradical in the phospho-organic acid used for a given type of bond (C-P or C-O-P). The presence of a benzene ring lowers the effectiveness of the organo-phosphorus compounds.

Ion-Exchange Properties

The preparation of modified cellulosic materials capable of forming stable complexes with metals, especially the heavy noble metals, considerably widens the field of application of cellulose derivatives having ion-exchange properties. These modified cellulose preparations were initially used for the manufacture of special quality papers for use in chromatography. Such materials, containing sulfografts or amino groups, are at present only being prepared on a small scale. In recent years, however, many new methods for obtaining ion-exchange materials have been devised and are being tested on a pilot plant scale. Results of these investigations indicate that graft polymerization methods seem to be potentially suitable for the industrial scale. Polymer reactions can also be carried out after the introduction of suitable functional groups to impart ion-exchange or complex-forming properties.

The different routes for imparting ion-exchange properties to cellulosic materials, which are of interest from the practical point of view, result in materials which can be classified into the following groups[127]:

1. Materials with cation-exchange properties:
   (a) mild cation activity; derivatives containing carboxyl groups.
   (b) strong cation-exchange capacity; derivatives containing sulfo-groups or residues of phosphorus-containing acids.
2. Materials with anion-exchange properties:
   (a) mild anion activity; derivatives containing methyl vinyl pyridine.
   (b) strong anion activity; derivatives containing quaternary ammonium groups.
3. Materials with electron exchange properties. These contain sulfydryl or thioamide groups.
4. Materials whose ion-exchange properties are acquired through complexing. These contain residues of hydroxamic acid or other complex-forming acids.

Ion-exchange properties are shown by low molecular mass hydroxamic acids as well as poly(acryl-hydroxamic acids) formed by treatment of polyacrylates with hydroxylamine[131]. However, the use of modified celluloses in the form of fibers or fabrics having ion-exchange properties has advantages over the usual synthetic-polymer ion-exchange materials which are used as granules. This is because, due to the enormous surface area of fibers, the rate of ion-exchange reactions, of sorption or desorption, is considerably higher. The higher hydrophilicity of modified cellulosic materials, as compared to synthetic polymers and their considerable swelling in aqueous media, tends to increase the rate of diffusion processes. Furthermore, the use of fabrics as ion-exchange materials lends itself to continuous processing. It must be noted that the hydrophilicity of a graft copolymer having ion-exchange properties usually decreases proportionally upon increasing the ion-exchange capacity. Hence, for high rates of ion-exchange, an optimum degree of grafting must be reached to get a balance between hydrophilicity and ion-exchange capacity.

The introduction of hydrophilic groups, (e.g., carboxyl groups) into the cellulose molecule influences the swelling capacity in water and may lead to a considerable decrease in wet strength and, in some cases, to the loss of physical form of the material. These factors also show the advisability of having the optimum, and not the maximum, quantity of synthetic polymer grafted to cellulose.

Although cellulosic ion-exchange materials can be used under many conditions, their use is more limited than that of synthetic materials due to the sensitivity of the glucosidic bond to acidic media. The successful use of any ion-exchange material requires that is should be useable over a large number of sorption-desorption cycles without loss of ion-exchange capacity or material properties. Hence, celluloses of high tensile strength and high molecular mass must be used as starting materials.

In some cases, in particular for modified cellulose having electron-exchange properties, the process of desorption presents severe difficulties. For example, in the case of sorption of heavy and noble metals, it is advisable and economic to use ion-exchange fibers for temporary use in one operation only. After reaching the limits of dynamic activity for adsorption of ions, the fiber is burned and the attached metal recovered. In such cases, cheap fibers, such as viscose, or pulp can be used as raw material.

Ion-exchange, using fiber-based supports, can be carried out in a series of connected columns loose-packed with fibers, through which the liquid is passed. When dynamic equilibrium is reached, the column is disconnected and subjected to desorption. The main difficulties with fibers as ion-exchange materials arise from the gradual tight-packing of fibers in the columns causing increased resistance to the flow of liquids.

Synthesis of graft copolymers of cellulose with acrylic and methacrylic acids is an important method of obtaining modified celluloses having ion-exchange properties. The grafting reaction can be carried out without producing appreciable amounts of homopolymer by introducing diazo groups into cellulose derivatives or by using the Fe^{II}-H_2O_2 redox system. Products containing 15–20% carboxyl groups and having a high ion-exchange capacity can be obtained. However, such copolymers are soluble in dilute sodium hydroxide which limits their application[128)].

Ion-exchange properties can also be achieved by introducing hydroxamic acid residues into graft polymers obtained from cellulose and polyacrylonitrile or poly(methyl acrylate). Graft copolymers of cellulose with poly(methyl metacrylate) can be treated with a 10–20% alcoholic solution of hydroxylamine at 20°–60 °C for 4–10 hours to introduce the hydroxamic acid residues. Thus,

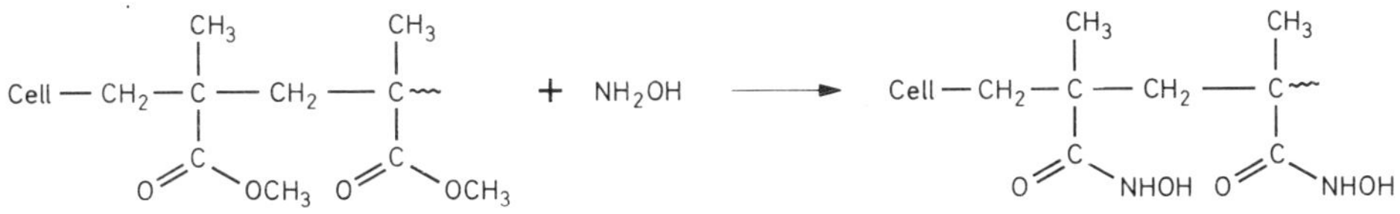

The product will bind 8–10% iron based on the weight of the graft copolymer[132)].

A graft copolymer of cellulose of similar type can also be obtained by the analogous reaction of hydroxylamine, carried out at 100 °C, with cellulose-polyacrylonitrile graft copolymers. Under these conditions an almost quantitative conversion of nitrile groups into amidoxime then hydroxamic acid takes place as follows:

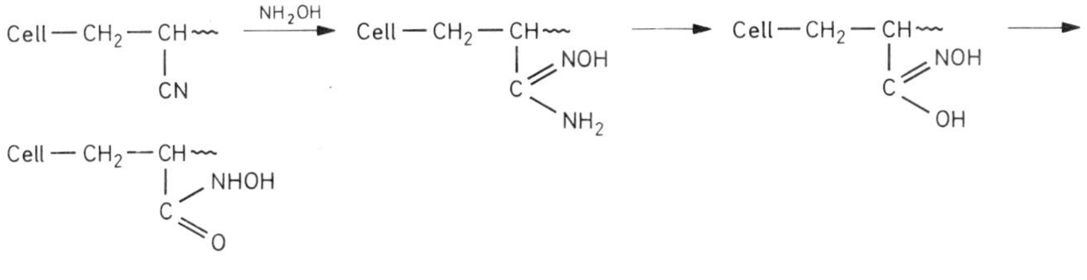

Synthesis of graft polymers of cellulose with dimethyl vinylethynyl carbinol results in a modified cellulose containing triple bonds, since the monomer has a triple bond in addition to a double bond. The grafting reaction is carried out by using a 3–4% aqueous solution of the monomer or 20% aqueous emulsions. The reaction scheme may be represented as follows[124]:

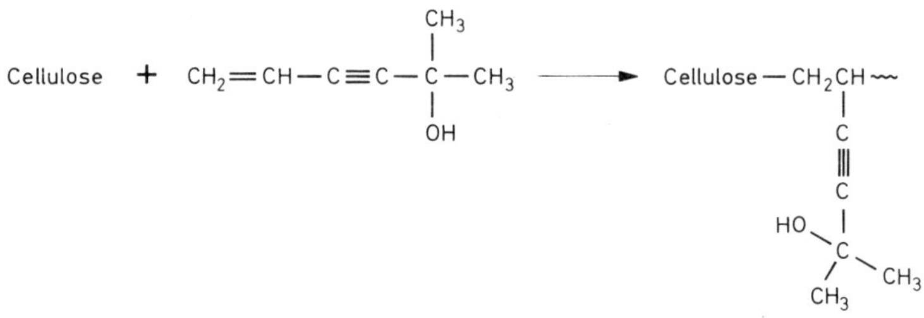

In addition to polymerization by opening up of the double bond as shown above, polymerization also takes place to some extent by the opening up of the triple bond. On treatment of the cellulose graft copolymer with a 5% aqueous solution of potassium hydroxide at 120°–130 °C, acetone is split off and a cellulose copolymer with vinyl acetylene is formed.

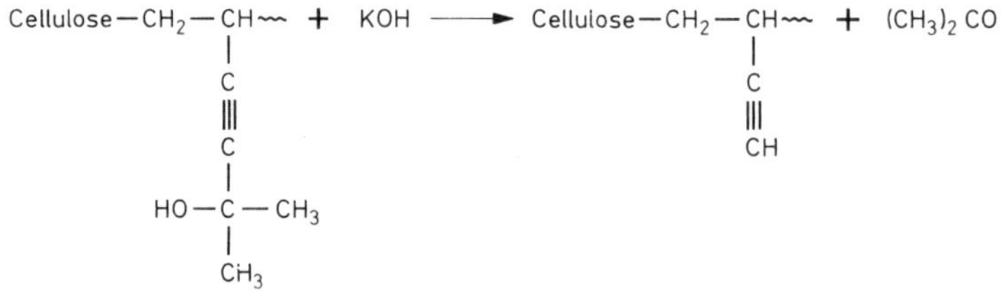

On treatment of the latter, copolymer with silver or copper salts, a modified cellulose contaning residues of silver or copper can be obtained, according to the following reaction scheme[124]:

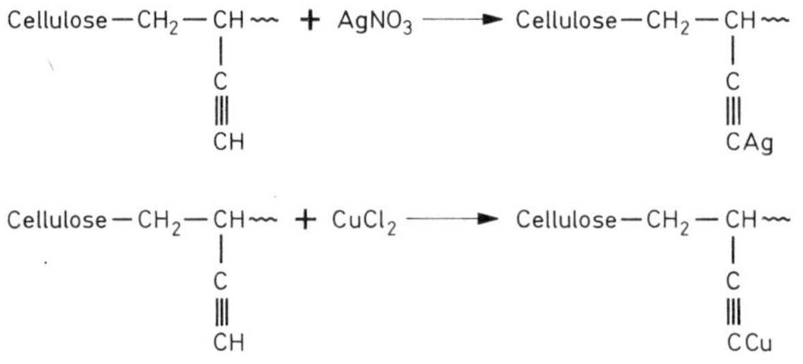

The introduction of nitrogenous heterocyclic compounds into the cellulose macromolecule gives materials having ion exchange properties which can be used for the preparation of chromatographic papers. Such modified celluloses can be prepared by grafting. For instance, graft polymerization of 2-methyl-5-vinyl pyridine with cellulose yields a poly(methyl vinyl pyridine)-cellulose graft copolymer. On treatment of the latter with an aqueous-alcoholic solution of dimethyl sulfate, the corresponding quaternary ammonium compounds are formed as follows[124]:

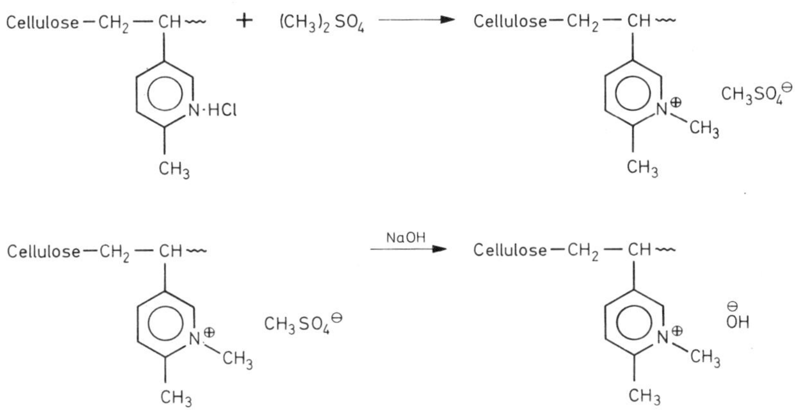

Such graft copolymers are strong anion-active complexes.

Anion-exchange fibers and cation-exchange fibers can be prepared by using the chemical reactivity of glycidyl methacrylate (GMA)-grafted cellulose (Fiber II). An anion-exchange fiber may be obtained by introducing a quaternary ammonium hydroxide and the cation-exchange fiber by introducing a sulfonic, a mercapto or a carbonyl group. Iwakura et al.[69] have examined the reaction of diethylamine with Fiber II (grafting content 100 wt.%) in anhydrous acetone, aqueous acetone, and water. The results are shown in Figure 84. Aqueous acetone gives the highest reaction ratio, but it decreases after 2 hours. The reason for this is that the rate of reaction was measured by the nitrogen content of the products, and hydrolysis of the ester is unavoidable.

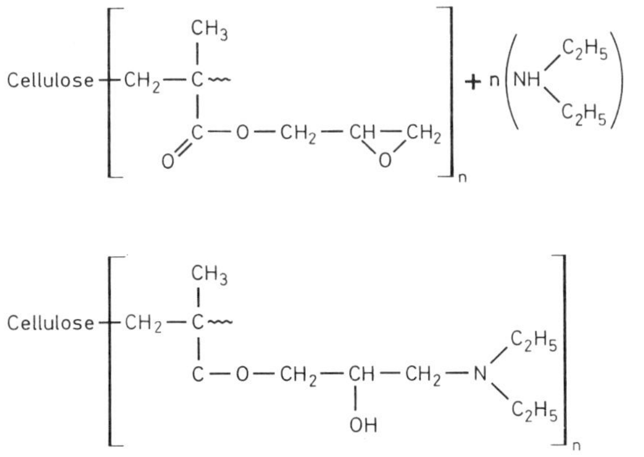

Water gave the highest reaction rate because cellulose has hydrophilic groups. Iwakura et al. further investigated the reaction of amines on rayon which had been graft-copolymerized with glycidyl methacrylate (GMA) and methyl methacrylate

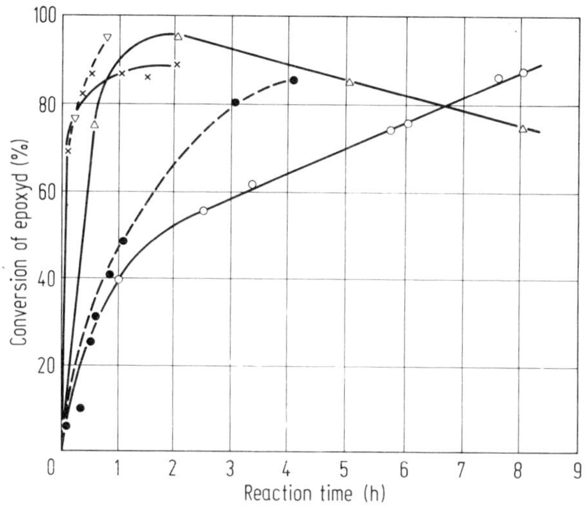

Fig. 84. Effect of reaction medium on amination of fiber II at 65 °C; (○) acetone-diethylamine; (△) acetone-water (20 wt.-% H_2O); (▽) water; (●) water-diethanolamine; (×) water-diethylamine (cograft). Reprinted from Ref.[69], p. 685, by courtesy of John Wiley & Sons, Inc.

(MMA) (grafting content GMA 57 wt. %, MMA 29 wt.-%), but the results were almost the same. On the other hand, an examination of the reaction velocity of (MMA-GMA)-cografted cellulose with amines in a homogeneous system showed that GMA-grafted cellulose had the higher reaction velocity. This was explained by the catalytic reaction of the group produced. No difference was observed in heterogeneous systems where the rate determining step is the diffusion of amines into the cellulose.

The quaternary ammonium chloride was prepared using benzyl chloride, on GMA-grafted, diethylamine-treated cellulose (grafting content, 94 wt. %, N. content 3.36 %)[69]. A range of reaction media, including dimethylformamide, acetone, and benzene, was assessed in this study.

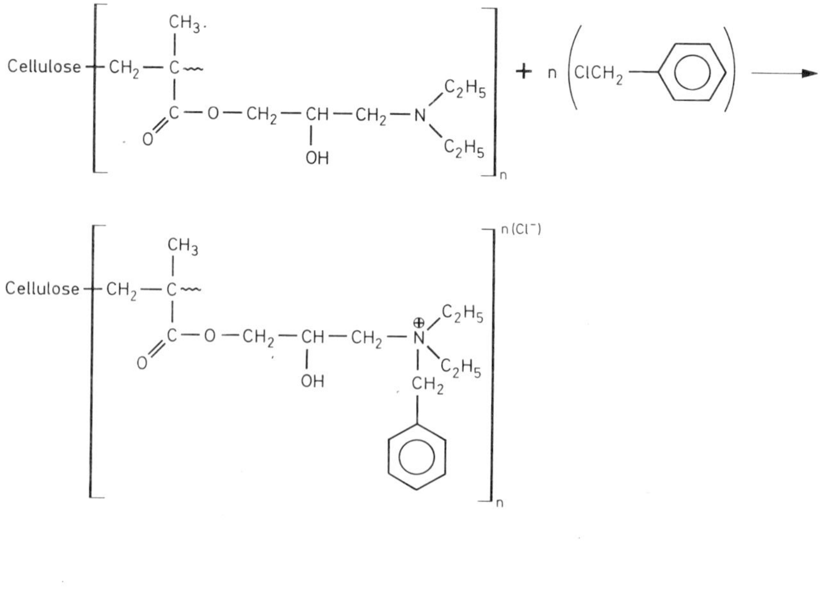

It is well known that a sodium hydroxysulfonate is produced by the ring-opening addition reaction of an epoxy group with Na_2SO_3:

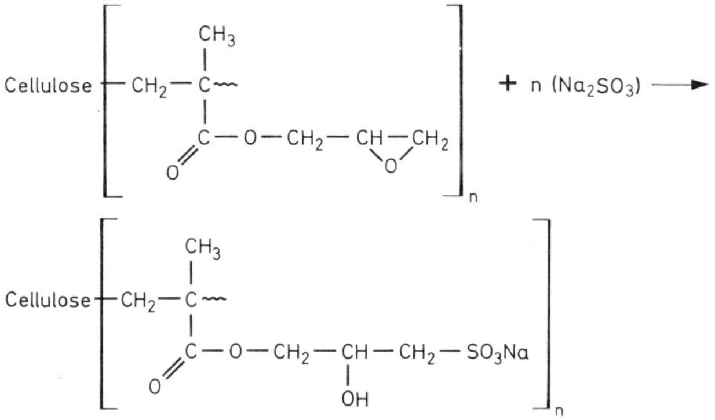

Table 85. The main types of cellulose graft copolymers, which have found practical application and their uses. Adapted from Ref.[55], p. 235, by courtesy of John Wiley & Sons, Inc.

Polymer used for grafting cellulose	Preparations of the cellulose graft copolymer	Field of Application
(1) Polyacrylonitrile	Wool-like appearance; improved resistance to microorganisms, light and wear	Rugs (modified hydrated cellulose fibers) Knitted goods, rain coats, tents.
(2) Poly(methyl vinyl pyridine)		
(a) phosphate salt	non-flammable	Decorating and facing materials, theatrical decorations
(b) quaternary salt (e.g., with dimethyl sulfate)	Strong anion exchanger and complex compound former	Adsorption of ions, complex formers, dyeing of food products, etc.
(c) Bases	Weak anion exchanger	
(3) Poly(acrylic acid), poly(methycrylic acid)		
(a) in free state	weak cation exchanger	Adsorption of metal ions, protein from food products etc.
(b) in form of calcium salt	Causes coagulation of proteins, and blood clotting	Blood-clotting cloth (hemostatic)
(c) in the form of silver and copper salts	bactericidal property (destroys bacteria)	Bed linen and underwear in clinics; lab coats for doctors; work clothes in antibiotics factories and other special plants
(4) Phosphorus-containing vinyl polymers	Non-flammable chelating agent	The same as 2 (a) The same as 2 (b)
(5) Fluorine-containing polymers	Waterproof and oil resistant properties	Work clothes for workers of different trades, tarpaulins, raincoats
(6) Polyisoprene and other polyolefins	Hydrophobic (waterproof), acid proof	Industrial goods

This reaction was applied to GMA-grafted cellulose (grafting content 101 wt.-%), and the polysulfonic acid was obtained[69]. For this addition reaction, Na_2SO_3-H_2O and $Na_2SO_3/NaHSO_3$—H_2O were effective in modifying the rate of ring opening.

GMA-grafted cellulose fiber can react with H_2S to form the polythiol:

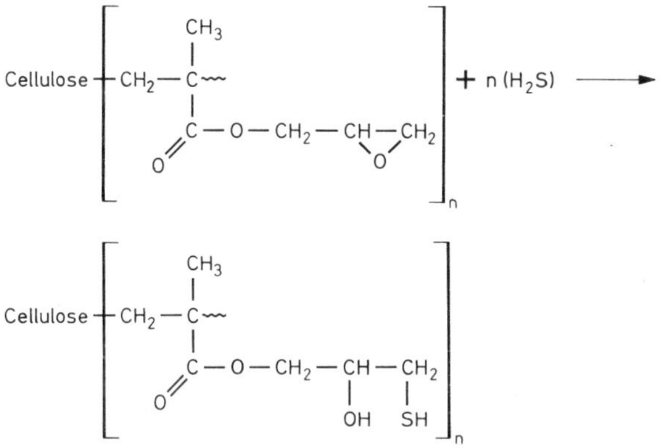

Two reaction products can be expected from the reaction of GMA-grafted cellulose and thioglycollic acid[69]:

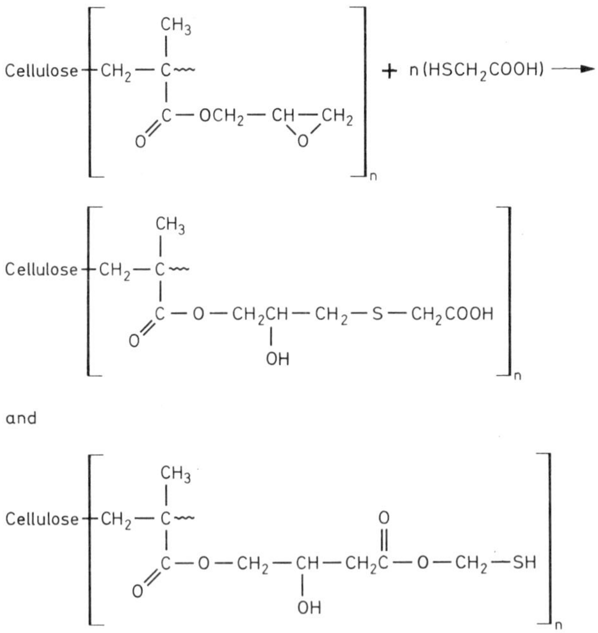

The main types of cellulose graft copolymers, which have found practical application, and their uses are summarized in Table 85.

The introduction of reactive groups of these types provides possible routes to bound reagents (e.g., immobilized enzymes or drugs), thus expanding the areas of application of cellulosic graft copolymers.

Dyeability

Cotton grafted with β-propiolactone[36)] can be dyed using dyes designed for acetate fibers, whereas an untreated control is not receptive to these dyes. The basic dye Methylene Blue 2B also dyes this grafted cotton very readily, whereas the ungrafted cotton is resistant to dyeing.

Grafting of acidic monomers such as acrylic and methacrylic acids to cellulose significantly improves its affinity to basic dyes. For example, the introduction of small amounts of carboxyl groups (1.8–2% on the weight of material) by grafting cellulose with acrylic acid considerably improves the absorption of basic dyes[128)].

Permanently dyed cellulose materials can be produced by first grafting rayon with glycidyl methacrylate and subsequently letting the epoxy group of the graft copolymer react with different dyes containing amino or phenolic hydroxyl groups[69)]. Alternatively, glycidyl methacrylate and aromatic amines or phenol give addition compounds by the ring-opening reaction of the epoxy group. The compounds obtained can be coupled with diazonium salts to give azo dyes with a vinyl group. When the addition products are grafted to cellulose, the grafted fibers acquire color on treatment with diazonium salts[133)].

The dye acceptability of cotton fabric grafted with a mixture of styrene and methyl methacrylate has been reported to be uniform and superior to that of ungrafted cotton (cf. Table 86). Uniformity is of special importance in matching dyed batches. The dyed samples have remarkable wash and light fastness[113)].

Kamel and Hebeish[134)] have reported on the behavior of cellulose, grafted with poly(methyl methacrylate) and polyacrylonitrile, toward some direct and reactive dyes. In this work, viscose rayon fibers were grafted and subsequently dyed in the usual manner. It was found that the exhaustion rate of dye onto the fibers was governed by the amount and nature of the polymer grafted. Figure 85 shows the effect of grafting of cellulose with polyacrylonitrile on dyeing with Solophenyl Red 6BL at 70 °C, in terms of the rates of dye exhaustion.

Table 86. Dye[a] absorption of styrene-methyl methacrylate grafted cotton fabric and color-fastness of the dyed material (standard: undyed, ungrafted blank — 100% reflectance). Reprinted from Ref.[113)], p. 345, by courtesy of John Wiley & Sons, Inc.

Percentage weight increase due to total grafting	Ratio of grafted polystyrene poly (methyl methacrylate) on the fabric	Amount of dye absorbed in mgm/g material (%)	Percentage Reflectance[b]	
			Unexposed	Exposed to fadeometer
Dye, ungrafted cotton	—	18.6	82, 60 Not uniform	Complete loss of color
12.8	$1:2^1/_2$	38.1	30	56
10.3	2:1	42.3	26	55
19.2	$3^1/_2:1$	60.0	20	50
24.7	1:1	60.0	22	49
32.0	$1:2^1/_2$	55.1	24	49

[a] Calcocid Fast red "A".

[b] A higher value of reflectance denotes lower intensity in colour.

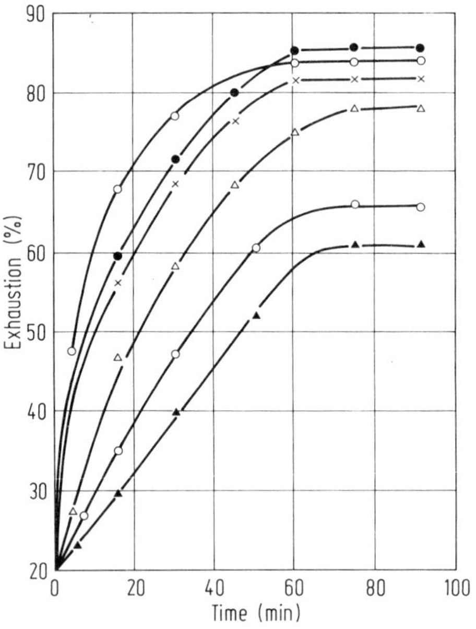

Fig. 85. Effect of grafting of cellulose with polyacrylonitrile on dyeing with Solophenyl Red 6 BL at 70 °C; (●) untreated cellulose; (×) oxidized cellulose; (○) 13.5% PAN; (△) 43.7% PAN; (○) 51.5% PAN; and (▲) 73.2% PAN. Reprinted from Ref.[134], p. 2411, courtesy of John Wiley and Sons, Inc.

It is clear that regardless of the substrate used, the exhaustion curve is characterized by an initially fast rate, followed by a slower rate, then a leveling off to give maximum exhaustion. In general, the maximum exhaustion level decreases with increasing graft, though the picture regarding initial exhaustion rates is far from clear. A similar situation is found with methyl methacrylate grafted cotton, Figure 86. Dyeings with reactive dyes (Remazol Gold Yellow G

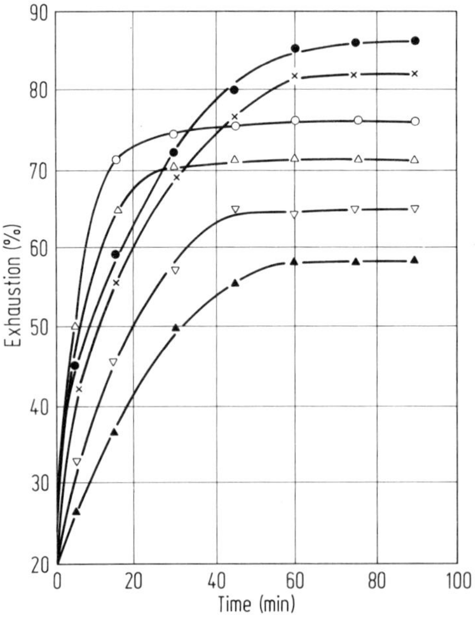

Fig. 86. Effect of grafting of cellulose with poly(methyl methacrylate) on dyeing with Solophenyl Red 6 BL at 70 °C; (●) untreated cellulose; (×) oxidized cellulose; (○) 14% PMMA; (△) 43.5% PMMA; (▽) 66.8% PMMA, and (▲) 93.2% PMMA. Reprinted from Ref.[134], p. 2412, courtesy of John Wiley and Sons, Inc.

and Cibacron Blue 3 GA) show that both acrylonitrile and methyl methacrylate grafted cotton give increased exhaustion relative to ungrafted fabrics.

Table 87 lists the affinity of three reactive dyes for different substrates. It is evident that the maximum exhaustion of Cibacron Brilliant Yellow 3G-E and Cibacron Orange T3G-E onto fibers is higher for polyacrylonitrile-cellulose grafted copolymers than for cellulose modified with poly(methyl methacrylate). Comparable dye exhaustion was found when dyeing was performed with Cibacron Rubine R-E. This reflects the nature of the dye. It is probably correct to say that the affinity of the reactive dyes for cellulose graft copolymers depends mainly on the (i) nature of the dye; (ii) the extent of disruption of the microstructure of the cellulose by grafting, which is dependent on the nature and amount of grafted polymers; and (iii), the affinity of the dye

Table 87. Effect of nature of substrate on dye affinity of some reactive dyes for cellulose. Reprinted from Ref.[134], p. 2415, by kind permission of John Wiley & Sons, Inc.

Substrate	Dye[a]	Exhaustion after different time periods, %				
		5 min	15 min	30 min	40 min	105 min
Untreated cellulose	Cibacron brilliant Yellow 3 G-E	52.38	62.86	69.52	78.10	78.10
Ceric-oxidized cellulose		17.40	23.82	23.81	23.81	23.81
PAN-cellulose graft copolymer (13.5% graft)		22.52	36.40	42.70	49.47	49.76
PAN-cellulose graft copolymer (73.2% graft)		15.24	26.00	37.48	45.95	45.95
PAN-cellulose graft copolymer (66.8% graft)		25.08	35.29	40.59	40.76	40.76
Untreated cellulose	Cibacron Orange T3G-E	88.79	90.52	90.52	90.52	90.52
Ceric-oxidized cellulose		73.28	81.90	81.90	81.90	81.90
PAN-cellulose graft copolymer (13.5% graft)		79.24	83.76	90.70	90.70	90.70
PAN-cellulose graft copolymer (73.2% graft)		77.54	87.07	87.07	87.07	87.07
PAN-cellulose graft copolymer (66.8% graft)		75.59	79.31	79.31	79.31	79.31
Untreated cellulose	Cibacron Rubine R-E	92.38	96.19	96.19	96.19	96.19
Ceric-oxidized cellulose		83.03	88.39	91.07	91.07	91.07
PAN-cellulose graft copolymer (13.5% graft)		74.29	79.05	86.67	89.95	89.95
PAN-cellulose graft copolymer (73.2% graft)		60.85	75.07	76.07	77.78	77.78
PAN-cellulose graft copolymer (66.8% graft)		62.20	74.50	77.93	77.93	77.93

[a] Dye concentration, 1% (on weight of fibers); temperature of dyeing, 70 °C.

for the grafted polymer. For a given dye[134)], increasing the magnitude of (ii) and (iii) leads to increased dye exhaustion into the fibers. However, the general trend is that an increase in the level of grafting causes a substantial reduction in dye exhaustion onto fibers, particularly with dyes having low affinity for cellulose. The large disruption of microstructure of the cellulose at high polymer contents seems to be outweighed by the low affinity of the dye for the grafted polymer. The dye acquires little affinity for PMMA, in contrast to PAN, for which the dye has considerable affinity. This would also explain the higher exhaustion onto cellulose modified with PAN relative to cellulose modified with PMMA. In this regard, however, differences in molecular mass, molecular mass distribution, and grafting frequency between the two copolymers must also be taken into account.

Dye fixation on untreated cellulose, PAN-cellulose graft copolymers and PMMA-cellulose graft copolymers is shown in Table 88. It is obvious that the dye fixation obtained with PAN-cellulose graft copolymers having up to 51% PAN was higher than that found with the untreated cellulose. The reverse is the case with a copolymer containing 73% PAN. However, when the dye fixation was calculated on the weight of the cellulose component only, the fixation increased significantly as the graft yield increased. The same situation was encountered when the dye fixation on PMMA-cellulose graft copolymers was determined. This leads to the conclusion that the grafted chains disrupt the

Table 88. Fixation of reactive dyes on various grafted substrates. Reprinted from Ref.[134)], p. 2416, by kind permission of John Wiley and Sons, Inc.

	Cibacron Brilliant Yellow 3GE[a] Dye-Reacted with Cellulose %		Remazol Golden Yellow[b] Dye-Reacted with Cellulose %	
	Based on weight of copolymer	Based on weight of cellulose only	Based on weight of copolymer	Based on weight of cellulose only
Untreated	—	0.86	—	0.48
Ce^{IV}ion-oxidized Cellulose	—	0.78	—	0.39
PAN-Cellulose Graft (13.5% graft)	0.94	1.04	0.57	0.63
PAN-Cellulose Graft (43.7% graft)	1.04	1.72	0.52	0.86
PAN-Cellulose Graft (51.5% graft)	1.17	2.34	0.51	1.02
PAN-Cellulose Graft (73.2% graft)	0.76	2.78	0.31	1.11
PMMA-Cellulose Graft (14.0% graft)	1.04	1.16	0.47	0.52
PMMA-Cellulose Graft (66.8% graft)	0.42	1.47	0.31	0.88

[a] Dye concentration, 1% (on weight of fibers); temperature of dyeing, 70 °C; duration of dyeing, 105 min.

[b] Dye concentration, 2% (on weight of fibers); temperature of dyeing, 60 °C; duration of dyeing, 60 min.

microstructure of cellulose and increase the accessibility of the cellulose hydroxyls to reaction with the dye molecules. For a given dye, the magnitude of this effect seems to depend on the amount and nature of the polymer grafted.

7.6.2. New Cellulosic Materials

During the last few years much technological research has been performed with the aim of achieving controlled changes in the properties of cellulosic materials. In general, cellulosic yarns or fabrics are treated with various reagents which often impart a new property of improve its performance in use. Another method followed to obtain cellulosic materials with novel properties is to synthesize new cellulose derivatives. For instance, reaction with ethylenimine gives improved affinity for reactive dyes as well as crease-resistance to fabrics from cotton or viscose rayon. Improved water-repellency and fire-proofing can be achieved by such secondary treatment[135].

The recently developed methods of chemical modification have considerably widened the scope of possible improvements in properties as well as fields of application. The new methods of chemical modification, and graft polymerization in particular, should be considered supplementary to the earlier methods of impregnation with specific reagents. In a few cases, such as the preparation of ion-exchange materials, fabrics with bactericidal action and substrates having medical uses, the chemical modification method alone is successful. However, for other uses, the method selected will depend on economic or technological factors, or on the type of equipment available.

In general, chemical modification methods, give products with better fastness to washing and rubbing and are preferred for prolonged used. Water-proofing can be achieved by application of paraffin wax based emulsions, but it is effective against small amounts of water only. If such materials are used for garments and exposed to continuous prolonged rain, the water-proof effect is not adequate. However, the water-repellency imparted by the use of hydrophobic graft copolymers is quite durable.

A surface application technique is, however, quite adequate in some cases, such as imparting antistatic properties, improving softness and handle, and reducing friction during fabrication of yarns and fabrics.

As a result of considerable research, the Moscow Textile Institute has developed products with novel properties for three different types of application.

I. Modified cellulosic materials for general use:
   (i) Fabrics from modified viscose staple (MTILON-B);
   (ii) Carpet materials from modified cuprammonium rayons (MTILON B & M);
   (iii) modified acetate silk (MTILON-A);
   (iv) materials of improved abrasion resistance.

II. Modified cellulosic materials for specific uses with improved properties:
   (i) fire-resistance;
   (ii) resistance to microorganisms;
   (iii) resistance to molds;
   (iv) resistance to light degradation and heat;

(v) water-repellency;
(vi) ion-and electron-exchange properties.

III. Cellulosic materials for medicinal use:
(i) bactericides;
(ii) blood-clotting materials;
(iii) antifungal materials;
(vi) materials for use in the radio tracer field.

Perhaps one of the greatest achievements of grafting in the field of chemical modification of cellulose is the development of Mtilon fibers. Graft copolymerization of the cellulose with 30–40% acrylonitrile leads to these fibers. Different types of Mtilon fiber have been produced.

A report by Achwal[136)] on Mtilon fibers reveals that Mtilon B is a modified viscose staple fiber which, due to the nitrile content, is characterized by a high resistance to weathering (photo-chemical degradation) as compared to normal viscose. Other advantages are increased resistance to abrasion and the possibility of dyeing with different classes of dyes used for dyeing cellulosic as well as wool fibers. Moreover, garments made of Mtilon B fibers acquire a wool-like appearance.

The count of the various types of cellulosic fiber modified by graft copolymerization is found to decrease somewhat as a result of grafting. This lowering in count is proportional to the amount of graft polymer present. Some degradation of cellulose usually accompanies grafting, resulting in a decrease of strength of the order of 10–15%. The index of the breaking load, the calculation of which involves the count of the fiber, naturally also decreases. The breaking load of Mtilon B obtained on a pilot plant is 10–20% lower than the value for the viscose staple fiber before grafting, though the breaking elongation is of the same order as for untreated rayon.

Mtilon B and Mtilon M are useful for making carpets; the lowering in the count of fibers on grafting is important in this case. These fibers used for carpet-making have a high resistance to the actions of light, high elasticity, high abrasion resistance, a wool-like appearance, and they can be dyed in bright shades. Apart from the improved properties of these modified rayons, their use is possible on a large scale since their cost compares favourably with that of ungrafted material, which is of importance to the consumer industry.

During the synthesis of graft copolymers of cellulose-rayon fibers, homopolymer formation can be almost completely prevented without other side reactions taking place. Hence, the specific cost of polyacrylonitrile, which forms the main part of the cost of grafting, consists of the amount of polymer grafted and small losses during the grafting or regeneration from used mixtures. In the USSR, the prices of viscose staple rayon and of acrylonitrile are of similar order, and the weight of the product increases on grafting. The commercial production price of Mtilon B, assuming almost complete regeneration from wash waters, is therefore not much higher than 10–15% above the price of viscose rayon itself.

The use of graft copolymerized acetate rayon material of improved properties is also of interest. Acetate fibers certainly have some advantages over the viscose fibers but possess the following drawbacks: (1) larger development of static charges, which creates difficulties in processing and wearing; (2) lower resistance to washing; and (3) dyeing is made possible only by the use of special classes of dyes.

Modified acetate fibers prepared by graft copolymerization can be processed on the same machines as those used for acetate rayon and following the same sequence of operations. The modified acetate rayon has a number of advantages, such as three to four times higher fastness to abrasion than normal acetate and a lower tendency for development of static electricity. Bhattacharjee and Perlin have reported interesting properties for their radiation-induced styrene-grafted, crosslinked carboxymethyl cellulose which apparently possesses significantly enhanced water retention[37)].

7.6.3. Properties of Grafted Paper

A number of studies have been reported on grafting paper with various vinyl monomers[21, 138, 139)]. It is claimed that the grafting reactions provide new ways of modifying paper. Such grafted papers may have quite different properties and different applications. They are often bulky and resilient, acid-resistant, and strong. It may also be possible, through the proper choice of monomers, to improve significantly the dimensional stability, particularly if crosslinking monomers are selected so that a combination of grafting and crosslinking is achieved[138)].

The results of one study on the properties of omitted grafted paper may be summarized as follows:

1. Water absorption is reduced by grafting hydrophobic polymers. The reduction, however, is not spectacular and indicates that the cellulose hydroxyls remain substantially accessible to water. Nevertheless, the grafted papers are quite water-repellent in many cases. In Figure 87 the rate of water pickup is

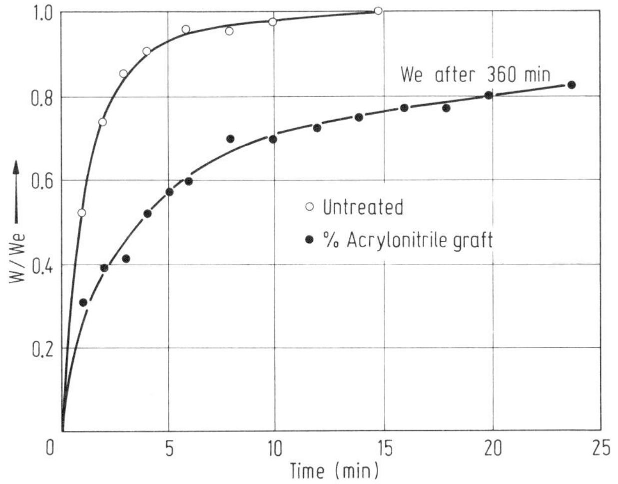

Fig. 87. The rate of water sorption (30–70%). W = regain at time t; We = regain at $t = \infty$. Reprinted from Ref.[21)], p. 254, by courtesy of the Technical Association of the Pulp and Paper Industry.

recorded for one of the best grafts from the point of view of moisture regain. It can be seen that the rate of sorption is greatly reduced, showing the potentiality of grafting as a means of producing dimensional stability.

2. The burst, tensile, and tear strengths can be increased by grafting in most cases. Figures 88 and 89 show thal all grafts have increased strengths at the

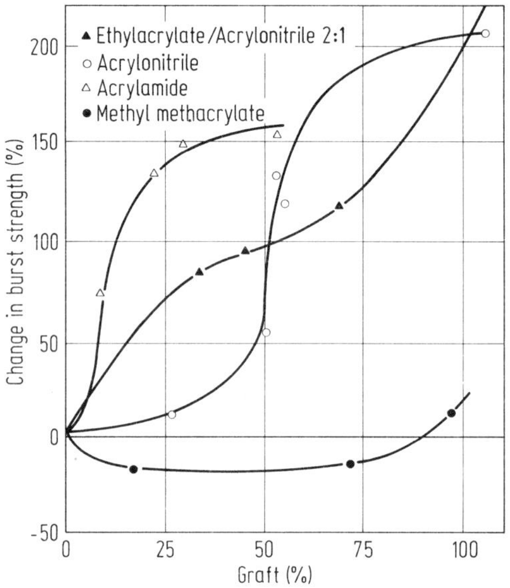

Fig. 88. Change in burst strength vs. graft percent. Reprinted from Ref.[21], p. 254, by courtesy of the Technical Association of the Pulp and Paper Industry.

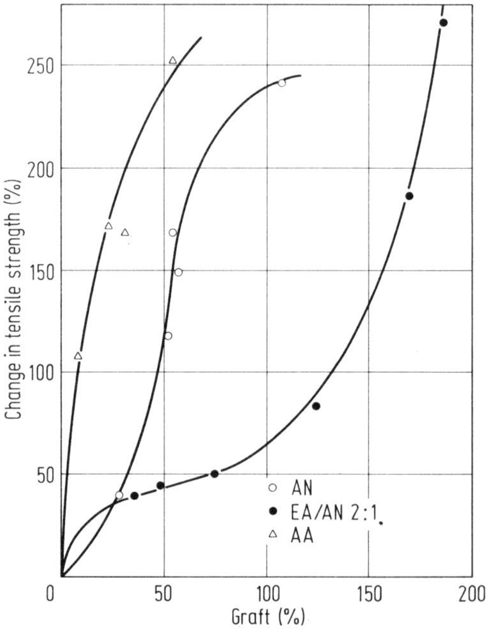

Fig. 89. Change in tensile strength vs. graft percent. AN: acrylonitrile; EA/AN 2:1 = ethyl acrylate-acrylonitrile 2:1 copolymer; AA = acrylamide. Reprinted from Ref.[21], p. 255, by courtesy of the Technical Association of the Pulp and Paper Industry.

highest percentage grafting. Acrylamide appears to be particularly beneficial with respect to the dry properties. The wet properties are also substantially improved in the case of grafted hydrophobic polymers. Comparisons with latex-treated papers show that grafting results in less improvement in the dry strength than the same amount of polymer added by latex treatment, but that it gives far greater wet strength.

3. The stiffness, particularly the wet stiffness, can be increased by the grafting of stiff polymers; fold endurance and softness can be improved by the grafting of soft rubbery polymers.
4. The resistance of paper to acids, bases, and chemicals, and to heat, light, and microorganisms is, in general, greatly improved by grafting.
5. Grafted paper can be rendered moldable to some extent through grafting with suitable monomers.
6. The ionic character of paper can be changed to improve dyeability and receptivity to certain top coats and resins.

Futhermore, it may be advantageous to use copolymerized papers in chromatography for the following reasons[140]:

(i) Specialized papers can be prepared in a one-step synthesis.
(ii) The grafted papers acquire improved wet strength.
(iii) They are not affected by acids, bases, organic solvents, and phenolic developing solutions.
(iv) The chromatogram development times are shorter.
(v) The sheets can be used again after washing.
(vi) Besides having unique ion-exchange properties, the grafted papers are useful for radioisotopes and labelled compounds.
(vii) They are applicable to reverse-phase chromatography.
(viii) Desired changes in the chromatography properties of the paper can be made by using monomers of different structures.
(ix) Because of the clear separation, detection and measurement of ion species in complex mixtures and extracts are possible.
(x) Since the grafted papers are stable to large doses of radiation, they are suitable for separation in radiation fields.

The potential of radiation-copolymerized celluloses in chromatography, particularly in the reverse-phase technique, has recently been discussed[140]. A range of styrenated celluloses has been prepared by copolymerizing styrene with Whatman 40, 41, and 42 papers. Significant differences in the chromatographic behavior of certain dyes were observed between the grafted and non-grafted papers. Similar changes in R_f-values between the two types of papers were also observed for the separation of a range of typical cations, such as Zn^{II}, Ni^{II}, and Cd^{II}. In the separation of fatty acids, stearic, myristic, and lauric, the grafted celluloses were shown to possess advantages over the untreated Whatman equivalent.

For column chromatography, a methyl methacrylate-cellulose radiation copolymer has been used quantitatively to separate a mixture of oil-soluble dyes. This same mixture could not be separated on a column of untreated cellulose powder[140].

Bardan et al.[141,142] have shown interest in methyl cellulose-polyacrylamide graft copolymers and their formation. This type of system has great potential as a substrate for immobilized enzyme studies owing to its inherent affinity for water and ease of handling[143]. In the grafting stages, Bardin et al. used potassium persulfate as the initiator, and various factors affecting the copolymerization were studied. The efficiency of grafting was found to be maximum at an initiator concentration of 3.7×10^{-4} mol/l. Grafting was found to be enhanced by increasing the monomer concentration and the temperature up to 40 °C. The system is also of interest because of the different effects of temperature change on the component parts of the copolymer. Thus, the solubility of methyl cellulose in water increases with cooling, whereas the solubility of polyacrylamide increases with an increase in temperature. In fact, the precipitation temperature of the methyl cellulose, as well as the copolymers, was found to be an increasing function of dilution, but the copolymers always showed a higher precipitation temperature than did the methyl cellulose at the same concentration.

Dimov et al. have studied the effects of reaction conditions on grafting itaconic acid and acrylamide onto cellulose fibers[144]. The properties of the fibers as ion exchangers were studied, as were the exchange capacities of the fibers for heavy metal cations. In addition, the value of such materials for extracting cationic dyes and treating waste waters from dyeing processes was examined. Chebotareva et al. have developed a method for determining the osmotic pressure of ion-exchangers[145]. This eliminates the need for contact with aggressive solutions. Ion-exchange fibers were prepared from rayon grafted with vinyl sulfonic acid, methacrylic acid, and methyl vinyl pyridine followed by alkylation with epichlorohydrin. Other fibers were prepared by heating poly(vinyl alcohol) with sodium bisulphite and grafting with acrylic acid. The ion-exchange fibers prepared from poly(vinyl alcohol) were found to possess greater osmotic stability.

The ion-exchange properties of grafted copolymers where the monomer contains a functional group, such as $—NH_2$; $—SO_3H$, or even a heterocyclic nitrogen (the vinyl pyridines), have also been discussed[140]. An interesting point is that the ion-exchange properties of papers grafted with vinylpyridine (VP) were connected with the behaviour of the copolymerized cellulose in the medium. Thus, in acid solutions, the nitrogen would be protonated while in alkaline solutions it would remain free. In acid solutions it might therefore behave as an active zone for anionic particles [e.g., $(ZnCl_4)^{2-}$] or similar species. In alkaline solutions (NH_3, pyridine, etc.) it may act as a donor to species such as $(ZnVP)^{2+}$ and thus be bound. The anions of strong acids may be held as PyH^+X^- in acid solutions, as might be the case with dyestuffs, while weak acids may have their anions held in alkaline solutions.

7.6.4. Properties of Grafted Pulp and Timber

Grafting reactions in wood have been reported by a number of investigators[146–150]. Most of these studies involved vinyl graft copolymerization induced by high energy radiation. Because of its heterogeneity, the effect of radiation upon the properties of wood is quite different from that on pure cellulose. This is probably due to the effect of the lignin, which is itself quite resistant to radiation damage and may reduce the scission of the cellulose chains[150,151]. It is therefore possible to modify the cellulose in wood by radiation grafting, with a resulting enhancement of physical and mechanical properties. Alteration in the physical properties of wood by grafting of wood with polystyrene is shown in Table 89.

In Figure 90 is shown a typical pair of load-deflection curves. In addition to a significantly higher bending strength (see Table 89), the mechanical nature of the material has been altered, and the load-deflection curves of this material are invariably linear right up to the point of failure. The alteration in the nature of

Table 89. Changes in physical properties of treated wood[a]. Reprinted from Ref.[150], p. 160, by courtesy of John Wiley & Sons, Inc.

Sample	Bending strength increase %	Shrinkage based on conditioned vol. %	Swelling based on swollen sample %	Vol. increase due to treatment %
Untreated		5.0	15.0	
Treated	45	0.9	1.1	8–9

[a] Treatment: impregnant 76 wt.-% styrene + 22.5 wt.-% methanol + 1.5 wt.-% Water: gamma dosage: 4.5 Mrad.; heating treatment; 105 °C for 24 hours.

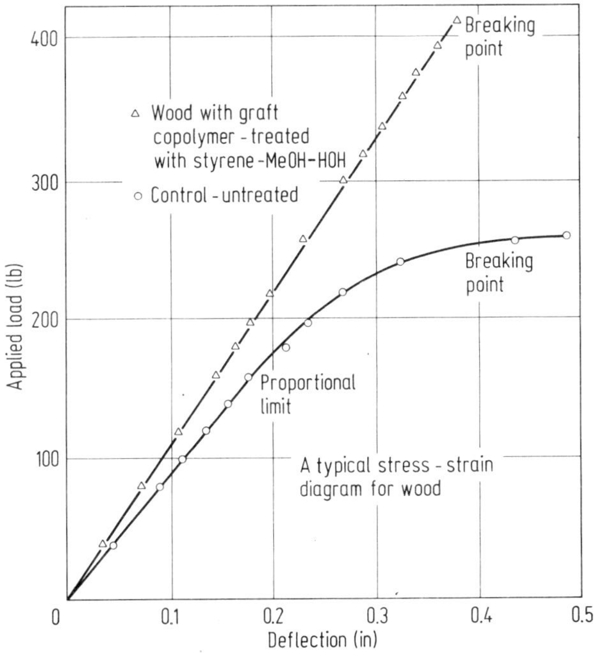

Fig. 90. Load-deflection characteristics of wood samples. Reprinted from Ref.[150], p. 160, by courtesy of John Wiley & Sons, Inc.

the load-deflection curves suggests that the chains of α-cellulose have been modified by the grafting of polystyrene. In the unmodified wood, the force necessary to move the cellulose chains with respect to one another, as in bending, is that required to overcome relatively weak Van der Waal's forces between adjacent chains. However, in the modified wood, there are also primary valence forces introduced by the polystyrene side chains which increase the bending strength[150].

As can be seen from Figure 91, penetration by water vapour and liquid is drastically reduced. Dimensional changes associated with water adsorption are correspondingly reduced (see Table 109). The resistance of wood to micro-biological attack is increased by grafting.

The water sorption kinetics and equilibria in wood pulp cellulose grafted with styrene and other vinyl monomers, using the mutual radiation techniques have recently been studied[152]. With respect to cellulose grafted with polystyrene, the equilibrium water content dropped monotonically with increasing percent graft.

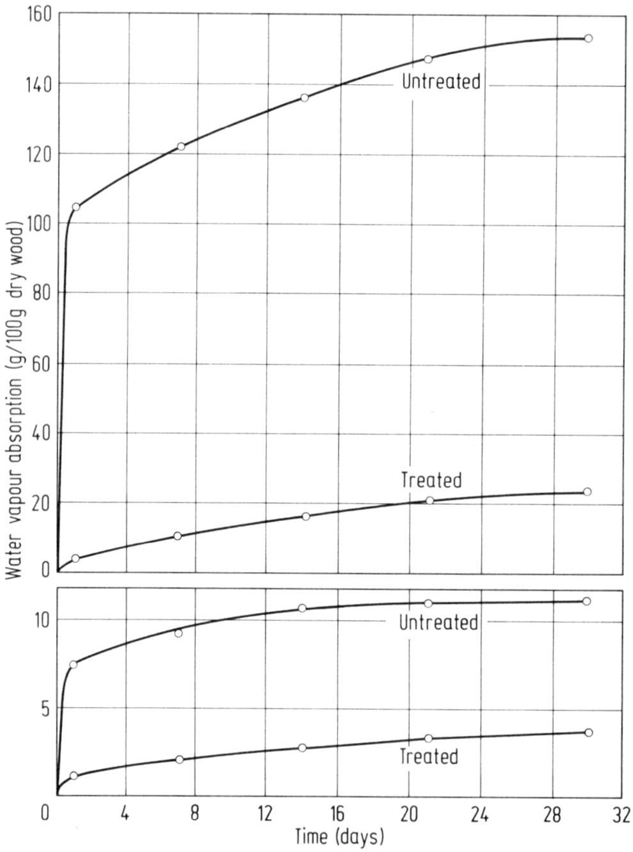

Fig. 91. Water absorption characteristics of wood. Samples immersed in distilled water at room temperature and under atmospheric pressure. Exposed to 75% R.H. at 25 °C. Reprinted from Ref.[150], p. 161, by courtesy of John Wiley & Sons, Inc.

This effect has been reported for similar systems[153], although when the water sorbed is calculated as a percentage of the pure hydrophilic backbone (e.g., cellulose acetate), there is no apparent effect on the specific sorption. This has prompted the suggestion of a "domain model" of hydrophobic (essentially non-sorbing) polymer domains dispersed in a seemingly unaltered matrix of hydrophilic backbone polymer.

It has been shown[152], however, that the water sorbed at equilibrium and calculated on the basis of pure cellulose is actually less than the specific sorption in ungrafted cellulose. The magnitude of the decrease increases with percent graft and is more pronounced at the very low water activities. Here, it appears that the discrete domain model is not applicable and that pendant polystyrene chains are admixed molecularly throughout the cellulose matrix, hindering site sorption which occurs at low activities.

The density of cellulose grafted with polystyrene by radiation-induced polymerization in the presence of dioxane has been measured and compared with the calculated density assuming physical mixing of the homopolymers, cellulose, and polystyrene, with no volume change on mixing. The results are presented in Table 90. In all cases, the measured values of density were significantly higher than the calculated values, assuming no volume change on mixing, which would be a reasonable assumption, if indeed there was no molecular intermingling of polystyrene and cellulose. These data support the arguments suggested by the sorption data that the grafted polystyrene is not solely deposited in isolated, discrete domains but is significantly intermingled with the cellulose matrix to affect not only the specific water sorption but the density as well.

The effect of styrene grafting (induced by the mutual irradiation technique in the presence of dioxane) on the diffusion coefficient of water in cellulose-polystyrene graft copolymers has been reported[152]. The results of this work reveal that grafting actually increases the magnitude of the diffusion coefficient, presumably as a consequence of subtle destructuring of the cellulose matrix by

Table 90. The effect of graft copolymerization on the density of styrene-cellulose grafts. Reprinted from Ref.[152], p. 318, by courtesy of John Wiley and Sons, Inc.

Cellulose sample	Percent graft based on original cellulose	Specific gravity	
		Measured	Calculated
84% Alpha wood pulp	0.0	1.527	(1.527)
	21.0	1.440	1.415
	29.0	1.430	1.385
	45.0	1.402	1.339
	68.0	1.361	1.289
Cotton	0.0	1.550	(1.550)
	37.0	1.415	1.372
Rayon	0.0	1.510	(1.510)
	34.0	1.368	1.355

the formation of graft copolymer. This means that, although styrene grafting decreases the amount of water sorbed, it increases the diffusion coefficient, characterizing the rate at which moisture is adsorbed.

At approximately the same level of grafting, the diffusivities of water in grafts of polar monomers, such as acrylonitrile, ethyl acrylate, and acrylic acid, were compared with those in grafts of styrene. Although all three monomers are quite polar in comparison with styrene, acrylic acid is the only monomer which is sufficiently hydrophilic to be water soluble. Indeed, acrylic acid grafts have been shown to behave quite differently from the other polar monomers, which in turn behave much like non-polar styrene monomer[152)]. This relative behaviour was also observed in the equilibrium sorption data presented in Table 91.

Presumably, the extremely hydrophilic acrylic acid grafts swell significantly within the cellulose matrix, causing decreased mobility of the neighbouring cellulose chain segments. The other slightly less hydrophilic monomers (ethyl acrylate and acrylonitrile) behave much more like styrene and are presumably confined largely to domains and do not significantly alter the mobility of the cellulose moieties[141)]. In a similar vein, Toneho has examined the high-energy, radiation-induced curing of resin-impregnated composites using various monomers, expecially styrene, as comonomers[154)]. Significant improvements to general tensile properties are recorded. Although dealt with in Chapter 8, the industrial application of grafting to timber products is worthy of inclusion here because of marked changes in tensile and resilience properties which can be achieved. Thus, Gamova et al.[155)] have reported the production of polymer grafted, wood-like materials obtained by grafting styrene onto sawdust using radiation initiation. It is found that the mechanical properties are dependent on the copolymer composition but independent of the degree of grafting above a specific minimum grafting level. Cunanan et al.[156)] report that significant improvements can be made to the physical properties of poorer quality timber by impregnation with styrene/methyl methacrylate, irradiating, and subsequent drying to remove unreacted monomer. Grafting yields of 86 to 96%, at an optimum 1–2 Mrad total dose, were achieved. Indications are that a specific comonomer composition is needed for a particular timber species in order to optimize the process.

It is sometimes expedient to sacrifice some of the mechanical/tensile character

Table 91. Equilibrium water sorption in cellulose grafted with various vinyl monomers. Reprinted from Ref.[152)], p. 320, by courtesy of John Wiley & Sons, Inc.

Sample description	Equilibrium water content at 5% relative humidity	Equilibrium water content at 75% relative humidity
9% ethyl acrylate graft	1.98	11.6
7% acrylonitrile graft	2.00	11.5
11% acrylic acid graft	2.04	14.4
8% styrene graft	1.95	10.7
Unmodified wood pulp	2.25	11.5

of a material, especially when improvements to other properties can be obtained. Gorskii and Zhonina have produced an electrically insulating box board by the radiation-induced grafting of styrene, vinyl acetate, or methyl methacrylate[157]. The products carry superior insulating properties when compared with their ungrafted counterparts. Meleschevich et al. have prepared chromatography and filter papers which have better thermal stability, rot resistance, weatherability, and acid/alkali resistance than unmodified papers[158]. Grafting is chemically or radiation-induced using a variety of monomer or comonomer systems. Vandana et al.[159] have examined the effect of moisture on the grafting of styrene and acrylonitrile onto beechwood. It was observed that grafting at the cell wall exceeded grafting in the cell cavity by a wide margin. While dry conditions yielded grafting in aromatic zones, moist conditions resulted in grafting in the softwood sections of the timber. These observations are supported by Paeglitis in his work on the same materials[160]. The styrene/acrylonitrile combination has been successfully employed by Hills and McGahen in their work on ashwood[161]. Here the efficient production of graftwood composites was achieved using a combination of high pressures and high-energy radiation initiation.

The chemical modification of timber through grafting has attracted recent attention. Lawniczak[162] used a mixed hydroperoxide/thermal induction process to achieve high levels of grafting to various wood pulps. A general observation is made that high initiator concentrations lead to excessive degradation. Hornoff et al.[163] have grafted styrene, acrylonitrile, methyl methacrylate, butyl acrylate, ethyl acrylate, vinyl acetate, methacrylic acid, and acrylamide onto spruce pulp in its xanthated form in the presence of H_2O_2, while Tirzina and Zoldners[164] have studied the effects of lignin on the grafting of styrene to birchwood and spruce. The effective chain transfer constants for the various timber types were measured and found to fall in the range of 7×10^{-2} and 6×10^{-3}. Young[165] has monitored the grafting of styrene/acrylonitrile mixtures onto xanthated high-yield wood pulps in the presence of Fe^{II} ions. It was seen that chemical pulps showed greater grafting levels than mechanical pulps. Horn[166] has investigated the formation, character, and reactivity of those mechanically formed radicals in pulp cellulose fibers which are capable of initiating graft copolymerization, a theme studied by Young and Nguyen through the grafting of mechanical pulps and assessment of the properties of the resultant composite sheets[167]. Dolinin et al. prepared an ion-exchange material with an ion-exchange capacity of 2.5–3.0 mequiv/g with respect to 0.1 M HCl by grafting spruce sawdust with 2-methyl-5-vinyl pyridine. The sawdust was pretreated with 1–10% NaOH or 1–3% HNO_3. Grafting was followed by alkylation with epichlorohydrin[168].

Zoldners et al.[169] have studied the effect of polar groups and polymer chemical bonding on the grafting of polystyrene onto birch wood sawdust. Grafting was radiation-induced with pure styrene, styrene-acrylonitrile, or styrene-acrylic acid mixtures. Molding of the grafted sawdust at 145–160 °C, under a pressure of 17–21 MPa, gave constructional loads having high flexural strength and water resistance.

For general information regarding the characterization of graft copolymers, the reader is advised to consult the recent excellent review of the subject which has been provided by Ikada[170].

References

1. Immergut, E. H., Mark, H.: Makromol Chem. *18/19*, 322 (1956)
2. Guthrie, J. T., Huglin, M. B., Phillips, G. O.: J. Appl. Polym. Sci. *15*; 1033 (1971)
3. Usmanov, Kh. U., Aikhodzhaev, B. I., Azizov, U.: J. Polymer Sci. *53*, 87 (1961)
4. Kobayashi, Y.: J. Polymer Sci. *51*, 539 (1961)
5. Stannett, V. T., Wellons, J. D., Yasuda, H.: J. Polymer Sci. *C 4*, 551 (1963)
6. Blouin, F. A., Morris, N. J., Arthur, J. C. Jr.: Textile Res. J. *36*, 309 (1966)
7. Gugliemelli, L. A., Weaver, M. O., Russell, C. R.: J. Polymer Sci. *B 6*, 599 (1968)
8. Huang, R. Y-M., Chandramouli, P.: J. Appl. Polym. Sci. *12*, 2549 (1968)
9. Brockway, C. E.: J. Polymer Sci. *2*, 3721 (1964)
10. Arai, K., Ogiwara, Y.: Sen-i Gakkaishi, *26*, 11 (1970)
11. Huang, R. Y-M., Immergut, B., Immergut, E. H., Rapson, W. H.: J. Polymer Sci. *A 1*, 1257 (1963)
12. Iwakura, Y., Kurosaki, T., Imai, Y.: J. Polymer Sci. *A 3*, 1185 (1965)
13. Stannett, V. T., Wellons, J. D., Yasuda, H.: J. Polymer Sci. *C 4*, 551 (1964)
14. Guthrie, J. T.: Polymer *16*, 134 (1975)
15. Toda, T.: J. Polymer Sci. *58*, 411 (1962)
16. Avny, Y., Rebenfeld, L.: Textile Res. J. *38*, 684 (1968)
17. Avny, Y., Schwenker, R. F. Jr.: Textile Res. J. *37*, 817 (1967)
18. Chaudhuri, D. K. R., Hermans, J. J.: J. Polymer Sci. *48*, 159 (1960); *51*, 373 (1961)
19. Kulkarni, A. Y., Chitale, A. G., Vaidya, B. K., Mehta, P. C.: J. Appl. Polym. Sci. *7*, 1581 (1963)
20. Geacintov, N., Stannett, V. T., Abrahamson, E. W., Hermans, J. J.: J. Appl. Polym. Sci. *3*, 35 (1960)
21. Schwab, E., Stannett, V. T., Hermans, J. J.: Tappi *44*, 251 (1961)
22. Yasuda, H., Wray, J. A., Stannett, V. T.: J. Polymer Sci. *C 2*, 387 (1963)
23. Melville, H. W., Stead, B. D.: J. Polymer Sci. *16*, 505 (1955)
24. Horowitz, C., M. Sc. Thesis: Polytechnic Inst. of Brooklyn 1963
25. Ende, H. A., Stannett, V. T.: J. Polymer Sci *A 2*, 4047 (1964)
26. Schwenker, R. F. Jr., Pacsu, E.: Tappi *46*, 665 (1963)
27. Kulkarni, A. Y., Mehta, P. C.: J. Polymer Sci. *B 1*, 509 (1963)
28. Arthur, J. C. Jr., Demint, R. J.: Textile Res. J. *31*, 988 (1961)
29. Marupov, R., Zhbankov, R. G., Kryazhev, Yu. G., Rogovin, Z. A.: Vysokomol. Soedin, Tssellyul Proizvodnye 150 (1963)
30. Imrisova, D., Maryska, S.: J. Appl. Polym. Sci. *11*, 901 (1967)
31. Guthrie, F. K.: Tappi *46*, 656 (1963)
32. Guthrie, J. T., Thesis, Ph. D.: University of Salford, U.K. 1971
33. Guthrie, J. T., Huglin, M. B., Phillips, G. O.: IUPAC Int. Symp. Macromolecules, Helsinki (1972) Prepr. Vol. 1. Paper II-26
34. Guthrie, J. T., Huglin, M. B., Phillips, G. O.: Polymer *18*, 521 (1977)
35. Arthur, J. C. Jr.: J. Macromol. Sci.-Chem. *4*, 1057 (1970)
36. Immergut, E. H., Rånby, B. G., Mark, H. F.: Ind. Eng. Chem. *42*, 2483 (1953)
37. Vink, H.: Svensk Papperstidn. *63*, 24 (1960)
38. Guthrie, J. T., Huglin, M. B., Phillips, G. O.: J. Appl. Polym. Sci. *16*, 1017 (1972)
39. Henley, D.: Arkiv Kemi *18*, 328 (1961)
40. Riande, E., Perena, J. M.: J. Macromol. Sci. Chem. *A 8*, 1413 (1974)
41. Jayme, G.: Papier *32*, 145 (1978)
42. Ohman, J.: Arkiv Kemi *31*, 125 (1970)
43. Shanbhag, V. P.: Arkiv Kemi *29*, 139 (1968)
44. Burchard, W.: Z. Phys. Chem. *42*, 293 (1964)
45. Janeschitz-Kriegl, H., Burchard, W.: J. Polymer Sci. *A-2*, *6*, 153 (1968)
46. Wellons, J. D., Williams, J. L., Stannett, V. T.: J. Polymer Sci. *A-1*, *5*, 1341 (1967)
47. Wellons, J. D., Stannett, V. T.: J. Polymer Sci. *A 3*, 847 (1965)
48. Morris, N. J., Blouin, F. A., Arthur, J. C. Jr.: J. Appl. Polym. Sci. *12*, 373 (1968)
49. Hebeish, A., Theses, Ph. D.: Gujarat University, India 1967
50. Schurz, J., Rebek, M., Spork, H.: Angew. Chem. *1*, 42 (1967)

51. Kobayashi, Y.: J. Polymer Sci. *51*, 359 (1961)
52. Kesting, R. E., Stannett, V. T.: Macromol. Chem. *55*, 1 (1962)
53. Neimo, L., Shitola, H.: Paperi Puu *47*, 369, 374 (1965)
54. Hebeish, A., Mehta, P. C.: Textile Res. J. *39*, 99 (1969)
55. Rogovin, Z. A.: J. Polymer Sci. *C 37*, 221 (1972)
56. Liushitz, R. M., Druzhinina, N. N., Penkova, M. P., Rogovin, Z. A.: Cell. Chem. Technol. *2*, 579 (1968)
57. Benoit, H., Froelich, D.: In Light Scattering from Polymer Solutions (Huglin, M. B., ed.) London-New York: Academic Press 1972, Ch. 11
58. Strazielle, C.: In Light Scattering from Polymer Solutions Huglin, M. B., ed.) London-New York: Academic Press 1972, Ch. 15
59. Tuzar, Z., Kratochvil, P., Strakova, P.: Europ. Polym. J. *6*, 1113 (1970)
60. Guthrie, J. T., Huglin, M. B., Phillips, G. O.: Makromol. Chem. *149*, 309 (1971)
61. Taga, T., Inagaki, H.: Angew. Makromol. Chem. *33*, 129 (1973)
62. Hamburger, C. J.: J. Polymer Sci. *A-1*, 7, 1023 (1969)
63. Moore, J. C.: J. Polymer Sci. *A 2*, 835 (1964)
64. Guthrie, J. T., Haq, Z.: Polymer *15*, 133 (1974)
65. Rollins, M. L., Cannizzaro, A. M., Blouin, F. A., Arthur, J. C. Jr.: J. Appl. Polym. Sci. *12*, 71 (1968)
66. Liggett, R. W., Hoffman, H. L. Jr., Tanquary, A. C.: Amer. Dyestuff Reptr. *58*, 25 (1969)
67. Reinhardt, R. M., Reid, J. D., Daul, G. C.: Textile Res. J. *26*, 1 (1956)
68. Kaizerman, S., Mino, G., Meinhold, L. F.: Textile Res. J. *32*, 136 (1962)
69. Iwakura, Y., Kurosaki, T., Uno, K., Imai, Y.: J. Polymer Sci. *C 4*, 673 (1964)
70. Immergut, E. H.: In Encyclopedia of Polymer Science and Technology, (Mark, H. F., Gaylord, N. G., Bikales, N. M., eds.) Vol. 3., New York: Interscience 1965, P. 242
71. Rollins, M. L., Moore, A. T., Goynes, W. R., Carra, J. H., de Gruy, I. V.: Amer. Dyestuff Reptr. *54*, 512 (1965)
72. Arthur, J. C. Jr.: Advances in Chem. Ser. *91*, 574 (1969)
73. Arthur, J. C. Jr., Daigle, D. J.: Textile Tes. J. *34*, 653 (1964)
74. Blouin, F. A., Arthur, J. C. Jr.: Textile Res. J. *33*, 727 (1963)
75. Blouin, F. A., Morris, N. J., Arthur, J. C. Jr.: Textile Res. J. *36*, 309 (1966)
76. Blouin, F. A., Cannizzaro, A. M., Arthur, J. C. Jr., Rollins, M. L.: Textile Res. J. *38*, 811 (1968)
77. Demint, R. J., Arthur, J. C. Jr., Markezich, A. R., McSherry, W. F.: Textile Res. J. *32*, 918 (1962)
78. Demint, R. J., Arthur, J. C. Jr., McSherry, W. F.: Textile Res. J. *31*, 821 (1961)
79. Arthur, J. C. Jr., Markezich, A. R., McSherry, W. F.: Textile Res. J. *33*, 896 (1963)
80. Blouin, F. A., Morris, N. J., Arthur, J. C. Jr.: Textile Res. J. *38*, 710 (1968)
81. Arthur, J. C. Jr., Grant, N. J.: Textile Res. J. *36*, 934 (1966)
82. Blouin, F. A., Arthur, J. C. Jr.: Polymer Preprints *6*, 359 (1965)
83. Arthur, J. C. Jr., Demint, R. J.: Textile Res. J. *30*, 505 (1960)
84. Arthur, J. C. Jr., Demint, R. J., McSherry, W. F., Jurgens, J. F.: Textile Res. J. *29*, 759 (1959)
85. Sakurada, I., Okada, T., Uchida, M., Kugo, E.: Doitati To Hoshasen *2*, 316 (1959)
86. Chitale, A. G., Kulkarni, A. Y.: J. Sci. Ind. Res. *20D*, 223 (1961)
87. Prahl, H. F., Tovey, H., Underwood, C. E.: Textile Res. J. *36*, 245 (1966)
88. Goynes, W. R., Harris, J. A.: J. Polymer Sci. *C 36*, 277 (1972)
89. Guthrie, J. T., Percival, J. A.: Polymer *18*, 531 (1977)
90. Noordermeer, J.W. M., Daryanini, R., Janeschitz-Kriegl, H.: Polymer *16*, 359 (1975)
91. Hearon, W., Hiatt, D., Fordyce, C. R.: J. Amer. Chem. Soc. *65*, 829 (1943)
92. Guthrie, J. T., Huglin, M. B., Phillips, G. O.: European Polym. J. *8*, 747 (1972)
93. Benson, R., Guthrie, J. T., Lartey, R. B.: Polymer *16*, 903 (1975)
94. Reich, L., Levi, D. W.: Makromol. Chem. *66*, 102 (1963)
95. Reich, L. J.: Polymer Sci. *B 3*, 231 (1965)
96. Reich, L., Levi, D. W.: Makromol. Rev. *1*, 173 (1967)
97. Wegner, J., Patat, F.: J. Polymer Sci., Polym. Symp. *31*, 121 (1971)
98. Patel, K. S., Patel, K. C., Patel, R. D.: Makromol. Chem. *132*, 7 (1970)
99. Huang, R. Y-M., Chandramouli, P.: J. Polymer Sci. *A-1*, *7*, 663 (1969)

100. Sharma, V. N., Daruwalla, E. H.: J. Appl. Polym. Sci. *21*, 331 (1977)
101. Brown, W. P., Tipper, C. F. H.: J. Appl. Polym. Sci. *22*, 1459 (1978)
102. Harris, J. A., Arthur, J. C. Jr., Carra J. H.: J. Appl. Polym. Sci. *22*, 905 (1978)
103. Reinhardt, R. M., Arthur, J. C. Jr.: Abstracts, A.C.S. 176th National Meeting, Maimi Beach, Florida Sept. 1978 (Cell. 94)
104. Arthur, J. C. Jr., Blouin, F. A.: Amer. Dyestuff. Reptr. *51*, 1024 (1962)
105. Arthur, J. C. Jr., Blouin, F. A.: J. Appl. Polym. Sci. *8*, 2813 (1964)
106. Sakurada, I.: Nippon Mengyo Gijutsu Kenkyusho Kenkyu Hokuku (Res. Rept. Japan Cotton Techn. Inst.) *56*, 1 (1962)
107. Tsuji, W., Imai, M., Kadono, Y.: Bull. Inst. Chem. Kyoto. Univ. *42*, 68 (1964)
108. Sakurada, I., Okada, T., Kaji, K.: J. Polymer Sci. *C 37*, 1 (1972)
109. Sakurada, I.: Chem. Technol. *2*, 376 (1972)
110. Tsuji, W., Imai, M., Kadono, Y., Ichikawa, K.: Bull. Inst. Chem. Res. Lyoto Univ. *43*, 94 (1964)
111. Mares, T., Arthur, J. C. Jr.: J. Polymer Sci. *C 37*, 349 (1972)
112. Byrne, G. A., Arthur, J. C. Jr.: Textile Res. J. *41*, 271 (1971)
113. DasGupta, S.: J. Polymer Sci. *C 37*, 333 (1972)
114. Kamogawa, H., Sekiya, T.: Textile Res. J. *31*, 585 (1961)
115. Gardon, J. L.: J. Appl. Polym. Sci. *5*, 734 (1961)
116. Walsh, W. K., Jin, C. R., Amstrong, A. A. Jr.: Textile Res. J. *35*, 648 (1965)
117. Kamel, M., Hebeish, A., Al-Aref, A.: J. Appl. Polym. Sci. *18*, 3463 (1974)
118. Kamel, M., Hebeish, A., Allam, M., Al-Aref, A.: J. Appl. Polym. Sci. *17*, 2725 (1973)
119. Gallagher, D. M.: Textile Res. J. *40*, 621 (1970)
120. Richards, G. N., White, E. F. T.: Textile Res. J. *34*, 810 (1964)
121. DasGupta: Can. Text. J. *18*, 57 (1964)
122. Negishi, M., Nakamura, Y., Kakinuma, T., Iizuka, Y.: J. Appl. Polym. Sci. *9*, 2227 (1965)
123. Hebeish, A.: L'Industrie Textile *No 991*, 441 (1970)
124. Achwal, W. B.: Colourage, Sept. 23, 32 (1971)
125. Kesting, R. E., Stannett, V. T.: Mackromol. Chem. *65*, 248 (1963)
126. Kulkarni, A. Y., Mehta, P. C.: Textile Res. J. *32*, 701 (1962)
127. Achwal, W. B.: Colourage Dec. 16, 34 (1971)
128. Achwal, W. B.: Colourage Dec. 31, 22 (1970)
129. Rogovin, Z. A.: Chemical Transformation and Modification of Cellulose, Khimiya 1967; Fasern *18*, 138 (1968); Lenzinger Ber. *30*, 20 (1970)
130. Rogovin, Z. A., Tuganova, M. A., Krjazhev, J. G., Zharova, T. J.: (to Moscow Textiljny Inst.) U.S. Pat. *3*, 391, 096 (1968); British Pat. *1*, 022, 083 (1966)
131. Kern, W., Schulz, R. C.: Angew. Chem. *69*, 153 (1957)
132. Achwal, W. B.: Colourage Oct. 21, 34 (1971)
133. Uno, K., Iwakura, Y., Makita, M., Ninomiya, T.: J. Polymer Sci. *A-1*, *5*, 2311 (1967)
134. Kamel, M., Hebeish, A.: J. Appl. Polym. Sci. *20*, 2407 (1976)
135. Einsele, U.: Melliand. Textilber. *45*, 641 (1964)
136. Achwal, W. B.: Colourage Nov. 18, 36 (1971)
137. Bhattacharjee, S. S., Perlin, A. S.: J. Polymer Sci. *B 13*, 113 (1975)
138. Stannett, V. T.: Tappi *47*, 53A (1964)
139. Putnam, E. S.: J. Polymer Sci. *C 2*, 377 (1963)
140. Dilli, S., Garnett, J. L., Martin, E. C.: J. Polymer Sci. *C 37*, 291 (1972)
141. Bardan, K., Mukhopadhyay, S., Chatterjee, S. R.: J. Polymer Sci. *15*, 141 (1977)
142. Mukhopadhyay, S., Prasad, J., Chatterjee, S. R.: Makromol. Chem. *176*, 1 (1975)
143. Maeda, H., Suzuki, H., Yamauchi, A., Sakimae, A.: Biochem. Biotech. *17*, 119 (1975)
144. Dimov, K., Semkova, M., Dimitrov, D., Terlemezyan, E.: Lenzinger Ber. *45*, 60 (1978)
145. Chebotareva, R. D., Grebenyuk, V. D., Shamolina, I. I.: Zh. Prikl. Khim. (Leningrad) *51*, 2197 (1978)
146. Anon.: Nucleonics *20*, 64 (1962)
147. Karpov, V. L., Malinsky, Y. M., Serenkov, V. I., Klimanova, R. S., Freiden, A. S.: Nucleonics *18*, 88 (1960)
148. Kenya, D. L., Fennessey, J. P., Stannett, V. T.: Forest Prod. J. *12*, 161 (1962)
149. Kent, J. A., Winston, A. W., Boyle, W. R., Harrison, L. W.: AEC Publication TID 7643, P. 355 (1962)

150. Ramalingam, K. V., Werezak, G. N., Hodgins, J. W.: J. Polymer Sci. *C 2*, 153 (1963)
151. Lawton, E. J., Bellamy, W. D., Hungate, R. E., Bryant, M. P., Hall, E.: Science *113*, 380 (1951)
152. LePoutre, P. F., Hopfenberg, H. B., Stannett, V. T.: J. Polymer Sci. *C 37*, 309 (1972)
153. Hopfenberg, H. B., Kimura, F., Rigney, P. T., Stannett, V. T.: J. Polymer Sci. *C 28*, 243 (1969)
154. Yoneho, T.: Kasen Geppo *29*, 60 (1976)
155. Gamova, I. A., Natkina, L. N., Shiryaeva, G. V., Prishchepa, N. D.: Isv. Uchebn. Zaved. Lesn. Zh. *19*, 113 (1976)
156. Cunanan, S. A., Bonoan, L. S., Verceluz, F. P., Azucena, E. A.: Phillipine A.E.C. Rept. (1976), PAEC(D)-76032
157. Goskii, G. M., Zhonina, T. A.: Tsellyul. Bum. Karton. *11*, 7 (1977)
158. Meleschevich, A. P., Vishev, Yu. V., Gordienko, L. L., Richko, N. T., Yurganova, L. G., Voitsekhovskii, R. V.: Khim. Technol. (Kiev) *4*, 59 (1977)
159. Vandana, I. Ya., Zolders, J., Surna, J.: Khim. Drev. *5*, 115 (1976)
160. Paeglitis, D.: Issled Obl. Khim. Drev. Tezisy Dokl. Konf. Mollodykhz Uch. 1st 56 1975
161. Hills, P. R., McGahen, D. J.: U.K.A.E.A., Brit. Pat. *1*, 449, 814 (1976)
162. Lawniczak, M.: Pr. Kom. Technol. Drewna, Poznan, Tav. Przyj Nauk. *6*, 49 (1976)
163. Hornoff, V., Kokta, B. V., Valade, J. L.: J. Appl. Polym. Sci. *21*, 477 (1977)
164. Tirzina, J., Zoldners, J.: Tezisy Dokl-Vses Konf. Khim. Ispol'z Lignina, 6th (1975) (Sergeeva, V. N., ed.) Riga, U.S.S.R.
165. Young, R. A.: J. Agric. Food. Chem. *25*, 738 (1977)
166. Hon, D. N. S.: Abstracts, ACS., 176th National Meeting, Miami Beach, Florida, Sept. 1978 (paper, Cell. 56)
167. Young, R. A., Naguyen, C.: Abstracts, A.C.S., 176th National Meeting, Miami Beach, Florida, Sept. 1978 (paper, Cell. 56)
168. Dolinin, V. R., Korovin, Yu. F., Tyuganov, M. A., Rogovin, Z. A.: Khim. Drev. *5*, 15 (1978)
169. Zolders, J., Pauls, R., Berisova, T. G.: Khim. Drev. *5*, 19 (1978)
170. Ikada, K.: In Advances in Polymer Science Vol. 29. (Cantow H. J., ed.). Berlin: Springer-Verlag 1978, P. 47

8. Industrial Application of Cellulose Graft Copolymers

8.1. Introduction

Those industries based on cellulosic raw materials were quick to see potential advantages in the chemical combination of a polar cellulose with a non-polar vinyl polymer. Projects to form and evaluate graft copolymers were therefore carried out with high hopes in research departments all over the world. There was a disappointing lack of success at first, frequently due to cost factors, and there are still only a few commercial products to be had, but the number seems to be increasing. Most of these products are based on either wood or textile fibers.

8.2. Wood-Plastic Composites (WPC)

The term "wood-plastic composites" or "WPC"[1,2] denotes the material formed when wood is impregnated with a monomer which is then polymerized. This process does not necessarily result in a graft copolymer, but some grafting may be expected from most methods of initiation. The polymerization may be initiated by treatment with high-energy radiation, either from a cobalt source or from an electron accelerator, or by the action of polymerization catalysts and heat. These treatments produce graft polymers of cellulose under the proper conditions, as has been discussed in earlier chapters. The conditions in forming wood-polymer combinations may or may not be favourable for true grafting to occur. The monomer fills the wood pores, but penetrates the cell walls only under swelling conditions[3], but there will probably be a certain amount of grafting wherever there is molecular contact between cellulose and monomer. This is less likely with simple peroxide initiation, because grafting would then occur by a chain-transfer mechanism, which does not operate too readily in the case of cellulose.

The properties of a solid cellulose matrix with the pores filled by a solid polymer would not be expected to change much if the two phases were connected by random covalent linkages, there is indeed little difference in the properties of radiation-polymerized products and those initiated with benzoyl peroxide, although the former are likely to contain more true grafting[4,5]. Still, since many investigators consider WPC's as examples of grafting, such composites should properly be mentioned here.

WPC's are, of course, more complicated than the preceding brief summary indicates, due to the very complex physical nature of wood and to the presence of other components. After all, cellulose is only half of the wood substance, and the other components also take part in the reaction, either by forming grafts themselves, as the hemicelluloses may, or by inhibition, as in the case of lignin[6].

Fibrous materials other than wood may serve as substrates for boards or similar composites, but wood was the first to become an industrial material. In November, 1967, an international study group met at Bangkok, their report, "Impregnated Fibrous Materials", published by the International Atomic Energy Agency, gives a detailed resumé of progress in the field at that time[7]. The initial paper gives a brief history of the impregnation of wood by phenolic or other resins[8], but most of the others deal with wood or other fibrous materials which are subsequently subject to copolymerization.

The number of substrates studied is vast, but the most promising woods seem to be oak, beech, poplar, birch, alder, maple, walnut, ash, and pine[9-12]. Naturally, the material of choice will depend upon the availability at the site of processing, as well as upon the properties desired.

An equally extensive study of monomers has been made. The choice here depends partly on availability, but still more so on economics, on the behavior of the process, and on the product's properties. Monomers of particular interest have been methyl methacrylate, styrene, and acrylonitrile[2,10-15]. Mixtures of monomers have proved particularly interesting[11,14-17]. It is also customary to add solvents, plasticizers, resins, or other modifiers to the monomer[16].

The properties of the product depend, of course, on the wood, the monomer, and the details of processing. However, one may make many generalizations.

The improvements in hardness and dimensional stability are particularly striking and very important for many uses[1,18-21]. Related properties are high compression resistance, abrasion resistance, and low water sorption. The attractive appearance is also an important attribute, especially since it lasts so well. WPC parquets in the Federal Pavilion at the New York World's Fair[22] and, for the past several years in the Air Terminal at Helsinki, Finland[19], have demonstrated the superior wearing qualities of these materials, although not all reports have revealed the same superior performance[23].

Indicative of the wide spread interest in WPC systems is the level of patent activity and associated literature. One process using radiation grafting has been described by Mott and Stannett[24]. This consists essentially of degassing under vacuum to remove air, nitrogen purging, degassing, exposure to the monomer, impregnation of the timber by the monomer under pressure, and subsequent irradiation. Low dose rates and total doses are favoured and a variety of monomers have been investigated. The results of grafting, depending on the monomer applied, can be related to those observed during textile modifications. Thus, improvements are observed in the moisture uptake, rot resistance, tensile strength, compressive strength, dimensional stability, elastic modulus, and overall wood hardness[25-27].

Attempts to fabricate 'wood-plastics' have been successful in certain instances with a wide range of monomers[28]. Much of the research along these lines arises from the utilization of natural resources as fully as possible. The outcome is of great interest to the developing nations. In a thorough treatment of the radiation-induced grafting of styrene and methyl methacrylate-impregnated birch and aspen, Shirayaeva et al.[29] noted improved compression strength, static bonding strength, shear strength, and static hardness. Grafting rates were accelerated by the presence of tributylphosphate and resulted in lesser doses being required. Alkenyl phosphates were observed to impart both fire and bio-organic decay resistance.

Another treatment involving impregnation followed by irradiation was used by Basson[30] in a study of grafting to pine and eucalyptus timbers. The degree of monomer uptake by the eucalyptus was less than that of pine. The investigation included microscopic examination, dose rate effects, and mechanical and weathering tests. Both types of treated timber were superior to their untreated counterparts in many respects.

Stannett and his group[31–34] have shown a general interest in grafting to wood and wood pulps. Here, however, the aim has been toward a greater understanding of the nature of the grafting process rather than toward improving the physical properties of the timber alone. This aspect was also studied by Sakata et al. using styrene in grafting to woody fibers[35]. In this instance, the study involved the distribution of grafted polystyrene branches between the lignin and carbohydrate fractions of the parent substrate (1.2:4 for lignin:carbohydrate, respectively). The branches were removed from lignin sections by treatment with acetic acid and from cellulose by hydrolysis with 72% sulphuric acid (see observations made earlier regarding the use of concentrated sulphuric acid in the acid hydrolysis of cellulose-polystyrene graft copolymers).

Birchwood has attracted attention recently because of considerable improvements which can be achieved in physical properties by low degrees of grafting. Irradiation of unitary and binary monomer-impregnated samples has been successfully carried out. Steiners et al.[36] have noted the absence of stress and deformation when birchwood is treated with styrene in carbon tetrachloride and then irradiated. Copolymerized samples have greater dynamic compliance and mechanical properties. Glukhov et al.[37] used styrene-acrylonitrile mixtures and observed that grafting was much easier than polymerization with either monomer alone. Similar observations were made by Oraby et al. in their study of the grafting of these same monomers to wood pulp cellulose[38].

The modification of timber or wood pulps by radiation-induced grafting offers a simple method of protection against a wide variety of decay-initiating systems. Improvements to physical properties, structural properties, and appearance may also be achieved. Once grafted, the timber or wood pulp composites are amenable to a much wider range of use than exists for ungrafted materials. Some of the major broad problems in graft polymerization on an industrial scale have been considered by Iwate in his review of this topic[39]. Mention should also be made of the use of grafting procedures to improve the flame retardency of timber through WPC formation as considered by Leipins et al.[40] and Needles[41] in their comprehensive reviews of radiation processing for flammability resistance. Consideration is given to the grafting of vinyl chloride monomer onto many types of wood using a Co-60 radiation facility[42]. Also noted is the non-flammability of woodplastic composites when grafted with Phosgard C-22-R (Monsanto flame resistant)/monomer mixtures[43].

8.2.1. The Geographical Distribution of WPC Development

The status of industrial development may be conveniently discussed country by country. Freĭdin et al.[44] published in 1959 what seems to be the first account of irradiation of monomer-impregnated wood to produce a WPC; the Russian patent application is dated 1958[45]. In the United States, Kenaga's 1963 patents[46] were also applied for in 1958. The U.S. Atomic Energy Commission initiated a very productive research program in 1961 at West Virginia University[12]. There has been a great deal of progress since these independant beginnings[1,18,47]. According to the Bangkok report, some eight or ten countries had reached various stages of industrial development and evaluation by 1967, while laboratory studies of such wood-polymer composites were being actively pursued in a dozen other countries[7].

1. United States

Rotariu and Mott[12] described the activities of four companies with production capacity for wood-plastic composites in 1967. Four years later, the situation was about the same, although there have been changes in ownership and activity.

The American Novawood Company, Lynchburg, Va., was founded and began small-scale production in 1964[12,22,48]. By the fall of 1966, their facilities had been expanded and various products were being evaluated for consumer use. These products — parquetry, stair treads and risers, handrails, thresholds, door sills, chair mats — are still being produced by the Applied Radiant Energy Corporation, which bought the original facilities and uses the same trade names[49]. Red oak is used for most of their products. The parquetry, "Gammapar", is also made from back ash or walnut. Methyl methacrylate, vinyl acetate, and styrene have been used as monomers.

Another parquet flooring of this type, "Perma Grain", is made by the ARCO Chemical Company, a division of the Atlantic Richfield Company[12,50]. Red oak is evacuated and impregnated with methyl methacrylate under high pressure. After irradiation with a cobalt source, the parquets are sanded and buffed. Vinyl chloride, styrene, and mixtures of styrene and acrylonitrile were also tested as impregnants.

Radiation Technology, Inc., of New Jersey and its Canadian subsidiary manufacture a composite called "Radwood" from red oak, birch, walnut, and wild cherry. When installed costs and maintenance costs are both considered, Radwood is said to be cheaper than conventional parquet flooring for periods over 5 years under heavy duty and over 10 years under medium duty[51].

The Bowling Division of the American Machine and Foundry Company[12,22,52] at Lowville, N.Y., has used a chemically catalyzed polymerization to make a WPC, originally as shafts for billiard cues, although other products have also been evaluated. The process was developed by Professor J. A. Meyer[5,53,54] at Syracuse University with benzoyl peroxide or azobisisobutyronitile as initiator. There may be little or no true grafting, but the properties are essentially the same as those of irradiated products. The catalytic process is advantageous for small units which can be built quickly and cheaply[55].

Lockwood is the registered trademark for the composite developed by the Lockheed-Georgia Nuclear Laboratories, Lockhead-Georgia Company, Marietta, Georgia[12)]. Over a hundred types of wood were processed, using a dozen different monomers. Methyl methacrylate seemed the most attractive; vinyl chloride was also promising. However, the company has discontinued its effort in the field and the Nuclear Laboratories are now closed[51)].

2. Canada

The Canadian subsidiary of Radiation Technology, Inc., which was just mentioned[51)], also markets Radwood.

There is also Barwood of Canada, Ltd., which, through an agreement with the Atomic Energy of Canada, fabricates and markets a red oak WPC impregnated with methyl methacrylate. The radiation source at the South March plant of the Atomic Energy of Canada has been in operation since early in 1970, and 10,000 square feet of flooring have been produced. A high-capacity plant for 10 million square feet annually is now being designed[53)].

3. United Kingdom

The U.K. Atomic Energy Authority at Harwell, Berkshire, has developed a WPC called Curifax[56)]. The wood is treated at Harwell and furnished on order to a number of companies now using it or investigating its use. The largest impregnator is 15 feet long; smaller ones are available for shorter lengths or for preformed blanks. Curifax is readily machined, preferably with carbide-tipped tools, and may be bonded with screws or adhesives; it is too hard to nail.

The most promising use is for flooring[2, 56)] and one of the first large-scale installations for Curifax has been a walkway complex in a chlorine factory, which used approximately 30 tons of WPC. Other applications have been cutlery handles, toilet brushes, and golf club heads. Textile machinery is being investigated experimentally. Drumsticks and musical instruments are on the market and the new organ console recently installed at Blackburn cathedral has used Curifax extensively[56)].

4. Russia

The research initiated by Freidin and Karpov[44, 57)] has continued in the Soviet Union[45, 58)]. Although there has been no announcement of large-scale production, a pilot plant with a capacity of one-half ton per day has been developed at the Karpov Institute in Moscow to produce wood-plastic composites. Among the woods used are beech, pine, and spruce; the polymers include styrene, methyl methacrylate, vinyl acetate, and vinyl chloride. A parquet sample of this sort is exhibited at the permanent Exhibition of the Achievements of the National Economy in the Pavilion of Atomic Energy in Moscow.

5. Finland

Research on radiation initiation and on chemically catalyzed initiation has been pursued concurrently in Finland since the early 1960's. The former research was carried out by the Department of Radiation Chemistry at the University of Helsinki[11, 14, 19] and the latter by Neste Oy[11, 19].

The best results with irradiation have been obtained with birch or alder impregnated with methyl methacrylate, or with a mixture of styrene and unsaturated polyesters. A test floor of irradiated birch and oak WPC parquets has been laid in one of the Helsinki elementary schools, and similar material has been tested as window frames in saunas and as a thin veneer on ski runners[19].

In the Neste Oy processes, peroxides are used as initiators, and styrene or styrene-polyester mixtures are the monomers used in most of the examples in the patents[59]. The floor of the reception room at the Helsinki Air Terminal is made of 8000 square meters of the Neste Oy parquetry[19]. The product "Neswood" has been licensed to Riihimaki Saha Oy and Paloheimo-Yhtyma in Riihimaki, Finland, for use in parquetry[60, 61]. This "Lamella-Neswood" is available either as parquet tiles or as strips of parquetry. It is recommended where heavy duty can be expected — public buildings, lobbies, concert halls, gymnasiums, and the like[60]. Neswood is also suitable for furniture parts, wall panels, tool handles, and many kinds of sporting goods, particularly ski runners[9].

6. *France*

Conservatoire Industrie in Courbevoie manufactures various articles from wood-plastic combinations, especially tool handles[10, 13]. In Courbevoie, too, the organization Saint-Gobain Techniques Nouvelles is working with beech, poplar, and oak woods and with methyl methacrylate, acrylonitrile, and styrene monomers[10].

The Centre Technique de Bois in Paris and the Centre d'Etudes Nucleaires in Saclay are cooperating in the study of a radiation-initiated combination of styrene with beechwood. France also collaborates with Japan in these studies and is particularly interested in density studies and fabrication procedures[10].

7. *Japan*

In their collaboration with France, the Japanese have paid particular attention to the dimensional stability and flame resistance of the products. Trial production batches of both chemical and radiation polymerizations have been made, and studies of marketability are underway[10, 62].

8. *Scandinavia*

There has been considerable research in Sweden on WPC, but no large-scale production and no plans for commercialization now exist. Samples have been evaluated for parquetry, tool handles, and plywood[10].

The Danish Atomic Energy Commission and the wood-manufacturing industries have cooperated in research, with special interest in the impregnation of veneer. Investigations have also begun at the Central Institute for Industrial Research in Norway[10].

9. West Germany

Some companies in the Federal Republic are interested in dimensionally stabilized wood and in surface-hardened wood. Heavily stressed products, like railway ties, have been suggested as the more economically feasible uses[13,63].

10. Other Countries

Many other countries are investigating wood-plastic combinations in universities and research centers, but at the time of the Bangkok meeting, no industrial developments had been indicated. These countries include Austria, Belgium, Holland, Italy, Czechoslovakia, and Hungary in Europe[10,25,64,65], and Australia[66], India[67], the Philippines[68], and Taiwan[69] in Asia and the Pacific regions. In Pakistan[70], jute has been investigated. In the Philippines and in Taiwan, WPC's are of interest to the wood-carving industry[71].

8.2.2. Uses and Properties

The properties and uses have already been indicated, but more detail on some points may be useful. The major types of utilization are reviewed below.

1. Construction Materials

The major emphasis of present utilization of wood-plastic composites is in this area, especially for tile or strip parquetry, stair treads, and the like[1,2,7,9,19,22,50,56,65]. They have improved hardness, dimensional stability, and resistance to abrasion[1,9,18,19,22,50,56,65]. Thin sheets have been suggested as facings for outer walls, doors, and window frames[68,72].

A significant difficulty, which must be recognized, is the need to educate construction workers in the uses and properties of plastic composite materials. Rotariu and Mott[12] have concluded that there are no major differences between the products from radiation polymerization and those from chemical polymerization without acid catalysts, of monomer systems which do not swell wood. However, if swelling systems are used and monomer is deposited within the fiber walls, there should be an advantage with the radiation product due to the greater likelihood of true grafting.

Capital costs would probably be higher for a radiation plant and operating costs lower. The heat-catalyst system would probably be more economical for low-volume production[12,22,52].

Several of the tables in Chapter 7 provide data which are relevant to the use and properties of wood-plastic composites. Studies of the mechanical properties (Tables 59–63, 66–69, 75 and 76), water repellency, and moisture regain properties (Table 77–81), resistance to fungal growths (Tables 83 and 84), and general physical properties (Tables 89–91) are covered. Some general areas of application can be inferred from Table 85. The information regarding relative ease of coloration and color fastness of composites (Tables 86 and 87) has relevance when composite identification is a requisite or when there is need of proof regarding a particular type of treatment.

The reader is also advised to consult those sections dealing with methods of developing cellulose composites such as those discussed in Chapters 3–6.

2. Furniture and Industrial Parts

The wood-plastic composites can be machined by equipment with hard metal cutters and glued after surface roughening, but not all adhesives are satisfactory[19,56]. Since nailing may cause splitting, pieces intended for nailing should be drilled beforehand, or they may be attached by screws. The uses most often mentioned are tool handles, furniture arms and legs, and other small articles[9,13,14,22,56].

Somewhat related to the use of WPC's in furniture and the like is the technique of high-energy irradiation to cure polymerizable lacquers or coatings on wood substrates. Hoffmann and Smith[73], for instance, have studied the use of the electron accelerator to cure mixtures of unsaturated polyesters with a vinyl monomer (styrene, ethyl acrylate, methyl methacrylate) applied as a coating or impregnant; they conclude that a cellulose graft polymer has been formed in this process. Much of the Finnish work on WPC's has dealt with monomer-polyester mixtures as impregnants for WPC's[11,17,19,74].

3. Sporting Goods

The toughness of wood-plastic composites has made them particularly attractive for use in sports. They are said to be particularly adapted for floors and furnishings of such areas as gymnasiums, sports halls, and basketball courts[66]. One of the most promising commercial outlets is as a base for cross-country skis, taking advantage of the excellent abrasion resistance and the affinity for ski wax[9,11,19]. Wood-polymer composites are used in the United States to manufacture billiard cues[12,52]. Other possibilities being investigated include gun stocks, archery equipment, golf clubs, chess sets, and musical instruments[9,56].

4. Wood Carving

Experience with the previously discussed products has led to the conclusion by many that it is not economically feasible at this stage to use the polymerization process to upgrade the properties of low-cost softwoods. On the other hand, there seems to be advantages in doing this in the specific field of wood carving[71].

In Taiwan, the use of a softwood, schefflera, is being considered to replace the conventionally used teak and camphor. The carving process is greatly accelerated and the final product is quite comparable to that of other woods — better, in fact, with regard to dimensional stability. Labor is 80% of the cost of producing an 8-inch carving in this way, and the total monomer treatment (80% vinyl chloride and 20% vinyl acetate) is 17.5%. Radiation polymerization is under consideration.

There is also interest in wood carving in the Philippines, but the availability and the cost of monomers are restricting features. Only methacrylates have been tested as monomers so far. An additional problem is that the present situation calls for rough carving before impregnation and finishing afterwards, which will increase handling costs in such a widely dispersed industry.

8.3. Grafted Textile Products

Textile products from grafted cellulose, like the wood-plastic composites, have appeared on the market in several countries. They include the rayons developed by Rogovin and his associates in Russia, the permanent-press cotton of Deering Milliken in the United States, and the fibers of the American Enka Corporation in the United States and of Chemiefaser Lenzing in Austria. The last two have both been developed by the Scott Paper Company. All of these will be discussed below. A grafted rayon was also manufactured in Japan on a small scale, but production was abandoned after a few years[75,76].

The two Russian materials most discussed in the literature are called Mtilon A and Mtilon B. Mtilon A has a cellulose acetate base copolymerized with poly(acrylic acid). Mtilon B is a viscose staple fiber with a polyacrylonitrile graft. Cotton may also be grafted in this way[77]. Graflon, the Japanese fiber formerly manufactured by Nisshin Spinning Company, was similar to Mtilon B with a higher polyacrylonitrile content. All of these were grafted by chemical initiation, probably using redox systems. The Japanese process used vapor-phase grafting of acrylonitrile[75,76].

Graflon was recommended for blankets, but the improvement in properties apparently failed to justify the expense of manufacture. For the same reason Grimm[78], in the Institute for Fiber-Technology in Dresden, decided that the superior stiffness and abrasion resistance did not justify further study of his process of grafting rayon with acrylonitrile by means of Fenton's reagent (hydrogen peroxide in the presence of Fe^{II}-ions). Grimm has compared the properties of his products with Graflon and Mtilon.

The Mtilons are still widely used in Russia[79-82]. Mtilon A, the modified acetate fiber, has three to four times the resistance to wear of the untreated acetate fiber and one tenth of the tendency to develop static during processing. It can be used to form a bactericidal fabric[80-83] and is being used as an ion-exchange material to recover precious metals, such as gold, from refining wastes[82]. Mtilon B, from viscose, has a wool-like appearance and is specially recommended especially for carpets[77,81,83]. It has improved resistance to micro-organisms, light, and abrasion, in addition to lowered hygroscopicity and less tendency to crease, wrinkle, or shrink[77,81,83].

Rogovin[80] has recently summarized the various uses in graft copolymers of cellulosic textiles in Russia. Earlier reviews in German and English also exist[81,82]. The Mtilon graft with acrylonitrile is produced (3 tons per day) for carpets and knitted wear[82,84,85]. Flame-proofed materials are produced by phosphorylation of a graft copolymer of poly(vinylpyrrolidone), or poly(methyl-vinylpyridine), or by grafting with a phosphorus-containing monomer[82,84,86]. Ion-exchange fibers are being used, as mentioned above, for recovery of heavy metals from solution and also in the dye and food industries[81,82]. They may be either cationic like Mtilon A or anionic (from nitrogeneous monomers, quaternized if necessary)[80,82,84,85]. The ion-exchange fibers are used as substrates to bind enzymes or other proteins, and the salts have special uses[83,84]. The calcium salt is a coagulant for blood and proteins and is used for medical gauze as a hemostatic agent[83,85,87,89]. The silver or copper salts are bactericidal and are used in hospital linen, surgical coats, and special laboratory and working clothes[81,83,85]. Fluorine-containing grafts, which are water- and oil-repellent, are also used in special work clothes[82]. Special technical uses have been found for isoprene and styrene grafts.

The durable-press cottons and polyester blends produced by Deering Milliken[88] are based on processes described in a series of patents[89]. These fabrics have been grafted with methylolacrylamide by radiation. They have several advantages over conventional permanent-press fabrics: durability to commercial laundering, superior soil release, and no development of formaldehyde odour. It appears that grafting (through the vinyl function) and crosslinking (through the methylol group) have occurred in the process.

Faessinger and Conte of the Scott Paper Company have developed very promising methods of grafting vinyl monomers to cellulose by initial xanthation[90-92] or by reaction of the cellulose with thio-compounds like ethylene sulfide[91,93]. The process is being developed by Scott Paper Company jointly with American Enka Corporation in the United States and with Chemiefaser Lenzing in Austria.

The Austrian group[84,85,94] is also making extensive pilot-plant studies on a flameproof fiber using a monomer which contains phosphorus, and also on a fiber for use in wet-web nonwovens. This latter product will be a long, thin fiber which can be easily and uniformly dispersed. It is formed by grafting the partly xanthated fiber with monomers which contain polar groups, such as the acrylates[95].

Considerable interest has been shown in the radiation-induced grafting of cellulosics with a view to enhancing their flammability resistance[40]. Early reports on this subject include those of Morris and Charlesby[96] and Miles and Delasanta[97]. Miles and Delasanta used triallyl phosphate and N-methylolacrylamide padded cotton, which was exposed to high-energy electrons at room temperature to produce a flame-retardant substrate. Cotton and various synthetic fibers have been made flame retardant by reaction with the following: vinyl phosphonate monomers[98]; N,N′,N″-triallyl phosphoric triamide and a crosslinking agent containing a vinyl group and a group reactive with cellulosic hydroxyl groups[99]; phosphonate acrylamides of general structure[100],

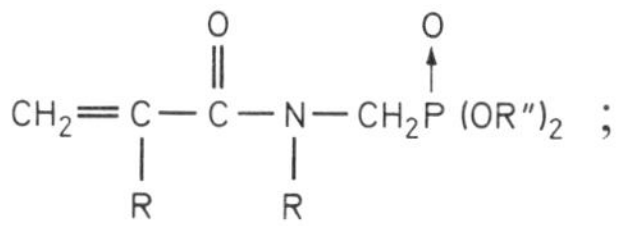

also cyclotri- and cyclotetraphosphazenes (for example, hexakis (methallyloxy) cyclo triphosphazine to retard cellophane film[101]) and organoboron compounds[102].

In the application of the flame retardant Fyrol 76, it has been noticed that the addition of N-methylolacrylamide gives increased fixation and also greater stability to flaming[40]. In certain applications, for example cotton printing, N-vinyl pyrrolidone makes a good substitute for N-methylolacrylamide. N-vinyl pyrrolidone has the advantage of being still liquid when water is removed, whereas N-methylolacrylamide is not[40]. With regard to flame retardancy, the greatest activity has been concerned with grafting to cotton, polyester, and cotton/polyester blends.

8.4. Grafted Pulps and Papers

There has been no evidence of the successful utilization of true grafting in the paper industry proper, although the Lenzing development of a fiber for non-woven webs[94,95] comes close to it. There is a grafted pulp on the market, however. This was a development of Adams and Hoftiezer of American Can Company[90,103] using the Bridgeford[104,105] method of grafting pulps with redox systems. Monomers are used containing hydrophobic groups which, after grafting, can be converted to hydrophilic groups. For instance, copolymers with poly(ethyl acrylate) or polyacrylonitrile can be readily changed to copolymers with poly(acrylic acid). These pulps are not at present used in paper, although their use as components of papermaking stocks has been patented[106], as have a number of other uses[107]. They are also being made and sold as a rooting medium for young plants[104,108].

Although it is not considered a true grafted product, the encapsulated fiber of the National Lead Company[109–111] should also be mentioned here. The pulp fibers are treated with titanium chloride or the like and then with an organometallic alkylating agent, like triethyl aluminium or a Grignard reagent. Ethylene is then polymerized on the reactive sites. Very likely, the growing polymer is a graft polymer with the cellulose and polyethylene linked by catalyst, but when the reaction is terminated, as, for instance, by methanol, these linkages will be cleaved and the resulting fiber will be simply encapsulated[110]. These fibers can be formed into a paper web and bonded by heat sealing. With a polymer level as low as 20%, the product has excellent wet strength, dimensional stability, and chemical resistance[112]. For some years, it was used in separator plates for lead-acetate storage batteries and is recommended for certain industrial filters and packaging applications[112,113].

Various approaches to paper boards have been made. A patent (to Kovrizhnykh and Flyate) has been granted which deals with a composite containing cellulose fiber (20–90%), vinyl fibers, (5–10%) and acrylonitrile grafted, rayon fibers (5–70%). This has been used to provide a filterboard having useful filtration, physical, mechanical, and chemical properties[114]. Sakamoto et al. have examined vinyl acetate grafted pulps with particular attention focused on the effects of the manufacturing method on the properties of paper prepared from the grafted pulp. Improvements were observed in tensile strength, surface smoothness, water regain, and heat bonding characteristics[115]. A recent Japanese patent deals with the manufacture of heat formable panels using waste aqueous pulp slurries grafted with olefinic monomers[116]. A redox initiator system, operating at pH = 4, is recommended with styrene as comonomer. The example quoted deals with an 8% aqueous waste, newspaper pulp slurry at 55 to 60 °C. This is treated with styrene (pulp/styrene ratio of 80:20) in the presence of 1% hydrogen peroxide and 0.1% Fe^{II} sulfate (based on weight of pulp in slurry) for 1 hour. Panels formed from the resulting copolymers, after hot pressing, have a tensile strength of 39 N mm^{-2}, bending modulus of 6500 N mm^{-2}, bending strength of 80 N mm^{-2}, and a compression strength of 50 N mm^{-2}. Grafting may be useful in specialty papers[6,117)], but Jayme[117)] considers that impregnation or reaction with the preformed polymer is better.

The paper industry is slow to adopt new processes and products requiring new equipment and conditions and raising unexpected questions of uniformity, speed, and cost. However, grafted textile and wood products are now in more than token use and cellulose grafting is no longer merely of academic interest.

8.5. Miscellaneous Examples

Taylor and Bagley have examined the hydrolyzed cellulose-polyacrylonitrile graft copolymer system as a polyelectrolyte gel forming highly viscous suspensions in water[118)]. Such a material could compete with existing thickeners since the rheological properties of the swollen, packed gel suspension are very similar to standard thickeners. The copolymer formulation provides stiff pastes in water suspensions even at concentrations as low as 1%[119)], and most of the material is easily centrifuged from suspension even at low speed. Patent coverage of this type of material has been taken by Adams and Tilloson[120)].

Bagley et al. have prepared a series of extrudable composites from graft-polymerized rigid fillers[121)]. Here a vinyl monomer is copolymerized with starch or cellulose to give an extrudable, biodegradable polymer. In a typically quoted example, 40 g of cellulose, 40 g styrene, and 10 cm^3 water were irradiated with Co 60, γ rays for 62 minutes (1 Mrad) under nitrogen before being allowed to stand at room temperature for 2 hours. The corresponding graft copolymer was mixed with 30 g of glycerol and the whole mixture extruded at 190 °C to give a product which, on repassage through the extruder at 190 °C, gave a translucent ribbon with a smooth, glossy surface and no die swell.

Absorbent materials (especially tampons) have been prepared from a 10–95% cellulose grafted with acrylic acid, alkyd acrylates and acrylamide. The average particle size of these copolymers is 50–500 μ[122)]. The remainder of the absorbent material consists of ungrafted wood pulp or regenerated cellulose. In a related field, Morin et al.[123)] have prepared graft copolymers of partly hydrolyzed cellulose triacetate and methacrylic acid (later converted to the sodium salt). Redox initiation at low temperatures was used in the grafting, which was achieved with high conversion (~100%) and efficiency (90–95%). The appearance of a disperse phase during grafting is attributed to the formation of spherical particles (0.5–1.0 μm in diameter) of insoluble graft copolymer on the surface of cellulose triacetate structures.

Interest has been shown in ion-exchange fibers. Yoshiboshi et al.[124)] have grafted substituted acrylic esters of type $CH_2{=}CRCOOCH_2CH_2NR_2$ (where $R{=}H, CH_3, C_2H_5, C_3H_7$) to provide high strength, ion-exchange systems. Such fibers were found to have a rapid Cr^{VI} adsorption rate and to require a lower sodium hydroxide consumption for their regeneration than does Amberlite IR 45. Oishi et al.[125)] have looked at similar series of compounds together with the corresponding phosphoric acid derivates. Here, electron beam initiation was used in grafting. The products were shown to have a high ion-exchange capacity and to exhibit little swelling in water or dioxan/water (90:100). Grafting levels of 740% were achieved. The preparation of mercaptoethylated graft fibers for the adsorption of heavy metal cations has been reported by Hayakawa and Yamakita[126].

In many ways, the industrial development of cellulosic graft copolymers is related to developments in associated technologies. An example of particular importance is provided by the advances made in electron beam accelerators[127] as considered by Abramyan. Electron beam facilities have found application in an ever-increasing number of polymerization processes and polymer degradation reactions. Equipment capable of generating energy in the range 0.3 to 5.0 MeV to a total dose of between 1 and 40 Mrad is readily available. Such sources, it may be recalled, have the advantages of high power, the possibility of inclusion in the production line (as seen in fabric, fiber, surface coating, or film modification), no radiation hazard when switched off, and high efficiency of utilization of the available radiation. The total power of industrial electron beam facilities produced every year is increasing rapidly[128–130]. The expansion of electron beam curing in industry owes much to the fact that significant modifications of properties can be made at a relatively low energy expenditure. Thus, a typical dose in grafting is 2.5 Mrad (i.e. 24 $J\ g^{-1}$), which corresponds to heating water by 6 °C.

Processes involving electron beam curing technologies, which either involve cellulosics directly or provide a guide to possible means of alteration, include the surface activation of polyethylene[131,132]; curing processes on various substrates including timber, cotton, paper and other substrates[131,133]; graft polymerization[131,134]; textile modification[135–136]; crosslinking reactions[132,137]; sterilization of medical products and associated packaging[138] foodstuffpreservation[139,140]; and leather goods processing[141].

References

1. Anonymous: Nucleonics *20* (3), 94 (1962)
2. Hills, P. R.: United Kingdom Atomic Energy Authority, AERE-M 2269 Wantage, U.K. (1969)
3. Timmons, T. K., Meyer, J. A., Cote, W. A. Jr.: Wood Science *4*, 13 (1971)
4. Juneja, S. C., Hodgins, J. W.: Forest Prod. J. *20* (12), 24 (1970)
5. Meyer, J. A.: Forest Prod. J. *15*, 362 (1965); Siau, J. F., Meyer, J. A.: Forest Prod. J. *16*, 47 (1966)
6. Phillips, R. B., Quere, J., Guiroy, G., Stannett, V. T.: Tappi *55*, 858 (1972)
7. International Atomic Energy Agency, Impregnated Fibrous Materials Vienna 1968
8. Takow, H.: Impregnated Fibrous Materials Vienna 1968
9. Neswood Neste Oy, Helsinki, undated, P. 3
10. Kinell, P.-Ö., Aagaard, P.: Impregnated Fibrous Materials I.A.E.A., Vienna 1968, P. 129
11. Miettinen, J. K.: Impregnated Fibrous Materials I.A.E.A., Vienna 152 1968
12. Rotariu, G. J., Mott, W. E.: Impregnated Fibrous Materials I.A.E.A., Vienna 1968, P. 190
13. Burmester, A.: Impregnated Fibrous Materials, I.A.E.A., Vienna, 1968, P. 209
14. Miettinen, J. K.: Paperi Puv *49*, 51 (1967); *49*, 63 (1967)
15. Mott, W. E., Rotariu, G. J.: Impregnated Fibrous Materials I.A.E.A., Vienna 83 1968; Stannett, V. T., ibid., P. 45

16. I.A.E.A. Report: Impregnated Fibrous Materials Vienna: 1968, P. 121
17. Strömberg, J., Miettinen, J. K.: Paperi Puu *51*, 785 (1969)
18. Anonymous: Nucleonics *23*, 17 (1965)
19. Autio, T., Miettinen, J. K.: Forest Prod. J. *20*, 36 (1970)
20. Gibson, E. J., Laidlow, R. A., Smith, G. A.: J. Appl. Chem. *16*, 58 (1966)
21. Aaagaard, P.: Svensk Kem. Tidskr. *79*, 501 (1967)
22. Anonymous: Modern Plastics *44*, 96 (1967)
23. Garratt, P. G., Hoigne, J., Heetman, J. G., Hess, W.: Holzforsch *26*, 96 (1972); Hills, P. R., U.K.A.E.A., A.E.R.E.-M. 2401 Harwell 1971
24. Mott, W. E., Stannett, V. T.: Isotopes Radiat. Tech. *6*, 323 (1969)
25. Czikovsky, T.: Atomic Energy Review *6*, 3 (1968)
26. Stamm, A. J.: Wood and Celluloe Science New York: Ronald Press 1964
27. Kenaga, D. L., Fennessey, J. P., Stannett, V. T.: Forest Prod. J. (1962), P. 161
28. Kandeel, S. A. E., Etman, M. H.: Alexandria J. Agric. Res. *19*, 365 (1973)
29. Shirayeva, G. V., Prishchepa, N. D., Glutiov, V. I., Lebedev, V. T., Karpov, V. L.: Proc. 3rd Tihany Symp. Rad. Chem. (1971), (ed., Dobo, J., Hedwig, P.); Akademiai Kaido Budapest, *1*, 783 (1972)
30. Basson, R. A.: Atom. Energy Board, Pretoria Rep. A/Conf. 49/P-668 (1971)
31. Radi, F. S., Hopfenberg, H. B., Stannett, V. T.: J. Appl. Polym. Sci. *16*, 2685 (1972)
32. Phillips, R. B., Brown, W., Stannett, V. T.: J. Appl. Polym. Sci. *15*, 2929 (1971)
33. Phillips, R. B., Brown, W., Stannett, V. T.: J. Appl. Polym. Sci. *16*, 1 (1972)
34. Phillips, R. B., Brown, W., Stannett, V. T.: J. Appl. Polym. Sci. *17*, 443 (1973)
35. Sakata, I., Miyata, N., Senjui, R.: Mokuzai Gakkaishi, *18*, 593 (1972)
36. Steiners, K., Goldsteins, A., Rocens, K., Dolacis, J., Rajavaa, E.: Latv. P.SR Zinat Akad. Vestis *11*, 35 (1972)
37. Glukov, V. I., Shirayaeva, G. V., Karpov, V. L.: Plass. Massy *11*, 28 (1972)
38. Oraby, W., Hopfenberg, H. B., Stannett, V. T.: J. Appl. Polym. Sci. *15*, 2987 (1971)
39. Iwate, F.: Kobunshi Kato *27*, 89 (1978)
40. Liepins, R., Stannett, V., Walsh, W. K.: Prog. Polym. Sci., *6*, 1 (1978)
41. Needles, H. L.: J. Consumer Prod. Flammability *4*, 156 (1977)
42. Freibush, A. M., Kieltyka, H. P.: U.S. patent *3*, 407,088 (1978)
43. Proksch, E.: Wood-plastic Work in Austria, presented at the Conference on Large Radiation Sources for Industrial Processes, Munich Germany 1969
44. Freĭdin, A. S., Malinskŭ, Yu, M., Karpov, V. L.: Gidroliz. i. Lesokhin. Prom. *12*, 4 (1959); Through Abstr. Bull. Inst. Paper Chem. *31* 2485 (1960)
45. Czvikovsky, T., Dobo, J.: Impregnated Fibrous Materials, Vienna: International Atomic Energy Agency 1968, P. 169
46. Kenaga, D. L.: (to Dow Chemical Company) U.S. Patents *3*, 077,417; *3* 077,418; *3*, 077,419 (1963); Kenaga, D. L., Fennessey, J. P., Stannett, V. T.: Forest Prod. J. *12*, 161 (1962)
47. Kent, J. A.: Proc. Nuclear Applns. Wood, Paper, Pulp Inds. Conf. 1964, P. 49; Kent, J. A.: U.S.A.E.C. Doc. ORNL 11C 7, 6 (1965); Kent, J. A., Loos, W. E., Walters, W. R., Boyle, W. R., Taylor, G. W., Winston, A. W.: U.S.A.E.C. Accession No. 33335 Rept. ORD-2954-4 1966; Kent, J. A., Winston, A. W., Boyle, W. R., Harmison, L. W.: U.S.A.E.C., TID-7643 1962; Kent, J. A., Winston, A., Boyle, W. R., Loos, W., Ayres, J. E.: Interim Report ORO-625 1965 Eng. Expt. Station, W. Va. Univ. Morgantown, W. Va: Klein, R. M.: U.S.A.E.C. Doc. ORNL 11C *7*, 15 (1965)
48. Anonymous: Chem. Eng. *72*, 64 (1965)
49. Gammapar: American Novawood Corp. 1971, R.D. Gillespie, personal communication
50. Witt, A. E., Morissey, J. E.: Modern Plastics *49*, 78 (1972)
51. Welt, M. A.: In U.K.A.E.A., AERE-M 2401, Harwell 1971, P. 4
52. Hebble, K., Golas, C., Schneider, D., Meyer, J. A., Moore, R.: Forest Prod. J. *17*, 19 (1967)
53. Langwig, J. E., Meyer, J. A., Davidson, R. W.: Forest Prod. J. *18*, 33 (1968), *19* 57 (1969); Meyer, J. A.: Forest Product J. *18*, 89 (1968); Meyer, J. A., Loos, W. E.: Forest Prod. J. *19*, 32 (1969)
54. McClary, L. R.: personal communication
55. Green, W. R.: U.K.A.E.A., AERE-M, 2401 Harwell: 1971, P. 1

56. Hills, P. R.: Composites 211 (1972)
57. Karpov, V. L., Malinsky, Y. M., Serenkov, V. I., Klimanova, R. S., Freidin, A. S.: Nucleonics *18*, 88 (1960)
58. Kashkina, N. A., Pormale, M. Ya., Surna, Ya. A.: In Modifikatsiya Drevesiny Akad. Nauk. Latv. SSR, Inst. Khim. Drevesiny, Riga (1967) P. 221; Kreitsberg, Z. N., Sergeeva, V. N., Grabovskii, Ya. K., Ore, V. N., ibid.: P. 209; Pormale, N. Ya., Kashkina, N. A., Surna, Ya. A.: Khim. Drevesiny S, 109 (1970); Through Abstr. Bull. Inst. Paper Chem. *41*, 7507 (1971)
59. Tammela, V.: Kem. Teolisuus, *26*, 9 (1969); Through CA 71, 39880 C (1969); Tammela, V. (to Neste Oy), German Pat. *1*, 642,190 (1970); Tammela, V., Harra, T. (to Neste Oy), German Pat. *2*, 013,186 (1970)
60. Lamella-Neswood, Paloheimo-Yhtymä, Riihimäki, Finland, undated.
61. Tanner, M., Valtari, H. A.: personal communication
62. Hirayama, T.: In Impregnated Fibrous Materials, Vienna: I.A.E.A. 1968, P. 250
63. Burmester, A.: Holz Roh-Werkstoff *25*, 11 (1967)
64. Kolosvany, G., Czvikovsky, T.: Holztechnol. *10*, 44 (1969)
65. Kolosvany, G., Czvikovsky, T.: Holztechnol. *10*, 104 (1969)
66. Clouston, J. C.: In Impregnated Fibrous Materials, Vienna: I.A.E.A. 1968, P. 217
67. Iya, V. K.: In Impregnated Fibrous Materials Vienna: I.A.E.A. 1968, P. 231
68. Bonoan, L. S.: In Impregnated Fibrous Materials, Vienna: I.A.E.A. 1968, P. 286
69. Wang, V.-P.: In Impregnated Fibrous Materials Vienna: I.A.E.A. 1968, P. 293
70. Awan, M.-H., Husain, A.: In Impregnated Fibrous Materials Vienna: I.A.E.A. 1968, P. 267; Awan, M.-H., Mohammed, D., Qadir, Q. A.: ibid. P. 278
71. Impregnated Fibrous Materials Vienna: I.A.E.A. 1968, P. 367
72. Arledter, H. F., Holweg, M.: Mitt. Deut. Ges. Holzforsch *57*, 105 (1971)
73. Hoffman, A. S., Smith, D. E.: Modern Plastics *43*, 111, 1966
74. Miettinen, J. K., Autio, T., Strömberg, J.: Britisch Pat. *1*, 178,215 (1970)
75. Matsuzaki, K. (1972), Ohno, Y. (1972); both personal communications
76. Puig, J.-R., Gaussens, G.: Bull. I.T.F. *21*, 133 (1967)
77. Khubitiya, M. M., Monastyrskiu, A. G.: Tekstil. Prom *26*, 10 (1966); Through CA *66*, 387,467 (1967)
78. Grimm, H.: Faserforsch, Textiltech. *19*, 20 (1968)
79. Phillip, B., Kukanoff, T.: Zellstoff Papier *9*, 292 (1970)
80. Rogovin, Z. A.: Cell. Chem. Technol. *6*, 17, 1972
81. Rogovin, Z. A.: Chemiefasern *18*, *738*; 742, 746 (1968)
82. Rogovin, Z. A.: Svensk. Papperstidn. *70*, 799 (1967)
83. Rogovin, Z. A., Virnik, A. D.: Faserforsch. Textiltech. *22*, 145 (1971)
84. Krässig, H.: Papier, *24*, 926 (1970)
85. Krässig, H.: Svensk. Papperstidn. *74*, 417 (1971)
86. Makhamov, K., Penenzhik, M. A., Virnik, A. D., Rogovin, Z. A.: Texstil. Prom. *24*, 62 (1964); Through Abstr. Bull. Inst. Paper Chem. *36*, 2390 (1965)
87. Virnik, A. D., Penenzhik, M. A., Rogovin, Z. A., Daurova, T. T., Khodakov, I. G.: Textil. Prom. *29*, 67 (1969); Through Abstr. Bull. Inst. Paper Chem. *41*, 2008 (1970)
88. Chem. Eng. News *44*, 19 (1966), 44, 42 (1966); Autio, T.: In Radioisotopes in the Pulp and Paper Industry, Vienna: I.A.E.A. 1968, P. 103, P. 107
89. Deering Milliken Res. Corp.: British Pat. *1*, 160,446 (1969); Gale, D. J. (to Deering Milliken Res. Corp) U.S. Pat. *3*, 617,196 (1971); U.S. Pat. *3*, 658,458 (1972); Lynch, J. A., Machell, G. (to Deering Milliken Res. Corp.) Canadian Pat. *797*, 724 (1968)
90. Conte, J. S., Faessinger, R. W.: (to Scott Paper Company), U.S. Pat. *3*, 505,257 (1970); Faessinger, R. W., Conte, J. S. (to Scott Paper Company), Canadian Pat. *823*, 261 and *823*, 262 (1969); (to Scott Paper Company), U.S. Pat. *3*, 359,224 (1967), British Pat. *1*, 255,184 (1971)
91. Rånby, B.: Pulp Paper Internatl. *11*, 52 (1969)
92. Scott Paper Company: British Pat. *1*, 137,624 (1968); Netherlands Pat. Appl. *6*, 514,877 (1966); Faessinger, R. W., Conte, J. S. (to Scott Paper Company), Belgian Pat., *646*, 284 (1964)
93. Conte, J. S., Faessinger, R. W.: (to Scott Paper Company), Canadian Pat. *824*, 725 (1964); Faessinger, R. W., Conte, J. S.: (to Scott Paper Company), U.S. Pat. *3*, 330,787 (1967), Belgian Pat. *646*, 285 (1964)

94. Schurz, J.: Papier *25*, 635 (1971)
95. Gotschy, F., Krässig, H.: Papier *26*, 813 (1972)
96. Morris, J., Charlesby, A.: Recent Developments in the Irradiation Treatment of Polymers in Applications of Radioactive Isotopes in Textile Research and Technology Lindau, W. Germany: Symposium 1964 (eds. Schmidt-Kuester, W. J., Weisner, L.), Munich Verlag Karl Thiemig KG: 1964
97. Miles, T. D., Delasanta, A. C.: Textile Res. J. *39*, 357 (1969)
98. Weil, E. D., Ger. Offen.: *2*, 115,050 October (1971)
99. Cahill, J. A., Myers, J. A.: U.S. Patent, *3*, 666,401, May (1972)
100. Duffy, J. J., Golborn, P.: (to Hooker Chemical Corp.) Ger. Offen. *2*, 215,434. Oct. (1972)
101. Hook, E. O., Nichols, L. D.: U.S. Patent, *3*, 711,389 Jan. (1973)
102. Liepins, R., Gilbert, S., Tibbetts, F., Kearney, J.: J. Appl. Polym. Sci. *17*, 2523 (1973)
103. Adams, J. W., Hoftiezer, H. W.: (to Tee-Pak), U.S. Pat. *3*, 194,727 (1965); Ranby, B., Svensk. Papperstidn *72*, 546 (1969)
104. Anonymous: Chem. Eng. News *46*, 58 (1968)
105. Bridgeford, D. J.: Ind. Eng. Chem. Prod. Res. Devel. 1, 45 (1962); Tappi, *46*, 673 (1963); Bridgeford, D. J. (to Tee-Pak) U.S. Pat. *3*, 083,118 (1963)
106. Adams, J. W., Hoftiezer, H. W.: (to Tee-Pak), Canadian Pat. *824*, 419 (1969); U.S. Pat. *3*, 395,070 (1969)
107. Adams, J. W., Hoftiezer, H. W.: (to American Can Company), U.S. Pat. *3*, 256,372 (1966); U.S. Pat. *3*, 366,582 (1968); Adams, J. W., Tilloson, A. H. (to American Can Company), U.S. Pat. *3*, 553,306 (1971)
108. Adams, J. W., Hoftiezer, H. W.: (to American Can Company) Canadian Pat. *838*, 039 (1970)
109. Carious Articles; Chem. Eng. News 41, 42 (1963); Chem. Process, *26*, 34 (1963); Ind. Eng. Chem. *55*, 11 (1963); Modern Plastics *40*, 97 (1963); Modern Textiles Mag. *44*, 37 (1963)
110. Herman, D. F., Dunlap I. R.: Tappi *48*, 418 (1965)
111. Herman, D. F., Kruse, U.: (to National Lead Company), U.S. Pat. *3*, 297,466 (1967)
112. Herman, D. F., Kruse, U., Brancato, J. J.: J. Polymer' Sci. *C 11*, 75 (1965)
113. Yanulis, M. J.: Tappi, *48* 61A (1965)
114. Kovrizhnijkh, L. P., Flyate, D. M.: U.S.S.R. Pat. *599*, 001 (1977)
115. Sakamoto, I., Sumi, M., Iwasaki, T., Tani, S., Inagaki, H.: Kamipa Gikyoshi *32*, 249 (1978)
116. P.R.B. S.A.: Japan Kokai *78*, 60, 958 (1978)
117. Jayme, G., Hebble, G. W.: Papier *25*, 113 (1971)
118. Taylor, N. W., Bagley, E. B.: J. Appl. Polym. Sci. *21*, 1607 (1977)
119. Le Poutre, P., Robertson, A. A.: Tappi *57*, 87 (1974)
120. Adams, J. W., Tillotson, A. H.: U.S. Pat. *3*, 682,856 (1972)
121. Bagley, E. B., Fanata, G. F., Doanne, W. M., Gugliemelli, L. A., Russell, C. R.: U.S. Pat. *4*, 026,849 (1977)
122. Personal Products Ltd. Bol. Inst. Prod. Florestais. Madeiras Deriv *3*, 31 (1977)
123. Morin, B. P., Smol'yakov, O. I., Tsarevskaya, I. Y., Rogina, M. V., Rogovin, Z. A.: Cell. Chem. Technol. *12*, 263 (1978)
124. Yoshiboshi, O., Takaharu, I. G., Serjii, G., Tomihiko, K.: Japan, Kokai, *77*, 144,389 (1977)
125. Oishi, Y., Hagaki, T., Goto, S., Kitada, T.: Japan, Kokai, *77*, 138,490, (1976)
126. Hayakawa, K., Yamakita, H.: J. Appl. Polym. Sci. *21*, 665 (1977)
127. Abramyan, E. A.: Atomic Energy Review (1978), 16 (3), 427 (1978)
128. Morganstern, K. H.: Radiation Processing with Electron Beam Accelerators — Present and Future Applications, presented at the World Electrotechnical Congress, Moscow June 1977
129. Morgenstern, K. H.: Appraisal of the Advantages and Disadvantages of Gamma, Electron Beam and X-ray Radiation Sterilization, at the Conference on the Radiosterilization of Medical Products, 1974: (Proc. Symp. Bombay 1974, IAEA, Vienna 269 (1975)
130. Glukhikh, V. A., Svin'in, M. P.: In Present Status and Future Development of High Voltage Accelerators, Proc. 2nd All-Union Meeting of Uses of Charged Particle Accelerators in National Economy, (Leningrad, 1975) 1, NIIEL FA 76 (1976)
131. Charlesby, A.: Atomic Radiation and Polymers, Oxford Pergamon Press: 1960

132. Finkel, E. L., Braginskij, R. P.: (Heat-resistant Wires and Cables with Radiation — Modified Insulation) Ehnergiya, Moscow 1975
133. Nablo, S. V.: Electron Beam Curing — a Cool Solution to the Energy Problem, Rep. 19th Annual Techn. Forum Flexographic Technical Association, Montreal April 1977
134. Hummel, D. O.: Some Aspects of Radiation — Initiated Copolymerization Large Radiation Sources for Industrial Processes (Proc. Sump. Munich, 1969) IAEA, Vienna 117 1969
135. Hoffman, A. S.: Radiation Modification of Textiles", Large Radiation Sources for Industrial Processes (Proc. Symp. Munich, 1969) IAEA, Vienna 301 1969
136. Chepel, L. V.: Use of Electron Accelerators in Radiation Chemistry Atomizdat, Moscow 1975
137. Kozlov, V. T., Evseev, A. G., Ponomarev, A. N. Os'kin, V. S., Tal'roze, V. L., Kozovrov, V. M., Smagin, E. N., Pashinin, V. I.: Proc. 2nd All-Union Meeting Use of Charged-Particle Accelerators in the National Economy (Leningrad, 1975) 1, NIIEhFA, 320 (1976)
138. Sterilization by Ionizing Radiation (Proc. Conf. Vienna, 1968) Multiscience Publication, Montreal
139. Metlitskij, L. V., Rogachev, V. I., Khrushchev, V. G.: Radiation Processing of Foodstuffs Ehkonomiya Moscow (1967)
140. Diehl, J. F.: Preservation of Food Radiation Physics and Chemistry, (Trans. 1st Int. Meeting Radiation Processing, Puerto Rico, 1976) NN1–3, (1977), 9,193
141. Strakhov, I. P., Shefrin, I. G.: (Ionizing Radiation Treatment in the Leather Industry) Legkaya Industriya, Moscow 1972

Subject Index